Role of Natural Bioactive Compounds in the Rise and Fall of Cancers

Role of Natural Bioactive Compounds in the Rise and Fall of Cancers

Volume 2

Editor

Claudio Luparello

MDPI • Basel • Beijing • Wuhan • Barcelona • Belgrade • Manchester • Tokyo • Cluj • Tianjin

Editor
Claudio Luparello
Università di Palermo
Italy

Editorial Office
MDPI
St. Alban-Anlage 66
4052 Basel, Switzerland

This is a reprint of articles from the Special Issue published online in the open access journal *Cancers* (ISSN 2072-6694) (available at: https://www.mdpi.com/journal/cancers/special_issues/cancers-NBC).

For citation purposes, cite each article independently as indicated on the article page online and as indicated below:

LastName, A.A.; LastName, B.B.; LastName, C.C. Article Title. *Journal Name* **Year**, *Article Number*, Page Range.

Volume 2
ISBN 978-3-03943-292-9 (Pbk)
ISBN 978-3-03943-293-6 (PDF)

Volume 1-2
ISBN 978-3-03943-294-3 (Pbk)
ISBN 978-3-03943-295-0 (PDF)

Contents

About the Editor

Claudio Luparello, Ph.D. (Cell Biology) is a Professor of Cytology and Histology at the University of Palermo, Italy. He has authored more than 40 publications and 3 book chapters and was the Editor of the book "Novel aspects in PTHrP physiopatology" edited by Nova Sci Pub. He serves as the Editorial Board member of Biological Procedure Online International Journal of Breast Cancer and TheScientificWorldJournal domains Cell Biology and Oncology. He is a member of AICC.

cancers

Review

Focus on Formononetin: Anticancer Potential and Molecular Targets

Samantha Kah Ling Ong [1], Muthu K. Shanmugam [1], Lu Fan [1], Sarah E. Fraser [2], Frank Arfuso [3], Kwang Seok Ahn [4,*], Gautam Sethi [1,*] and Anupam Bishayee [2,*]

1. Department of Pharmacology, Yong Loo Lin School of Medicine, National University of Singapore, Singapore 117600, Singapore; samantha.ong@u.nus.edu (S.K.L.O.); phcsmk@nus.edu.sg (M.K.S.); phcfanl@nus.edu.sg (L.F.)
2. Lake Erie College of Osteopathic Medicine, Bradenton, FL 34211, USA; SFraser78637@med.lecom.edu
3. Stem Cell and Cancer Biology Laboratory, School of Pharmacy and Biomedical Sciences, Curtin Health Innovation Research Institute, Curtin University, Perth, WA 6102, Australia; frank.arfuso@curtin.edu.au
4. Department of Science in Korean Medicine, Kyung Hee University, 24 Kyungheedae-ro, Dongdaemun-gu, Seoul 02447, Korea

* Correspondence: ksahn@khu.ac.kr (K.S.A.); phcgs@nus.edu.sg (G.S.); abishayee@lecom.edu or abishayee@gmail.com (A.B.); Tel.: +82-2-961-2316 (K.S.A.); +65-6516-3267 (G.S.); +1-(941)-782-5950 (A.B.); Fax: +65-6873-7690 (G.S.)

Received: 22 March 2019; Accepted: 28 April 2019; Published: 1 May 2019

Abstract: Formononetin, an isoflavone, is extracted from various medicinal plants and herbs, including the red clover (*Trifolium pratense*) and Chinese medicinal plant *Astragalus membranaceus*. Formononetin's antioxidant and neuroprotective effects underscore its therapeutic use against Alzheimer's disease. Formononetin has been under intense investigation for the past decade as strong evidence on promoting apoptosis and against proliferation suggests for its use as an anticancer agent against diverse cancers. These anticancer properties are observed in multiple cancer cell models, including breast, colorectal, and prostate cancer. Formononetin also attenuates metastasis and tumor growth in various in vivo studies. The beneficial effects exuded by formononetin can be attributed to its antiproliferative and cell cycle arrest inducing properties. Formononetin regulates various transcription factors and growth-factor-mediated oncogenic pathways, consequently alleviating the possible causes of chronic inflammation that are linked to cancer survival of neoplastic cells and their resistance against chemotherapy. As such, this review summarizes and critically analyzes current evidence on the potential of formononetin for therapy of various malignancies with special emphasis on molecular targets.

Keywords: formononetin; cancer; preclinical models; cell signaling; angiogenesis

1. Introduction

Cancer refers to the disease that develops when abnormal cells proliferate uncontrollably, followed by invasion into the surrounding tissues and eventually spreads to the organs or other parts of the body via the circulatory and lymphatic systems [1]. The fundamental process that leads to the development of cancer is the process of continuous, unregulated proliferation of cancer cells, resulting in tumor development [2]. Cancer has claimed the lives of 8.8 million individuals in 2015, making it the second leading cause of death worldwide after cardiovascular diseases [3,4]. In the past few decades, numerous treatment methods have been developed against cancer after acquiring a deeper understanding of multiple underlying signals and mechanisms that contribute to the survival and progression of neoplastic cells. These treatment modalities include both adjuvant and neoadjuvant chemotherapy, targeted therapy, immunotherapy, surgery, and radiotherapy. The incidence of cancer still remains high, with increasing mortality due to the disease progression despite significant progress in treatment regimens [3,4]. This phenomenon can be largely attributed to the limited effects exhibited by existing cancer therapies and the high cost of treatment, coupled with significant adverse effects [5]. Furthermore, conventional cytotoxic agents usually embody life-threatening toxicities [6]. There have been nearly 500 cases of withdrawal of medicinal products from the market as a result of adverse drug reactions over the past six decades, with the most common reason being hepatotoxicity [7]. In addition, certain cancers, such as breast cancer, can still resurface after a dormant period of 15 years following successful treatment [8], suggesting that there is a need to develop new and safe treatment options that can be proven to be more efficacious.

Natural products and dietary agents have gained enormous popularity over the years to be used in the clinical setting. They can be obtained from plants, microorganisms, or animals and have been used medicinally for decades, usually in complementary and alternative medicine [9]. An example would be Veregen, a topical ointment containing green tea extract sinecatechins, which is used to treat external genital and perianal warts, and the herb *Artemisia annua*, from which pharmaceutical scientist Tu Youyou managed to isolate artemisinin that is now used as a potent antimalarial drug [10]. These studies indicate the promising use of natural products for future discovery and development of cancer preventive and anticancer drugs. In the past three decades, nearly 80% of the drugs approved by the United States Food and Drug Administration for cancer therapy contain natural products or imitate their functions [11,12]. Natural products have since then made their appearance and represent a large portion of today's pharmaceutical agents used in cancer therapy, including taxol, vinblastine and camptothecin [13]. Over 60% of the current anticancer drugs are derived from natural sources [13], and the impetus to discover more natural products for chemotherapy and chemoprevention has become evident in the past decade as more agents are now in clinical trials.

Antitumorigenic activity can be found in compounds with different structural groups, including isoprenoids (including terpenoids and carotenoids), isoflavones, etoposide, and teniposide [5,11,14–35]. They function through various mechanisms, including the induction of apoptosis via DNA cleavage by inhibiting the activity of topoisomerase I or II, mitochondrial permeabilization, inhibiting crucial enzymes in signal transduction (i.e., proteases), cellular metabolism, or by inhibiting tumor-induced angiogenesis [9,36]. Certain isoflavones, such as isoliquiritigenin, have been shown to possess antitumorigenic properties, including pro-apoptotic effect on human cancer cells [37]. In recent years, there has been increasing evidence in preclinical and clinical studies that suggest that a dysregulated inflammatory response plays a crucial role in cancer development and may even be a key driver of cancer [38,39].

Chronic inflammation is also one of the hallmarks of cancer, and can drive the development of cancer through increasing the production of pro-inflammatory mediators involved in various signaling mechanisms, including cytokines, chemokines, reactive oxygen species/intermediates, increased expression of oncogenes, cyclooxygenase-2, 5-lipoxygenase, matrix metalloproteinases (MMPs), and pro-inflammatory transcription factors, such as nuclear factor-κB (NF-κB), signal transducer and activator of transcription 3 (STAT3), activator protein 1 (AP-1) and hypoxia-inducible factor 1α (HIF-1α) [38–66]. These factors mediate the basis of cancer progression, such as the proliferation of tumor cells, metastasis, survival, invasion, angiogenesis, chemoresistance and radio-resistance [38,47,61,64,67–71].

Targeting selected transcription factors and pathways increases the sensitivity of cancer cells to chemotherapeutics and radiation, resulting in apoptosis [60]. Furthermore, the use of natural products, such as farnesol, curcumin and celastrol, have been found to significantly enhance the anticancer effects of chemotherapeutics, such as bortezomib and thalidomide, in multiple myeloma [57,72]. This suggests that natural products can be used as adjunct cancer therapy. Despite the steady improvement of current anticancer therapeutics, developing novel drugs remains a priority of cancer treatment due to an overwhelming increase in resistance to chemotherapeutic drugs [9,73–76]. With greater knowledge and constant advances in technology, there is a great prospect for safer and more efficacious treatment options for cancer.

The red clover, *Trifolium pratense* (family: Fabaceae), is a legume known for its numerous health benefits, and can hold a crucial role in the prevention and management of certain health conditions, including type 2 diabetes, hyperlipidemia and hypertension [77]. It is a perennial herb that is commonly found in Asia, Europe, and North America, and has been traditionally used to treat skin and respiratory conditions, such as eczema, psoriasis, asthma and pertussis [78]. The isoflavones present in red clover have estrogen-like activities and have been subjected to an intense research over the past two decades due to their potential cancer-preventive, cardio-protective and anti-osteoporosis effects [78].

The extract from the red clover plant contains genistein, daidzein, formononetin (biochanin B), and biochanin A. Formononetin [7-hydroxy-3-(4-methoxyphenyl)-4H-1-benzopyran-4-one], (Figure 1), one of the main bioactive components extracted from the red clover, has been found to be the principal compound that contributes the therapeutic effects observed in the extract. It is exclusively produced by the Fabaceae family, and can be extracted from the roots of *Astragalus membranaceus*, *T. pratense* and *Glycyrrhiza glabra* [79]. Since it is structurally similar to 17-estrogen, formononetin's bioactivity mimics the effect of estrogen and this compound is considered to be a phytoestrogen [80]

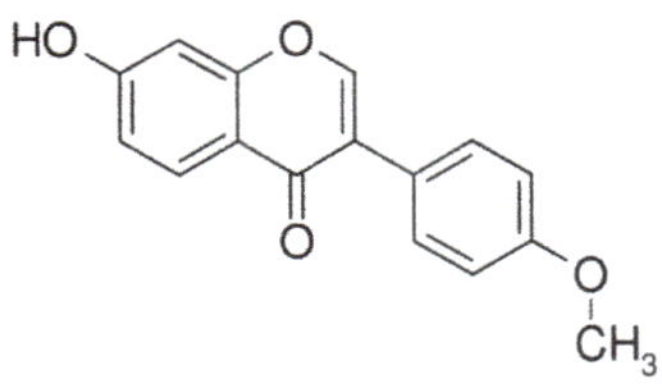

Figure 1. The chemical structure of formononetin.

Formononetin has shown beneficial effects in clinical trials for menopausal relief [81,82], reduction in bone loss [83], and improved arterial compliance [84]. Formononetin has been used clinically in China, in traditional medicine, as one of the fundamental herbs for treating carcinomas due to its protective effects against certain malignant tumors [85]. Formononetin has become the subject of intense research over the past decade due to its estrogenic effects and antitumorigenic properties. This review will summarize and critically analyze current evidence on the potential of formononetin for anticancer therapy with special emphasis on molecular targets.

2. Toxicity and Pharmacokinetics of Formononetin

A water-soluble derivative of formononetin, formononetin-3′-sulphonate (Sul-F, $C_{16}H_{12}O_7SNa$), has been shown to provide significant neuro- and cardio-protective effects both in vitro and in vivo [86,87]. Pro-estrogenic isoflavones, such as formononetin, can potentially be converted to more potent phytoestrogens in the human body. Incubation of formononetin in human liver microsomes caused demethylation, resulting in the production of formononetin derivatives and metabolites, including daidzein [88]. Due to the fact that formononetin is a naturally occurring isoflavone and phytoestrogen, it is associated with estrogen receptor (ER) binding. Phytoestrogens, especially isoflavones, can be classified as endocrine disruptors and are known to modify or interfere with the endocrine function [80].

Formononetin and its metabolites can significantly enhance pro-inflammatory cytokines and induce an allergic immune response. Interleukin-4 (IL-4) is closely associated with the CD4^{+} T helper cells and EL4 T lymphoma cells (*Mus musculus*). An increase in IL-4 production is observed upon administration of formononetin, daidzein, and equol due to an elevation in the activation of activator protein 1 (AP-1) via the phosphoinositide 3-kinase (PI3K)/protein kinase C (PKC)/p38 mitogen activated protein kinase (MAPK) signaling pathway. This suggests that formononetin and its metabolites may potentially cause allergic responses through amplifying the production of IL-4 in T-cells [89]. Hence, it may be a reasonable approach to restrict or limit the usage of formononetin in order to prevent allergic responses.

3. In Vitro Anticancer Pharmacological Properties of Formononetin

3.1. Antiproliferative Effects

The anticancer potential of formononetin has been explored in numerous in vitro models as shown in Table 1. First and foremost, formononetin showed potential in inhibiting tumor growth and proliferation. Uncontrolled proliferation is one of the primary hallmarks of cancer and represents one of the most prominent factors associated with malignancy. This antiproliferative effect of formononetin has been observed in multiple cancer models and is one of the main anticancer properties of formononetin. As compared to other isoflavones, formononetin has been proven to possess the greatest antiproliferative activity [90]. The antiproliferative property of formononetin has been observed in ER-positive breast cancer cells, such as MCF-7 and T-47D [91,92], and displayed minimal effect against ER-negative breast cancer cells, namely MDA-MB-231 and MDA-MB-435 [93]. It also demonstrated antiproliferative effects against prostate cancer (PC-3, DU-145 and LNCaP) [94,95], non-small cell lung cancer (A549 and NCI-H23) [96], cervical cancer (HeLa) [97], bladder cancer (T24) [98], osteosarcoma (U2OS), ovarian cancer (ES2 and OV90) [99], glioma (C6), and colorectal cancer cells (HCT-116, SW1116 and RKO) [37,85].

For most cell lines, formononetin has been found to possess a concentration- and time-dependent effect against tumor proliferation [85,94,96,98]. Antiproliferative effects of formononetin were also demonstrated in cytokine-induced cancer models, such as multiple myeloma, where formononetin attenuated the expression of inflammatory cytokines, including tumor necrosis factor-α (TNF-α), transforming growth factor-β1 (TGF-β1), interleukin-6 (IL-6), and interleukin-8 (IL-8) by downregulating hypoxia-inducible factor 1α (HIF-1α) [39,100] and nuclear factor-κB (NF-κB) [101]. Most studies conducted thus far indicate that formononetin has been able to prevent the proliferation of tumor cells without causing serious adverse effects as compared to other chemotherapeutic drugs.

3.2. Proapoptotic Effects

For progression of abnormal cells into cancerous ones, it is crucial that the apoptosis pathway is hijacked to allow promotion of growth and development of the damaged or abnormal cells. Cancer cells have the ability to evade the apoptotic checkpoints, allowing them to proliferate uncontrollably [102]. As a result, the proapoptotic property of formononetin elevates its potential to be used as an anticancer agent in cancer therapy. However, the underlying mechanisms of formononetin that promote cell apoptosis differ among different cell lines as well as cancer models. Apoptosis observed in cells can be classified into two different stages—early and late apoptosis—which can be differentiated through the presence of propidium iodide [102]. Formononetin largely elevated the proportion of early apoptotic cells in DU-145 prostate and U2SO osteosarcoma cell lines, and was found to be dose-dependent for prostate PC-3 cells (25–100 μM) [103–105].

The influence of formononetin on apoptosis of different cancer cell lines may involve the upregulation of specific transcription factors. The apoptotic mechanism in prostate DU-145 cells is activated by upregulating dexamethasone-induced retrovirus associated DNA sequences (Ras)-related protein 1 (maximum for 48 h before decreasing rapidly after) and Bcl-2-associated protein (Bax), and simultaneously reducing B-cell lymphoma 2 (Bcl-2) levels, thereby causing the DU-145 cells to display morphological changes indicative of the early apoptotic stage, and trigger apoptosis via the mitochondrial apoptotic pathway [106]. On the contrary, high concentrations of formononetin (>12.5 μM) have been found to effectively inhibit proliferation and trigger apoptosis of PC-3 prostate cancer cells by inhibiting the insulin-like growth factor 1 (IGF-1) receptor androgen-independent pathway [104]. T24 human bladder cancer cells displayed morphological changes of apoptosis when treated with formononetin, and there was a significant reduction in the expression of miR-21 and phosphorylated protein kinase B (AKT). In addition, phosphatase and tensin homolog (PTEN), a notable tumor suppressor gene, was upregulated in T24 cells after formononetin treatment, which suppressed uncontrolled tumor proliferation [98]. Furthermore, a study by Zhang and colleagues [107] suggested that formononetin did not elicit toxic effects on non-cancerous cell lines, indicating that it may be a safe choice to halt cancerous cell growth.

Table 1. In vitro anticancer effects of formononetin.

Cancer Type/Cell Line Used	Concentration	Anticancer Effect	Mechanisms of Action	References
		Bladder cancer		
T24 cell line	50–200 μM	Antiproliferative Anti-invasion	↑Apoptosis; ↑PTEN; ↓miR-21; ↓pAKT	[98]
MCF-7 cell line	30–100 μM	Antiproliferative	↑Apoptosis; ↑G0/G1 cell cycle arrest; ↓IGF-1/IGFR-PI3K/AKT pathway	[91]
		Breast cancer		
ER-positive MCF-7 cells and T47D cell	25–100 μM	Antiproliferative	↑Apoptosis; ↓p38MAPK pathway	[92]
ER-positive MCF-7 cells and T47D cell	25–100 μM	Antiproliferative	↑Caspase-3; ↓IGF1R; ↓miR375	[93]
MDA-MB-231 4TI	2.5–40 μmol/L	Antiproliferative	↓MMP-2; ↓MMP-9, ↓TIMP1; ↓TIMP2; ↓PI3K/AKT pathway	[108]
Cervical cancer				
HeLa cells	Not available	Antiproliferative	↑Apoptosis; ↓PI3K/AKT pathway; ↓ERK pathway	[97]
		Colon cancer		
LoVo	50 μM	Anti-invasion	↑Apoptosis; ↓VEGF; ↓MMP	[109]
		Colorectal cancer		
HCT116 cell line	6.25–200 μM	Antiproliferative	↑Apoptosis; ↑Bax; ↑NAG-1; ↓Bcl-2; ↓Bcl-xL	[37]
SW1116 cell line HCT116 cell line	20–200 μM	Antiproliferative	↑miR-149; ↓EphB3; ↓PI3K/AKT pathway; ↓STAT3 pathway	[85]
RKO cell line	20–80 μM	Antiproliferative	↑Apoptosis; ↓ERK pathway	[101]
		Glioma		
Glioma C6 cell line	20–320 μM	Antiproliferative	↑Apoptosis; ↑Bax; ↑cleaved caspase-3 & caspase-9; ↓Bcl-2; ↓MMP-2; ↓MMP-9	[110]
		Glioblastoma		
U87MG cell line U251MG cell line T98G cell line	50–200 μM	Antiproliferative	↓HDAC5; ↓doxorubicin-induced EMT	[111]
		Multiple myeloma		
U266 cell line	5–60 μM	Antiproliferative	↓HIF-1α; ↓inflammatory cytokines; ↓AKT pathway	[100]
		Nasopharyngeal carcinoma		
CNE1 cell line CNE2 cell line	5–40 μM	↓Cell viability	↑Apoptosis; ↓PI3K/AKT pathway; ↓ERK pathway	[112]

Table 1. *Cont.*

Cancer Type/Cell Line Used	Concentration	Anticancer Effect	Mechanisms of Action	References
			Non-small cell lung cancer	
A549 cell line NCI-H23 cell line	100–200 μM	Antiproliferative	↑Apoptosis; ↑caspase-3; ↑Bax; ↓Bcl-2; ↑P21; ↓cyclin A; ↓cyclin D1; ↑G1 cell cycle arrest; ↑p53 at Ser15 and Ser20	[96]
			Osteosarcoma	
U2OS	20–80 μM	Antiproliferative	↑Apoptosis; ↑caspase-3; ↑Bax; ↓Bcl-2; ↓PI3K/AKT pathway; ↓ERK pathway	[90]
			Ovarian cancer	
ES2 cell line OV90 cell line	20–40 μM	Antiproliferative	↑Apoptosis; ↑G0/G1 cell cycle arrest	[99]
			Prostate cancer	
LNCaP cell line PC-3 cell line	20–80 μM	Antiproliferative	↑Apoptosis; ↑G1 cell cycle arrest; ↓AKT/cyclin D1/CDK4; ↓ERK1/2 pathway	[94]
PC-3 cell line DU-145 cell line	10–100 μM	Antiproliferative	↑Apoptosis; ↑G1/S cell cycle arrest; ↓IGF/IGFR1 pathway	[95]
PC-3 cell line	25–100 μM	Antiproliferative	↑Apoptosis; ↓IGF/IGFR1 pathway	[104]
DU-145 cell line	6.25–200 μM	Antiproliferative	↑Apoptosis; ↑Bax; ↑RASD1; ↑caspase-3; ↑PARP; ↓Bcl-2	[106]
PC-3 cell line	25–100 μM	Antiproliferative	↑Bax/Bcl-2 ratio; ↓p38MAPK/AKT pathway	[107]

AKT, protein kinase B; Bax, Bcl-2-associated protein; Bcl-2, B-cell lymphoma 2; Bcl-xL, B-cell lymphoma-extra-large; CDK, cyclin-dependent kinase; EMT, epithelial-mesenchymal transition; ERK, extracellular signal regulated kinase; EphB3, ephrin type-B receptor 3; HDAC5, histone deacetylase 5; HIF-1α, hypoxia-inducible factor 1α; IGF-1, insulin-like growth factor 1; IGF-1R, insulin-like growth factor 1 receptor; MAPK, mitogen-activated protein kinase; miR, microRNA; MMP, matrix metalloproteinase; NAG-1, NSAID-activated gene; PI3K, phosphatidylinositol 3-kinase; PTEN, phosphatase and tensin homolog; p21, cyclin-dependent kinase inhibitor; PARP, poly-ADP ribose polymerase; RASD, ras-related dexamethasone induced; STAT3, signal transducer and activator of transcription 3; TIMP, tissue inhibitor of metalloproteinase; VEGF, vascular endothelial growth factor.

One of the primary factors associated with cell apoptosis is the Bax and Bcl-2 protein levels, usually coupled with the AKT and extracellular signal regulated kinase (ERK) pathway. Bax protein is crucial for apoptosis in normal cells to prevent excessive proliferation and possible tumor formation. However, Bcl-2 functions in the opposite manner: enhancing cell survival by suppressing apoptosis [113]. Formononetin treatment reduced Bcl-2 protein levels and upregulated the pro-apoptotic proteins, such as Bax and caspase-3, thereby increasing the Bax/Bcl-2 ratio and inducing apoptosis. This effect was observed in several cancers, including prostate cancer (PC-3) [107], osteosarcoma (U2OS) [105], non-small cell lung cancer (A549 and NCI-H23) [96], and colon carcinoma (HCT-116) [37], and was sometimes coupled with caspase and phosphorylation activity. In PC-3 cells, increased phosphorylation of p38 and blocked AKT phosphorylation accompanied the growth in the Bax/Bcl-2 ratio to induce apoptosis [107]. An elevated activation and cleavage of caspase-3 was observed in A549 and NCI-H23 non-small cell lung cancer cell lines [96], and antiapoptotic proteins Bcl-2 and Bcl-xL were downregulated in HCT-116 colon cancer cells, along with caspase activation [37]. Furthermore, the upstream regulator and a novel pro-apoptotic protein, non-steroidal anti-inflammatory drug (NSAID)-activated gene (NAG-1) was found to be overexpressed in formononetin-treated HCT-116 cells, potentially promoting the apoptotic effects of formononetin; however, it failed to induce phase-specific cell cycle arrest [37]. In a recent study formononetin was found to inhibit the growth of osteogenic sarcoma U2OS cells and induce apoptosis by modulating the intracellular miR-375/ERα-PI3K/AKT signaling pathway [114].

Studies conducted thus far indicate that formononetin supports the apoptotic process of cancer cells through intrinsic pathways, but the mechanism and pathways may differ between different cancers. The most evident pathways suppressed by formononetin to induce apoptosis are the PI3K/AKT and the extracellular signal regulated kinase (ERK) pathways, which are the main regulatory pathways found dysregulated in many cancers. These pathways regulate the cell cycle and transmit signals from the surface of the cell to the nuclear DNA [115,116]. Some examples include cervical cancer (HeLa), osteosarcoma (U2OS) [90,105], and colorectal cancer (RKO) [101]. Other studies showed that formononetin successfully promoted apoptosis in prostate cancer (DU-145 and PC-3) [104,106,107], breast cancer (MCF-7 and MDA-MB-231) [92,93,108], non-small cell lung cancer (A549 and NCI-H23) [96], colon cancer (HCT116 and SW1116) [37,85], and nasopharyngeal carcinoma (CNE1 and CNE2) [112] via modulation of aforementioned pathways. Formononetin inactivated the MAPK signaling pathway, causing apoptosis in MCF-7 breast cancer cells and increased expression levels of Bax in LNCaP and PC-3 prostate cancer cells, leading to apoptosis [94]. Another commonly dysregulated pathway in cancer is the STAT3 signaling pathway. Emerging evidence suggests that abnormal STAT3 signaling drives the initiation and progression of human cancers through the inhibition of apoptosis and driving multiple pro-oncogenic functions [57,66,72,117–120]. Formononetin has been found to suppress fibroblast growth factor 2 (FGF2)-induced STAT3 activation, successfully suppressing multiple myeloma, leukemia, lymphoma and solid tumors that display constitutive STAT3 activation [121]. Furthermore, formononetin has been found to elicit greater antiproliferative and proapoptotic activity as compared to other isoflavones, such as calycosin [90].

3.3. Induction of Cell Cycle Arrest

Treatment of various cancer cell lines using formononetin has been shown to attenuate cancer cell proliferation via cell cycle arrest. Formononetin has been demonstrated to be highly effective in suppressing the oncogenic PI3K/AKT pathway [122–132] and inducing cell cycle arrest in numerous cell lines, including breast cancer (MCF-7), colorectal cancer (HCT-116 and SW1116) [85], non-small cell lung cancer [96], and prostate cancer (PC-3 and DU-145) [95]. Formononetin can induce cell cycle arrest at the G0/G1 phase by modifying the level of expression of cell cycle arrest-associated proteins and growth factors, as observed in DU-145 and PC-3 prostate cancer cells [95], A549 and NCI-H23 lung adenocarcinoma cells [96], ES2 and OV90 ovarian cancer cells [99], and MCF-7 breast cancer cells [91]. In non-small cell lung cancer, proliferation of A549 and NCI-H23 cells was significantly inhibited with treatment with formononetin. In human non-small cell lung cancer cells, formononetin induced cell cycle arrest in the G1-phase and promoted apoptosis via increasing p21 expression and reducing cyclin A and D1 expression in a time- and concentration-dependent manner [96].

Similarly, formononetin induced G1 arrest in PC-3 and DU-145 prostate cancer cells by downregulating cyclin D1, protein kinase B (AKT), and cyclin dependent kinase 4 (CDK4) in a concentration-dependent manner, and this phenomenon was observed to be more significant in PC-3 cells than DU-145 cells [95]. Inactivation of AKT facilitated the phosphorylation of glycogen synthase kinase-3β (GSK-3β), which, coupled with the reduction of cyclin D1, can deter progression into the G1/S phase [133]. Likewise, formononetin induced G0/G1 cell cycle arrest in MCF-7 cells by inactivating the IGF-1/IGFR-PI3K/AKT pathway [91]. On the contrary, cell cycle arrest in SW1116 and HCT-116 colon carcinoma cells at the G0/G1 checkpoint was induced through the suppression of cyclin D1 in a concentration-dependent manner, without changes in the expression levels of cyclin B1, which propagated the number of cells undergoing G1 phase cell cycle arrest to 79.7% upon treatment with 100 μM formononetin [85]. Collectively, these results suggest that the administration of formononetin modulated the expression levels of crucial cell cycle regulators through different pathways, consequently leading to cell cycle arrest at the G1 phase and apoptosis in various cancer cell lines.

3.4. *Antioxidant Effects*

Formononetin has been investigated for its antioxidant properties and is known for being a potent and effective natural antioxidant capable of protecting cells from most free radicals. This is a crucial property of formononetin that may support its use as an anticancer agent because free radicals can lead to cancer development [134]. Formononetin is also the most effective isoflavone in inhibiting lipid peroxidation as it possesses the strongest antioxidant activity [135].

3.5. *Angiogenesis-Modulating Effects*

Tumor angiogenesis is an essential pathway for the development and progression of malignant tumors [136], and studies have shown that the PI3K pathway may hold a vital role in this process [137]. It is a crucial process that leads to the growth and spread of cancer. The effect of formononetin on angiogenesis differs for different cancers. In one study, formononetin downregulated the expression of key pro-angiogenic factors, such as vascular endothelial growth factor (VEGF) and matrix metalloproteinases (MMPs), in LoVo human colorectal adenocarcinoma cells and reduced xenografted tumor size and the number of proliferating cells in the tumor tissues with decreased serum VEGF level [109]. MMP-2 and MMP-9 are known to be directly associated with tumor angiogenesis [138].

The most well-known pathway involved in antiangiogenic therapies currently under evaluation in clinical trials targets the VEGF pathway. However, there is a chance of the tumor acquiring resistance to the VEGF-targeted therapy by shifting to other angiogenesis mechanisms [121], therefore rendering the antiangiogenic treatment ineffective. Consequently, there is a need to develop alternative therapeutic agents that inhibits other non-VEGF angiogenic pathways. Other angiogenesis inducers, apart from VEGF, are the fibroblast growth factors basic fibroblast growth factor (b-FGF/FGF-2), making them a potential drug target for cancer, such as melanoma [137,139]. Formononetin has been found to be a novel FGF receptor 2 inhibitor as it suppressed the sprouting of FGF2-induced micro-vessel in rat aortic rings and angiogenesis, specifically targeting the FGF receptor 2-mediated AKT signaling pathway, resulting in the attenuation of tumor growth and angiogenesis [121]. As a result, formononetin could be investigated in future studies as an angiogenesis inhibitor through targeting FGFs.

3.6. *Metastasis-Regulatory Effects*

When highly metastatic MDA-MB-231-luc and 4TI breast cancer cells were exposed to formononetin, no significant difference in the cell viability was observed as compared to untreated cells [108]. However, treatment with formononetin (2.5–40 μmol/L) reduced the migration of MDA-MB-231 and 4TI cells in a concentration-dependent manner. Furthermore, formononetin reduced the invasion of MDA-MB-231 and 4T1 cells. Invasiveness of metastatic cancer was also attenuated in the LoVo human colon cancer cell line [109]. This phenomenon may have been due to the reduced expression levels of MMPs. MMP-2 and MMP-9 levels were reduced concentration-dependently with formononetin treatment, and formononetin elevated the expression levels of tissue inhibitor of MMP-1 and MMP-2, which are negative regulators of MMPs [108], suggesting that formononetin has the ability to influence the expression levels of proteins and genes associated with proteolytic activation.

3.7. Anti-Inflammatory Effects

One of the hallmarks of cancer development is chronic inflammation. It can drive tumor progression by regulating proliferation, invasion, metastasis, angiogenesis, chemoresistance, and radio-resistance of tumor cells [1,2,38,60,61,102]. NF-κB, a pro-inflammatory transcription factor, is associated with the inflammation and suppression of apoptosis, and is a key driver of different cellular processes in multiple cancer models [21,50,60,62,64]. NF-κB can translocate into the nucleus and activate numerous genes that are involved in multiple processes crucial for multistage carcinogenesis, including proliferation, invasion, and angiogenesis, upon activation by cytokines or chemotherapeutic agents. As a result, chemotherapeutics that can potentially inhibit NF-κB have a significant role in anticancer therapy [5,12,21,38,50,58,60,62,102,129]. Formononetin demonstrated an inhibitory effect on NF-κB activation, and NF-κB is a significant transcription factor for the induction of nitric oxide synthase to decrease the production of nitric oxide in vitro [140]. The inhibition of NF-κB suggests the potential crucial and therapeutic effect that formononetin may have in protecting against inflammation, which may result in cancer development. Overall, the important oncogenic and pro-inflammatory pathways affected by formononetin have been depicted in Figure 2.

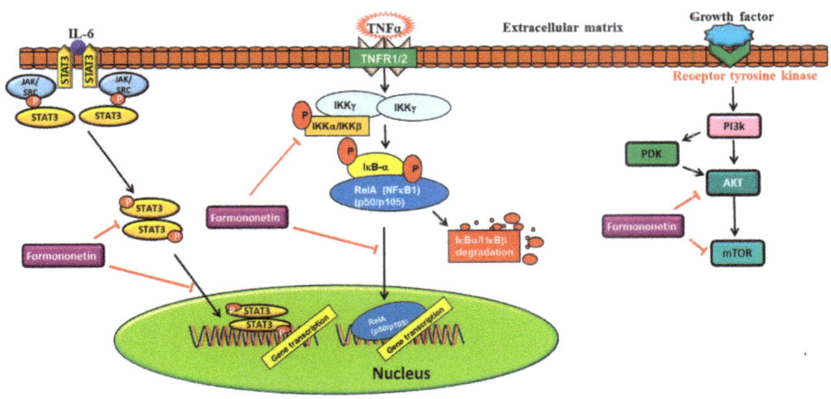

Figure 2. Formononetin regulates major oncogenic pathways involved in cancer progression. ⊥, inhibition/downregulation; ↑ upregulation/activation.

3.8. Combinatorial Studies with Selected Chemotherapeutics

Previous clinical studies suggest that polysaccharides obtained from *Astragalus* plants can counteract the adverse effects of chemotherapeutic drugs, including a significant reduction of myelosuppression in cancer patients [141]. In addition to the antiproliferative and proapoptotic properties of formononetin, the expression level of p53 was concentration-dependently upregulated after treatment with formononetin through increased phosphorylation of p53 at Ser15 and Ser20, enhancing its transcriptional activity [96]. For example, formononetin demonstrated synergy when coupled with the use of other chemotherapeutic drugs. Temozolomide (TMZ) is an oral chemotherapy drug often used in the treatment for certain brain cancers, such as glioblastoma multiforme. However, TMZ is a chemotherapeutic drug that is known to causes adverse side effects, including hematologic complications and both intrinsic and acquired resistance [110,142]. Although results found that both formononetin and TMZ alone were sufficient to inhibit the growth of C6 glioma cells concentration-dependently, when formononetin is used in combination with TMZ, it displayed a synergistic effect on C6 cells. The combination of both drugs increased Bax protein expression and cleaved caspase-3 and caspase-9, attenuated Bcl-2 expression, and promoted tumor cell apoptosis [110]. Furthermore, the drug combination prevented the migration of C6 glioma cells due to the down-regulated expression of MMP-2 and MMP-9. This suggests the potential use of formononetin as a combination drug during chemotherapy or as a post-operative adjuvant therapy to curb the adverse effects displayed by other chemotherapeutic drugs, such as TMZ. Co-treatment consisting of formononetin with synthetic inhibitors for ovarian cancer, such as LY294002 (PI3K inhibitor) or U0126 [mitogen activated protein kinase kinase (MEK) inhibitor], further prevented the proliferative effects on ovarian cancer cells (ES2 and OV90), thus increasing the occurrence of apoptosis in both cell lines [99].

Furthermore, a potent chemotherapy drug doxorubicin has been identified to induce epithelial-mesenchymal transition (EMT) in glioma cells via elevated expression of vimentin and reduced expression of E-cadherin in U87MG glioma cells [111]. Numerous studies have suggested the important role that EMT holds in carcinogenicity, metastasis, progression, and acquired chemoresistance [129,143–146]. Formononetin has been proven to be able to sensitize glioma cells to doxorubicin, and combination therapy using doxorubicin and formononetin is able to successfully reverse the induction of EMT by doxorubicin. In addition, histone deacetylase 5 (HDAC5) has been identified to enhance glioma cell proliferation, and doxorubicin-treated glioma cells have been identified to have significantly increased HDAC5 levels. However, co-treatment with formononetin reduced the expression of HDAC5 in glioma cells. These results implied that co-treatment with formononetin and other chemotherapeutic drugs, such as doxorubicin, potentially sensitizes cancer cells, such as glioma cells, through the prevention of EMT and the inhibition of HDAC5 [111]. This further suggests that formononetin could be considered as an adjuvant agent with existing chemotherapeutic drugs.

3.9. Novel Semi-Synthetic Hybrids of Formononetin

Several new bioactive derivatives of formononetin have been shown to have potent anti-cancer activity. Ren et al., 2012 showed that formononetin nitrogen mustard derivative (IC50-3.8 μM) exhibited potent antitumor activity against colorectal HCT-116 cells and was associated with G2/M phase cell cycle arrest and induction of apoptosis [147]. In another study, formononetin-dithiocarbamate hybrid (IC50-1.97 μM) inhibited androgen independent prostate cancer PC3 cell growth and induced apoptosis by modulating the MAPK and wingless (Wnt) signaling pathway [148]. Several lines of evidence suggest that epidermal growth factor receptor (EGFR) is an attractive target for non-small cell lung cancer (NSCLC) therapy. Lin et al 2017 reported on a new series of formononetin derivatives following the binding model of lapatinib to EGFR. Formononetin derivatives exhibited potent anti-proliferative activity against triple negative breast cancer MDA-MB-231 cells and induced apoptosis by down-regulating multiple EGFR/PI3K/Akt/Bcl-2-associated death promoter (Bad), EGFR/ERK and EGFR/PI3K/Akt/β-catenin signaling pathways in breast cancer cells [149]. In another study a formononetin 7-phosphoramidate derivative significantly induced early apoptosis in HepG-2 cells [150]. In a recent study by Chengli et al. [151] three new derivatives of formononetin were shown to inhibit the growth, invasion and migration of A549 lung cancer cells. Furthermore, Bohong et al. [152] demonstrated that multiwalled carbon nanotube–formononetin (MWCNT-FMN) composite for sustained delivery induced apoptosis by activating reactive oxygen species (ROS) production in cervical carcinoma HeLa cells.

4. In Vivo Anticancer Pharmacological Activities of Formononetin

The anticancer potential of formononetin has been demonstrated in numerous in vivo models as shown in Table 2. First, formononetin has demonstrated the potential to inhibit tumor proliferation in various murine models of cancers. As compared to controls, formononetin is able to significantly and dose-dependently inhibit local tumor growth in nude mice bearing MCF-7 human breast cancer [91]. The weight of the tumor in the formononetin-treated group was reduced significantly when compared to the control, and this was substantiated with a reduction of 39.6% in tumor weight. Similar effects have been observed using human colorectal cancer cell lines. HCT-116 nude mice xenografts (CCL-247) treated with formononetin exhibited a large reduction in tumor volume and the number of proliferating cells as compared to the vehicle-treated group [109]. Mortality and significant changes in body weight were not observed in the formononetin-treated HCT-116 nude mice xenografts, indicating that formononetin displayed tolerable toxicity while being able to significantly reduce tumor size and mass in vivo. This observation was substantiated with comparable white blood cell counts between the formononetin-treated and control group, suggesting that formononetin did not exhibit neutropenic effects in preclinical models, unlike other common chemotherapeutic drugs [109]. Furthermore, formononetin has been proven to successfully inhibit the growth of tumors in vivo through targeting the TNF-α/NF-κB pathway in colorectal tumor bearing nude mice [101], and both tumor weight and volume decreased dose-dependently after treatment with formononetin.

The antitumor potential of formononetin is also observed in human multiple myeloma xenografts in nude mice, with a decrease in Ki-67 expression levels in tumor tissues [49]. Oxidative stress induced through formononetin treatment impeded the expression of phosphorylated STAT3 and STAT5 through the STAT3 and STAT5 signaling axis by successfully removing the binding ability of both transcription factors; thereby reversing their activation and consequently suppressing tumor growth in multiple myeloma [49]. Similarly, the antiproliferative effect was also observed in vivo in U2OS osteosarcoma and PC-3 and DU-145 prostate cancer nude mouse xenografts in a dose-dependent manner through a significant reduction in tumor weight [95,105]. This further affirms the potential application of formononetin as a cancer therapeutic agent. However, it is important to note that varying concentrations of phytoestrogens may exude different effects on the proliferative and apoptotic properties of different cancer models.

In vivo mouse metastasis models were used to study the inhibitory effects of formononetin on metastatic breast cancer cells (MDA-MB-231-luc and 4T1) for lung metastasis [108]. Formononetin is able to successfully diminish the development of lung metastasis in mouse xenografts with metastatic breast cancer, which suggests that formononetin may possess antimigration and invasion properties on breast cancer cells and therefore, can increase survival time in preclinical models. This is an important finding that could promote the use of formononetin in clinical settings for chemotherapy because it plays a protective role against breast cancer metastasis.

Formononetin has been investigated for its ability to exhibit antioxidant properties in in vivo models. Antioxidant and estrogenic properties are important factors that could catalyze the development of cancers as they are found to be involved in oxidative damage to the cellular macromolecules. Several antioxidant enzymes, including superoxide dismutase (SOD), catalase (CAT), and glutathione peroxidase (GSH-Px), are essential to counter the reactive oxygen species in vivo [153]. Preclinical studies using ovariectomized mice supported the use of formononetin as a potent free radical scavenging molecule that can prevent lipid peroxidation, which is a crucial factor in the progression of cancer [154].

The pharmacological property of formononetin in vivo is dependent on the level of estrogen present, which can be proestrogenic or antiestrogenic [92]. As previously mentioned, free radicals are highly reactive chemicals that can damage several major components of cells, especially the DNA, playing a pivotal role in the development of cancer. The oxidative effects of formononetin were monitored through the levels of SOD, GSH-Px, CAT, and malondialdehyde (MDA). A high and low formononetin intake for 6 months was found to increase uterine weight and the levels of SOD, GSH-Px, and CAT, and reduced MDA content in preclinical conditions, suggesting that formononetin displayed obvious antioxidant and estrogenic effects in in vivo conditions, and the estrogenic property of formononetin is not dose-dependent [153].

Table 2. In vivo anticancer effects of formononetin.

Cancer Model	Dose, Duration and Route of Administration	Observed Effects	Mechanisms	References
MCF-7 cells-induced xenograft in Balb/c nude mice	60 mg/kg/day; 20 days; i.p.	↓Tumor growth	↓IGF-1/IGFR-PI3K/AKT pathway	[91]
MDA-MB-231 cells-induced xenografts in Balb/c nude mice	100 mg/kg/day; 25 days; intra-gastric (i.g.)	↓Tumor growth (synergistic effect with sunitinib)	↓FGF2-induced angiogenesis; ↓PI3K/AKT pathway; ↓STAT3 pathway	[121]
MDA-MB-231-luc cells-induced xenografts in Balb/c nude mice	10 or 20 mg/kg/day; once every 2 days for 35 days; i.p.	↓Tumor growth; ↓lung metastasis; ↑overall survival	↓PI3K/AKT pathway	[108]
HCT-116 cells-induced xenografts in Balb/c nu/nu mice	20 mg/kg/day; 2 weeks; i.p.	↓Tumor growth	↓Tumor angiogenesis; ↓VEGF	[109]
Human multiple myeloma U266 xenograft in Balb/c nude mice	20 and 50 mg/kg/day; 25 days; i.g.	↓Tumor growth	↓PI3K/AKT pathway	[92]
Human multiple myeloma tumor tissues implanted in athymic nu/nu mice	40 mg/kg; thrice/week for 3 weeks; i.p.	↓Tumor growth	↓Tumor angiogenesis; ↓STAT3/5 pathway; ↓VEGF	[49]
PC-3 cells-induced prostate xenograft in nude mice	60 mg/kg/day; 20 days; i.p.	↓Tumor growth	↑Apoptosis; ↓IGF/IGFR1 pathway	[95]
RKO tumor-bearing Balb/c nude mice	5, 10 or 20 mg/kg/day; 14 days; i.g.	↓Tumor growth	↓IL6; ↓TNF-α; ↓NF-κB pathway	[101]

AKT, protein kinase B; FGF2, fibroblast growth factor 2 receptor; IGF-1, insulin-like growth factor 1; IGF-1R, insulin-like growth factor 1 receptor; IL-6, interleukin-6; NF-κB, nuclear factor-κB; PI3K, phosphatidylinositol 3-kinase; STAT3, signal transducer and activator of transcription 3; TNF-α, tumor necrosis factor-α; VEGF, vascular endothelial growth factor.

Further studies conducted to analyze the estrogenic effects of formononetin showed that it significantly elevated the expression of atrial ER subtype ERβ in ovariectomized mice [155]. Although it has not been yet demonstrated that formononetin is able to negatively impact the ER in preclinical models, this study provides us with insight that formononetin could likely affect the regulation of ERs, which are essential in the development of hormone-sensitive cancers, such as breast and ovarian cancer.

Tumor angiogenesis is a crucial process for tumor growth and metastasis for the development of cancer. The most well-established effect is the role of VEGF and VEGFR2. However, as mentioned earlier, there is evidence that the tumor can become resistant to VEGF-targeted therapy and acquire resistance against monotherapy with VEGFR inhibitors that manipulate other angiogenesis mechanisms [156]. The antiangiogenic property of formononetin is observed in nude mouse xenografts through the reduced levels of VEGF in drug-treated animals when compared against controls [109]. In preclinical models, formononetin displayed antiangiogenic properties in a myriad of cancers, especially colorectal cancer (HCT-116) [109] and breast cancer (MDA-MB-231-luc) [108]. The HCT-116 human metastatic colorectal adenocarcinoma cell line is known for its invasiveness and metastatic properties. Following treatment with formononetin, HCT-116 tumor xenografts in mice displayed a notable reduction in the number of invaded cells [109]. In another study, formononetin at doses of 25, 50 or 100 mg/kg administered for 14 days significantly suppressed the growth of subcutaneously implanted osteogenic sarcoma U2OS tumor growth and was associated with the downregulation of miR-375 with concomitant upregulation of Bax, Caspase-3, and Apaf-1 [114]. Further studies in preclinical models with small molecules have been of interest in recent years as a therapeutic measure against cancer since multi-kinase inhibitors targeting VEGFR have been proven effective in clinical settings against breast tumors. However, these agents displayed signs of toxicity and did not exhibit high response rates. Formononetin derivative when administered intraperitoneally (i.p.) for 21 days (5 mg/kg) was shown to significantly inhibit breast tumor growth in a nude mice model when compared to EGFR inhibitors gefitinib or lapatinib and was found to be well tolerated with no significant change in the body weight of the mice [149].

Further research in the preclinical setting has found evidence that formononetin could be effective for antiangiogenesis treatment against human breast cancer (MDA-MB-231) xenografted in nude mice by preventing tumor growth via inhibition of tumor angiogenesis [121]. Furthermore, formononetin supplemented the effect of sunitinib, a receptor tyrosine kinase inhibitor that targets VEGFR2, on tumor growth inhibition through largely decreasing the invasiveness of cancer cells stimulated by tumor growth in vivo through the fibroblast growth factor receptor 2 (FGFR2)-mediated AKT signaling pathway, attenuating tumor growth and angiogenesis [121,157]. The inhibition of angiogenesis in xenografted human breast cancer (MDA-MB-231) by formononetin was found to modulate FGF2-induced micro-vessel growth that suppressed the emergence of rat aortic rings and angiogenesis through the repression of FGF2-initiated activation of FGFR2 and AKT signaling [121].

5. Formononetin in Clinical Studies

Formononetin has been tested in preclinical settings for other diseases, such as Alzheimer's [158], but clinical studies for the use of formononetin as a treatment method for cancer have yet to be conducted. Jarred and colleagues [159] identified that treatment with a red clover-derived isoflavone mixture was able to induce apoptosis in low- to moderate-grade human prostate carcinoma. Further clinical research needs to be performed to ascertain the anti-cancer properties of formononetin in multiple cancer types.

6. Conclusion and Future Perspectives

Evidence presented in this review provides a comprehensive summary on the potential anticancer properties of formononetin in both in vitro and in vivo studies and the current progress of clinical studies. Numerous molecular targets and mechanisms of actions are involved in the antitumorigenic property (primarily on the induction of cell apoptosis and the inhibition of cell proliferation) of formononetin as evidenced from numerous in vitro studies, whereas the safety and efficacy of formononetin and its metabolites in biological systems are further confirmed in in vivo studies. The tumor-inhibitory effects of formononetin have been associated with the modulation of PI3K/AKT and STAT3 signaling pathways in both in vitro and in vivo models. The various anticancer molecular targets of formononetin are briefly summarized in Figure 3. In addition, formononetin has been found to possess additive and synergistic effects with chemotherapeutic drugs, such as Sunitinib. The potential role for formononetin to be used as an adjunct therapy and future prospective drug development for cancer patients are supported through these findings. However, further studies need to be conducted (both in vivo and clinical studies) to allow for further assessment of the efficacy and safety of formononetin for prevention and treatment of various cancer types. This is crucial as various formononetin derivatives and metabolites have varying pharmacokinetic properties and activities that need to be fully elucidated, and this requires further investigation to ensure that this bioactive phytochemical is safe for clinical development. The significant antitumor properties of formononetin make it a novel candidate for anticancer drug development.

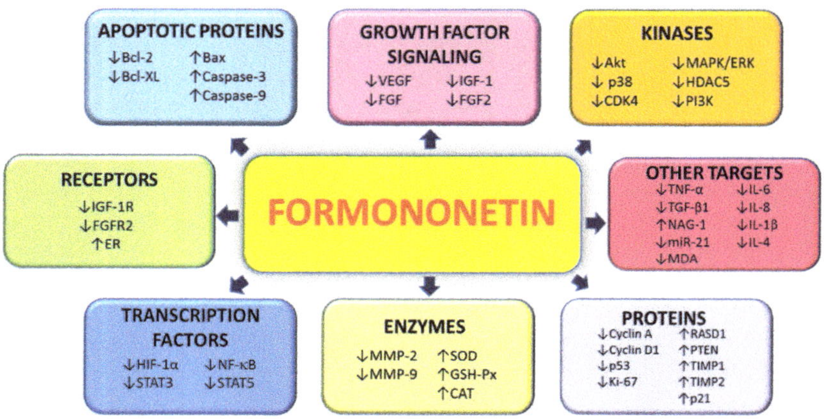

Figure 3. Molecular targets influenced by formononetin in cancer.

Author Contributions: Conceptualization, S.K.L.O., M.K.S., L.F., K.S.A., G.S. and A.B.; writing—original draft preparation, S.K.L.O., L.F. and M.K.S.; writing—review and editing, G.S., K.S.A., S.E.F., F.A. and A.B.; funding acquisition, G.S. and K.S.A.

Funding: This work was supported by grants from National Research Foundation (NRF)-Korean government awarded to K.S.A (NRF-2015R1A4A1042399, 2017R1A6A3A11031224 and 2018R1D1A1B07042969).

Conflicts of Interest: The authors declare that they have no competing interests.

Abbreviations

AKT	protein kinase B
AP-1	activator protein 1
Bax	Bcl-2-associated protein
Bcl-2	B-cell lymphoma 2
Bcl-xL	B-cell lymphoma-extra-large
b-FGF	basic fibroblast growth factor
CAT	catalase
CDK	cyclin-dependent kinase
CD4	cluster of differentiation 4
CREB	cAMP response element-binding protein
DMSO	dimethyl sulfoxide
EGR3	early growth response protein 3

EMT	epithelial-mesenchymal transition
ERK	extracellular signal regulated kinase
FGF	fibroblast growth factor
FGF2	fibroblast growth factor 2 receptor
GSH-Px	glutathione peroxidase
GSK-3β	glycogen synthase kinase 3β
HDAC5	histone deacetylase 5
HIF-1α	hypoxia-inducible factor 1α
IGF-1	insulin-like growth factor 1
IGF-1R	insulin-like growth factor 1 receptor
IL-1β	interleukin 1β
IL-4	interleukin-4
IL-6	interleukin-6
IL-8	interleukin-8
MAPK	mitogen-activated protein kinase
MDA	malondialdehyde
miR-21	microRNA 21
MMP	matrix metalloproteinase
NAG-1	NSAID-activated gene
NF-κB	nuclear factor-κB
PI3K	phosphatidylinositol 3-kinase
PTEN	phosphatase and tensin homolog
ROCK	Rho-associated protein kinase
siRNA	small interfering ribonucleic acid
SOD	superoxide dismutase
STAT3	signal transducer and activator of transcription 3
Sul-F	formononetin-3′-sulphonate
TMZ	temozolomide
TNF-α	tumor necrosis factor-α
VCAM-1	vascular cell adhesion protein 1
VEGF	vascular endothelial growth factor
VEGFR	vascular endothelial growth factor receptor
Wnt	Wingless

References

1. Martin, T.A.; Ye, L.; Sanders, A.J.; Lane, J.; WG, J. Cancer invasion and metastasis: Molecular and cellular perspective. In *Metastatic Cancer Clinical Biological Perspectives*; Landes Bioscience 2000–2013: Austin, TX, USA, 2013. [CrossRef]
2. Cooper, G.M. The development and causes of cancer. In *The Cell: A Molecular Approach*, 2nd ed.; Sinauer Associates: Sunderland, MA, USA, 2000.
3. Ferlay, J.; Colombet, M.; Soerjomataram, I.; Mathers, C.; Parkin, D.M.; Pineros, M.; Znaor, A.; Bray, F. Estimating the global cancer incidence and mortality in 2018: Globocan sources and methods. *Int. J. Cancer* **2018**. [CrossRef]
4. Bray, F.; Ferlay, J.; Soerjomataram, I.; Siegel, R.L.; Torre, L.A.; Jemal, A. Global cancer statistics 2018: Globocan estimates of incidence and mortality worldwide for 36 cancers in 185 countries. *CA Cancer J. Clin.* **2018**, *68*, 394–424. [CrossRef]
5. Shanmugam, M.K.; Kannaiyan, R.; Sethi, G. Targeting cell signaling and apoptotic pathways by dietary agents: Role in the prevention and treatment of cancer. *Nutr. Cancer* **2011**, *63*, 161–173. [CrossRef] [PubMed]
6. Clarke, G.; Johnston, S.; Corrie, P.; Kuhn, I.; Barclay, S. Withdrawal of anticancer therapy in advanced disease: A systematic literature review. *BMC Cancer* **2015**, *15*, 892. [CrossRef] [PubMed]
7. Onakpoya, I.J.; Heneghan, C.J.; Aronson, J.K. Post-marketing withdrawal of 462 medicinal products because of adverse drug reactions: A systematic review of the world literature. *BMC Med.* **2016**, *14*, 10. [CrossRef] [PubMed]

8. Pan, H.; Gray, R.; Braybrooke, J.; Davies, C.; Taylor, C.; McGale, P.; Peto, R.; Pritchard, K.I.; Bergh, J.; Dowsett, M.; et al. 20-year risks of breast-cancer recurrence after stopping endocrine therapy at 5 years. *N. Engl. J. Med.* **2017**, *377*, 1836–1846. [CrossRef] [PubMed]
9. Demain, A.L.; Vaishnav, P. Natural products for cancer chemotherapy. *Microb. Biotechnol.* **2011**, *4*, 687–699. [CrossRef]
10. Neill, U.S. From branch to bedside: Youyou tu is awarded the 2011 lasker~debakey clinical medical research award for discovering artemisinin as a treatment for malaria. *J. Clin. Investig.* **2011**, *121*, 3768–3773. [CrossRef]
11. Bishayee, A.; Sethi, G. Bioactive natural products in cancer prevention and therapy: Progress and promise. *Semin. Cancer Biol.* **2016**, *40–41*, 1–3. [CrossRef]
12. Newman, D.J.; Cragg, G.M. Natural products as sources of new drugs over the 30 years from 1981 to 2010. *J. Nat. Prod.* **2012**, *75*, 311–335. [CrossRef]
13. Cragg, G.M.; Pezzuto, J.M. Natural products as a vital source for the discovery of cancer chemotherapeutic and chemopreventive agents. *Med. Princ. Pract.* **2016**, *25*, 41–59. [CrossRef] [PubMed]
14. Deorukhkar, A.; Krishnan, S.; Sethi, G.; Aggarwal, B.B. Back to basics: How natural products can provide the basis for new therapeutics. *Exp. Opin. Invest. Drugs* **2007**, *16*, 1753–1773. [CrossRef] [PubMed]
15. Yang, S.F.; Weng, C.J.; Sethi, G.; Hu, D.N. Natural bioactives and phytochemicals serve in cancer treatment and prevention. *Evid. Based Complement Altern. Med.* **2013**, *2013*, 698190. [CrossRef] [PubMed]
16. Tang, C.H.; Sethi, G.; Kuo, P.L. Novel medicines and strategies in cancer treatment and prevention. *BioMed. Res. Int.* **2014**, *2014*, 474078. [CrossRef] [PubMed]
17. Hsieh, Y.S.; Yang, S.F.; Sethi, G.; Hu, D.N. Natural bioactives in cancer treatment and prevention. *BioMed. Res. Int.* **2015**, *2015*, 182835. [CrossRef]
18. Yarla, N.S.; Bishayee, A.; Sethi, G.; Reddanna, P.; Kalle, A.M.; Dhananjaya, B.L.; Dowluru, K.S.; Chintala, R.; Duddukuri, G.R. Targeting arachidonic acid pathway by natural products for cancer prevention and therapy. *Semin. Cancer Biol.* **2016**, *40–41*, 48–81. [CrossRef]
19. Hasanpourghadi, M.; Looi, C.Y.; Pandurangan, A.K.; Sethi, G.; Wong, W.F.; Mustafa, M.R. Phytometabolites targeting the warburg effect in cancer cells: A mechanistic review. *Curr. Drug Targets* **2017**, *18*, 1086–1094. [CrossRef]
20. Shanmugam, M.K.; Warrier, S.; Kumar, A.P.; Sethi, G.; Arfuso, F. Potential role of natural compounds as anti-angiogenic agents in cancer. *Curr. Vasc. Pharmacol.* **2017**, *15*, 503–519. [CrossRef]
21. Aggarwal, B.B.; Sethi, G.; Baladandayuthapani, V.; Krishnan, S.; Shishodia, S. Targeting cell signaling pathways for drug discovery: An old lock needs a new key. *J. Cell Biochem.* **2007**, *102*, 580–592. [CrossRef]
22. Jung, Y.Y.; Hwang, S.T.; Sethi, G.; Fan, L.; Arfuso, F.; Ahn, K.S. Potential anti-inflammatory and anti-cancer properties of farnesol. *Molecules* **2018**, *23*, 2827. [CrossRef]
23. Merarchi, M.; Sethi, G.; Fan, L.; Mishra, S.; Arfuso, F.; Ahn, K.S. Molecular targets modulated by fangchinoline in tumor cells and preclinical models. *Molecules* **2018**, *23*, 2538. [CrossRef] [PubMed]
24. Sethi, G.; Shanmugam, M.K.; Warrier, S.; Merarchi, M.; Arfuso, F.; Kumar, A.P.; Bishayee, A. Pro-apoptotic and anti-cancer properties of diosgenin: A comprehensive and critical review. *Nutrients* **2018**, *10*, 645. [CrossRef] [PubMed]
25. Ko, J.H.; Sethi, G.; Um, J.Y.; Shanmugam, M.K.; Arfuso, F.; Kumar, A.P.; Bishayee, A.; Ahn, K.S. The role of resveratrol in cancer therapy. *Int. J. Mol. Sci.* **2017**, *18*, 2589. [CrossRef] [PubMed]
26. Tewari, D.; Nabavi, S.F.; Nabavi, S.M.; Sureda, A.; Farooqi, A.A.; Atanasov, A.G.; Vacca, R.A.; Sethi, G.; Bishayee, A. Targeting activator protein 1 signaling pathway by bioactive natural agents: Possible therapeutic strategy for cancer prevention and intervention. *Pharmacol. Res.* **2018**, *128*, 366–375. [CrossRef] [PubMed]
27. Shanmugam, M.K.; Lee, J.H.; Chai, E.Z.; Kanchi, M.M.; Kar, S.; Arfuso, F.; Dharmarajan, A.; Kumar, A.P.; Ramar, P.S.; Looi, C.Y.; et al. Cancer prevention and therapy through the modulation of transcription factors by bioactive natural compounds. *Semin. Cancer Biol.* **2016**, *40–41*, 35–47. [CrossRef] [PubMed]
28. Shanmugam, M.K.; Nguyen, A.H.; Kumar, A.P.; Tan, B.K.; Sethi, G. Targeted inhibition of tumor proliferation, survival, and metastasis by pentacyclic triterpenoids: Potential role in prevention and therapy of cancer. *Cancer Lett.* **2012**, *320*, 158–170. [CrossRef] [PubMed]
29. Shrimali, D.; Shanmugam, M.K.; Kumar, A.P.; Zhang, J.; Tan, B.K.; Ahn, K.S.; Sethi, G. Targeted abrogation of diverse signal transduction cascades by emodin for the treatment of inflammatory disorders and cancer. *Cancer Lett.* **2013**, *341*, 139–149. [CrossRef]

30. Ranaware, A.M.; Banik, K.; Deshpande, V.; Padmavathi, G.; Roy, N.K.; Sethi, G.; Fan, L.; Kumar, A.P.; Kunnumakkara, A.B. Magnolol: A neolignan from the magnolia family for the prevention and treatment of cancer. *Int. J. Mol. Sci.* **2018**, *19*, 2362. [CrossRef]
31. Ko, J.H.; Arfuso, F.; Sethi, G.; Ahn, K.S. Pharmacological utilization of bergamottin, derived from grapefruits, in cancer prevention and therapy. *Int. J. Mol. Sci.* **2018**, *19*, 4048. [CrossRef]
32. Parikh, N.R.; Mandal, A.; Bhatia, D.; Siveen, K.S.; Sethi, G.; Bishayee, A. Oleanane triterpenoids in the prevention and therapy of breast cancer: Current evidence and future perspectives. *Phytochem. Rev.* **2014**, *13*, 793–810. [CrossRef]
33. Shanmugam, M.K.; Dai, X.; Kumar, A.P.; Tan, B.K.; Sethi, G.; Bishayee, A. Oleanolic acid and its synthetic derivatives for the prevention and therapy of cancer: Preclinical and clinical evidence. *Cancer Lett.* **2014**, *346*, 206–216. [CrossRef] [PubMed]
34. Rajendran, P.; Li, F.; Shanmugam, M.K.; Kannaiyan, R.; Goh, J.N.; Wong, K.F.; Wang, W.; Khin, E.; Tergaonkar, V.; Kumar, A.P.; et al. Celastrol suppresses growth and induces apoptosis of human hepatocellular carcinoma through the modulation of stat3/jak2 signaling cascade in vitro and in vivo. *Cancer Prev. Res. (Phila.)* **2012**, *5*, 631–643. [CrossRef] [PubMed]
35. Kannaiyan, R.; Hay, H.S.; Rajendran, P.; Li, F.; Shanmugam, M.K.; Vali, S.; Abbasi, T.; Kapoor, S.; Sharma, A.; Kumar, A.P.; et al. Celastrol inhibits proliferation and induces chemosensitization through down-regulation of nf-kappab and stat3 regulated gene products in multiple myeloma cells. *Br. J. Pharmacol.* **2011**, *164*, 1506–1521. [CrossRef] [PubMed]
36. Akincilar, S.C.; Low, K.C.; Liu, C.Y.; Yan, T.D.; Oji, A.; Ikawa, M.; Li, S.; Tergaonkar, V. Quantitative assessment of telomerase components in cancer cell lines. *FEBS Lett.* **2015**, *589*, 974–984. [CrossRef]
37. Auyeung, K.K.; Ko, J.K. Novel herbal flavonoids promote apoptosis but differentially induce cell cycle arrest in human colon cancer cell. *Invest. New Drugs* **2010**, *28*, 1–13. [CrossRef]
38. Sethi, G.; Shanmugam, M.K.; Ramachandran, L.; Kumar, A.P.; Tergaonkar, V. Multifaceted link between cancer and inflammation. *Biosci. Rep.* **2012**, *32*, 1–15. [CrossRef] [PubMed]
39. Liu, D.; Miao, H.; Zhao, Y.; Kang, X.; Shang, S.; Xiang, W.; Shi, R.; Hou, A.; Wang, R.; Zhao, K.; et al. NF-κB potentiates tumor growth by suppressing a novel target lpts. *Cell Commun. Signal.* **2017**, *15*, 39. [CrossRef] [PubMed]
40. Arora, L.; Kumar, A.P.; Arfuso, F.; Chng, W.J.; Sethi, G. The role of signal transducer and activator of transcription 3 (STAT3) and its targeted inhibition in hematological malignancies. *Cancers* **2018**, *10*. [CrossRef]
41. Lee, J.H.; Kim, C.; Lee, S.G.; Sethi, G.; Ahn, K.S. Ophiopogonin d, a steroidal glycoside abrogates STAT3 signaling cascade and exhibits anti-cancer activity by causing GSH/GSSG imbalance in lung carcinoma. *Cancers* **2018**, *10*. [CrossRef]
42. Wong, A.L.A.; Hirpara, J.L.; Pervaiz, S.; Eu, J.Q.; Sethi, G.; Goh, B.C. Do stat3 inhibitors have potential in the future for cancer therapy? *Expert Opin. Investig. Drugs* **2017**, *26*, 883–887. [CrossRef]
43. Lee, J.H.; Kim, C.; Baek, S.H.; Ko, J.H.; Lee, S.G.; Yang, W.M.; Um, J.Y.; Sethi, G.; Ahn, K.S. Capsazepine inhibits jak/stat3 signaling, tumor growth, and cell survival in prostate cancer. *Oncotarget* **2017**, *8*, 17700–17711. [CrossRef]
44. Zhang, J.; Ahn, K.S.; Kim, C.; Shanmugam, M.K.; Siveen, K.S.; Arfuso, F.; Samym, R.P.; Deivasigamanim, A.; Lim, L.H.; Wang, L.; et al. Nimbolide-induced oxidative stress abrogates stat3 signaling cascade and inhibits tumor growth in transgenic adenocarcinoma of mouse prostate model. *Antioxid. Redox Signal.* **2016**, *24*, 575–589. [CrossRef] [PubMed]
45. Chai, E.Z.; Shanmugam, M.K.; Arfuso, F.; Dharmarajan, A.; Wang, C.; Kumar, A.P.; Samy, R.P.; Lim, L.H.; Wang, L.; Goh, B.C.; et al. Targeting transcription factor stat3 for cancer prevention and therapy. *Pharmacol. Ther.* **2016**, *162*, 86–97. [CrossRef] [PubMed]
46. Subramaniam, A.; Shanmugam, M.K.; Ong, T.H.; Li, F.; Perumal, E.; Chen, L.; Vali, S.; Abbasi, T.; Kapoor, S.; Ahn, K.S.; et al. Emodin inhibits growth and induces apoptosis in an orthotopic hepatocellular carcinoma model by blocking activation of stat3. *Br. J. Pharmacol.* **2013**, *170*, 807–821. [CrossRef] [PubMed]
47. Subramaniam, A.; Shanmugam, M.K.; Perumal, E.; Li, F.; Nachiyappan, A.; Dai, X.; Swamy, S.N.; Ahn, K.S.; Kumar, A.P.; Tan, B.K.; et al. Potential role of signal transducer and activator of transcription (stat)3 signaling pathway in inflammation, survival, proliferation and invasion of hepatocellular carcinoma. *Biochim. Biophys. Acta* **2013**, *1835*, 46–60. [CrossRef] [PubMed]

48. Jung, Y.Y.; Lee, J.H.; Nam, D.; Narula, A.S.; Namjoshi, O.A.; Blough, B.E.; Um, J.Y.; Sethi, G.; Ahn, K.S. Anti-myeloma effects of icariin are mediated through the attenuation of jak/stat3-dependent signaling cascade. *Front. Pharmacol.* **2018**, *9*, 531. [CrossRef]
49. Kim, C.; Lee, S.G.; Yang, W.M.; Arfuso, F.; Um, J.Y.; Kumar, A.P.; Bian, J.; Sethi, G.; Ahn, K.S. Formononetin-induced oxidative stress abrogates the activation of stat3/5 signaling axis and suppresses the tumor growth in multiple myeloma preclinical model. *Cancer Lett.* **2018**, *431*, 123–141. [CrossRef]
50. Li, F.; Zhang, J.; Arfuso, F.; Chinnathambi, A.; Zayed, M.E.; Alharbi, S.A.; Kumar, A.P.; Ahn, K.S.; Sethi, G. Nf-kappab in cancer therapy. *Arch. Toxicol.* **2015**, *89*, 711–731. [CrossRef]
51. Manu, K.A.; Shanmugam, M.K.; Ramachandran, L.; Li, F.; Fong, C.W.; Kumar, A.P.; Tan, P.; Sethi, G. First evidence that gamma-tocotrienol inhibits the growth of human gastric cancer and chemosensitizes it to capecitabine in a xenograft mouse model through the modulation of nf-kappab pathway. *Clin. Cancer Res.* **2012**, *18*, 2220–2229. [CrossRef]
52. Shanmugam, M.K.; Ahn, K.S.; Lee, J.H.; Kannaiyan, R.; Mustafa, N.; Manu, K.A.; Siveen, K.S.; Sethi, G.; Chng, W.J.; Kumar, A.P. Celastrol attenuates the invasion and migration and augments the anticancer effects of bortezomib in a xenograft mouse model of multiple myeloma. *Front. Pharmacol.* **2018**, *9*, 365. [CrossRef]
53. Manu, K.A.; Shanmugam, M.K.; Ramachandran, L.; Li, F.; Siveen, K.S.; Chinnathambi, A.; Zayed, M.E.; Alharbi, S.A.; Arfuso, F.; Kumar, A.P.; et al. Isorhamnetin augments the anti-tumor effect of capeciatbine through the negative regulation of nf-kappab signaling cascade in gastric cancer. *Cancer Lett.* **2015**, *363*, 28–36. [CrossRef] [PubMed]
54. Li, F.; Shanmugam, M.K.; Siveen, K.S.; Wang, F.; Ong, T.H.; Loo, S.Y.; Swamy, M.M.; Mandal, S.; Kumar, A.P.; Goh, B.C.; et al. Garcinol sensitizes human head and neck carcinoma to cisplatin in a xenograft mouse model despite downregulation of proliferative biomarkers. *Oncotarget* **2015**, *6*, 5147–5163. [CrossRef] [PubMed]
55. Mohan, C.D.; Bharathkumar, H.; Dukanya; Rangappa, S.; Shanmugam, M.K.; Chinnathambi, A.; Alharbi, S.A.; Alahmadi, T.A.; Bhattacharjee, A.; Lobie, P.E.; et al. N-substituted pyrido-1,4-oxazin-3-ones induce apoptosis of hepatocellular carcinoma cells by targeting nf-kappab signaling pathway. *Front. Pharmacol.* **2018**, *9*, 1125. [PubMed]
56. Raghunath, A.; Sundarraj, K.; Arfuso, F.; Sethi, G.; Perumal, E. Dysregulation of NRF2 in hepatocellular carcinoma: Role in cancer progression and chemoresistance. *Cancers* **2018**, *10*. [CrossRef] [PubMed]
57. Lee, J.H.; Kim, C.; Ko, J.H.; Jung, Y.Y.; Jung, S.H.; Kim, E.; Kong, M.; Chinnathambi, A.; Alahmadi, T.A.; Alharbi, S.A.; et al. Casticin inhibits growth and enhances ionizing radiation-induced apoptosis through the suppression of stat3 signaling cascade. *J. Cell Biochem.* **2019**, *120*, 9787–9798. [CrossRef] [PubMed]
58. Shanmugam, M.K.; Ahn, K.S.; Hsu, A.; Woo, C.C.; Yuan, Y.; Tan, K.H.B.; Chinnathambi, A.; Alahmadi, T.A.; Alharbi, S.A.; Koh, A.P.F.; et al. Thymoquinone inhibits bone metastasis of breast cancer cells through abrogation of the cxcr4 signaling axis. *Front. Pharmacol.* **2018**, *9*, 1294. [CrossRef]
59. Liu, L.; Ahn, K.S.; Shanmugam, M.K.; Wang, H.; Shen, H.; Arfuso, F.; Chinnathambi, A.; Alharbi, S.A.; Chang, Y.; Sethi, G.; et al. Oleuropein induces apoptosis via abrogating nf-kappab activation cascade in estrogen receptor-negative breast cancer cells. *J. Cell Biochem.* **2019**, *120*, 4504–4513. [CrossRef]
60. Li, F.; Sethi, G. Targeting transcription factor nf-kappab to overcome chemoresistance and radioresistance in cancer therapy. *Biochim. Biophys. Acta* **2010**, *1805*, 167–180.
61. Sethi, G.; Tergaonkar, V. Potential pharmacological control of the nf-kappab pathway. *Trends. Pharmacol. Sci.* **2009**, *30*, 313–321. [CrossRef]
62. Ahn, K.S.; Sethi, G.; Aggarwal, B.B. Nuclear factor-kappa b: From clone to clinic. *Curr. Mol. Med.* **2007**, *7*, 619–637. [CrossRef]
63. Puar, Y.R.; Shanmugam, M.K.; Fan, L.; Arfuso, F.; Sethi, G.; Tergaonkar, V. Evidence for the involvement of the master transcription factor nf-kappab in cancer initiation and progression. *Biomedicines* **2018**, *6*. [CrossRef]
64. Chai, E.Z.; Siveen, K.S.; Shanmugam, M.K.; Arfuso, F.; Sethi, G. Analysis of the intricate relationship between chronic inflammation and cancer. *Biochem. J.* **2015**, *468*, 1–15. [CrossRef] [PubMed]
65. Loh, C.Y.; Arya, A.; Naema, A.F.; Wong, W.F.; Sethi, G.; Looi, C.Y. Signal transducer and activator of transcription (stats) proteins in cancer and inflammation: Functions and therapeutic implication. *Front. Oncol.* **2019**, *9*, 48. [CrossRef] [PubMed]
66. Lee, M.; Hirpara, J.L.; Eu, J.Q.; Sethi, G.; Wang, L.; Goh, B.C.; Wong, A.L. Targeting stat3 and oxidative phosphorylation in oncogene-addicted tumors. *Redox Biol.* **2018**, 101073. [CrossRef] [PubMed]

67. Liu, Y.; Tergaonkar, V.; Krishna, S.; Androphy, E.J. Human papillomavirus type 16 e6-enhanced susceptibility of l929 cells to tumor necrosis factor alpha correlates with increased accumulation of reactive oxygen species. *J. Biol. Chem.* **1999**, *274*, 24819–24827. [CrossRef]
68. Akincilar, S.C.; Khattar, E.; Boon, P.L.; Unal, B.; Fullwood, M.J.; Tergaonkar, V. Long-range chromatin interactions drive mutant tert promoter activation. *Cancer Discov.* **2016**, *6*, 1276–1291. [CrossRef]
69. Khattar, E.; Kumar, P.; Liu, C.Y.; Akincilar, S.C.; Raju, A.; Lakshmanan, M.; Maury, J.J.; Qiang, Y.; Li, S.; Tan, E.Y.; et al. Telomerase reverse transcriptase promotes cancer cell proliferation by augmenting trna expression. *J. Clin. Investig.* **2016**, *126*, 4045–4060. [CrossRef]
70. Mishra, S.; Verma, S.S.; Rai, V.; Awasthee, N.; Chava, S.; Hui, K.M.; Kumar, A.P.; Challagundla, K.B.; Sethi, G.; Gupta, S.C. Long non-coding rnas are emerging targets of phytochemicals for cancer and other chronic diseases. *Cell Mol. Life Sci.* **2019**, *76*, 1947–1966. [CrossRef]
71. Deng, S.; Shanmugam, M.K.; Kumar, A.P.; Yap, C.T.; Sethi, G.; Bishayee, A. Targeting autophagy using natural compounds for cancer prevention and therapy. *Cancer* **2019**, *125*, 1228–1246. [CrossRef]
72. Siveen, K.S.; Sikka, S.; Surana, R.; Dai, X.; Zhang, J.; Kumar, A.P.; Tan, B.K.; Sethi, G.; Bishayee, A. Targeting the stat3 signaling pathway in cancer: Role of synthetic and natural inhibitors. *Biochim. Biophys. Acta* **2014**, *1845*, 136–154. [CrossRef]
73. Yang, M.H.; Jung, S.H.; Sethi, G.; Ahn, K.S. Pleiotropic pharmacological actions of capsazepine, a synthetic analogue of capsaicin, against various cancers and inflammatory diseases. *Molecules* **2019**, *24*, 995. [CrossRef] [PubMed]
74. Girisa, S.; Shabnam, B.; Monisha, J.; Fan, L.; Halim, C.E.; Arfuso, F.; Ahn, K.S.; Sethi, G.; Kunnumakkara, A.B. Potential of zerumbone as an anti-cancer agent. *Molecules* **2019**, *24*, 734. [CrossRef] [PubMed]
75. Lee, J.H.; Kim, C.; Um, J.Y.; Sethi, G.; Ahn, K.S. Casticin-induced inhibition of cell growth and survival are mediated through the dual modulation of akt/mtor signaling cascade. *Cancers* **2019**, *11*, 254. [CrossRef] [PubMed]
76. Aggarwal, V.; Kashyap, D.; Sak, K.; Tuli, H.S.; Jain, A.; Chaudhary, A.; Garg, V.K.; Sethi, G.; Yerer, M.B. Molecular mechanisms of action of tocotrienols in cancer: Recent trends and advancements. *Int. J. Mol. Sci.* **2019**, *20*, 656. [CrossRef] [PubMed]
77. Polak, R.; Phillips, E.M.; Campbell, A. Legumes: Health benefits and culinary approaches to increase intake. *Clin. Diabetes* **2015**, *33*, 198–205. [CrossRef] [PubMed]
78. Cassileth, B. Complementary therapies, herbs, and other otc agents: Red clover (trifolium pratense). *Oncology (Williston Park)* **2010**, *24*, 960. [PubMed]
79. Kaczmarczyk-Sedlak, I.; Wojnar, W.; Zych, M.; Ozimina-Kaminska, E.; Taranowicz, J.; Siwek, A. Effect of formononetin on mechanical properties and chemical composition of bones in rats with ovariectomy-induced osteoporosis. *Evid. Based Complement. Altern. Med.* **2013**, *2013*, 457052. [CrossRef]
80. Patisaul, H.B.; Jefferson, W. The pros and cons of phytoestrogens. *Front Neuroendocrinol.* **2010**, *31*, 400–419. [CrossRef] [PubMed]
81. Lukaczer, D.; Darland, G.; Tripp, M.; Liska, D.; Lerman, R.H.; Schiltz, B.; Bland, J.S. Clinical effects of a proprietary combination isoflavone nutritional supplement in menopausal women: A pilot trial. *Altern. Ther. Health Med.* **2005**, *11*, 60–65.
82. Hidalgo, L.A.; Chedraui, P.A.; Morocho, N.; Ross, S.; San Miguel, G. The effect of red clover isoflavones on menopausal symptoms, lipids and vaginal cytology in menopausal women: A randomized, double-blind, placebo-controlled study. *Gynecol. Endocrinol.* **2005**, *21*, 257–264. [CrossRef]
83. Atkinson, C.; Compston, J.E.; Day, N.E.; Dowsett, M.; Bingham, S.A. The effects of phytoestrogen isoflavones on bone density in women: A double-blind, randomized, placebo-controlled trial. *Am. J. Clin. Nutr.* **2004**, *79*, 326–333. [CrossRef]
84. Nestel, P.J.; Pomeroy, S.; Kay, S.; Komesaroff, P.; Behrsing, J.; Cameron, J.D.; West, L. Isoflavones from red clover improve systemic arterial compliance but not plasma lipids in menopausal women. *J. Clin. Endocrinol. Metab.* **1999**, *84*, 895–898. [CrossRef]
85. Wang, A.L.; Li, Y.; Zhao, Q.; Fan, L.Q. Formononetin inhibits colon carcinoma cell growth and invasion by microrna149mediated ephb3 downregulation and inhibition of pi3k/akt and stat3 signaling pathways. *Mol. Med. Rep.* **2018**, *17*, 7721–7729. [PubMed]

86. Zhang, S.; Tang, X.; Tian, J.; Li, C.; Zhang, G.; Jiang, W.; Zhang, Z. Cardioprotective effect of sulphonated formononetin on acute myocardial infarction in rats. *Basic Clin. Pharmacol. Toxicol.* **2011**, *108*, 390–395. [CrossRef] [PubMed]
87. Zhu, H.; Zou, L.; Tian, J.; Lin, F.; He, J.; Hou, J. Protective effects of sulphonated formononetin in a rat model of cerebral ischemia and reperfusion injury. *Planta Med.* **2014**, *80*, 262–268. [CrossRef] [PubMed]
88. Tolleson, W.H.; Doerge, D.R.; Churchwell, M.I.; Marques, M.M.; Roberts, D.W. Metabolism of biochanin a and formononetin by human liver microsomes in vitro. *J. Agric. Food Chem.* **2002**, *50*, 4783–4790. [CrossRef] [PubMed]
89. Park, J.; Kim, S.H.; Cho, D.; Kim, T.S. Formononetin, a phyto-oestrogen, and its metabolites up-regulate interleukin-4 production in activated t cells via increased ap-1 DNA binding activity. *Immunology* **2005**, *116*, 71–81. [CrossRef] [PubMed]
90. Liu, Y.; He, J.; Chen, X.; Li, J.; Shen, M.; Yu, W.; Yang, Y.; Xiao, Z. The proapoptotic effect of formononetin in human osteosarcoma cells: Involvement of inactivation of erk and akt pathways. *Cell Physiol. Biochem.* **2014**, *34*, 637–645. [CrossRef] [PubMed]
91. Chen, J.; Zeng, J.; Xin, M.; Huang, W.; Chen, X. Formononetin induces cell cycle arrest of human breast cancer cells via igf1/pi3k/akt pathways in vitro and in vivo. *Horm. Metab. Res.* **2011**, *43*, 681–686. [CrossRef] [PubMed]
92. Chen, J.; Sun, L. Formononetin-induced apoptosis by activation of ras/p38 mitogen-activated protein kinase in estrogen receptor-positive human breast cancer cells. *Horm. Metab. Res.* **2012**, *44*, 943–948. [CrossRef]
93. Chen, J.; Zhao, X.; Ye, Y.; Wang, Y.; Tian, J. Estrogen receptor beta-mediated proliferative inhibition and apoptosis in human breast cancer by calycosin and formononetin. *Cell Physiol. Biochem.* **2013**, *32*, 1790–1797. [CrossRef]
94. Ye, Y.; Hou, R.; Chen, J.; Mo, L.; Zhang, J.; Huang, Y.; Mo, Z. Formononetin-induced apoptosis of human prostate cancer cells through erk1/2 mitogen-activated protein kinase inactivation. *Horm. Metab. Res.* **2012**, *44*, 263–267. [CrossRef]
95. Li, T.; Zhao, X.; Mo, Z.; Huang, W.; Yan, H.; Ling, Z.; Ye, Y. Formononetin promotes cell cycle arrest via downregulation of akt/cyclin d1/cdk4 in human prostate cancer cells. *Cell Physiol. Biochem.* **2014**, *34*, 1351–1358. [CrossRef] [PubMed]
96. Yang, Y.; Zhao, Y.; Ai, X.; Cheng, B.; Lu, S. Formononetin suppresses the proliferation of human non-small cell lung cancer through induction of cell cycle arrest and apoptosis. *Int. J. Clin. Exp. Pathol.* **2014**, *7*, 8453–8461.
97. Jin, Y.M.; Xu, T.M.; Zhao, Y.H.; Wang, Y.C.; Cui, M.H. In vitro and in vivo anti-cancer activity of formononetin on human cervical cancer cell line HELA. *Tumour Biol.* **2014**, *35*, 2279–2284. [CrossRef]
98. Wu, Y.; Zhang, X.; Li, Z.; Yan, H.; Qin, J.; Li, T. Formononetin inhibits human bladder cancer cell proliferation and invasiveness via regulation of mir-21 and pten. *Food Funct.* **2017**, *8*, 1061–1066. [CrossRef]
99. Park, S.; Bazer, F.W.; Lim, W.; Song, G. The o-methylated isoflavone, formononetin, inhibits human ovarian cancer cell proliferation by sub g0/g1 cell phase arrest through PI3K/AKT and ERK1/2 inactivation. *J. Cell Biochem.* **2018**, *119*, 7377–7387. [CrossRef] [PubMed]
100. Wu, X.L.; Li, H.Y.; Wang, R.H.; Ma, X.X.; Yue, B.; Yan, J.; Yang, N.L.; Xing, X.H.; Sheng, Z.F.; Wang, G.H.; et al. Formononetin suppresses hypoxia inducible factor-1α/inflammatory cytokines expression via inhibiting akt signal pathway in multiple myeloma cells. *Int. J. Clin. Exp. Med.* **2016**, *9*, 1117–1127.
101. Huang, J.; Xie, M.; Gao, P.; Ye, Y.; Liu, Y.; Zhao, Y.; Luo, W.; Ling, Z.; Cao, Y.; Zhang, S.; et al. Antiproliferative effects of formononetin on human colorectal cancer via suppressing cell growth in vitro and in vivo. *Proc. Biochem.* **2015**, *50*, 912–917. [CrossRef]
102. Fouad, Y.A.; Aanei, C. Revisiting the hallmarks of cancer. *Am. J. Cancer Res.* **2017**, *7*, 1016–1036. [PubMed]
103. Aapro, M.S.; Plezia, P.M.; Alberts, D.S.; Graham, V.; Jones, S.E.; Surwit, E.A.; Moon, T.E. Double-blind crossover study of the antiemetic efficacy of high-dose dexamethasone versus high-dose metoclopramide. *J. Clin. Oncol.* **1984**, *2*, 466–471. [CrossRef]
104. Huang, W.J.; Bi, L.Y.; Li, Z.Z.; Zhang, X.; Ye, Y. Formononetin induces the mitochondrial apoptosis pathway in prostate cancer cells via downregulation of the igf-1/igf-1r signaling pathway. *Pharm. Biol.* **2013**, *52*, 466–470. [CrossRef] [PubMed]
105. Hu, W.; Xiao, Z. Formononetin induces apoptosis of human osteosarcoma cell line u2os by regulating the expression of bcl-2, bax and mir-375 in vitro and in vivo. *Cell Physiol. Biochem.* **2015**, *37*, 933–939. [CrossRef]

106. Liu, X.J.; Li, Y.Q.; Chen, Q.Y.; Xiao, S.J.; Zeng, S.E. Up-regulating of rasd1 and apoptosis of du-145 human prostate cancer cells induced by formononetin in vitro. *Asian Pac. J. Cancer Prev.* **2014**, *15*, 2835–2839. [CrossRef] [PubMed]
107. Zhang, X.; Bi, L.; Ye, Y.; Chen, J. Formononetin induces apoptosis in pc-3 prostate cancer cells through enhancing the bax/bcl-2 ratios and regulating the p38/akt pathway. *Nutr. Cancer* **2014**, *66*, 656–661. [CrossRef] [PubMed]
108. Zhou, R.; Xu, L.; Ye, M.; Liao, M.; Du, H.; Chen, H. Formononetin inhibits migration and invasion of mda-mb-231 and 4t1 breast cancer cells by suppressing mmp-2 and mmp-9 through pi3k/akt signaling pathways. *Horm. Metab. Res.* **2014**, *46*, 753–760. [CrossRef] [PubMed]
109. Auyeung, K.K.; Law, P.C.; Ko, J.K. Novel anti-angiogenic effects of formononetin in human colon cancer cells and tumor xenograft. *Oncol. Rep.* **2012**, *28*, 2188–2194. [CrossRef] [PubMed]
110. Zhang, X.; Ni, Q.; Wang, Y.; Fan, H.; Li, Y. Synergistic anticancer effects of formononetin and temozolomide on glioma c6 cells. *Biol. Pharm. Bull* **2018**, *41*, 1194–1202. [CrossRef]
111. Liu, Q.; Sun, Y.; Zheng, J.M.; Yan, X.L.; Chen, H.M.; Chen, J.K.; Huang, H.Q. Formononetin sensitizes glioma cells to doxorubicin through preventing EMT via inhibition of histone deacetylase 5. *Int. J. Clin. Exp. Pathol.* **2015**, *8*, 6434–6441.
112. Qi, C.; Xie, M.; Liang, J.; Li, H.; Li, Z.; Shi, S.; Yang, X.; Wang, Z.; Tang, J.; Tang, A. Formononetin targets the mapk and pi3k/akt pathways to induce apoptosis in human nasopharyngeal carcinoma cells in vitro and in vivo. *Int. J. Clin. Exp. Med.* **2016**, *9*, 1180–1189.
113. Rosse, T.; Olivier, R.; Monney, L.; Rager, M.; Conus, S.; Fellay, I.; Jansen, B.; Borner, C. Bcl-2 prolongs cell survival after bax-induced release of cytochrome c. *Nature* **1998**, *391*, 496–499. [CrossRef] [PubMed]
114. Hu, W.; Wu, X.; Tang, J.; Xiao, N.; Zhao, G.; Zhang, L.; Ou, L. In vitro and in vivo studies of antiosteosarcoma activities of formononetin. *J. Cell. Physiol.* **2019**. [CrossRef]
115. Montagut, C.; Settleman, J. Targeting the RAF-MEK-ERK pathway in cancer therapy. *Cancer Lett.* **2009**, *233*, 125–134. [CrossRef] [PubMed]
116. Luo, J.; Manning, B.D.; Cantley, L.C. Targeting the pi3k-akt pathway in human cancer: Rationale and promise. *Cancer Cell* **2003**, *4*, 257–262. [CrossRef]
117. Gelain, A.; Mori, M.; Meneghetti, F.; Villa, S. Signal transducer and activator of transcription protein 3 (stat3): An update on its direct inhibitors as promising anticancer agents. *Curr. Med. Chem.* **2018**. [CrossRef]
118. Avalle, L.; Camporeale, A.; Camperi, A.; Poli, V. Stat3 in cancer: A double edged sword. *Cytokine* **2017**, *98*, 42–50. [CrossRef] [PubMed]
119. Lee, J.H.; Kim, C.; Lee, J.; Um, J.Y.; Sethi, G.; Ahn, K.S. Arctiin is a pharmacological inhibitor of stat3 phosphorylation at tyrosine 705 residue and potentiates bortezomib-induced apoptotic and anti-angiogenic effects in human multiple myeloma cells. *Phytomedicine* **2018**, *55*, 282–292. [CrossRef]
120. Jung, Y.Y.; Shanmugam, M.K.; Narula, A.S.; Kim, C.; Lee, J.H.; Namjoshi, O.A.; Blough, B.E.; Sethi, G.; Ahn, K.S. Oxymatrine attenuates tumor growth and deactivates stat5 signaling in a lung cancer xenograft model. *Cancers* **2019**, *11*. [CrossRef]
121. Wu, X.Y.; Xu, H.; Wu, Z.F.; Chen, C.; Liu, J.Y.; Wu, G.N.; Yao, X.Q.; Liu, F.K.; Li, G.; Shen, L. Formononetin, a novel fgfr2 inhibitor, potently inhibits angiogenesis and tumor growth in preclinical models. *Oncotarget* **2015**, *6*, 44563–44578. [CrossRef]
122. Baek, S.H.; Ko, J.H.; Lee, J.H.; Kim, C.; Lee, H.; Nam, D.; Lee, J.; Lee, S.G.; Yang, W.M.; Um, J.Y.; et al. Ginkgolic acid inhibits invasion and migration and tgf-beta-induced emt of lung cancer cells through pi3k/akt/mtor inactivation. *J. Cell. Physiol.* **2017**, *232*, 346–354. [CrossRef]
123. Mohan, C.D.; Srinivasa, V.; Rangappa, S.; Mervin, L.; Mohan, S.; Paricharak, S.; Baday, S.; Li, F.; Shanmugam, M.K.; Chinnathambi, A.; et al. Trisubstituted-imidazoles induce apoptosis in human breast cancer cells by targeting the oncogenic pi3k/akt/mtor signaling pathway. *PLoS ONE* **2016**, *11*, e0153155. [CrossRef]
124. Singh, S.S.; Yap, W.N.; Arfuso, F.; Kar, S.; Wang, C.; Cai, W.; Dharmarajan, A.M.; Sethi, G.; Kumar, A.P. Targeting the pi3k/akt signaling pathway in gastric carcinoma: A reality for personalized medicine? *World J. Gastroenterol.* **2015**, *21*, 12261–12273. [CrossRef] [PubMed]
125. Siveen, K.S.; Ahn, K.S.; Ong, T.H.; Shanmugam, M.K.; Li, F.; Yap, W.N.; Kumar, A.P.; Fong, C.W.; Tergaonkar, V.; Hui, K.M.; et al. Y-tocotrienol inhibits angiogenesis-dependent growth of human hepatocellular carcinoma through abrogation of akt/mTOR pathway in an orthotopic mouse model. *Oncotarget* **2014**, *5*, 1897–1911. [CrossRef] [PubMed]

126. Kim, S.W.; Kim, S.M.; Bae, H.; Nam, D.; Lee, J.H.; Lee, S.G.; Shim, B.S.; Kim, S.H.; Ahn, K.S.; Choi, S.H.; et al. Embelin inhibits growth and induces apoptosis through the suppression of akt/mtor/s6k1 signaling cascades. *Prostate* **2013**, *73*, 296–305. [CrossRef] [PubMed]
127. Park, K.R.; Nam, D.; Yun, H.M.; Lee, S.G.; Jang, H.J.; Sethi, G.; Cho, S.K.; Ahn, K.S. Beta-caryophyllene oxide inhibits growth and induces apoptosis through the suppression of pi3k/akt/mtor/s6k1 pathways and ros-mediated mapks activation. *Cancer Lett.* **2011**, *312*, 178–188. [CrossRef] [PubMed]
128. Lee, J.H.; Kim, C.; Lee, S.G.; Yang, W.M.; Um, J.Y.; Sethi, G.; Ahn, K.S. Ophiopogonin d modulates multiple oncogenic signaling pathways, leading to suppression of proliferation and chemosensitization of human lung cancer cells. *Phytomedicine* **2018**, *40*, 165–175. [CrossRef] [PubMed]
129. Ko, J.H.; Nam, D.; Um, J.Y.; Jung, S.H.; Sethi, G.; Ahn, K.S. Bergamottin suppresses metastasis of lung cancer cells through abrogation of diverse oncogenic signaling cascades and epithelial-to-mesenchymal transition. *Molecules* **2018**, *23*, 1601. [CrossRef] [PubMed]
130. Lee, H.; Baek, S.H.; Lee, J.H.; Kim, C.; Ko, J.H.; Lee, S.G.; Chinnathambi, A.; Alharbi, S.A.; Yang, W.M.; Um, J.Y.; et al. Isorhynchophylline, a potent plant alkaloid, induces apoptotic and anti-metastatic effects in human hepatocellular carcinoma cells through the modulation of diverse cell signaling cascades. *Int. J. Mol. Sci.* **2017**, *18*, 1095. [CrossRef]
131. Kannaiyan, R.; Manu, K.A.; Chen, L.; Li, F.; Rajendran, P.; Subramaniam, A.; Lam, P.; Kumar, A.P.; Sethi, G. Celastrol inhibits tumor cell proliferation and promotes apoptosis through the activation of c-jun n-terminal kinase and suppression of pi3 k/akt signaling pathways. *Apoptosis* **2011**, *16*, 1028–1041. [CrossRef]
132. Sethi, G.; Ahn, K.S.; Sung, B.; Kunnumakkara, A.B.; Chaturvedi, M.M.; Aggarwal, B.B. Sh-5, an akt inhibitor potentiates apoptosis and inhibits invasion through the suppression of anti-apoptotic, proliferative and metastatic gene products regulated by ikappabalpha kinase activation. *Biochem. Pharmacol.* **2008**, *76*, 1404–1416. [CrossRef]
133. Ma, X.; Hu, Y. Targeting pi3k/akt/mtor cascade: The medicinal potential, updated research highlights and challenges ahead. *Curr. Med. Chem.* **2013**, *20*, 2991–3010. [CrossRef] [PubMed]
134. Dreher, D.; Junod, A.F. Role of oxygen free radicals in cancer development. *Eur. J. Cancer* **1996**, *32A*, 30–38. [CrossRef]
135. Arora, A.; Nair, M.G.; Strasburg, G.M. Antioxidant activities of isoflavones and their biological metabolites in a liposomal system. *Arch. Biochem. Biophys.* **1998**, *356*, 133–141. [CrossRef]
136. Kubota, Y. Tumor angiogenesis and anti-angiogenic therapy. *Keio J. Med.* **2012**, *61*, 47–56. [CrossRef]
137. Xiao, L.; Yang, S.; Hao, J.; Yuan, X.; Luo, W.; Jiang, L.; Hu, Y.; Fu, Z.; Zhang, Y.; Zou, C. Endostar attenuates melanoma tumor growth via its interruption of b-fgf mediated angiogenesis. *Cancer Lett.* **2015**, *359*, 148–154. [CrossRef]
138. Egeblad, M.; Werb, Z. New functions for the matrix metalloproteinases in cancer progression. *Nat. Rev. Cancer* **2002**, *2*, 161–174. [CrossRef]
139. Li, Y.; Cheng, H.S.; Chng, W.J.; Tergaonkar, V. Activation of mutant tert promoter by ras-erk signaling is a key step in malignant progression of braf-mutant human melanomas. *Proc. Natl. Acad. Sci. USA* **2016**, *113*, 14402–14407. [CrossRef] [PubMed]
140. Wang, Y.; Zhu, Y.; Gao, L.; Yin, H.; Xie, Z.; Wang, D.; Zhu, Z.; Han, X. Formononetin attenuates il-1beta-induced apoptosis and NF-κB activation in ins-1 cells. *Molecules* **2012**, *17*, 10052–10064. [CrossRef]
141. Duan, P.; Wang, Z.M. Clinical study on effect of astragalus in efficacy enhancing and toxicity reducing of chemotherapy in patients of malignant tumor. *Zhongguo Zhong Xi Yi Jie He Za Zhi* **2002**, *22*, 515–517.
142. Van den Bent, M.J.; Taphoorn, M.J.; Brandes, A.A.; Menten, J.; Stupp, R.; Frenay, M.; Chinot, O.; Kros, J.M.; van der Rijt, C.C.; Vecht Ch, J.; et al. Phase ii study of first-line chemotherapy with temozolomide in recurrent oligodendroglial tumors: The european organization for research and treatment of cancer brain tumor group study 26971. *J. Clin. Oncol.* **2003**, *21*, 2525–2528. [CrossRef]
143. Liu, X.; Fan, D. The epithelial-mesenchymal transition and cancer stem cells: Functional and mechanistic links. *Curr. Pharm. Des.* **2015**, *21*, 1279–1291. [CrossRef] [PubMed]
144. Su, Y.; Wang, J.; Zhang, X.; Shen, J.; Deng, L.; Liu, Q.; Li, G. Targeting sim2-s decreases glioma cell invasion through mesenchymal–epithelial transition. *J. Cell Biochem.* **2014**, *115*, 1900–1907.
145. Kalluri, R.; Weinberg, R.A. The basics of epithelial-mesenchymal transition. *J. Clin. Investig.* **2009**, *119*, 1420–1428. [CrossRef] [PubMed]

146. watsuki, M.; Mimori, K.; Yokobori, T.; Ishi, H.; Beppu, T.; Nakamori, S.; Baba, H.; Mori, M. Epithelial-mesenchymal transition in cancer development and its clinical significance. *Cancer Sci.* **2010**, *101*, 293–299. [CrossRef] [PubMed]
147. Ren, J.; Xu, H.J.; Cheng, H.; Xin, W.Q.; Chen, X.; Hu, K. Synthesis and antitumor activity of formononetin nitrogen mustard derivatives. *Eur. J. Med. Chem.* **2012**, *54*, 175–187. [CrossRef]
148. Fu, D.J.; Zhang, L.; Song, J.; Mao, R.W.; Zhao, R.H.; Liu, Y.C.; Hou, Y.H.; Li, J.H.; Yang, J.J.; Jin, C.Y.; et al. Design and synthesis of formononetin-dithiocarbamate hybrids that inhibit growth and migration of pc-3 cells via mapk/wnt signaling pathways. *Eur. J. Med. Chem.* **2017**, *127*, 87–99. [CrossRef]
149. Lin, H.Y.; Sun, W.X.; Zheng, C.S.; Han, H.W.; Wang, X.; Zhang, Y.H.; Qiu, H.Y.; Tang, C.Y.; Qi, J.L.; Lu, G.H.; et al. Synthesis, characterization and biological evaluation of formononetin derivatives as novel egfr inhibitors via inhibiting growth, migration and inducing apoptosis in breast cancer cell line. *RSC Adv.* **2017**, *7*, 48404–48419. [CrossRef]
150. Li, Y.Q.; Yang, F.; Wang, L.; Cao, Z.; Han, T.J.; Duan, Z.A.; Li, Z.; Zhao, W.J. Phosphoramidate protides of five flavones and their antiproliferative activity against hepg2 and l-o2 cell lines. *Eur. J. Med. Chem.* **2016**, *112*, 196–208. [CrossRef]
151. Yang, C.; Xie, Q.; Zeng, X.; Tao, N.; Xu, Y.; Chen, Y.; Wang, J.; Zhang, L. Novel hybrids of podophyllotoxin and formononetin inhibit the growth, migration and invasion of lung cancer cells. *Bioorg. Chem.* **2019**, *85*, 445–454. [CrossRef]
152. Guo, B.; Liao, C.; Liu, X.; Yi, J. Preliminary study on conjugation of formononetin with multiwalled carbon nanotubes for inducing apoptosis via ros production in hela cells. *Drug Des. Dev. Ther.* **2018**, *12*, 2815–2826. [CrossRef]
153. Mu, H.; Bai, Y.H.; Wang, S.T.; Zhu, Z.M.; Zhang, Y.W. Research on antioxidant effects and estrogenic effect of formononetin from trifolium pratense (red clover). *Phytomedicine* **2009**, *16*, 314–319. [CrossRef] [PubMed]
154. Barrera, G. Oxidative stress and lipid peroxidation products in cancer progression and therapy. *ISRN Oncol.* **2012**, *2012*, 137289. [CrossRef] [PubMed]
155. Xing, D.X.; Liu, X.L.; Xue, C.K.; Huang, Q.; Liu, Z.G.; Xiong, L. The estrogenic effect of formononetin and its effect on the expression of rats' atrium estrogen receptors. *Zhong Yao Cai* **2010**, *33*, 1445–1449. [PubMed]
156. Bertolini, F.; Marighetti, P.; Martin-Padura, I.; Mancuso, P.; Hu-Lowe, D.D.; Shaked, Y.; D'Onofrio, A. Anti-vegf and beyond: Shaping a new generation of anti-angiogenic therapies for cancer. *Drug Discov. Today* **2011**, *16*, 1052–1060. [CrossRef] [PubMed]
157. Hao, Z.; Sadek, I. Sunitinib: The antiangiogenic effects and beyond. *Oncol. Targets Ther.* **2016**, *9*, 5495–5505. [CrossRef] [PubMed]
158. Fei, H.X.; Zhang, Y.B.; Liu, T.; Zhang, X.J.; Wu, S.L. Neuroprotective effect of formononetin in ameliorating learning and memory impairment in mouse model of alzheimer's disease. *Biosci. Biotechnol. Biochem.* **2018**, *82*, 57–64. [CrossRef]
159. Jarred, R.A.; Keikha, M.; Dowling, C.; McPherson, S.J.; Clare, A.M.; Husband, A.J.; Pedersen, J.S.; Frydenberg, M.; Risbridger, G.P. Induction of apoptosis in low to moderate-grade human prostate carcinoma by red clover-derived dietary isoflavones. *Cancer Epidemiol. Biomark. Prev.* **2002**, *11*, 1689–1696.

cancers

MDPI

Article

Malva pseudolavatera Leaf Extract Promotes ROS Induction Leading to Apoptosis in Acute Myeloid Leukemia Cells In Vitro

Marianne El Khoury [1,†], Tony Haykal [1,†], Mohammad H. Hodroj [1], Sonia Abou Najem [1], Rita Sarkis [2], Robin I. Taleb [1] and Sandra Rizk [1,*]

1 Department of Natural Sciences, Lebanese American University, Byblos 1401, Lebanon; marianne.elkhoury01@lau.edu (M.E.K.); tony.haykal@lau.edu (T.H.); mohammadhassan.hodroj@lau.edu (M.H.H.); sonia.abounajem@lau.edu.lb (S.A.N.); robin.taleb@lau.edu.lb (R.I.T.)

2 Laboratory of Regenerative Hematopoiesis, Swiss Institute for Experimental Cancer Research (ISREC) & Institute of Bioengineering (IBI), School of Life Sciences, Ecole Polytechnique Fédérale de Lausanne (EPFL), 1015 Lausanne, Switzerland; Rita.sarkis@epfl.ch

* Correspondence: Sandra.rizk@lau.edu.lb; Tel.: +961-9944851

† These authors contributed equally to this work.

Received: 30 November 2019; Accepted: 6 February 2020; Published: 13 February 2020

Abstract: *Malva pseudolavatera* Webb & Berthel. is a plant from the Malvaceae family that has long been included in the human diet due to its various curative effects. Many plant leaf extracts from the various species of *Malva* genus have been reported to possess anti-cancer properties, however, studies on *M.pseudolavatera* Webb & Berthel. leaves have documented anti-inflammatory and anti-oxidant effects with no emphasis on their possible anti-cancer potential. The present study explores the anti-cancer properties of *Malva pseudolavatera* Webb & Berthel. leaf extract on acute myeloid leukemia (AML) cell lines in vitro and deciphers the underlying molecular mechanism. Treatment of AML cell lines with *M. pseudolavatera* methanolic leaf extract showed a dose- and time-dependent inhibition of proliferation and a dose-dependent increase in apoptotic hallmarks such as an increase in phosphatidylserine on the outer membrane leaflet and membrane leakage in addition to DNA fragmentation. The pro-apoptotic effect was induced by reactive oxygen species (ROS) as well as an upregulation of cleaved poly(ADP-ribose) polymerase (PARP), increase in Bax/Bcl-2 ratio, andrelease of cytochrome-c from the mitochondria. Major compounds of the extract included methyl linolenate, phytol, γ-sitosterol, and stigmasterol as revealed by gas chromatography coupled with mass spectrometry, and amino acids, amino acid derivatives, tiliroside, 13-hydroxyperoxyoctadecadienoic, and quercitrin as detected by liquid chromatography coupled to mass spectrometry.

Keywords: *Malva pseudolavatera* Webb & Berthel.; apoptosis; acute myeloid leukemia; reactive oxygen species

1. Introduction

Malva pseudolavatera Webb & Berthel. is an annual or biennial subshrub that grows in fields and roadsides in coastal areas and low-altitude mountain regions [1]. Commonly known as "tree mallow" in North America and "khubbaza" in the Middle East, it is a plant of the Malvaceae family [2]. Previously named *Lavatera cretica* (Malvaceae family), the species was transferred to the *Malva* genus and is currently called *M. pseudolavatera* or *Malva linnaei* or *Malva multiflora* [3]. *Malva pseudolavatera* Webb & Berthel. is the accepted name of the species as included on http://www.theplantlist.org [4].

Apart from being used as food in some regions such as in Turkey, Spain, and Pakistan, leaves of the *Malva* genus plants have been traditionally used in folk medicine all around the world to treat a multitude of diseases, most commonly diarrhea, arthritis, inflammation, cough, and respiratory infections [5]. Among the 50 species of Malva, *M. pseudolavatera* Webb & Berthel. is known for its versatile uses [6]. In Spain, it is considered as a remedy for influenza, upper respiratory tract infections, and cough [7], whereas in Portugal, it is used for its laxative, analgesic, and antiseptic effects [8]. In fact, studies showed that *M. pseudolavatera* Webb & Berthel. aqueous leaf extracts were able to scavenge free radicals and inhibit lipoxygenase activity in vitro, indicating its potent antioxidant and anti-inflammatory activities [9].

Malva species leaves have a lot of similarities in the overall morphology and anatomy, and only differ in some characteristics such as the number of lobes, size of the blade, or margin dentation. Leaves of many Malva species such as *M. sylvestris* and *M. parviflora* share a common basic chemical composition as they all contain anthocyanins, flavonoids, essential oils, and tocopherols [10,11]. In fact, many studies have shown the inhibitory effects of leaf extracts from Malva species on cancer cell lines. The methanolic extract from *M. sylvestris* leaves exhibited a dose-dependent cytotoxicity on melanoma and lymphoma cell lines in vitro [12]. Moreover, the ethanolic extract from *M. parviflora* leaves showed mild cytotoxicity against the MCF-7 human breast adenocarcinoma cell line [13]. Other studies have reported that the hexane and methanol extracts from *M. parviflora* leaves inhibited the proliferation of HeLa cervical carcinoma cells, but the aqueous extract did not cause any inhibition of cell proliferation [14]. However, no studies have examined the anti-cancer potential of extracts from *M. pseudolavatera* Webb & Berthel. leaves.

Acute myeloid leukemia (AML) is a type of cancer that starts in the bone marrow and quickly moves undifferentiated myeloblasts into the blood [15]. It is an aggressive malignancy with incidence levels still on the rise in several countries including Canada, UK, and Australia [16]. Chemotherapy is the main treatment for AML [17] and plants have long been used as important sources for novel chemotherapeutic drug characterization [18,19].

In the present study, we investigated the potential anti-cancer properties of *M. pseudolavatera* Webb & Berthel. methanolic leaf extract (MMLE) on AML cell lines in vitro and deciphered the underlying molecular mechanism.

2. Materials and Methods

2.1. Cell Culture

Acute myeloid leukemia cell lines, namely Mono-Mac-1, U937, and KG-1 (American Type Culture Collection), were cultured in Roswell Park Memorial Institute medium (RPMI, Sigma-Aldrich, St. Louis, MO, USA)supplemented with 10% fetal bovine serum (FBS) (Gibco^{TM}, Dublin, Ireland) and 100 U/mL penicillin and 100 µg/mL streptomycin (Lonza, Basel, Switzerland) in a humidified incubator at 5% CO_2 at 37 °C. Trypan blue exclusion method was used to count the cells before experimentation.

2.2. Isolation and Culture of Mesenchymal Stem Cells (MSCs) from Rat Bone Marrow

A single, 12-week-old rat was provided by the animal facility at the Lebanese American University. The animal was maintained under optimal laboratory conditions and received food and water ad libidum. All experiments were approved by the university's Animal Care and Use Committee (ACUC) and complied with the Guide for the Care and Use of Laboratory Animals (Committee for the Update of the Guide for the Care and Use of Laboratory Animals, 2010) [20,21]. MSCs were isolated from rat bone marrow according to a modified procedure. Briefly, the rat was sacrificed by CO_2 asphyxiation and both hind legs were aseptically removed. Femoral and tibial bones were then isolated and washed with 70% ethanol and placed in sterile phosphate buffered saline (PBS, Lonza) supplemented with 100 U/mL penicillin and 100 µg/mL streptomycin (Lonza). After removing the bone epiphyses with sterilized scissors, the bone marrows were flushed out using a needle filled with Dulbecco's modified Eagle medium (DMEM, Sigma-Aldrich) supplemented with 10% fetal bovine serum (Gibco^{TM}) and 100 U/mL penicillin and 100 µg/mL streptomycin (Lonza). The cells collected were then incubated in vented flasks at 37 °C with 5% CO_2. After 5 days of daily medium change, MSCs were identified by their spindle-shaped morphology as observed using the ZOE fluorescent cell imager (Bio-Rad, Irvine, CA, USA) [22–24].

2.3. Isolation and Culture of Normal Mononuclear Cells (MNCs) from Human Bone Marrow (BM)

Normal mononuclear cells were offered by Dr. Marwan El-Sabban's Lab at the American University of Beirut (AUB) as a kind gift. The normal MNCs were obtained originally from bone marrow aspirate leftovers of healthy patients attending AUB Medical center (AUB-MC). BM aspirates were centrifuged on Ficoll/Hypaque (GE Healthcare Life Sciences, Uppsala, Sweden), a density gradient step to separate MNCs from red blood cells and neutrophils. Then the buffy coat was aspirated and seeded in petri dishes using Dulbecco's modified Eagle's medium (DMEM)-low glucose (Sigma-Aldrich, Saint Louis, MO, USA) supplemented with 10% FBS (Gibco, Dublin, Ireland) and 100 U/mL penicillin and 100 µg/mL streptomycin (Lonza, Basel, Switzerland) in a humidified incubator at 37 °C and 5% CO_2. One week later, the cells in suspension were collected as a purified MNC population and cultured in the same conditions as detailed by Zibara et al. [25].

2.4. Plant Material

Malva pseudolavatera Webb & Berthel. leaves were collected from Batroun, Lebanon (34.2498° N, 35.6643° E. 20 m above sea level), during January 2018, and identified according to the indications and characteristics described by Edgecombe [2], and then identified by Dr. Nisrine Machaka-Houri, plant researcher and expert on Lebanese flora [26]. A voucher specimen was deposited in the Beirut Arab University Herbarium (ID-RCED2019-361).

2.5. Preparation of Crude Leaf Extract (MMLE)

Leaves were washed with distilled water, stored between paper towel sheets at 4 °C for 2 weeks to dry out, then ground and left to shake in absolute methanol at 200 rpm for 1 week. The extract was later filtered through a cheesecloth and centrifuged at 15,000 rpm to discard the pellet. Methanol was evaporated using a rotary evaporator. The methanolic crude extract was weighed, then dissolved in dimethyl sulfoxide (DMSO) and diluted with RPMI to a final concentration of 9 mg/mL. When applied on the cell lines, the DMSO level maximally reached 0.8% at 360 µg/mL for KG-1 and Monomac-1 and 1% at 450 µg/mL for U937.

2.6. Cytotoxicity Assay

AML cells and MSCs were seeded in 96-well plates at a density of 0.5×10^5 cells/well and incubated overnight before treatment of triplicates of wells with increasing concentrations of *M. pseudolavatera* Webb & Berthel. methanolic leaf extract (MMLE). After 24 or 48 h of incubation, WST-1 cell proliferation reagent (Roche, Mannheim, Germany) was used to estimate cell viability according to the manufacturer's guidelines. Absorbance of each well was detected at 450 nm using a Multiskan™ FC microplate photometer to quantify metabolically-active cells before calculating the percent proliferation relative to the control untreated cells.

2.7. Cell Cycle Analysis

Monomac-1 cells were seeded in 6-well plates at a density of 1×10^5 cells/well and incubated overnight before treatment with increasing concentrations of MMLE for 24 h. Cells were then fixed overnight with ethanol, and the DNA was stained with propidium iodide (PI, Sigma-Aldrich) after the enzymatic removal of RNA using RNase (Roche). DNA content was measured using an Accuri C6 flow cytometer to determine the distribution of cells in each cell cycle phase: pre-G0/G1 phase cells had <2n, G0/G1 phase cells had 2n, S phase cells had between 2n and 4n, and G2/M phase cells had 4n.

2.8. Apoptosis Detection Using Fluorescent Annexin V Staining

Monomac-1 and KG-1 cells were seeded in 24-well plates at a density of 1×10^5 cells/well and incubated overnight before treatment with increasing concentrations of the MMLE for 24 h. Cells were then stained with annexin V-FITC (Abcam, Cambridge, UK)and visualized with the ZOE fluorescent cell imager under bright-field conditions then the filter was set for FITC before merging the images.

2.9. Apoptosis Quantification by Dual Annexin V/PI Staining

Monomac-1 cells were seeded in 6-well plates at a density of 2×10^5 cells/well and incubated overnight before incubation with increasing concentrations of the MMLE for 24 h. Cells were then stained with annexin V-FITC and PI (Abcam) according to the manufacturer's instructions and analyzed by the Accuri C6 flow cytometer. Annexin V binds to phosphatidylserine molecules translocated to the outer layer of the cell membrane upon apoptosis induction. PI reaches the cellular DNA in cells that have lost the cellular membrane integrity, so it stains late apoptotic and necrotic cells but not viable and early apoptotic cells. This allowed for the discrimination between viable, early apoptotic, late apoptotic, and necrotic cells.

2.10. Cell Death ELISA

Monomac-1 and KG-1 cells were seeded in 24-well plates at a density of 2×10^5 cells/well and incubated overnight before treatment of duplicates of wells with increasing concentrations of MMLE. Treatment with the chemotherapeutic drug, etoposide (Abcam) at a concentration of 100 μM (58.85 μg/mL) was used as positive control. After 24 h, cells were collected and lysed in incubation buffer before quantification of fragmented cytosolic histone-associated-DNA content using the Cell Death ELISA kit according the manufacturer's instructions (Roche). Extracted DNA was then incubated in wells coated with biotin-associated anti-histone antibodies, followed by incubation with anti-DNA antibodies linked to peroxidase enzyme, then washed with washing buffer before the addition of the peroxidase substrate. Absorbance at 405 nm was measured by spectrophotometry using a Multiskan™ FC microplate photometer and the DNA fragmentation enrichment factor (absorbance of treated cells/absorbance of non-treated cells) was calculated as the ratio of absorbance in the treated samples to that of the untreated controls.

2.11. Western Blot

Monomac-1 cells were plated in 6-well plates at a density of 5×10^5 cells/mL before treatment with two increasing concentrations of MMLE for 24 h. The concentrations used were the closest to the IC50. Total proteins were extracted using the Qproteome mammalian protein prep kit (Qiagen, Hilden, Germany) and quantified using the Lowry method. Proteins were then separated by SDS-PAGE (10%) and transferred to PVDF membranes that were blocked with 5% skimmed milk, then incubated with primary antibodies: anti-β-actin (Santa Cruz Biotechnology, Dallas, Tx, USA), anti-cytochrome-c and anti-cleaved poly(ADP-ribose) polymerase (PARP) (Abcam), anti-Bax and anti-Bcl2 (Elabscience, Houston, TX, USA). β-actin was used as a loading control. Membranes were then washed and incubated with a secondary antibody (Bio-Rad, Irvine, CA, USA) followed by exposure for image development using Clarity™ Western ECL substrate (Abcam) on a ChemiDoc machine (Bio-Rad). Quantification using the ImageJ program allowed us to calculate the relative expression of proteins, as compared to the loading control.

2.12. Reactive Oxygen Species Detection

Using the DCFDA cellular ROS detection assay kit (Abcam), levels of ROS were quantified in Monomac-1 and KG-1 cells treated with increasing concentrations of MMLE. Tert-butyl hydrogen peroxide (TBHP) is a potent ROS inducer and was used as a positive control. DCFDA (2′,7′-dichlorodihydrofluorescein diacetate) oxidative conversion to H_2DCFDA upon ROS reduction was quantified by spectrofluorometry on the Varioskan™ LUX multimode microplate reader (Thermo Fisher Scientific, Bremen, Germany).

2.13. Gas ChromatographyMass Spectrometry Analysis of the Methanolic Extract of M. Pseudolavatera Webb & Berthel. Leaves

MMLE composition was analyzed using gas chromatography coupled with mass spectrometry (GC-MS). The carrier gas used was helium with splitless injection and a flow rate of 1.2 mL/min was applied. A temperature program consisted of 2 min at 70 °C, from 70 °C to 130 °C at 8 °C/min and hold for 5 min, from 130 °C to 180 °C at 2 °C/min and hold for 10 min, from 180 °C to 220 °C at 15 °C/min and hold for 2 min, and then from 220 °C to 280 °C at 15 °C/min and hold for 22 min. Preliminary identification of the various compounds was performed by comparing their mass spectra with the literature (NIST11 and Wiley9). Percentage composition was computed from GC peak areas.

2.14. Liquid ChromatographyMass Spectrometry Analysis of the Methanolic Extract of M. Pseudolavatera Webb & Berthel. Leaves

A 2.5 µg sample was injected into C18 Gravity-SB Nucleodur (300 Å, 1.8 µm, 2 × 100 mm, Macherey-Nagel, Düren, Germany) using a Dionex Ultimate 3000 analytical RSLC system (Dionex, Germering, Germany) coupled to a heated electrospray source HESI source (Thermo Fisher Scientific, Bremen, Germany). The separation was performed with flow rate of 300 µl/min by applying a gradient of solvent B from 3% to 50% within 35 min, followed by column washing and re-equilibration steps. Solvent A was composed of water with 0.1% formic acid, while solvent B consisted of acetonitrile with 0.1% formic acid. Eluting compounds were analyzed on a QExactive HF-HT-Orbitrap-FT-MS benchtop instrument (Thermo Fisher Scientific, Bremen, Germany). MS1 scan was performed with 60,000 resolution, AGC (automatic gain control) of 3e6 and maximum injection time of 200 ms. MS2 scan was performed in Top10 mode with 2 m/z isolation window, AGC of 5e5, 15 000 resolution, maximum injection time of 50 ms, and averaging 2 µscans. Higher-energy collisional dissociation (HCD) was used as the fragmentation method with normalized collision energy of 28%. For compound analysis, mzCloud and ChemSpider database for chemicals were used.

2.15. Statistical Analysis

All experiments were repeated three times (n=3). Statistical analyses were performed using GraphPad Prism 8. The data was reported as mean ± SEM and the *p*-values were calculated by *t*-tests or two-way ANOVA depending on the experiment. Significant differences were reported with * indicating a *p*-value of $0.01 < p < 0.05$, ** indicating a *p*-value of $0.001 < p < 0.01$, *** indicating a *p*-value of $0.0001 < p < 0.001$, and **** indicating a *p*-value of $p < 0.0001$.

3. Results

3.1. M. pseudolavatera Leaf Extract Exhibits Selective Anti-Proliferative Effects on AML Cell Lines

In order to detect the percent proliferation of AML cell lines, MSCs, and MNCs treated with MMLE, WST-1 cell proliferation reagent was used. A dose-dependent and time-dependent significant decrease in proliferation of the three AML cell lines, Monomac-1, KG-1, and U937 was observed with an IC50 of 200 µg/mL and 86.80 µg/mL for Monomac-1 (Figure 1A), 207.9 µg/mL and 89.47 µg/mL for KG-1 (Figure 1B), and 402 µg/mL and 229 µg/mL for U937 (Figure 1C) after 24 h and 48 h, respectively. The extract had no significant cytotoxic effect on MSCs and MNCs (Figure 1D,E). This indicates that MMLE exhibits selective anti-proliferative effects on all AML cancer cell lines used.

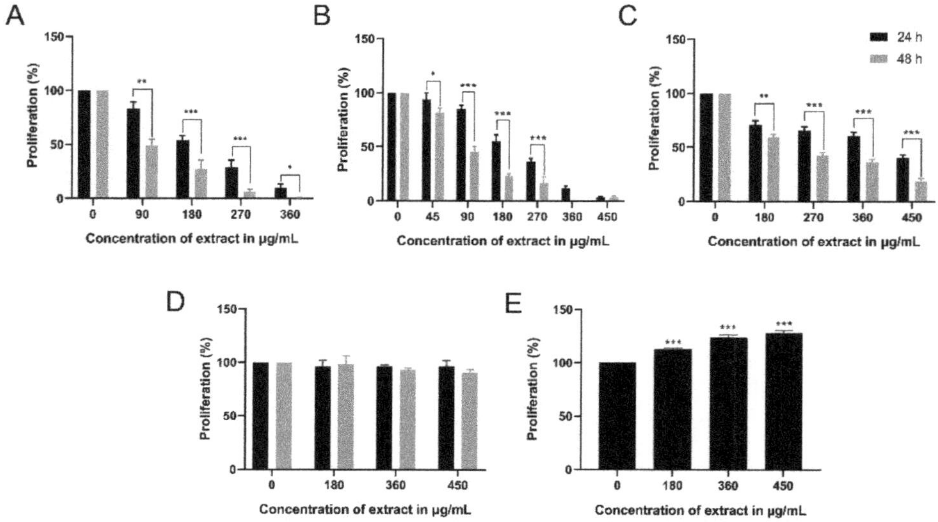

Figure 1. Proliferation of Monomac-1 (**A**), KG-1 (**B**), U937 (**C**), mesenchymal stem cells (MSCs) (**D**), and normal mononuclear cells MNCs (**E**), after 24 h and 48 h of treatment with methanolic leaf extract (MMLE). A significant dose- and time-dependent inhibition of proliferation of the three AML cell lines was noticed with increasing concentrations of MMLE. Significant differences were reported with * indicating a p-value: $0.01 < p < 0.05$, ** indicating a p-value: $0.001 < p < 0.01$ and *** indicating a p-value: $0.0001 < p < 0.001$.

3.2. *M. pseudolavatera Leaf Extract Induces Cellular Fragmentation in AML Cell Lines*

In order to elucidate the mechanism by which MMLE exerted its cytotoxic effect, PI staining was performed. To check for any cell cycle arrest and analyze the cell cycle distribution of Monomac-1 cells treated with MMLE, DNA content was quantified by PI staining followed by cytometric analysis. A dose-dependent increase in cellular fragmentation was detected as the cells gradually shifted from the G0/G1, S, and G2/M stages to the pre-G0/G1 stage where cells are fragmented and contain DNA <2n. In fact, the proportion of Monomac-1 cells in the pre-G0/G1 stage increased significantly from 8.05% in the untreated cells to 74.9% in cells treated with 270 μg/mL (after IC50) (Figure 2). This shows that MMLE does not induce a cell cycle arrest, but rather activates a mechanism leading to cellular fragmentation.

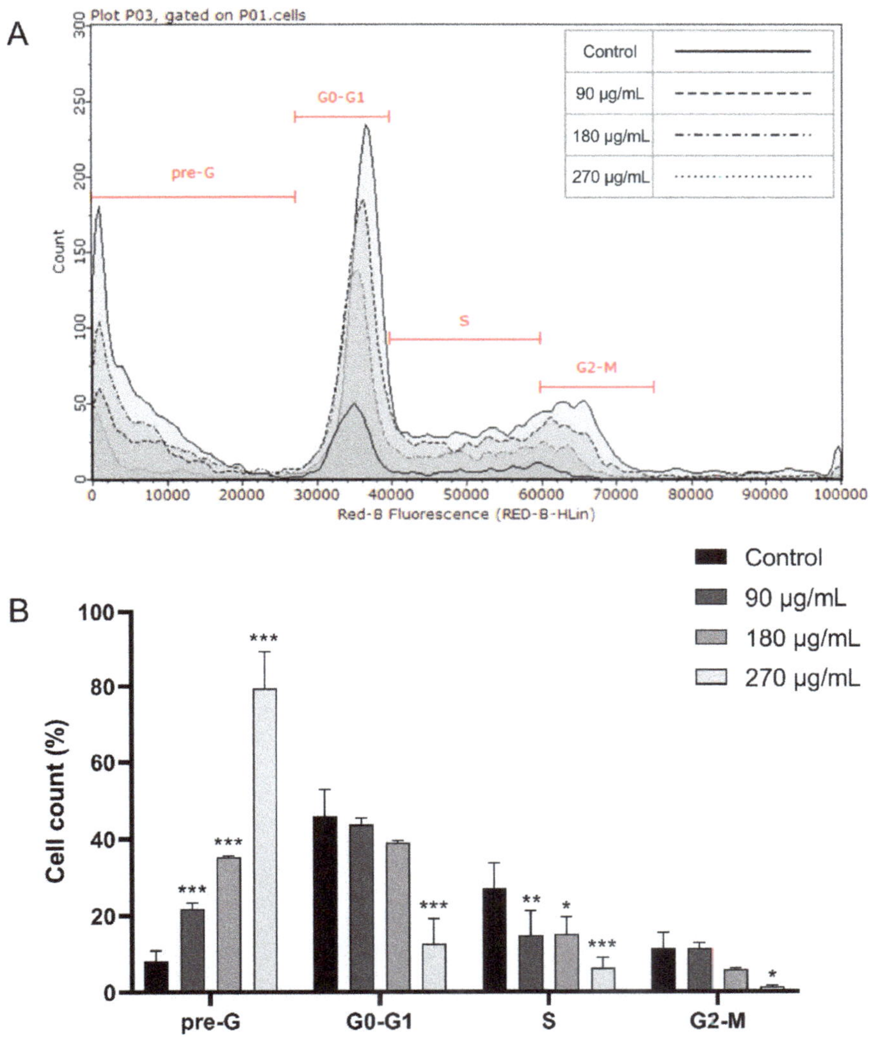

Figure 2. Cell cycle analysis of Monomac-1 treated with MMLE for 24 h (**A**). The percentage of the cells in different phases of cell cycle was determined by C Flow software (**B**) A significant increase in the pre-G and a decrease in G0/G1, S, and G2/M in a dose-dependent manner was obtained and indicated an increase in DNA fragmentation in Monomac-1 cells upon MMLE treatment.

3.3. *M. pseudolavatera Leaf Extract Significantly Induces Apoptosis in AML Cell Lines*

To explore whether cell death is induced by apoptosis, annexin V staining was followed by fluorescence microscopy. Upon treatment with increasing concentrations of MMLE, a marked increase in annexin binding on Monomac-1 cells was observed, indicating a shift of phosphatidylserine from the inner leaflet to the outer leaflet of the cell membrane, a major apoptotic event (Figure 3A).

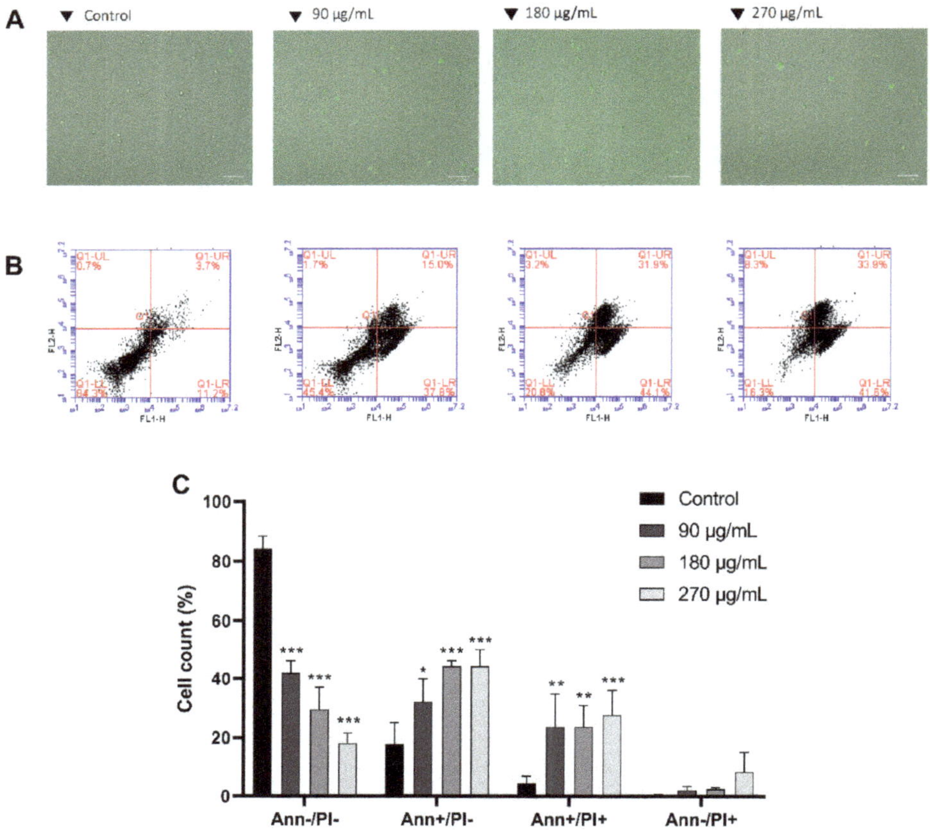

Figure 3. Annexin V (**A**) and annexin V/PI staining (**B**,**C**) of Monomac-1 cells treated for 24 h with increasing concentrations of MMLE. A significant increase in positively stained cells was observed upon 24 h treatment with increasing concentrations of MMLE (A). A decrease in annexin-negative/PI-negative stained cells and an increase in annexin-positive/PI-negative, annexin-positive/PI-positive, and annexin-negative/PI-positive stained cells were noted in Monomac-1 treated with 90, 180, and 270 µg/mL for 24 h.

To quantitatively assess the induction of apoptosis, flow cytometry analysis was carried out after annexin V/PI staining and cells were distributed into four quadrants where the lower left quadrant represents normal healthy cells, negatively staining for both annexin V and PI. The lower right quadrant represents early apoptotic cells which stain positively only for annexin V. The upper right quadrant represents the late apoptotic cells, staining positively for both annexin V and PI. The upper left quadrant represents necrotic cells which stain positively for PI only. After 24 h of treatment with increasing concentrations of MMLE, a decrease in Monomac-1 healthy cells from 84.3% in the control group to 18.2% at 270 µg/mL was coupled to a significant increase in early apoptotic cells from 17.8% in the control to 44.9% when treated with 270 µg/mL MMLE. A significant increase in late apoptotic cells from 4.5% to 27.7% was also observed upon treatment (Figure 3B,C). This shows that apoptosis is the likely mechanism by which MMLE inhibits the proliferation of AML cell lines.

To validate apoptosis induction, DNA fragmentation was quantified by Cell Death ELISA kit. A 5.0-fold and 12.8-fold significant increase in DNA fragmentation was observed upon 24 h treatment of Monomac-1 with 180 µg/mL and 270 µg/mL, respectively (Figure 4A), which correspond to concentrations below and above the IC50. A similar pattern of DNA fragmentation was noted in

KG-1 cells with a 1.7-fold and 2.5-fold increase after 24 h treatment with 135 μg/mL and 270 μg/mL respectively (Figure 4B). The increase of the dual stain fluorescence using annexin V/PI dual staining and DNA fragmentation confirms that MMLE induces apoptosis in Monomac-1 and KG-1 cell lines in a dose-dependent manner.

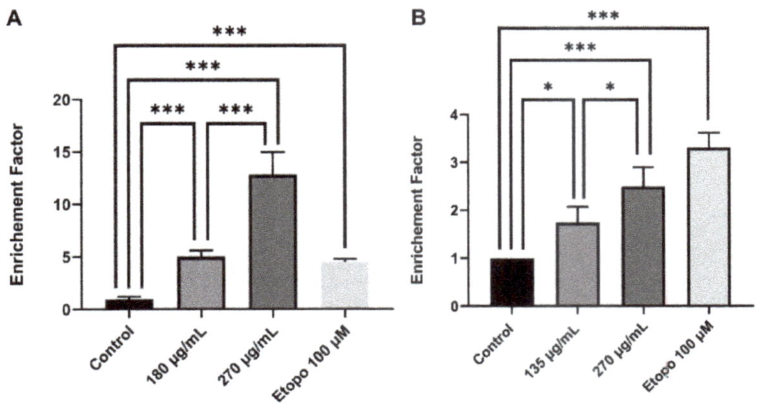

Figure 4. Cell Death ELISA on Monomac-1 (**A**) and KG-1 cells (**B**) treated with increasing concentrations of MMLE and a positive control treated with etoposide for 24 h. A significant dose-dependent increase in the enrichment factor was observed in both Monomac-1 and KG-1 cells when treated with concentrations before and after IC50. Significance relative to the control was reported with * indicating a *p*-value: $0.01 < p < 0.05$ and *** indicating a *p*-value: $0.0001 < p < 0.001$.

3.4. *M. pseudolavatera Leaf Extract Causes Upregulation of Pro-Apoptotic Proteins*

Western blot analysis was then performed to determine the apoptotic signaling pathway induced by MMLE on Monomac-1 cell line treated with 2 different concentrations, before and after the IC50, for 24 h. Upon treatment of Monomac-1 cells with 270 μg/mL of MMLE, an upregulation of cleaved PARP (c-PARP) was observed (Figure 5). This treatment also induced a downregulation of the anti-apoptotic protein Bcl-2, coupled to an upregulation of the pro-apoptotic protein Bax indicating an increase in the Bax/Bcl-2 ratio (Figure 5) leading to apoptosis. This was further confirmed by the observed dose-dependent upregulation of cytochrome-c upon treatment of Monomac-1 with MMLE (Figure 5).

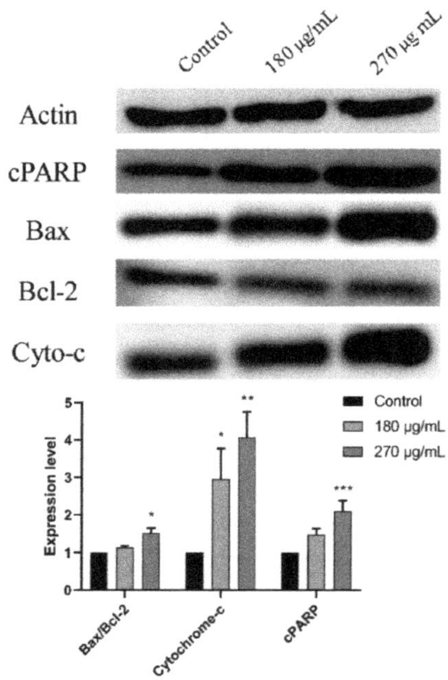

Figure 5. Western blot analysis and quantification of expression levels of apoptosis-regulating proteins in Monomac-1 cells treated with MMLE for 24 h. Significant upregulation of pro-apoptotic proteins such as cleaved poly(ADP-ribose) polymerase (PARP), Bax, and cytochrome-c and downregulation of anti-apoptotic proteins such as Bcl-2 were observed upon treatment of Monomac-1 cells with 180 µg/mL and 270 µg/mL. Significance relative to the control was reported with * indicating a *p*-value: $0.01 < p < 0.05$, ** indicating a *p*-value: $0.001 < p < 0.01$, and *** indicating a *p*-value: $0.0001 < p < 0.001$. Detailed information of western blot can be found at Figure S1.

3.5. *M. pseudolavatera Leaf Extract Induces Oxidative Stress in AML Cell Lines*

Reactive oxygen species levels (ROS) in Monomac-1 and KG-1 treated with increasing concentrations of MMLE were quantified using the DCFDA Cellular ROS Detection Assay kit. The recorded ROS levels showed a significant upregulation reaching 1.608-fold increase and 1.351-fold increase at 360 µg/mL for Monomac-1 and KG-1, respectively (Figure 6). This indicates that the extract is inducing oxidative stress by production of excess ROS in AML cell lines.

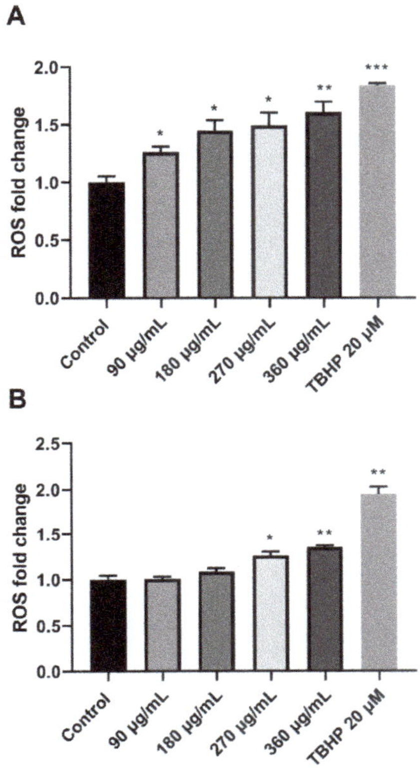

Figure 6. Fold change of ROS in Monomac-1 (**A**) and KG-1 cells (**B**) treated with increasing concentrations of MMLE and a positive control treated with 20 μM of TBHP. ROS levels increased significantly with increasing concentration of MMLE for both Monomac-1 (A) and KG-1 cells (B). Significance relative to the negative control was reported with * indicating a *p*-value: $0.01 < p < 0.05$, ** indicating a *p*-value: $0.001 < p < 0.01$, and *** indicating a *p*-value: $0.0001 < p < 0.001$.

3.6. Chemical Elucidation of *M. pseudolavatera* Leaf Extract Using GC-MS

The chemical composition of the extract was assessed using gas chromatography coupled to mass spectrometry (GC-MS) (Figure 7). Table 1 shows the major and minor constituents of the extract, some of which have been identified. The major compound (Peak 5: 31.7912%) was (Z,Z,Z)-9,12,15-octadecatrienoic acid methyl ester and another omega-3 fatty acid ester that os hexadecatrienoic acid was identified Peaks 7,10,13: (0.0463%). The second most abundant compound was phytol (Peak 6: 19.3447%). γ-sitosterol also constituted an important portion of the extract (Peak 25: 13.2396%) along with stigmasterol (Peak 24: 5.3751%).

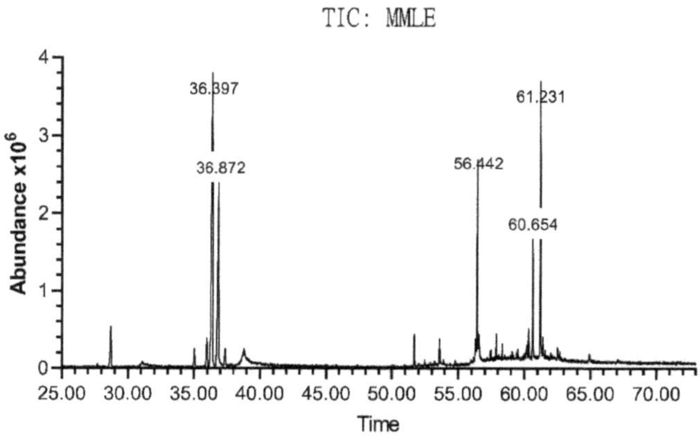

Figure 7. Chromatogram of *M. pseudolavatera* Webb & Berthel. methanolic leaf extract elucidated by GC-MS. Analysis of the different peaks obtained at different retention times with varying areas under the peak showed the presence of many compounds in different amounts.

Table 1. Table showing the composition of the *M. pseudolavatera* Webb & Berthel. methanolic leaf extract as elucidated by GC-MS.

Peak	RT	Compound	%MMLE
1	5.751	Unidentified A	0.2141
2	28.7101	Hexadecanoic acid methyl ester	3.4385
3	35.0314	Unidentified B	1.4208
4	35.9573	(E,E)-9,11-Octadecadienoic acid methyl ester	2.7188
5	36.3974	Methyl linolenate	31.7912
6	36.8717	Phytol	19.3447
7	37.3804	Octadecanoic acid methyl ester	1.3903
8	38.7407	Unidentified C	0.124
9	38.7864	7,10,13-Hexadecatrienoic acid methyl ester	0.0463
10	38.8093	Unidentified C	0.0408
11	51.7034	2,2′-methylenebis [6-(1,1-dimethylethyl)-4-methyl-phenol	1.6741
12	53.607	Unidentified D	1.47
13	56.3158	Unidentified E	1.4302
14	56.4415	(Z,Z,Z)-9-(3-hexenylidenecyclopropylidene)-, 2-hydroxy-1-(hydroxymethyl)nonanoic acid, ethyl ester	8.0675
15	56.584	Unidentified F	1.5
16	57.4074	Unidentified G	0.2354
17	57.505	Unidentified H	0.3
18	57.8875	Cyclotetracosane	0.6203
19	58.322	Unidentified I	0.31
20	59.5393	α-Tocopherol	0.3324
21	60.168	Unidentified J	0.3748
22	60.2195	Unidentified K	0.9207
23	60.3509	3-β-Ergost-5-en-3-ol	1.5331
24	60.6538	Stigmasterol	5.3751
25	61.2311	γ-Sitosterol	13.2396
26	61.4254	3-methoxy-19-Norpregna-1,3,5(10)-trien-17-ol	1.2401
27	64.94	Unidentified L	0.39

3.7. Chemical Elucidation of M. Pseudolavatera Leaf Extract Using LC-MS

The chemical composition of the extract was assessed using liquid chromatography coupled with mass spectrometry (LC-MS). Table 2 shows the major identified constituents of the extract. The two major compounds detected were the amino acids DL-phenylalanine and DL-tryptophan (RT = 2.609 min and 5.583 min, respectively). Other important abundant compounds were tiliroside (RT = 23.857 min), 13-hydroperoxyoctadecadienoic acid (RT = 35.512 min), and quercitrin (RT = 18.598 min). Many other compounds were identified as well.

Table 2. Table showing the composition of the *M. pseudolavatera* Webb & Berthel. methanolic leaf extract as elucidated by liquid chromatography coupled with mass spectrometry (LC-MS).

RT	Compound	Area Max
2.609	DL-phenylalanine	1.21×10^6
5.583	DL-tryptophan	1.10×10^6
23.857	Tiliroside	4.25×10^5
10.893	5-[(6,7,8-trimethoxy-4-quinazolinyl)amino]pentyl nitrate	3.12×10^5
35.512	13-hydroperoxyoctadecadienoic acid	1.95×10^5
14.915	*N*-acetyl-*L*-phenylalanine	1.12×10^5
18.865	3-amino-2-pyrazinecarboxylate	1.04×10^5
18.598	quercitrin	7.35×10^4
13.603	*N*-(4-{methyl[(1-methyl-1*H*-pyrazol-4-yl) methyl]sulfamoyl}phenyl)acetamide	7.21×10^4
26.441	Decyl hydrogen sulfate	6.97×10^4
14.775	Suberic acid	6.73×10^4
19.007	9-hydroxynonanoic acid	6.10×10^4
16.791	*L*-acetyltryptophan	5.12×10^4
14.082	5-(benzyloxy)-2-piperazinopyrimidine	5.05×10^4

4. Discussion

Natural products from plants have been widely considered an important source for identifying drugs with anti-cancer properties since they are rich in bioactive components having multiple targets with minimal side-effects [18,27]. In fact, of all pharmaceutical drugs present on the market, one-third are plant-derived [28], and many chemotherapeutic drugs which have been isolated from plants are now used as standard-of-care in cancer treatment regimens. In this study, the methanolic leaf extract of *M. pseudolavatera* Webb & Berthel. was examined for its anti-proliferative and pro-apoptotic effects on AML cell lines since leaves from other Malva species have shown cytotoxic effects on other cancer types [12–14]. In a previous study by Solowey et al. (2014), the ethanolic extract from *Urtica membranacea* showed potent anti-cancer effects at 750 µg/mL and 1500 µg/mL and these concentrations showed to be therapeutically correlated to a mouse breast cancer model with no side-effects [29]. These concentrations are higher than the concentrations of MMLE used to treat AML cells for 24 and 48 h. For the rest of this study, all experiments were performed with 24 h incubation of the extract in order to elucidate the mechanism of action of the extract in inhibiting cancer cell proliferation, although therapeutic levels are best reached after 48 h of incubation.

Moreover, the inhibitory effect of MMLE on the proliferation of the AML cell lines was significantly stronger than its effect on the growth of the MSCs and MNCs which exhibited resistance to the extract with no significant toxicity. These results supported the promising effects of the extract in selectively targeting cancerous cells with minimal if any effects on normal healthy cells, which is a major advantage of an effective chemotherapeutic drug exhibiting target selectivity [30].

An important mechanism by which chemotherapeutic drugs achieve their selective cytotoxicity is by activating apoptosis or programmed cell death [31]. Many hallmarks of apoptosis were detected in AML cell lines treated with MMLE among which were both membrane and nuclear changes typically detected by flow cytometry and protein blots [32].

Another characteristic of apoptosis alongside DNA fragmentation and membrane moieties flipping is the cleavage of poly(ADP-ribose) polymerase (PARP) [33]. PARP was previously reported to play an important role in salvaging cells suffering from DNA damage because it is involved in DNA repair. In fact, the cleaved fragment of PARP-1 binds to double strand breaks in the DNA preventing DNA repair machinery from accessing the damage leading to apoptosis. So, upregulation of cleaved PARP (c-PARP) in Monomac-1 treated with MMLE implies that DNA repair is no longer occurring, which promotes cell death via apoptosis induction [34].

Bcl-2 and Bax are also involved in the control of cell survival by decreasing and increasing the permeability of the outer mitochondrial membrane, respectively [35,36]. The reported increase in the Bax/Bcl-2 ratio upon MMLE treatment confirms the increase in mitochondrial membrane permeability. This promotes the release of cytochrome-c from the mitochondria, explaining its upregulation, which is essential for the activation of caspases leading to apoptosis [37]. In fact, exploiting chemotherapeutic effects on mitochondrial membrane leakage was shown to be effective in selectively triggering apoptosis in cancer cells since metabolic reprogramming is an inherent step required for hyperproliferation of cancer cells [38].

Another aspect of apoptotic cell death involves reactive oxygen species (ROS). Excess cellular ROS levels cause oxidative stress which damages proteins, DNA, and cellular membranes and activates death-receptor-mediated or mitochondrial apoptotic pathways [39]. However, a previous study showed that aqueous extract from *M. pseudolavatera* Webb & Berthel. possesses antioxidant properties by scavenging free radicals [9]. The differential effect of MMLE can be attributed to variation in its concentration, implying its dose-dependent activity. In fact, in similar medicinal plants with anti-cancer properties, extracts were shown to exhibit antioxidant activity at low concentration of the extract, without achieving any cytotoxicity. At high concentrations, that same extract was shown to be cytotoxic and induced ROS [40]. In MMLE, an antioxidant compound(s) may be present in high concentrations and hence act to destroy the mitochondrial membrane and generate ROS. This mechanism was described in green tea extract with the phenolic compound epigallocatechin gallate that can decrease lipid peroxidation and enhance antioxidant capacity in hepatocytes at low concentrations and destroy the mitochondrial membrane and generate intracellular oxidative stress at high concentrations [41].

The examination of the extract composition by GC-MS and LC-MS provided insight into some of the potential compounds in MMLE responsible for its pro-apoptotic effects. Abundant identified molecules included amino acids like-phenylalanine (Phe) and tryptophan (Trp) (detected by LC-MS) as well as some of their acetylated derivatives. These two essential amino acids were not previously shown to possess such activity. Omega-3 fatty acid esters (detected by GC-MS) and these two amino acids (phe and trp) were previously shown to slow the growth of many types of cancers and increase patient outcomes when included in a standard regimen of chemotherapy since they increase the sensitivity of the cells to the conventional therapies. They also exhibit selective toxicity on cancer cells of various types but not on normal cells [42,43]. This could explain the selective toxicity seen on AML cell lines and not on MSCs and MNCs. Also, (13S)-hydroperoxyoctadecadienoic (13-HPODE) detected by LC-MS, is a linoleic acid derivative previously shown to inhibit growth of a chronic myeloid leukemia cell line K-562 [44]. In fact, the mechanism of action described was ROS-mediated and caspase-dependent apoptosis which could explain the effects observed by MMLE on AML cell lines.

Phytol, another major compound detected by GC-MS, is a diterpene alcohol and it was previously found to inhibit the growth of many cancer cell types, among which acute T-cell lymphoblastic leukemia (Molt-4) cells in a dose- and time-dependent manner. The effects of phytol were attributed to apoptosis demonstrated by DNA fragmentation and formation of apoptotic bodies. In other studies, phytol was found to have a synergistic effect with some used chemotherapeutic drugs like β-caryophyllene, in addition to inducing apoptosis in epidermoid carcinoma cells (A431) and immortalized keratinocytes (HaCaT cells) by ROS induction, activation of the apoptotic pathway involving the release of cytochrome-c, the activation of the caspase pathway, cleavage of PARP, and an increase in the Bax/Bcl-2 ratio [45,46]. These previously reported effects are consistent with the current observed morphological and molecular changes in AML cell lines upon treatment with MMLE. Another detected alcohol is quercitrin (by LC-MS). It is a plant-derived polyphenol found to enhance the effect of topotecan in breast cancer cell lines [47,48]. It was also found to reduce the cytotoxicity and genotoxicity of topotecan in bone marrow cells of mouse models in a dose-dependent manner [49].

Another important phytochemical detected by LC-MS is tiliroside. It is a glycosidic flavonoid present in many edible plants [50]. It was found to be cytotoxic against human CML cell line K-562 [51] and to inhibit cell proliferation and induce apoptosis via the extrinsic pathway in breast cancer cell lines MCF-7 and T47D [52].

Phytosterols, particularly γ-sitosterol and stigmasterol, are also important constituents of the extract (detected by GC-MS). γ-sitosterol was previously shown to be cytotoxic against colon and liver cancer cell lines by downregulating c-myc and inducing apoptosis [53]. Moreover, stigmasterol was formerly studied for activating apoptosis in hepatocellular carcinoma cell lines through upregulation of the Bax protein and downregulation of the Bcl-2 protein [54]. In another study, stigmasterol inhibited the proliferation of gastric cancer cell lines through a mitochondrial pathway [55], in accordance with the effects MMLE exhibited on AML cell lines in this study.

All these findings suggest that MMLE contains many compounds which can potentially be acting together as cancer growth inhibitors through inducing ROS and activating apoptosis in AML cell lines. Many compounds detected by GC-MS and LC-MS were not previously known for their effects on cancer cell lines and many other compounds were not identified. This should also be taken into account when explaining the pro-apoptotic effect of MMLE on AML cells.

5. Conclusions

In conclusion, *Malva pseudolavatera* Webb & Berthel. methanolic leaf extract showed a promising selective anti-proliferative and pro-apoptotic effect on acute myeloid leukemia cell lines, by cleaving PARP, releasing cytochrome-c, and increasing the Bax/Bcl-2 ratio. Chemical analysis of the extract showed that it is a ROS inducer and that it contains many compounds that are potentially anti-cancer compounds. Future work aims at exploring the effect of the extract on other types of cancer cells, fractionating the extract to identify the compounds with highest biological therapeutic activity and confirming the efficacy of the extract in vivo.

Supplementary Materials: The following are available online at http://www.mdpi.com/2072-6694/12/2/435/s1. Figure S1: Western blot analysis and quantification of expression levels of apoptosis-regulating proteins in Monomac-1 cells treated with MMLE for 24 h.

Author Contributions: Preparing plant extracts, performing experiments, interpreting the results of all the cell lines and writing the first version of the manuscript: M.E.K. and T.H. Dual annexin V/PI staining and revising of the manuscript: M.H.H. Performing the cell cycle analysis by flow cytometry and revising the manuscript: S.A.N. Performing the chemical characterization of the extract: R.I.T. and R.S. Developing the concept of the study, supervising the work, interpreting the results, and generating the final version of the manuscript: S.R. All authors have read and agreed to the published version of the manuscript.

Funding: This research was financially funded by intramural funds from the School Research Development Council (School of Arts and Sciences, Lebanese American University) and the Department of Natural Sciences (Lebanese American University).

Conflicts of Interest: The authors declare no conflict of interest.

References

1. The Jebson Herbarium University of California, Berkley. Available online: http://ucjeps.berkeley.edu/eflora/eflora_display.php?tid=89042 (accessed on 25 June 2019).
2. Edgecombe, W.S. *Weeds of Lebanon*; American University of Beirut: Beirut, Lebanon, 1964; pp. 244–247.
3. Webb, P.B.; Berthelot, S. *Histoire Naturelle des Iles Canaries, Phytographia Canariensis*; Bethune, editeur: Paris, France, 1836; pp. 1836–1850.
4. The Plant List. Malva pseudolavatera Webb & Berthel. Available online: http://theplantlist.org/tpl1.1/record/tro-50337435 (accessed on 7 April 2019).
5. Abdullatif, A. Malva: Food, medicine and chemistry. *Eur. Chem. Bull.* **2017**, *6*, 295–320.
6. Singorini, M.A.; Piredda, M.; Bruschi, P. Plants and traditional knowledge: An ethnobotanical investigation on Monte Ortobene. *J. Ethnobiol. Ethnomed.* **2009**, *5*, 6.
7. Rivera, N.D.; Obon, D.C.C. Medicaments et Aliments: l'approche ethnopharmacologique. Acte du 2eme Colloque Europeen d'Ethnopharamcologie et de la 1ere Coference Internationale d'Ethnomedicine. *Heidelberg* **1993**, *2*, 223.
8. Gaspar, N.; Godinho, J.; Vasconcelos, T.; Caldas, D.; Mendes, P.; Barros, O. Ethnobotany in the center of Portugal. In *Natural Products in the New Millennium: Prospects and Industrial Applications*; Springer: Dordercht, The Netherlands, 2002; pp. 271–284.
9. Ben-Nasr, S.; Aazza, S.; Mnif, W.; da Graca Costa Miguel, M. Antioxidant and anti-lipoxygenase activities of extracts from different parts of Lavatera cretica L. grown in Algarve (Portugal). *Pharm. Mag.* **2015**, *11*, 48–54.
10. Commitee of Herbal Medicinal Products. *Assessment Report on Malva Sylvestris L. and/or Malva Neglecta Wallr., Follium and Malva Sylvestris L., Flos*; European Medicines Agency: London, UK, 2018.
11. Farhan, H.; Rammal, H.; Hijazi, A. Chemical composition, in vitro cytotoxicity and anti-free radical properties of six extracts from Lebanese Trigonella berythea boiss. *Pak. J. Pharm. Sci.* **2013**, *26*, 1157–1163.
12. Raghda Rayssan, S.; Muayad, S. Cytotoxicity Assessment of Malva Sylvestris Crude Extract. *Int. J. Pharm. Sci. Res.* **2019**, *11*, 70–74.
13. Ali, M.; Abul Farah, M.; Al-Hemaid, F.; Abou-Tarboush, F. In vitro cytotoxicity screening of wild plant extracts from Saudi Arabia on human breast adenocarcinoma cells. *Genet. Mol. Res.* **2014**, *13*, 3981–3990. [CrossRef] [PubMed]
14. Donaldson, J.; Cates, R. Screening for anticancer agents from Sonoran desert plants: A chemical ecology approach. *Pharm. Biol.* **2014**, *42*, 478–487. [CrossRef]
15. American Cancer Society. What Is Acute Myeloid Leukemia (AML). Available online: https://www.cancer.org/cancer/acute-myeloid-leukemia/about/what-is-aml.html (accessed on 3 April 2019).
16. Shysh, A.C. The incidence of acute myeloid leukemia in Calgary, Alberta, Canada: A retrospective cohort study. *BMC Public Health* **2018**, *18*, 94. [CrossRef] [PubMed]
17. Cancer Research UK. Chemotherapy for Acute Myeloid Leukaemia (AML). Available online: https://www.cancerresearchuk.org/about-cancer/acute-myeloid-leukaemia-aml/treating-aml/chemotherapy/chemotherapy-for-aml (accessed on 3 April 2019).
18. Atanasov, A. Discovery and resupply of pharmacologically active plant-derived natural products: A review. *Biotechnol. Adv.* **2015**, *33*, 1582–1614. [CrossRef]
19. Rates, S. Plants as source of drugs. *Toxicon* **2001**, *39*, 603–613. [CrossRef]
20. National Research Council (US). Committee for the Update of the Guide for the Care and Use of Laboratory Animals. In *Guide for the Care and Use of Laboratory Animals*, 8th ed.; National Academies Press (US): Washington, DC, USA, 2011.
21. Zeeni, N.; Daher, C.; Fromentin, G.; Tome, D.; Darcel, N.; Chaumontet, C. A cafeteria diet modifies the response to chronic variable stress in rats. *Stress* **2013**, *16*, 211–219. [CrossRef] [PubMed]
22. Najar, M.; Fayyad-Kazan, H.; Faour, W.H.; Merimi, M.; Sokal, E.M.; Lombard, C.A.; Fahmi, H. Immunological modulation following bone marrow-derived mesenchymal stromal cells and Th17 lymphocyte co-cultures. *Inflamm. Res.* **2019**, *68*, 203–213. [CrossRef]
23. Soleimani, M.; Nadri, S. A protocol for isolation and culture of mesenchymal stem cells from mouse bone marrow. *Nat. Protoc.* **2009**, *4*, 102–106. [CrossRef] [PubMed]

24. Haykal, T.; Nasr, P.; Hodroj, M.H.; Taleb, R.I.; Sarkis, R.; Moujabber, M.N.E.; Rizk, S. *Annona cherimola* Seed Extract Activates Extrinsic and Intrinsic Apoptotic Pathways in Leukemic Cells. *Toxins* **2019**, *11*, 506. [CrossRef]
25. Zibara, K.; Hamdan, R.; Dib, L.; Sindet-Pedersen, S.; Kharfan-Dabaja, M.; Bazarbachi, A.; El-Sabban, M. Acellular Bone Marrow Extracts Significantly Enhance Engraftment Levels of Human Hematopoietic Stem Cells in Mouse Xeno-Transplantation Models. *PLoS ONE* **2012**, *7*, e40140. [CrossRef] [PubMed]
26. Nisrine Machaka-Houri. Available online: http://www.nisrinemachaka.com/ (accessed on 24 April 2019).
27. Aung, T.N.; Qu, Z.; Kortschak, R.D.; Adelson, D.L. Understanding the effectiveness of natural compound mixtures in cancer through their molecular mode of action. *Int. J. Mol. Sci.* **2017**, *18*, 656. [CrossRef]
28. Howitz, K.T.; Sinclair, D.A. Xenohormesis: Sensing the chemical cues of other species. *Cell* **2008**, *133*, 387–391 [CrossRef]
29. Solowey, E.; Lichtenstein, M.; Sallon, S.; Paavilainen, H.; Solowey, E.; Lorberboum-Galski, H. Evaluating medicinal plants for anticancer activity. *Sci. World J.* **2014**, *2014*, 721402. [CrossRef]
30. Fattoruso, S.I.; Di Lauro, L.; Conti, F.; Amodio, A.; Lopez, M. Target selectivity of anticancer drugs. *Clin. Ter.* **2008**, *3*, 189–206.
31. Pfeffer, C.M.; Singh, A.T.K. Apoptosis: A target for Anti-cancer therapy. *Int. J. Mol. Sci.* **2018**, *19*, 448. [CrossRef] [PubMed]
32. Luparello, C.; Asaro, D.M.L.; Cruciata, I.; Hassell-Hart, S.; Sansook, S.; Spencer, J.; Caradonna, F. Cytotoxic Activity of the Histone Deacetylase 3-Selective Inhibitor Pojamide on MDA-MB-231 Triple-Negative Breast Cancer Cells. *Int. J. Mol. Sci.* **2019**, *20*, 804. [CrossRef] [PubMed]
33. Skidmore, C.J.; Davies, M.I.; Goodwin, P.M.; Halldorsson, H.; Lewis, P.J.; Shall, S.; Zia'ee, A.A. The Involvement of Poly(ADP-ribose) Polymerase in the Degradation of NAD Caused by Radiation and N-Methyl-N-Nitrosourea. *Eur. J. Biochem.* **1979**, *101*, 135–142. [CrossRef] [PubMed]
34. Smulson, M.E.; Pang, D.; Jung, M.; Dimtchev, A.; Chasovskikh, S.; Spoonde, A.; Simbulan-Rosenthal, C.; Rosenthal, D.; Yakovlev, A.; Dritschilo, A. Irreversible binding of poly(ADP)ribose polymerase cleavage product to DNA ends revealed by atomic force microscopy: Possible role in apoptosis. *Cancer Res.* **1998**, *58*, 3495–3498. [PubMed]
35. Chen, Q.; Xu, H.; Xu, A.; Ross, T.; Bowler, E.; Hu, Y.; Lesnefsky, E.J. Inhibition of Bcl-2 Sensitizes Mitochondrial Permeability Transition Pore (MPTP) Opening in Ischemia-Damaged Mitochondria. *PLoS ONE* **2015**, *10*, 3. [CrossRef] [PubMed]
36. Edlich, F.; Banerjee, S.; Suzuki, M.; Cleland., M.M.; Arnoult, D.; Wang, C.; Neutzner, A.; Tjandra, N.; Youle, R.J. Bcl-xL retrotranslocates Bax from the mitochondria into the cytosol. *Cell* **2011**, *145*, 104–116. [CrossRef] [PubMed]
37. Cosulich, S.C.; Worrall, V.; Hedge, P.J.; Green, S.; Clarke, P.R. Regulation of apoptosis by BH3 domains in a cell-free system. *Curr. Biol.* **1997**, *7*, 913–920. [CrossRef]
38. Nguyen, C.; Pandey, S. Exploiting Mitochondrial Vulnerabilities to Trigger Apoptosis Selectively in Cancer Cells. *Cancers* **2019**, *11*, 916. [CrossRef]
39. Redza-Dutordoir, M.; Averill-Bates, D.A. Activation of apoptosis signalling pathways by reactive oxygen species. *Biochim. Biophys. Acta* **2016**, *1863*, 2977–2992. [CrossRef]
40. Sun, C.; Zhang, H.; Ma, X.F.; Zhou, X.; Gan, L.; Liu, Y.Y.; Wang, Z.H. Isoliquiritigenin enhances radiosensitivity of HepG2 cells via disturbance of redox status. *Cell Biochem. Biophys.* **2013**, *65*, 433–444. [CrossRef]
41. Galati, G.; Lin, A.; Sultan, A.M.; O'Brien, P.J. Cellular and in vivo hepatotoxicity caused by green tea phenolic acids and catechins. *Free Radic. Biol. Med.* **2006**, *40*, 570–580. [CrossRef] [PubMed]
42. Hardman, W.E. (n-3) Fatty Acids and Cancer Therapy. *J. Nutr.* **2004**, *134*, 3427–3430. [CrossRef] [PubMed]
43. D'Eliseo, D.; Velotti, F. Omega-3 Fatty Acids and Cancer Cell Cytotoxicity: Implications for Multi-Targeted Cancer Therapy. *J. Clin. Med.* **2016**, *5*, 15. [CrossRef] [PubMed]
44. Mahipal, S.V.; Subhashini, J.; Reddy, M.C.; Reddy, M.M.; Anilkumar, K.; Roy, K.R.; Reddy, G.V.; Reddanna, P. Effect of 15-lipoxygenase metabolites, 15-(S)-HPETE and 15-(S)-HETE on chronic myelogenous leukemia cell line K-562: Reactive oxygen species (ROS) mediate caspase-dependent apoptosis. *Biochem. Pharmacol.* **2007**, *74*, 202–214. [CrossRef] [PubMed]
45. Komiya, T.; Kyohkon, M.; Ohwaki, S.; Eto, J.; Katsuzaki, H.; Imai, K.; Kataoka, T.; Yoshioka, K.; Ishii, Y.; Hibasami., H. Phytol induces programmed cell death in human lymphoid leukemia Molt 4B cells. *Int. J. Mol. Med.* **1999**, *4*, 377–380. [CrossRef] [PubMed]

46. Pavithra, P.S.; Mehta, A.; Verma, R.S. Synergistic interaction of β-caryophyllene with aromadendrene oxide-2 and phytol induces apoptosis on skin epidermoid cancer cells. *Phytomedicine* **2018**, *47*, 121–134. [CrossRef] [PubMed]
47. Li, Y.; Yao, J.; Han, C.; Yang, J.; Chaudhry, M.T.; Wang, S.; Liu, H.; Yin, Y. Quercetin, Inflammation and Immunity. *Nutrients* **2016**, *8*, 167. [CrossRef]
48. Akbas, S.H.; Timur, M.; Ozben, T. The Effect of Quercetin on Topotecan Cytotoxicity in MCF-7 and MDA-MB 231 Human Breast Cancer Cells. *J. Surg. Res.* **2005**, *125*, 49–55. [CrossRef]
49. Bakheet, S.A. Assessment of anti-cytogenotoxic effects of quercetin in animals treated with topotecan. *Oxid. Med. Cell. Longev.* **2011**, *2011*, 824597. [CrossRef]
50. Goto, T.; Teraminami, A.; Lee, J.Y.; Ohyama, K.; Funakoshi, K.; Kim, Y.I.; Hirai, S.; Uemura, T.; Yu, R.; Takahashi, N.; et al. Tiliroside, a glycosidic flavonoid, ameliorates obesity-induced metabolic disorders via activation of adiponectin signaling followed by enhancement of fatty acid oxidation in liver and skeletal muscle in obese-diabetic mice. *J. Nutr. Biochem.* **2012**, *23*, 768–776. [CrossRef]
51. Esteves-Souza, A.; Silva, T.; Alves, C.; Carvalho, M.; Braz-Filho, R.; Echevarria, A. Cytotoxic activities against Ehrlich carcinoma and human K562 leukaemia of alkaloids and flavonoid from two Solanum Species. *J. Brazil. Chem. Soc.* **2002**, *13*, 838–842. [CrossRef]
52. Dai, M.; Wikantyasning, E.; Wahyuni, A.; Kusumawati, I.; Saifudin, A.; Suhendi, A. Antiproliferative properties of tiliroside from Guazuma ulmifolia lamk on T47D and MCF7 cancer cell lines. *Natl. J. Physiol. Pharm. Pharmacol.* **2016**, *6*, 627. [CrossRef]
53. Endrini, S.; Rahmat, A.; Ismail, P.; Taufiq-Yap, Y. Cytotoxic effect of γ-Sitosterol from kejibeling (Strobilanthes crispus) and its mechanism of action towards c-myc gene expresiion and apoptotic pathway. *Med. J. Indones.* **2014**, *23*, 203–208. [CrossRef]
54. Kim, Y.S.; Li, X.F.; Kang, K.H.; Ryu, B.; Kim, S.K. Stigmasterol isolated from marine microalgae *Navicula incerta* induces apoptosis in human hepatoma HepG2 cells. *BMB Rep.* **2014**, *47*, 433–438. [CrossRef]
55. Li, K.; Yuan, D.; Yan, R.; Meng, L.; Zhang, Y.; Zhu, K. Stigmasterol exhibits potent antitumor effects in human gastric cancer cells mediated via inhibition of cell migration, cell cycle arrest, mitochondrial mediated apoptosis and inhibition of JAK/STAT signaling pathway. *J. BUON.* **2018**, *23*, 1420–1425.

cancers

MDPI

Article

Combined Effects of Eicosapentaenoic Acid and Adipocyte Renin–Angiotensin System Inhibition on Breast Cancer Cell Inflammation and Migration

Fahmida Rasha [1,2], Chanaka Kahathuduwa [2,3], Latha Ramalingam [1,2], Arelys Hernandez [1], Hanna Moussa [2,4] and Naima Moustaid-Moussa [1,2,*]

1 Department of Nutritional Sciences, Texas Tech University, 1301 Akron Ave, Lubbock, TX 79409, USA; fahmida.rasha@ttu.edu (F.R.); latha.ramalingam@ttu.edu (L.R.); arelys.hernandez@ttu.edu (A.H.)

2 Obesity Research Institute, Texas Tech University, Lubbock, TX 79409, USA; chanaka.kahathuduwa@ttuhsc.edu (C.K.); hanna.moussa@ttu.edu (H.M.)

3 Department of Psychiatry, School of Medicine, Texas Tech University Health Science Center, Lubbock, TX 79430, USA

4 Department of Mechanical Engineering, Texas Tech University, Lubbock, TX 79409, USA

* Correspondence: naima.moustaid-moussa@ttu.edu; Tel.: +1-806-834-7946

Received: 30 November 2019; Accepted: 11 January 2020; Published: 16 January 2020

Abstract: Obesity is a major risk factor for breast cancer (BC). Obesity-related metabolic alterations such as inflammation and overactivation of the adipose renin–angiotensin system (RAS) may contribute to the progression of BC. Clinically used antihypertensive drugs such as angiotensin-converting enzyme inhibitors (ACE-I) and dietary bioactive components such as eicosapentaenoic acid (EPA) are known for their anti-inflammatory and adipose RAS blocking properties. However, whether EPA enhances the protective effects of ACE-I in lessening adipocyte inflammation on BC cells has not been studied. We hypothesized that combined EPA and ACE-I would attenuate BC cell inflammation and migration possibly via adipose RAS inhibition. To test our hypothesis, we examined the (i) direct effects of an ACE-I (captopril (CAP)) or EPA, individually and combined, on MCF-7 and MDA-MB-231 human BC cells, and the (ii) effects of conditioned medium (CM) from human adipocytes pretreated with the abovementioned agents on BC cells. We demonstrated that CM from adipocytes pretreated with EPA with or without captopril (but not direct treatments of BC cells) significantly reduced proinflammatory cytokines expression in both BC cell lines. Additionally, cell migration was reduced in MDA-MB-231 cells in response to both direct and CM-mediated CAP and/or EPA treatments. In summary, our study provides a significant insight into added benefits of combining anti-inflammatory EPA and antihypertensive ACE-I to attenuate the effects of adipocytes on breast cancer cell migration and inflammation.

Keywords: obesity; breast cancer; renin–angiotensin system; eicosapentaenoic acid; adipocyte inflammation

1. Introduction

Breast cancer (BC) is the most common type of cancer among U.S. women and has a lifetime risk of more than 12% [1]. Obesity is one of the major modifiable risk factors for BC, especially in postmenopausal women, and is associated with poor cancer outcomes and survival in patients with BC [2,3]. Moreover, as inflammation is an important underlying basis for both diseases, it is critical to understand its involvement in mechanism(s) of obesity-related BC [4]. In obesity, adipose tissue expands and becomes hypertrophic and hypoxic, with invasion of proinflammatory M1-type macrophages and exaggerated secretion of protumor adipocytokines [5]. Various proinflammatory adipocytokines such as interleukin (IL)-6, IL-8, and monocyte chemoattractant protein-1 (MCP-1) can promote the activation and transcription of many protumor signaling pathways such as nuclear factor kappa B (NF-κB) and signal transducer and activator of transcription 3 (STAT3), resulting in increased migratory and invasive capacities in the adjacent cancer cells [6]. Additionally, cancer-associated adipocytes form a proinflammatory microenvironment via induced macrophage invasion and thus synergistically activate STAT3 and NF-κB pathways [7], which in turn leads to apoptosis inhibition and cancer cell proliferation [8]. Moreover, aberrant STAT3 signaling also upregulates angiogenic vascular endothelial growth factor (VEGF), leading to higher tumor angiogenesis [9]. Hence, obesity-associated proinflammation increases cancer cell progression via their interaction with adjacent adipocytes [6]. Furthermore, obese adipocytes also overexpress metabolic markers such as fatty acid synthase (FASN), which is associated with adipocyte hypertrophy and increased proinflammatory adipocytokine secretion in obesity [10,11]. FASN provides an alternative energy source to proliferating cancer cells in the form of free fatty acids helping their growth and proliferation [12]. This, in turn, facilitates the formation of a toxic tumor microenvironment (TME) in the breast, which contributes to obesity-associated BC progression [13].

The renin–angiotensin system (RAS) is classically known to regulate fluid balance and blood pressure. Angiotensinogen (Agt) is the main precursor protein for the bioactive hormone angiotensin II (Ang II). Agt is first cleaved by renin into Ang I, which is converted into Ang II by the angiotensin-converting enzyme (ACE). Ang II mediates its effect via binding to two major G-protein-coupled receptors, namely, type 1 receptor (AT1R) and type 2 receptor (AT2R) [14,15]. Adipose RAS activation has multiple roles in obesity, such as increasing inflammatory, prothrombotic, and angiogenic markers [16] that can also promote tumor growth, proliferation, and invasion in the breast [17,18]. In addition, Ang II/AT1R signaling triggers production and infiltration of tumor-associated macrophages (TAM) in different tumor models, while RAS inhibitors can restrain tumor growth and TAM response [19]. Thus, RAS is a potential key player in adipocyte–breast-cancer-cell interactions. RAS inhibitors such as angiotensin-converting enzyme inhibitors (ACE-I) and AT1R blockers (ARB) have recently gained interest as possible anticancer agents in both clinical and preclinical models of BC [17,18,20,21]. Some ACE-I and ARBs were found to reduce BC risk in a time-dependent manner in BC patients [21].

RAS can also be downregulated by anti-inflammatory dietary bioactive compounds such as omega-3 polyunsaturated fatty acids (n-3 PUFA) [22], which we have also shown to reduce Agt secretion from adipocytes [23]. The dietary n-3 PUFAs eicosapentaenoic acid (EPA)-20:5n-3 and docosahexaenoic acid (DHA)-22:5n-3 have potent anti-inflammatory properties that include reducing obesity-associated systemic and adipose tissue inflammation, thus alleviating TME formation in adipocyte–BC-cell crosstalk [24–27]. Therefore, given the promising anti-inflammatory and anticancer properties of ACE-I and EPA, we hypothesized that ACE-I (captopril (CAP)) and EPA alone or in combination would attenuate BC cell inflammation and migration, in part through inhibition of adipocyte RAS. We examined both direct and adipose CM-mediated effects of captopril and EPA and their combination on cultured BC cell genes and proteins as well as cell migration using wound healing analyses.

2. Results

2.1. Effect of Captopril and EPA on Markers of Fatty Acid Synthesis and Inflammation in BC Cells and Role of Human Adipocyte-Conditioned Media (CM)

MDA-MB-231 and MCF-7 cells were treated with 100 μm of EPA or 100 μm of CAP with or without EPA either directly or using CM from CAP, EPA, or both pretreated adipocytes for 48 h. Direct treatments with EPA, CAP, or CAP + EPA did not alter most measured markers of cancer cell growth and inflammation in MDA-MB-231 cells, with the exception of IL-8, which was significantly reduced by EPA and CAP + EPA treatments for 48 h in MDA-MB-231 cells ($p < 0.05$) compared with control (CT), while CAP alone had no effect. However, EPA and CAP + EPA had comparable effects, indicating no additional effect of direct CAP + EPA combination on BC cell inflammatory markers

(Figure 1E). Treatment of MDA-MB-231 cells with human adipocyte CM significantly increased all tested markers of cell growth and inflammation after 48 h, as shown by increased mRNA levels of FASN, STAT3, NF-κB, IL-6, and IL-8 compared with control (Figure 1A–E; also see Tables S1–S7) ($p < 0.05$). However, treatment of MDA-MB-231 cells with CM from adipocytes pretreated with EPA, CAP, and their combination significantly reduced the mRNA content of all measured markers of BC cell growth and inflammation compared with treatment with CM derived from untreated adipocytes (Figure 1A–E) ($p < 0.05$). However, no changes were observed in FASN, STAT3, NF-κB, and IL-8 mRNA transcription levels in MDA-MB-231 BC cells treated with CAP + EPA pretreated adipocyte CM, compared with CAP-CM or EPA-CM (Figure 1A–C,E) ($p < 0.05$). Interestingly, CAP + EPA pretreated CM reduced IL-6 mRNA levels to a greater extent in MDA-MB-231 cells compared with CAP-CM and/or EPA-CM treatments, indicating potential additive anti-inflammatory effects of CAP and EPA combination (Figure 1D) ($p < 0.05$). Exploratory factorial regression analyses performed to examine the interactions of CAP and EPA when administered as a combination resulted in significant negative regression coefficients for CM-CAP and CM-EPA factors but significant positive CM-CAP × EPA interaction for mRNA levels of all measured markers of MDA-MB-231 cell growth and inflammation (Tables S1–S5). This result suggests that CAP and EPA may act via a common pathway in reducing mRNA expression in CM-treated MDA-MB-231 cells.

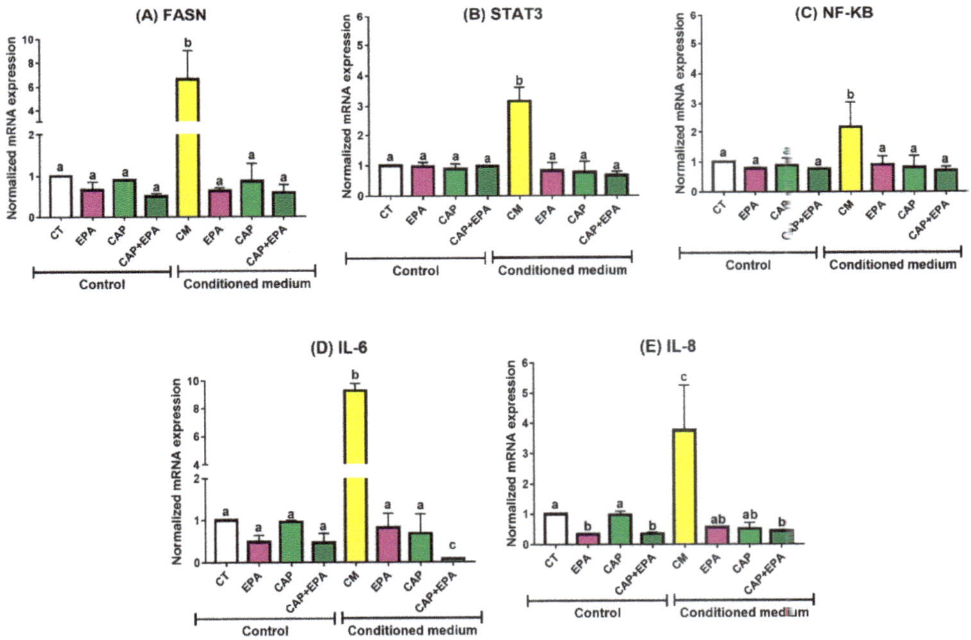

Figure 1. Eicosapentaenoic acid (EPA) and captopril (CAP) (angiotensin-converting enzyme inhibitors; ACE-I) effects on mRNA expression in MDA-MB-231 cells. MDA-MB-231 cells were treated with 100 µm of CAP with or without 100 µm of EPA for 48 h. Human mesenchymal stem cells (HMSCs) were differentiated into adipocytes and treated with 100 µm of CAP with or without 100 µm of EPA for 24 h. Conditioned media (CM) was collected and transferred to breast cancer (BC) cells for 48 h. Cells were harvested and changes in mRNA levels of fatty acid synthase (FASN) (**A**), signal transducer and activator of transcription 3 (STAT3) (**B**), nuclear factor kappa B (NF-κB) (**C**), interleukin (IL)-6 (**D**), and IL-8 (**E**) were measured ($p < 0.05$; N = 3; three replicates under each treatment group; bars with different letters (a, b, c) indicate significance).

On the other hand, CM from human adipocytes significantly increased markers of cell growth and inflammation in MCF-7 cells after 48 h, as shown by increased mRNA levels of FASN, STAT3, NF-κB, and IL-8 compared with CT ($p < 0.05$), while CM from adipocytes pretreated with EPA, CAP, and CAP + EPA significantly reduced the abovementioned markers of cell growth and inflammation after 48 h compared with CM-control (Figure 2A–C,E; Tables S8–S14) ($p < 0.05$). However, no changes in the mRNA levels of the respective markers were observed between CAP and EPA treated groups with or without CAP–EPA combination. Additionally, direct treatments with EPA and CAP + EPA significantly reduced MCF-7 BC cell inflammation, as demonstrated by significantly lower IL-6 and IL-8 mRNA transcription levels, while direct treatments with CAP reduced only IL-6 mRNA levels after 48 h compared with control in MCF-7 cells (Figure 2D,E) ($p < 0.05$). However, the changes were not significant between EPA and CAP + EPA treated groups, indicating no additional effects of CAP–EPA combination in MCF-7 cells compared with EPA alone or CAP alone.

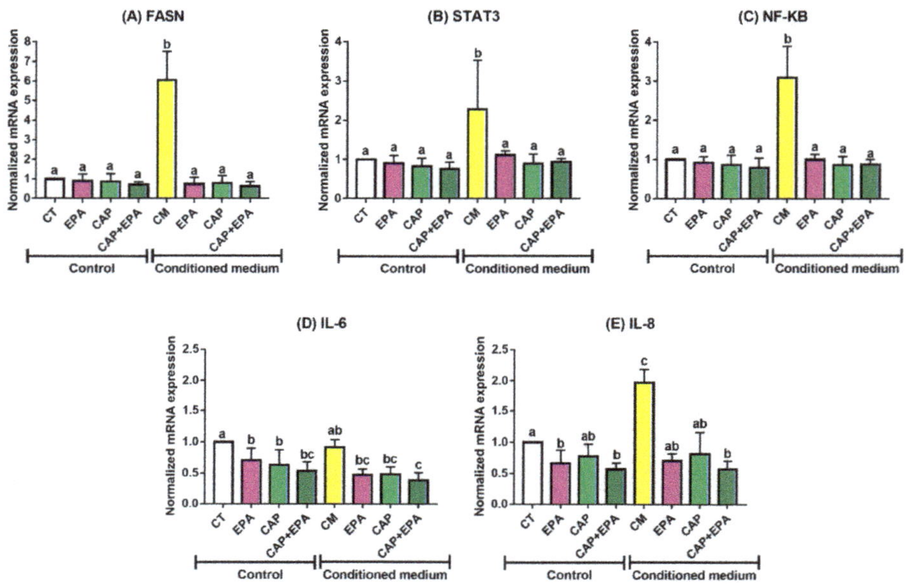

Figure 2. EPA and captopril (ACE-I) effects on mRNA expression in MCF-7 cells. MCF-7 cells were treated with 100 μm of CAP with or without 100 μm of EPA for 48 h. HMSCs were differentiated into adipocytes and treated with 100 μm of CAP with or without 100 μm of EPA for 24 h. CM was collected and transferred to BC cells for 48 h. Cells were harvested and mRNA level changes of FASN (**A**), STAT3 (**B**), NF-κB (**C**), IL-6 (**D**), and IL-8 (**E**) were measured ($p < 0.05$; N = 3; three replicates under each treatment group; bars with different letters (a, b, c) indicate significance).

Next, we measured proinflammatory IL-6 and IL-8 protein levels secreted by both MDA-MB-231 and MCF-7 cells after 48 h of EPA and CAP treatments. Medium was collected from both cell lines after direct and adipocyte CM (with or without EPA, CAP, or combined EPA and CAP) treatments for 48 h. Consistent with the above-described results (Figure 3A,B), no changes were observed in secreted IL-6 and IL-8 levels in MDA-MB-231 cells with direct treatments of EPA, CAP, CAP + EPA for 48 h. In addition, increased IL-6 and IL-8 secretion in MDA-MB-231 cells were observed in response to adipocyte CM treatments compared with CT and direct CAP ± EPA treatments ($p < 0.05$). Moreover, significant reductions in both IL-6 and IL-8 protein levels were identified in MDA-MB-231 cells in response to CM-EPA and CM-CAP treatments (with or without EPA) compared with CM-control ($p < 0.05$) (Figure 3A,B). However, the changes were not significant between CM-EPA and CM-CAP groups as well as compared to CT and direct CAP ± EPA treatments after 48 h ($p < 0.05$). Additionally, both IL-6 and IL-8 protein levels were undetectable (* nd) without CM treatments in MCF-7 cells, as reported earlier by others [28] (Figure 3C,D). Interestingly, adipocyte CM with CAP + EPA significantly reduced IL-6 secretion in MCF-7 cells compared with CM-control treatment, but no changes were observed compared to CM-EPA or CM-CAP individual treatments (Figure 3C) ($p < 0.05$). However, no significant changes were observed in secreted IL-8 levels by MCF-7 cells when treated with adipocyte CM-EPA or CM-CAP (with or without EPA) (Figure 3D). Therefore, these results suggest possible inflammatory-reducing benefits of CAP–EPA combination in triple-negative breast cancer (TNBC) MDA-MB-231 cells but not in MCF-7 cells, indicating the potential role of n-3 PUFAs and antihypertensive ACE inhibitors in attenuating the tumor-promoting proinflammatory effects of adipocytes on breast cancer cells.

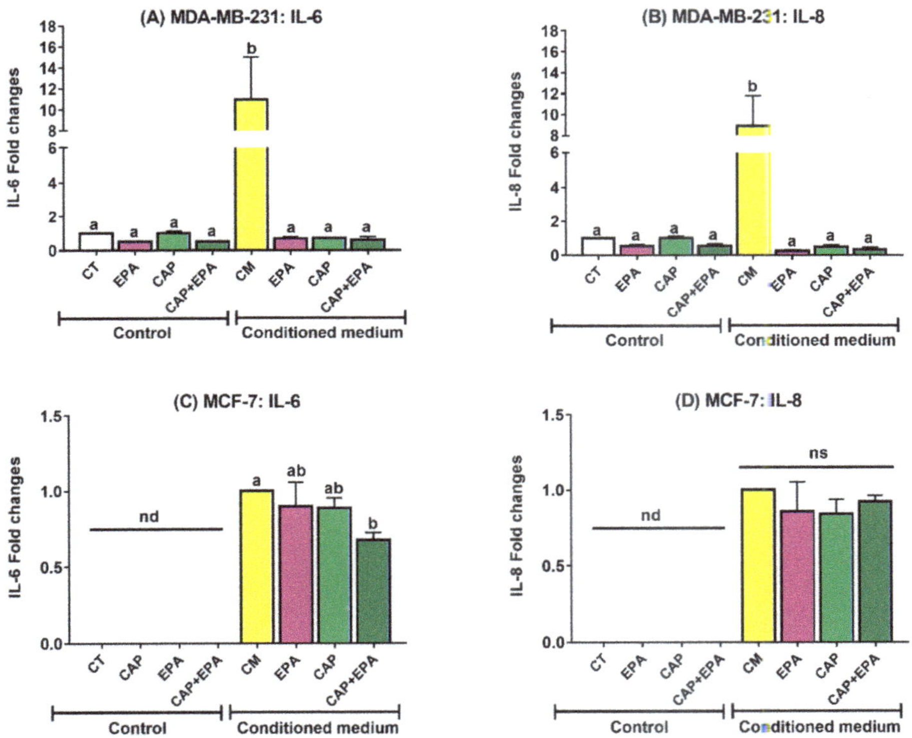

Figure 3. Adipose-CM with ACE-I with or without EPA reduced proinflammatory interleukins secretion in BC cells. Mature human adipocytes were pretreated with 100 µm of CAP with or without 100 µm of EPA for 24 h followed by CM collection and transferred to MDA-MB-231 (**A**,**B**) and MCF-7 (**C**,**D**) cells for 48 h. Medium was collected, and secreted IL-6 and IL-8 levels were measured in response to direct (no adipocyte CM) or CM-mediated CAP and EPA effects ($p < 0.05$; N = 3; and in each experiment, three replicates were used for each treatment group; bars with different letters (a, b) indicate significance;); nd: not detectable; ns: not significant).

2.2. Combined Effect of Captopril and EPA on Breast Cancer Cell Migration in Response to Treatment with Adipocyte CM, Measured by a Wound Healing Assay

Since inflammation is associated with BC cell migration, we performed in vitro wound healing assays to dissect the effects of both direct and human adipocyte-mediated effects of CAP ± EPA treatments in MDA-MB-231 cells. First, MDA-MB-231 cells were treated directly with CAP with or without EPA and images of wound closure were taken at various times up to 48 h (Figure 4A,B; also see Tables S15 and S16). No significant effects of direct CAP treatment were found in MDA-MB-231 cells at any of the time points tested ($p < 0.05$) (Figure 4A,B). Intriguingly, a significantly higher percent wound area (or lower percent wound healing) was observed in response to direct EPA and CAP + EPA treatments at 4, 8, 12, 24, and 36 h time points when compared with bovine serum albumin (BSA)-control ($p < 0.05$), indicating reduced wound healing cell migration with EPA and CAP + EPA. However, the reductions in wound healing were not significant between EPA and CAP + EPA treatments at the respective time points ($p < 0.05$), indicating no additional effect of CAP when combined with EPA on MDA-MB-231 cell wound-healing capacity (Figure 4A,B).

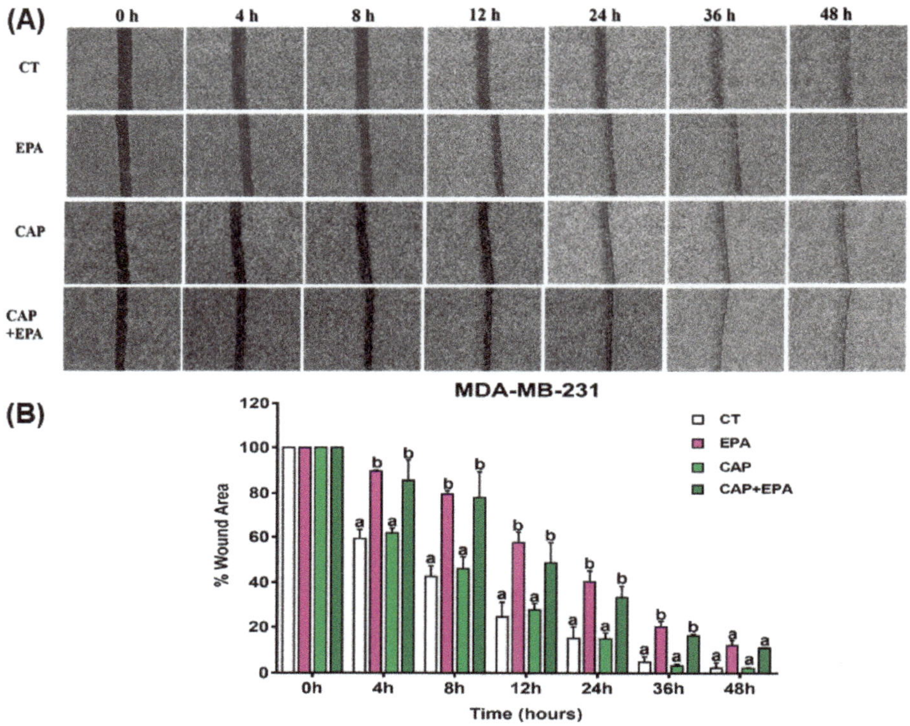

Figure 4. Direct effect of captopril and EPA on cell migration in MDA-MB-231 cells. Representative wound healing images at 0, 4, 8, 12, 24, 36, and 48 h. Wounds were inflicted with a 200 uL pipette tip on MDA-MB-231 (**A**,**B**) cells in confluent monolayers. Graphical representation of migrating MDA-MB-231 (**A**,**B**) cells in response to CAP with or without EPA treatments of BC cells. Cells treated with regular Dulbecco's Modified Eagle's Medium (DMEM) with 1% bovine serum albumin (BSA) conjugation were used as controls (BSA). Bars represent mean of the percent wound area in MDA-MB-231 (**A**,**B**) cells in the two independent experiments ± SEM. $p < 0.05$ (comparison between 0 and 48 h; N = 2 combined experiments, in which each had three replicates per treatment group). Different letters (a, b) indicate significance.

Next, we identified the adipocyte-CM-mediated effects of CAP ± EPA treatments in MDA-MB-231 cells. Mature human adipocytes were pretreated with CAP with or without EPA for 24 h and medium was transferred to MDA-MB-231 cells, followed by capturing time-dependent images, as described above (Figure 5A,B). Cell migration, as denoted by the changes in percent wound area due to closure, was significantly increased in response to CM-mediated EPA, CAP, and CAP + EPA treatments compared with CM-control in MDA-MB-231 cells at all time points tested, such as 4, 8, 12, 24, 36, and 48 h ($p < 0.05$) (Figure 5A,B), whereas CM-CAP + EPA effects were not significantly different from

CM-EPA or CM-CAP, indicating no additional effect of CAP and EPA combination on MDA-MB-231 cell migration (Figure 5A,B). When time was modeled as a continuous variable, the wound area with CM treatment decreased by 2.13% per hour. This rate of reduction of wound area was hindered by treatment of adipocyte CM with each of EPA, CAP, and their combination as evidenced by significant positive interactions (Table S16). The subsequent factorial model revealed a significant positive effect of EPA and CAP but a significant negative interaction of CAP and EPA; these findings suggest that EPA and CAP may act via a common pathway in slowing down cell MDA-MB-231 cell migration (Tables S17 and S18). We did not perform wound healing assays in MCF-7 cells since the inflammatory markers were expressed and secreted at a lower level in MCF-7 compared with the MDA-MB-231 cell line we used. In addition, we observed greater changes in these markers, such as IL-6 in MDA-MB-231 cells, in response to both direct and CM-mediated treatments of CAP and EPA. These results further confirm our previous observation that CAP and EPA effects may be mediated by a common pathway to reduce expression of mRNA biomarkers of MDA-MB-231 cell proliferation.

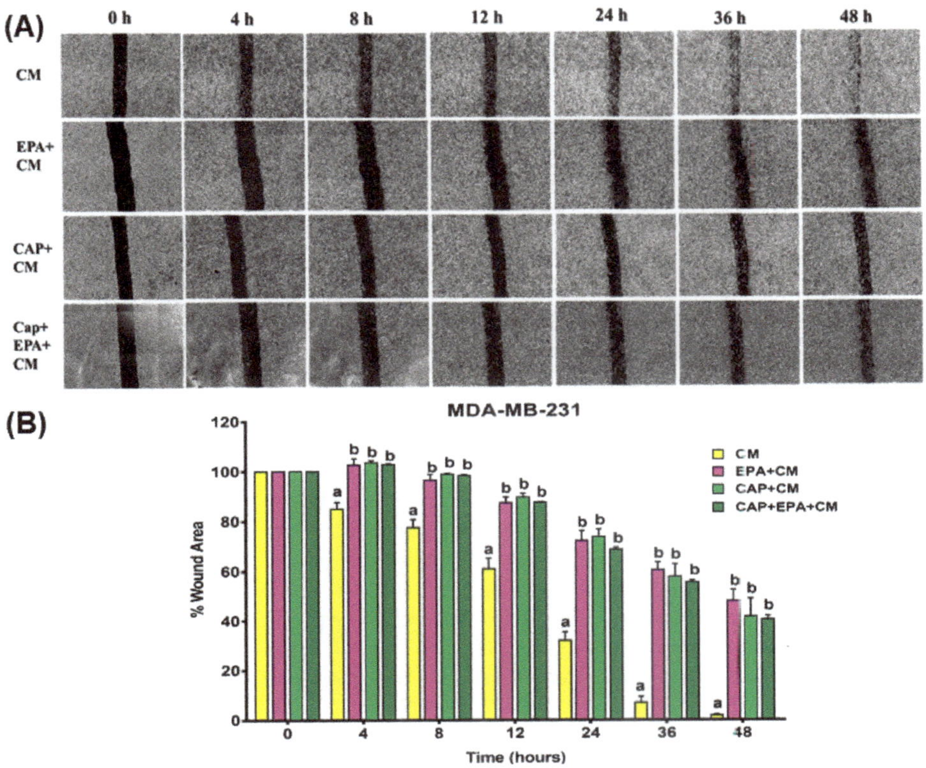

Figure 5. CM-mediated effect of captopril and EPA on cell migration in MDA-MB-231 cells. Representative wound healing images at 0, 4, 8, 12, 24, 36, and 48 h. Wounds were inflicted with a 200 uL pipette tip on MDA-MB-231 (**A**,**B**) cells in confluent monolayers. Graphical representation of migrating MDA-MB-231 (**A**,**B**) cells in response to CM-EPA and CM-CAP with or without EPA treatments of BC cells. Cells treated with human adipose conditioned medium were used as controls (CM). Bars represent mean of the percent wound area in MDA-MB-231 (**A**,**B**) cells in the two independent experiments ± SEM. $p < 0.05$ (comparison between 0 and 48 h; N = combined experiments, in which each had three replicates per treatment group). Different letters (a, b) indicate significance.

3. Discussion

The aim of the current study was to provide new insights into adipocyte–BC interaction through inhibition of RAS in adipocytes. We previously showed protective effects of n-3 PUFAs in adipocyte–BC cell interactions [27]. Moreover, we demonstrated that n-3 PUFAs reduce Agt secretion from adipocytes and reduce both systemic and adipose tissue inflammation [29]. Hence, we proposed to determine whether an n-3 PUFA (EPA), individually or in combination with ACE-I, would reduce BC cell inflammation and migration directly on BC cells or via adipose RAS inhibition. In this study, we demonstrated reduced BC cell inflammation and motility in response to CM-derived EPA, CAP, and CAP + EPA treatments while no changes were observed among these treatments, indicating that the combination of CAP and EPA had no synergistic effects in attenuating adipocyte RAS effects on BC cells.

The doses of EPA and CAP that we used were within or lower than ranges used in human studies, making our findings translatable to humans. For omega-3 fatty acids, doses ranging from 1 to 6 g of EPA per day have been used in various human cancer studies and are associated with both improved cardiovascular and cancer outcomes [30,31]. Researchers reported that a ≥150 µg/mL (~150 µm) plasma EPA level was linked to benefits in preventing cardiovascular disease (CVD) outcomes [32]. In addition, plasma EPA concentrations in healthy subjects taking 2–4 g/day of EPA ethyl ester for 28 days increased up to a mean of 366 µg/mL, without any adverse events [33]. Although it is difficult to translate in vitro doses into actual human intake, the above mentioned reports clearly indicate that our 100 µm of EPA dose was lower than doses that can be reached in human plasma levels with prescribed fish oil or EPA. Moreover, other cell studies used doses of fatty acids up to 5 times higher than our dose [34,35]. For captopril, it has been used safely at doses up to 450 mg/day in humans [36], which may lead to plasma doses likely higher than the 100 µm dose we used for cell treatments. Captopril was given at 150 mg/day to lung cancer patients [37] and lisinopril (another ACE-I) given as 10 mg tablets to breast cancer patients to prevent cardiotoxicity [38]. We were unable to find data regarding this drug's plasma level concentration in cancer patients since ACE-I is only prescribed as antihypertensive medication for cancer patients and not as an anticancer agent. Hence, the suitable dose for preventing breast cancer is not known. However, earlier research reported plasma captopril levels in normal versus chronic renal failure subjects as 364 and 347 µm, respectively, after 1 h, but it was undetectable in blood after 6 h [39]. In addition, we previously performed a cell viability (MTT) assay with both 100 µm of EPA and 100 µm of captopril and found no toxic effects against breast cancer cell viability.

Adipose tissue is a major component of the breast tissue and is actively involved in forming TME; thus, adipose tissue through its inflammatory and lipogenic activities may increase the risk of BC development [3,40]. Human adipose CM is a source of various growth factors, several adipocyte secretory hormones, cytokines, and metabolites [41], which are immensely responsible for inducing expression of markers associated with breast cancer cell signaling, motility, and metastases [42,43]. Consistent with this, we have found increased inflammatory, IL-6, IL-8, NF-κB, and STAT3 mRNA expression as well as induced expression of the lipogenic enzyme FASN in both MCF-7 and MDA-MB-231 cells in response to human adipocyte CM. Overexpression of FASN plays an important role in tumorigenesis and is thus a vital target of obesity-mediated BC therapy [44]. Additionally, selective FASN inhibitors are potent antitumor agents showing antiproliferative and apoptosis inhibition effects in in vitro and in vivo models of receptor-positive (ER/HER2+) BC [45].

Moreover, two metabolically different BC cell lines, MCF-7 and MDA-MB-231, were used in the current study to identify the presence of any metabolic responses of these cell lines to n-3 PUFA or ACE-I. While the observed changes in mRNA levels between two cell lines were almost identical, their cytokine secretory profiles varied in response to direct or CM treatments. Both IL-6 and IL-8 levels were below detection in response to direct EPA/CAP treatments in MCF-7 cells, which is in agreement with other studies investigating the basal protein secretion levels by different BC cell lines [46,47]. Furthermore, we observed higher IL-6/IL-8 secretion by MDA-MB-231 cells compared with MCF-7 cells, which is consistent with previous studies indicating a stronger secretory profile in MDA-MB-231

TNBC cells compared with its less aggressive counterpart MCF-7 cells [48]. Another possible reason for the differential proinflammatory profile of MCF-7 versus MDA-MB-231 could be the intrinsically reduced transcriptional/post-transcriptional and/or translational/post-translational IL-6/IL-8 profile of MCF-7 cells in vitro compared with MDA-MB-231 cells [48].

Moreover, IL-6 activates STAT3 and NF-κB by (i) autocrine and paracrine signaling [8] and (ii) via synergistic crosstalk between the pathways in BC cells [7], resulting in cancer cell proliferation and apoptosis evasion [8]. By contrast, EPA is a natural anti-inflammatory bioactive compound previously reported to reduce IL-6 production, in part by inhibiting NF-κB activation [49] and through its downstream anti-inflammatory lipid mediators such as E-resolvins [50]. Consistent with this, we have also shown reduced mRNA levels of IL-6/STAT3/NF-κB in response to EPA-CM treatments. On the other hand, EPA also reduces both inflammation and Agt secretion from adipocytes both in vivo and in vitro [22,29]. Hence, when combining EPA with ACE-I (captopril), we observed reduced IL-6 inflammation in both MDA-MB-231 and MCF-7 BC cells as indicated by lower IL-6 mRNA and protein levels, respectively. Interestingly, the effect of EPA and ACE-I combination was more pronounced in reducing TNBC cell inflammation as shown by reduced IL-6 levels in MDA-MB-231 cells, which could be attributed to the more aggressive, inflammatory, and metastatic nature of MDA-MB-231 cells compared with MCF-7 cells [51]. Similarly, EPA and EPA-derived ethanol amines were more potent in reducing cell proliferation, migration, and invasion of MDA-MB-231 cells compared with non-invasive MCF-7 cells in vitro, which is in agreement with our present study [52]. We did not observe any synergistic effects of ACE-I and EPA combination in BC cell proliferation and migration, possibly due to their competitive anti-inflammatory in vitro mechanisms of action via EGFR and/or NF-κB (Rel-A) signaling pathways [19,53]. Moreover, many preclinical studies reported antihypertensive and anti-inflammatory roles of n-3 PUFAs respectively via ACE-2 upregulation, Ang II downregulation, and partly via GPCR-mediated NF-κB modulation [22,54,55]. Importantly, to our knowledge, we are the first to report the potential anti-inflammatory roles of ACE-I and EPA combination in targeting adipocyte and BC cell interactions. Despite this knowledge, potential mechanisms of ACE-I and EPA interactions in lessening the effects of adipocytes on BC cells remain unknown. Clinical studies are also lacking to understand the crosstalk between RAS inhibitors and n-3 PUFAs or other anti-inflammatory bioactive compounds to improve obesity-mediated cancer outcomes.

Inflammation is positively associated with BC cell motility and wound healing, although their mechanistic inter-relationship requires further studies [56]. Adipose CM contains various cytokines, including IL-6 and IL-8, which could be responsible for increased BC cell motility and wound healing, as reported by others [57]. We found a similar induced wound-healing capacity in MDA-MB-231 cells in response to human adipose CM. On the contrary, given the anti-inflammatory and antiproliferative effects of EPA on MDA-MB-231 cell motility [27] and the proposed similar effects for ACE-I, we combined them to identify their synergistic or additive effects in MDA-MB-231 TNBC. Surprisingly, both direct and CM-mediated EPA and ACE-I combination were ineffective in reducing TNBC cell motility compared with either EPA or ACE-I alone. This is in disagreement with Krusche et al., who reported the synergistic inhibitory effect of ACE-I and a Chinese herb (artesunate) on human umbilical vein endothelial cells (HUVEC) motility [58]. This might be due to the competitiveness between EPA and captopril to target common inflammatory mechanisms of action such as NF-κB downregulation and/or inactivation [59]. Taken together, despite some protective anti-inflammatory effects of EPA and ACE-I combination, such as attenuating IL-6, the underlying mechanisms require further studies in both in vitro and in vivo models of obesity-induced breast cancer.

4. Materials and Methods

4.1. Cell Culture Experiments

Human breast cancer cells such as MDA-MB-231 (HTB-26; Lot: 700792) TNBC and ER/PR positive MCF-7 (HTB-22; Lot: 64125078) BC cells were purchased from the American Type Culture collection

(ATCC) (Manassas, VA, USA). MDA-MB-231 and MCF-7 cells were seeded at a density of 50,000 cells per well and maintained under standard culture conditions at 37 °C in a humid atmosphere of 5% CO_2 in Dulbecco's Modified Eagle's Medium (DMEM). The medium was supplemented with 10% fetal bovine serum (FBS) (Thermo Fisher Scientific, Waltham, MA, USA) and antibiotics (50 μg/mL penicillin, 50 μg/mL streptomycin, and 100 ug/mL neomycin) (Thermo Fisher Scientific, Waltham, MA, USA). Bone-marrow-derived human mesenchymal stem cells (HMSCs) (PT-2501; Lot: 0000483199) were purchased from Lonza (Allendale, NJ, USA). These cells were differentiated into mature human adipocytes according to the optimized published protocol [60] and maintained under standard culture conditions at 37 °C and 5% CO_2 in DMEM: Nutrient Mixture F-12 (DMEM/F-12) (Thermo Fisher Scientific, Waltham, MA USA). Following differentiation, cells were transferred to regular DMEM with 10% FBS for 24 h and CM experiments were performed. EPA (Nu-Check Prep, Waterville, MN, USA) and/or captopril (Sigma-Aldrich, St. Louis, MO, USA) was conjugated with 1% BSA (Sigma-Aldrich, St. Louis, MO, USA) for 2 h in a shaking water bath at 37 °C prior to cell treatment. Since fatty acids are found in the circulation as attached to albumin, many studies used BSA as a complex medium for EPA and other fatty acids in in vitro culture experiments [61]. For consistency, all treatments were performed in a BSA complexed medium.

4.2. Treatment with ACE Inhibitor, Captopril, and Eicosapentaenoic Acid for Conditioned Medium Experiments

MDA-MB-231 and MCF-7 cells were starved with 1% BSA for 2 h prior to CAP ± EPA experiments. The BC cells were treated directly with the ACE inhibitor CAP (100 μm) with or without EPA (100 μm) combination for 48 h to identify their individual/combined effect in modulating BC cell metabolism (Figure S1). The time and dose of our current treatments (CAP and EPA) were based on our previous work using EPA alone [27] or CAP alone (manuscript under review) against adipocyte–BC-cell interactions. CM experiments were performed between differentiated human adipocytes (HMSCs) and BC cells. Prior to the treatments, differentiated mature adipocytes were starved with 1% BSA for 2 h. Following starvation, mature adipocytes were treated with 100 μm of CAP ± 100 μm of EPA complexed with BSA, and BSA alone was used as the control. Conditioned media from differentiated adipocytes were collected after 24 h and centrifuged at 10,000× *g* for 10 min, then filter-sterilized to remove any cell debris. MDA-MB-231 and MCF-7 cells were seeded in six-well plates and then at 70% confluency exposed to CM from human adipocytes for 48 h. BC cells and media were collected and stored at −80 °C for further analyses (Figure S1).

4.3. Enzyme-Linked Immunosorbent Assay (ELISA)

Quantification of secreted IL-6 and IL-8 levels were determined by ELISA (R&D Systems, Minneapolis, MN, USA) according to the manufacturer's protocol.

4.4. RNA Isolation and Real-Time Quatitative Polymerase Chain Reaction (RT-qPCR)

RNA was purified using the Quick RNA mini kit (Zymo Research, Irvine, CA, USA) according to the established manufacturer's protocol followed by cDNA synthesis. cDNA was reverse-transcribed using Maxima reverse transcriptase (Thermo Fisher Scientific, Waltham, MA, USA). mRNA transcription levels were assessed with RT-qPCR using the Sybr green master mix (Thermo Fisher Scientific, Waltham, MA, USA). All genes were normalized to two housekeeping genes (18S ribosomal RNA and TBP (TATA box binding protein)). We used both 18S and TBP as our reference genes for normalization of RT-qPCR data, both were normalized to each other, and we did not see any regulatory effects in response to our proposed treatment conditions.

Sequences of the primers (Sigma Aldrich, St. Louis, MO, USA) are listed as follows (forward, reverse):

IL-6 (5′-AGACAGCCACTCACCTCTTCAG-3′, 5′-TTTCTGCCAGTGCCTCTTTGC-3′),
IL-8 (5′-AGGACAAGAGCCAGGAAGAA-3′, 5′-GGGTGGAAAGGTTTGGAGTATG-3′),
NF-κB (5′-ATGGCTTCTATGAGGCTGAG-3′, 5′-GTTGTTGTTGGTCTGGATGC-3′),
STAT3 (5′-AGAAGGACATCAGCGGTAAGA-3′, 5′-GGATAGAGATAGACCAGTGGAGAC-3′),
FASN (5′-TCGTGGGCTACAGCATGGT-3′, 5′-GCCCTCTGAAGTCGAAGAAGAA-3X),
18S (5′-CTACCACATCCAAGGAAGCA-3′, 5′-TTTTTCGTCACTACCTCCCCG-3′), and
TBP (5′-ATGGTGGTGTTGTGAGAAGATG-3′, 5′-CAGATAGCAGCACGGTATGAG-3′).

4.5. Wound Healing Assay

MDA-MB-231 cells were seeded (50,000 cells per well) in six-well plates. At 95% confluence, cells were washed with phosphate-buffered saline (PBS) (Thermo Fisher Scientific, Waltham, MA, USA) and starved with 0.5% FBS in DMEM overnight to inhibit cell proliferation, thereby ensuring wound closure was due to cell migration only. Using a sterilized 200 μL pipette tip, two straight scratches were made through the cell's monolayer, stimulating wound. Cells were first rinsed once gently with PBS, then subjected to treatments described in the results, such as CAP with or without EPA. Images were taken at 0, 4, 8, 12, 24, 36, and 48 h using the EVOS Cell Imaging System (Thermo Fisher Scientific, Waltham, MA, USA) and data were analyzed using the in-house Cellular Growth Analyzer for Windows software (version 2.2). The in-house analyzer program was used to change the original image into black (the scratch) and the cell area into white (inside cells). The detection parameter assigned in the program was the pixel size, and the program checked along the x direction and later along the y direction, moving from left to right and counting the number of black pixels it came across. This stopped when it encountered a white pixel. This worked on the basis that the wound area (black) was the largest gap visible in the scratch assay.

4.6. Statistical Analysis

Data obtained from the experiments were normalized and analyzed using IBM SPSS (version 20, 2015) and R statistical software (version 3.5.3), and graphs were made using GraphPad Prism (Version 8.0). Statistical significance between groups (control vs. treatments) was tested using one-way ANOVA and subsequent pairwise post hoc comparisons adjusted using Tukey's correction. Differences were considered statistically significant for a Tukey-corrected p-value < 0.05. All experiments were performed at least two times and results are expressed as mean ± SEM of three independent biological replicates. Additional exploratory factorial regression analyses were performed to examine the interactions between EPA and CAP on all measured biomarkers.

For the wound healing assays, the data were analyzed separately for BSA and CM-treated conditions using two-way ANOVA models. The two-way ANOVA models were constructed to examine for the main effects of treatment (i.e., EPA, CAP, and combination), time modeled as a categorical variable, and their interaction. To determine the effect of each treatment on each biomarker per unit time, additional exploratory linear regression analyses were performed to modeling time as a continuous variable and the treatment condition as a categorical variable and their interaction. The interactions of EPA and CAP on temporal trends of biomarkers were examined in factorial regression analyses.

5. Conclusions

In conclusion, our study reported significant protective roles of EPA with and without ACE-I combination in attenuating adipocyte-induced proinflammation and cancer cell migration. To our knowledge, our study is the first to report inflammatory IL-6 downregulation effects of captopril and EPA combination in both MDA-MB-231 and MCF-7 BC cells. One acknowledged limitation of our study is that we only established a proof of principle for protective effects of CAP–EPA combination against adipocyte–BC-cell interaction, possibly via modulation of inflammation. This provided us a basis for future more in-depth mechanistic studies to further dissect effects of adipocytes on cancer cell proliferation and motility in response to CAP and EPA treatments in the tumor microenvironment.

Additionally, we did not see any differences between individual and combined ACE-I and EPA treatments except for IL-6. This indicates the need for more in-depth mechanistic studies possibly via knockdown of inflammatory signaling pathways in adipocytes and BC cells such as RAS and IL-6/STAT3/NF-κB pathways. Moreover, comparison of the efficacy of other ACE-I such as lisinopril, ARBs such as telmisartan, losartan, and/or n-3 PUFAs such as DHA or fish oil is necessary to better understand how these affect adipocyte–BC-cell interactions. Hence, our present study provides a novel significant validation for future mechanistic studies to combine diet and antihypertensive medication as a potential therapeutic approach for obesity-associated BC. More research is warranted to identify the possible mechanism of action and synergy of this unique combination of ACE-I and EPA or other n-3 PUFAs in obesity and breast cancer.

Supplementary Materials: The following are available online at http://www.mdpi.com/2072-6694/12/1/220/s1, Figure S1: Experimental design; Table S1: Exploratory factorial regression analyses to examine CAP and EPA interactions of mRNA markers associated with MDA-MB-231 cell growth and inflammation; Table S2: Exploratory factorial regression analyses to examine CAP and EPA interactions of mRNA markers associated with MDA-MB-231 cell growth and inflammation; Table S3: Exploratory factorial regression analyses to examine CAP and EPA interactions of mRNA markers associated with MDA-MB-231 cell growth and inflammation; Table S4: Exploratory factorial regression analyses to examine CAP and EPA interactions of mRNA markers associated with MDA-MB-231 cell growth and inflammation; Table S5: Exploratory factorial regression analyses to examine CAP and EPA interactions of mRNA markers associated with MDA-MB-231 cell growth and inflammation; Table S6: Exploratory factorial regression analyses to examine CAP and EPA interactions of protein markers associated with MDA-MB-231 cell inflammation; Table S7: Exploratory factorial regression analyses to examine CAP and EPA interactions of protein markers associated with MDA-MB-231 cell inflammation; Table S8: Exploratory factorial regression analyses to examine CAP and EPA interactions of mRNA markers associated with MCF-7 cell growth and inflammation; Table S9: Exploratory factorial regression analyses to examine CAP and EPA interactions of mRNA markers associated with MCF-7 cell growth and inflammation; Table S10: Exploratory factorial regression analyses to examine CAP and EPA interactions of mRNA markers associated with MCF-7 cell growth and inflammation; Table S11: Exploratory factorial regression analyses to examine CAP and EPA interactions of mRNA markers associated with MCF-7 cell growth and inflammation; Table S12: Exploratory factorial regression analyses to examine CAP and EPA interactions of mRNA markers associated with MCF-7 cell growth and inflammation; Table S13: Exploratory factorial regression analyses to examine CAP and EPA interactions of protein markers associated with MCF-7 cell inflammation; Table S14: Exploratory factorial regression analyses to examine CAP and EPA interactions of protein markers associated with MCF-7 cell inflammation; Table S15: Exploratory time-series regression analyses to examine CAP and EPA interactions in MDA-MB-231 cell migration; Table S16: Exploratory time-series regression analyses to examine CAP and EPA interactions in MDA-MB-231 cell migration; Table S17: Exploratory time-series factorial regression analyses to examine CAP and EPA interactions in MDA-MB-231 cell migration; Table S18: Exploratory time-series factorial regression analyses to examine CAP and EPA interactions in MDA-MB-231 cell migration.

Author Contributions: N.M.-M. conceived this project, and N.M.-M. and F.R. designed the experiments with input from and discussions with L.R. and C.K.; F.R. conducted most experiments with assistance from A.H.; H.M. developed the analysis software for the wound healing assays. F.R. and H.M. analyzed wound healing experimental data with input from N.M.-M., C.K. and L.R.; C.K. assisted with statistical analyses. All coauthors were involved in discussions of this work during its execution, reviewed the paper, and provided feedback. All authors have read and agreed to the published version of the manuscript.

Funding: This work was supported by the Texas Tech University (TTU) Transdisciplinary Research Academy (N.M.-M.), TTU startup funds (N.M.-M.), and the Obesity Research Institute. Internal funds are used to cover the costs to publish in open access.

Conflicts of Interest: The authors declare no conflict of interest.

Abbreviations

Agt	Angiotensinogen
Ang I	Angiotensin I
Ang II	Angiotensin II
ACE	Angiotensin-converting enzyme
ACE-I	Angiotensin-converting enzyme inhibitor
ARB	Angiotensin receptor type I blocker
AT1R	Angiotensin receptor type I
AT2R	Angiotensin receptor type II
ATCC	American Type Culture Collection
BC	Breast cancer
CAA	Cancer-associated adipocytes
CAP	Captopril
CM	Conditioned medium
CVD	Cardiovascular disease
DHA	Docosahexaenoic acid
DMEM	Dulbecco's Modified Eagle's Medium
ELISA	Enzyme-linked immunosorbent assay
EPA	Eicosapentaenoic acid
FBS	Fetal bovine serum
FASN	Fatty acid synthase
ER	Estrogen receptor
ER+	Estrogen receptor positive
ER-	Estrogen receptor negative
GAPDH	Glyceraldehyde-3-phosphate dehydrogenase
HER2+	Human epidermal growth factor receptor-2 positive
HMSC	Human mesenchymal stem cells
HUVEC	Human umbilical vein endothelial cells
IL-6	Interleukin-6
IL-8	Interleukin-8
MCP-1	Monocyte chemoattractant protein 1
MMP	Matrix metalloproteinase
n-3 PUFA	Omega-3 polyunsaturated fatty acid
NF-κB	Nuclear factor kappa B
PBS	Phosphate-buffered saline
PR	Progesterone receptor
RAS	Renin–angiotensin system
RT-qPCR	Real-time quantitative polymerase chain reaction
STAT3	Signal transducer and activator of transcription 3
TAM	Tumor-associated macrophage
TME	Tumor microenvironment
TNBC	Triple-negative breast cancer
VEGF	Vascular endothelial growth factor

References

1. Stewart, B.; Wild, C.P. *World Cancer Report 2014*; World Health Organization (WHO): Geneva, Switzerland, 2017.
2. Carmichael, A.R. Obesity as a risk factor for development and poor prognosis of breast cancer. *BJOG Int. J. Obstet. Gynaecol.* **2006**, *113*, 1160–1166. [CrossRef] [PubMed]
3. Ahn, J.; Schatzkin, A.; Lacey, J.V.; Albanes, D.; Ballard-Barbash, R.; Adams, K.F.; Kipnis, V.; Mouw, T.; Hollenbeck, A.R.; Leitzmann, M.F. Adiposity, Adult Weight Change, and Postmenopausal Breast Cancer Risk. *Arch. Intern. Med.* **2007**, *167*, 2091. [CrossRef] [PubMed]

4. Simone, V.; D'avenia, M.; Argentiero, A.; Felici, C.; Rizzo, F.M.; De Pergola, G.; Silvestris, F. Obesity and breast cancer: Molecular interconnections and potential clinical applications. *Oncologist* **2016**, *21*, 404–417. [CrossRef]
5. Deng, T.; Lyon, C.J.; Bergin, S.; Caligiuri, M.A.; Hsueh, W.A. Obesity, Inflammation, and Cancer. *Annu. Rev. Pathol. Mech. Dis.* **2016**, *11*, 421–449. [CrossRef]
6. Dirat, B.; Bochet, L.; Dabek, M.; Daviaud, D.; Dauvillier, S.; Majed, B.; Wang, Y.Y.; Meulle, A.; Salles, B.; Le Gonidec, S.; et al. Cancer-associated adipocytes exhibit an activated phenotype and contribute to breast cancer invasion. *Cancer Res.* **2011**, *71*, 2455–2465. [CrossRef]
7. De Simone, V.; Franze, E.; Ronchetti, G.; Colantoni, A.; Fantini, M.C.; Di Fusco, D.; Sica, G.S.; Sileri, P.; MacDonald, T.T.; Pallone, F.; et al. Th17-type cytokines, IL-6 and TNF-alpha synergistically activate STAT3 and NF-kB to promote colorectal cancer cell growth. *Oncogene* **2015**, *34*, 3493–3503. [CrossRef]
8. Hodge, D.R.; Hurt, E.M.; Farrar, W.L. The role of IL-6 and STAT3 in inflammation and cancer. *Eur. J. Cancer* **2005**, *41*, 2502–2512. [CrossRef]
9. Banerjee, K.; Resat, H. Constitutive activation of STAT3 in breast cancer cells: A review. *Int. J. Cancer* **2016**, *138*, 2570–2578. [CrossRef]
10. Wang, Y.; Kuhajda, F.P.; Li, J.N.; Pizer, E.S.; Han, W.F.; Sokoll, L.J.; Chan, D.W. Fatty acid synthase (FAS) expression in human breast cancer cell culture supernatants and in breast cancer patients. *Cancer Lett.* **2001**, *167*, 99–104. [CrossRef]
11. Berndt, J.; Kovacs, P.; Ruschke, K.; Kloting, N.; Fasshauer, M.; Schon, M.R.; Korner, A.; Stumvoll, M.; Bluher, M. Fatty acid synthase gene expression in human adipose tissue: Association with obesity and type 2 diabetes. *Diabetologia* **2007**, *50*, 1472–1480. [CrossRef] [PubMed]
12. Wang, D.; Dubois, R.N. Associations between obesity and cancer: The role of fatty acid synthase. *J. Natl. Cancer Inst.* **2012**, *104*, 343–345. [CrossRef] [PubMed]
13. Iyengar, P.; Combs, T.P.; Shah, S.J.; Gouon-Evans, V.; Pollard, J.W.; Albanese, C.; Flanagan, L.; Tenniswood, M.P.; Guha, C.; Lisanti, M.P.; et al. Adipocyte-secreted factors synergistically promote mammary tumorigenesis through induction of anti-apoptotic transcriptional programs and proto-oncogene stabilization. *Oncogene* **2003**, *22*, 6408–6423. [CrossRef] [PubMed]
14. Kalupahana, N.S.; Moustaid-Moussa, N. The adipose tissue renin-angiotensin system and metabolic disorders: A review of molecular mechanisms. *Crit. Rev. Biochem. Mol. Biol.* **2012**, *47*, 379–390. [CrossRef]
15. Jing, F.; Mogi, M.; Horiuchi, M. Role of renin–angiotensin–aldosterone system in adipose tissue dysfunction. *Mol. Cell. Endocrinol.* **2013**, *378*, 23–28. [CrossRef] [PubMed]
16. Ramalingam, L.; Menikdiwela, K.; LeMieux, M.; Dufour, J.M.; Kaur, G.; Kalupahana, N.; Moustaid-Moussa, N. The renin angiotensin system, oxidative stress and mitochondrial function in obesity and insulin resistance. *Biochim. et Biophys. Acta Mol. Basis Dis.* **2017**, *1863*, 1106–1114. [CrossRef] [PubMed]
17. Namazi, S.; Rostami-Yalmeh, J.; Sahebi, E.; Jaberipour, M.; Razmkhah, M.; Hosseini, A. The role of captopril and losartan in prevention and regression of tamoxifen-induced resistance of breast cancer cell line MCF-7: An in vitro study. *Biomed. Pharmacother.* **2014**, *68*, 565–571. [CrossRef]
18. Muscella, A.; Greco, S.; Elia, M.G.; Storelli, C.; Marsigliante, S. Angiotensin II stimulation of Na+/K+ATPase activity and cell growth by calcium-independent pathway in MCF-7 breast cancer cells. *J. Endocrinol.* **2002**, *173*, 315–323. [CrossRef]
19. Pinter, M.; Jain, R.K. Targeting the renin-angiotensin system to improve cancer treatment: Implications for immunotherapy. *Sci. Transl. Med.* **2017**, *9*, eaan5616. [CrossRef]
20. Rodrigues-Ferreira, S.; Nahmias, C. G-protein coupled receptors of the renin-angiotensin system: New targets against breast cancer? *Front. Pharmacol.* **2015**, *6*, 24. [CrossRef]
21. Ni, H.; Rui, Q.; Zhu, X.; Yu, Z.; Gao, R.; Liu, H. Antihypertensive drug use and breast cancer risk: A meta-analysis of observational studies. *Oncotarget* **2017**, *8*, 62545. [CrossRef]
22. Ulu, A.; Harris, T.R.; Morisseau, C.; Miyabe, C.; Inoue, H.; Schuster, G.; Dong, H.; Iosif, A.M.; Liu, J.Y.; Weiss, R.H.; et al. Anti-inflammatory effects of omega-3 polyunsaturated fatty acids and soluble epoxide hydrolase inhibitors in angiotensin-II-dependent hypertension. *J. Cardiovasc. Pharmacol.* **2013**, *62*, 285–297. [CrossRef] [PubMed]
23. Fabian, C.J.; Kimler, B.F.; Hursting, S.D. Omega-3 fatty acids for breast cancer prevention and survivorship. *Breast Cancer Res.* **2015**, *17*, 62. [CrossRef]

24. Kalupahana, N.S.; Claycombe, K.; Newman, S.J.; Stewart, T.; Siriwardhana, N.; Matthan, N.; Lichtenstein, A.H.; Moustaid-Moussa, N. Eicosapentaenoic Acid Prevents and Reverses Insulin Resistance in High-Fat Diet-Induced Obese Mice via Modulation of Adipose Tissue Inflammation. *J. Nutr.* **2010**, *140*, 1915–1922. [CrossRef] [PubMed]
25. Kalupahana, N.S.; Claycombe, K.J.; Moustaid-Moussa, N. (n-3) Fatty Acids Alleviate Adipose Tissue Inflammation and Insulin Resistance: Mechanistic Insights. *Adv. Nutr.* **2011**, *2*, 304–316. [CrossRef] [PubMed]
26. Al-Jawadi, A.; Moussa, H.; Ramalingam, L.; Dharmawardhane, S.; Gollahon, L.; Gunaratne, P.; Layeequr Rahman, R.; Moustaid-Moussa, N. Protective properties of n-3 fatty acids and implications in obesity-associated breast cancer. *J. Nutr. Biochem.* **2018**, *53*, 1–8. [CrossRef] [PubMed]
27. Al-Jawadi, A.; Rasha, F.; Ramalingam, L.; Alhaj, S.; Moussa, H.; Gollahon, L.; Dharmawardhane, S.; Moustaid-Moussa, N. Protective effects of eicosapentaenoic acid in adipocyte-breast cancer cell cross talk. *J. Nutr. Biochem.* **2020**, *75*, 108244. [CrossRef]
28. Freund, A.; Jolivel, V.; Durand, S.; Kersual, N.; Chalbos, D.; Chavey, C.; Vignon, F.; Lazennec, G. Mechanisms underlying differential expression of interleukin-8 in breast cancer cells. *Oncogene* **2004**, *23*, 6105–6114. [CrossRef]
29. Siriwardhana, N.; Kalupahana, N.S.; Fletcher, S.; Xin, W.; Claycombe, K.J.; Quignard-Boulange, A.; Zhao, L.; Saxton, A.M.; Moustaid-Moussa, N. n-3 and n-6 polyunsaturated fatty acids differentially regulate adipose angiotensinogen and other inflammatory adipokines in part via NF-κB-dependent mechanisms. *J. Nutr. Biochem.* **2012**, *23*, 1661–1667. [CrossRef]
30. Brinton, E.A.; Mason, R.P. Prescription omega-3 fatty acid products containing highly purified eicosapentaenoic acid (EPA). *Lipids Health Dis.* **2017**, *16*, 23. [CrossRef]
31. Superko, H.R.; Superko, A.R.; Lundberg, G.P.; Margolis, B.; Garrett, B.C.; Nasir, K.; Agatston, A.S. Omega-3 Fatty Acid Blood Levels Clinical Significance Update. *Curr. Cardiovasc. Risk Rep.* **2014**, *8*, 407. [CrossRef]
32. Itakura, H.; Yokoyama, M.; Matsuzaki, M.; Saito, Y.; Origasa, H.; Ishikawa, Y.; Oikawa, S.; Sasaki, J.; Hishida, H.; Kita, T.; et al. Relationships between plasma fatty acid composition and coronary artery disease. *J. Atheroscler. Thromb.* **2011**, *18*, 99–107. [CrossRef]
33. Braeckman, R.A.; Stirtan, W.G.; Soni, P.N. Pharmacokinetics of Eicosapentaenoic Acid in Plasma and Red Blood Cells After Multiple Oral Dosing With Icosapent Ethyl in Healthy Subjects. *Clin. Pharmacol. Drug Dev.* **2014**, *3*, 101–108. [CrossRef]
34. Song, J.; Li, C.; Lv, Y.; Zhang, Y.; Amakye, W.K.; Mao, L. DHA increases adiponectin expression more effectively than EPA at relative low concentrations by regulating PPARγ and its phosphorylation at Ser273 in 3T3-L1 adipocytes. *Nutr. Metab.* **2017**, *14*, 52. [CrossRef]
35. Mansara, P.P.; Deshpande, R.A.; Vaidya, M.M.; Kaul-Ghanekar, R. Differential Ratios of Omega Fatty Acids (AA/EPA+DHA) Modulate Growth, Lipid Peroxidation and Expression of Tumor Regulatory MARBPs in Breast Cancer Cell Lines MCF7 and MDA-MB-231. *PLoS ONE* **2015**, *10*, e0136542. [CrossRef]
36. Cunha, J.P. Consumer_Captopril_Capoten. Available online: https://www.rxlist.com/consumer_captopril_capoten/drugs-condition.htm (accessed on 3 January 2020).
37. Small, W., Jr.; James, J.L.; Moore, T.D.; Fintel, D.J.; Lutz, S.T.; Movsas, B.; Suntharalingam, M.; Garces, Y.I.; Ivker, R.; Moulder, J.; et al. Utility of the ACE Inhibitor Captopril in Mitigating Radiation-associated Pulmonary Toxicity in Lung Cancer: Results From NRG Oncology RTOG 0123. *Am. J. Clin. Oncol.* **2018**, *41*, 396–401. [CrossRef]
38. Guglin, M.; Munster, P.; Fink, A.; Krischer, J. Lisinopril or Coreg CR in reducing cardiotoxicity in women with breast cancer receiving trastuzumab: A rationale and design of a randomized clinical trial. *Am. Heart J.* **2017**, *188*, 87–92. [CrossRef]
39. Onoyama, K.; Hirakata, H.; Iseki, K.; Fujimi, S.; Omae, T.; Kobayashi, M.; Kawahara, Y. Blood concentration and urinary excretion of captopril (SQ 14,225) in patients with chronic renal failure. *Hypertension* **1981**, *3*, 456–459. [CrossRef]
40. Iyengar, N.M.; Gucalp, A.; Dannenberg, A.J.; Hudis, C.A. Obesity and Cancer Mechanisms: Tumor Microenvironment and Inflammation. *J. Clin. Oncol.* **2016**, *34*, 4270–4276. [CrossRef] [PubMed]
41. Sagaradze, G.; Grigorieva, O.; Nimiritsky, P.; Basalova, N.; Kalinina, N.; Akopyan, Z.; Efimenko, A. Conditioned Medium from Human Mesenchymal Stromal Cells: Towards the Clinical Translation. *Int. J. Mol. Sci.* **2019**, *20*, 1656. [CrossRef] [PubMed]

42. Nieman, K.M.; Romero, I.L.; Van Houten, B.; Lengyel, E. Adipose tissue and adipocytes support tumorigenesis and metastasis. *Biochim. Biophys. Acta* **2013**, *1831*, 1533–1541. [CrossRef] [PubMed]
43. Carter, J.C.; Church, F.C. Mature breast adipocytes promote breast cancer cell motility. *Exp. Mol. Pathol.* **2012**, *92*, 312–317. [CrossRef] [PubMed]
44. Vazquez-Martin, A.; Colomer, R.; Brunet, J.; Lupu, R.; Menendez, J.A. Overexpression of fatty acid synthase gene activates HER1/HER2 tyrosine kinase receptors in human breast epithelial cells. *Cell Prolif.* **2008**, *41*, 59–85. [CrossRef] [PubMed]
45. Alwarawrah, Y.; Hughes, P.; Loiselle, D.; Carlson, D.A.; Darr, D.B.; Jordan, J.L.; Xiong, J.; Hunter, L.M.; Dubois, L.G.; Thompson, J.W.; et al. Fasnall, a Selective FASN Inhibitor, Shows Potent Anti-tumor Activity in the MMTV-Neu Model of HER2(+) Breast Cancer. *Cell Chem. Biol.* **2016**, *23*, 678–688. [CrossRef] [PubMed]
46. Faggioli, L.; Costanzo, C.; Merola, M.; Bianchini, E.; Furia, A.; Carsana, A.; Palmieri, M. Nuclear factor kappa B (NF-kappa B), nuclear factor interleukin-6 (NFIL-6 or C/EBP beta) and nuclear factor interleukin-6 beta (NFIL6-beta or C/EBP delta) are not sufficient to activate the endogenous interleukin-6 gene in the human breast carcinoma cell line MCF-7. Comparative analysis with MDA-MB-231 cells, an interleukin-6-expressing human breast carcinoma cell line. *Eur. J. Biochem.* **1996**, *239*, 624–631. [PubMed]
47. Chavey, C.; Muhlbauer, M.; Bossard, C.; Freund, A.; Durand, S.; Jorgensen, C.; Jobin, C.; Lazennec, G. Interleukin-8 expression is regulated by histone deacetylases through the nuclear factor-kappaB pathway in breast cancer. *Mol. Pharmacol.* **2008**, *74*, 1359–1366. [CrossRef]
48. Bravata, V.; Minafra, L.; Forte, G.I.; Cammarata, F.P.; Russo, G.; Di Maggio, F.M.; Augello, G.; Lio, D.; Gilardi, M.C. Cytokine profile of breast cell lines after different radiation doses. *Int. J. Radiat. Biol.* **2017**, *93*, 1217–1226. [CrossRef]
49. Trebble, T.; Arden, N.K.; Stroud, M.A.; Wootton, S.A.; Burdge, G.C.; Miles, E.A.; Ballinger, A.B.; Thompson, R.L.; Calder, P.C. Inhibition of tumour necrosis factor-a and interleukin-6 production by mononuclear cells following dietary fish-oil supplementation in healthy men and response to antioxidant co-supplementation. *Br. J. Nutr.* **2003**, *90*, 405–412. [CrossRef]
50. Duvall, M.G.; Levy, B. DHA- and EPA-derived resolvins, protectins, and maresins in airway inflammation. *Eur. J. Pharmacol.* **2016**, *785*, 144–155. [CrossRef]
51. Illan-Cabeza, N.A.; Jimenez-Pulido, S.B.; Hueso-Urena, F.; Ramirez-Exposito, M.J.; Sanchez-Sanchez, P.; Martinez-Martos, J.M.; Moreno-Carretero, M.N. Effects on estrogen-dependent and triple negative breast cancer cells growth of Ni(II), Zn(II) and Cd(II) complexes with the Schiff base derived from pyridine-2-carboxaldehyde and 5,6-diamino-1,3-dimethyluracil explored through the renin-angiotensin system (RAS)-regulating aminopeptidases. *J. Inorg. Biochem.* **2018**, *185*, 52–62. [CrossRef]
52. Brown, I.; Lee, J.; Sneddon, A.A.; Cascio, M.G.; Pertwee, R.G.; Wahle, K.W.; Rotondo, D.; Heys, S.D. Anticancer effects of n-3 EPA and DHA and their endocannabinoid derivatives on breast cancer cell growth and invasion. *Prostaglandins Leukot. Essent. Fat. Acids* **2019**. [CrossRef] [PubMed]
53. Weng, W.H.; Leung, W.H.; Pang, Y.J.; Kuo, L.W.; Hsu, H.H. EPA significantly improves anti-EGFR targeted therapy by regulating miR-378 expression in colorectal cancer. *Oncol. Lett.* **2018**, *16*, 6188–6194. [CrossRef] [PubMed]
54. Niazi, Z.R.; Silva, G.C.; Ribeiro, T.P.; Leon-Gonzalez, A.J.; Kassem, M.; Mirajkar, A.; Alvi, A.; Abbas, M.; Zgheel, F.; Schini-Kerth, V.B.; et al. EPA:DHA 6:1 prevents angiotensin II-induced hypertension and endothelial dysfunction in rats: Role of NADPH oxidase- and COX-derived oxidative stress. *Hypertens. Res. Off. J. Jpn. Soc. Hypertens.* **2017**, *40*, 966–975. [CrossRef] [PubMed]
55. Ulu, A.; Stephen Lee, K.S.; Miyabe, C.; Yang, J.; Hammock, B.G.; Dong, H.; Hammock, B.D. An omega-3 epoxide of docosahexaenoic acid lowers blood pressure in angiotensin-II-dependent hypertension. *J. Cardiovasc. Pharmacol.* **2014**, *64*, 87–99. [CrossRef] [PubMed]
56. Coussens, L.M.; Werb, Z. Inflammation and cancer. *Nature* **2002**, *420*, 860–867. [CrossRef]
57. Chang, Q.; Bournazou, E.; Sansone, P.; Berishaj, M.; Gao, S.P.; Daly, L.; Wels, J.; Theilen, T.; Granitto, S.; Zhang, X.; et al. The IL-6/JAK/Stat3 Feed-Forward Loop Drives Tumorigenesis and Metastasis. *Neoplasia* **2013**, *15*, 848–862. [CrossRef]
58. Krusche, B.; Arend, J.; Efferth, T. Synergistic inhibition of angiogenesis by artesunate and captopril in vitro and in vivo. *Evid. Based Complement. Alternat. Med.* **2013**, *2013*, 454783. [CrossRef]

59. Miguel-Carrasco, J.L.; Zambrano, S.; Blanca, A.J.; Mate, A.; Vazquez, C.M. Captopril reduces cardiac inflammatory markers in spontaneously hypertensive rats by inactivation of NF-kB. *J. Inflamm.* **2010**, *7*, 21. [CrossRef]
60. Lee, M.J.; Fried, S.K. Optimal protocol for the differentiation and metabolic analysis of human adipose stromal cells. *Methods Enzymol.* **2014**, *538*, 49–65. [CrossRef]
61. Wortman, P.; Miyazaki, Y.; Kalupahana, N.S.; Kim, S.; Hansen-Petrik, M.; Saxton, A.M.; Claycombe, K.J.; Voy, B.H.; Whelan, J.; Moustaid-Moussa, N. n3 and n6 polyunsaturated fatty acids differentially modulate prostaglandin E secretion but not markers of lipogenesis in adipocytes. *Nutr. Metab.* **2009**, *6*, 5. [CrossRef]

cancers

MDPI

Article

Isobolographic Analysis Demonstrates the Additive and Synergistic Effects of Gemcitabine Combined with Fucoidan in Uterine Sarcomas and Carcinosarcoma Cells

Marcin Bobiński [1,*], Karolina Okła [1], Jarogniew Łuszczki [2], Wiesława Bednarek [1], Anna Wawruszak [3], Gema Moreno-Bueno [4], Magdalena Dmoszyńska-Graniczka [3], Rafał Tarkowski [1,†] and Jan Kotarski [1,†]

1 I Chair and Department of Gynaecological Oncology and Gynaecology, Medical University of Lublin, 20-081 Lublin, Poland; karolinaokla@gmail.com (K.O.); wbed@wp.pl (W.B.); rafaltar@yahoo.com (R.T.); jan.kotarski.gabinet@gmail.com (J.K.)

2 I Chair and Department of Pathophisiology, Medical University of Lublin, 20-081 Lublin, Poland; jarogniewluszczki@umlub.pl

3 Chair and Department of Biochemistry and Molecular Biology, Medical University of Lublin, 20-081 Lublin, Poland; anna.wawruszak@umlub.pl (A.W.); magdalena.dmoszynska-graniczka@umlub.pl (M.D.-G.)

4 Laboratorio de Investigación Traslacional, MD Anderson Cancer Centre Madrid, Calle de Arturo Soria, 270 28033 Madrid, Spain; gmoreno@iib.uam.es

* Correspondence: m.s.bobinski@gmail.com; Tel.: +48-81-53-27-847

† These authors contributed equally to the research.

Received: 29 October 2019; Accepted: 30 December 2019; Published: 31 December 2019

Abstract: Background: Uterine sarcomas and carcinosarcoma are associated with unfavorable prognosis. The regimens that are used in chemotherapy are associated with high incidence of side effects and usually do not significantly increase patients' survival rates. In this study we investigated the activity and interactions between gemcitabine and fucoidan, the natural compound known for its anti-tumor properties, in human sarcomas and carcinosarcoma cell models. Methods: SK-UT-1, SK-UT1-B (carcinosarcoma), MES-SA (leiomyosarcoma), and ESS-1 (endometrial stromal sarcoma) cell lines were used for the experiments. Cells were incubated in the presence of gemcitabine, fucoidan, and mixtures, after the incubation the MTT tests were performed. In order to assess the interactions between tested compounds isobolographic analysis was performed. Additional assessments of apoptosis and cell cycle were done. Results: Additive effect of combined treatment with gemcitabine and fucoidan was observed in ESS-1 and SK-UT-1 cell line. Although the supra-additive (synergistic) effect noticed in SK-UT-1B cell line. It was not possible to determine the interactions of fucoidan and gemcitabine in MES-SA cell line due to insufficient response to treatment. Addition of fucoidan to gemcitabine enhances its proapoptotic activity, what was observed especially in ESS-1 and SK-UT-1B cell lines. The arrest of cell cycle induced by mixture of gemcitabine and fucoidan, superior comparing gemcitabine alone was observed in SK-UT-1B. Conclusions: Obtained data showed that a combination of fucoidan and gemcitabine in uterine endometrial stromal sarcoma and carcinosarcoma cell lines has additive or even synergistic effect in decreasing cell viability. Furthermore, this drug combination induces apoptosis and arrest of cell cycle. The resistance of uterine leiomyosarcoma cell line, justifies searching for other drugs combinations to improve therapy efficacy.

Keywords: uterine sarcoma; gemcitabine; fucoidan; isobolography

1. Introduction

Uterine sarcomas are a group of malignancies consist of various types of tumors arising from mesenchymal tissue. The most common type of uterine sarcomas is leiomyosarcoma (about 60% of cases), less common types are endometrial stromal sarcoma (low and high grade), liposarcoma, rhabdomyosarcoma, and many other rare types [1]. The incidence of uterine sarcomas remains low respectively to epithelial malignancies (3–7% of all uterine malignancies) [2]. The diagnosis of uterine sarcoma is associated with bad prognosis and low rate of response to chemotherapy. Carciosarcoma is a type of tumor consisting of both mesenchymal and epithelial cells, that nowadays is considered as a type of endometrial cancer. These tumors are associated with bad prognosis and, similarly to uterine sarcomas, the results of its systemic treatment remain unsatisfactory. Systemic treatment of uterine mesenchymal tumors is based on cytostatic agents such as doxorubicine, gemcitabine, docetaxel, dacrbazine, and ifosfamde [3].

The exceptions from the above characteristics are endometrial stromal sarcomas, associated with indolent clinical behavior and favorable prognosis. The majority of these tumors express estrogen receptors and usually respond to hormonal treatment [2].

The retrospective trial comparing gemcitabine plus docetaxel, ifosfamide plus cisplatin, doxorubicin plus ifosfamide, ifosfamide alone, topotecan alone, and observation only in stage I and II leiomyosarcoma revealed no significant differences between groups in recurrence rate [4]. Due to the rarity of uterine sarcomas, the number of completed, randomized, prospective trials is limited. The response rates among patients suffering from uterine sarcomas, in prospective trials assessing the efficacy of various combinations of gemcitabine, docetaxel, and bevacizumab were between 25% and 35.8% [5–7].

The results obtained in systemic therapy of sarcomas are unsatisfactory. Most authors strongly recommend conducting clinical and pre-clinical trials in this field. The situation described above led to a search for new agents to be active against uterine sarcomas. Fucoidan is sulphated polysaccharide derived from brown seaweed, that recently gained attention due to its biological activities. It is known to affect multiple pathways in cancer cells including PI3K/AKT, MAPK, PTEN, VEGF, and caspases. Its effectiveness was proven in various models including lymphoma, leukemia, prostate cancer, breast cancer, hepatic cancer [8]. Recently we reported the anti-proliferative and pro-apoptotic activity of fucoidan in monotherapy among uterine sarcoma and carcinosarcoma cell lines. Simultaneously we confirmed previous observations that fucoidan do not affect normal (benign) human cells. Additionally, fucoidan is widely using as a dietary supplement. The characteristics listed above allow it to be considered as a safe product [8–10].

Obtained results led us to question the activity of fucoidan in combination with standard chemotherapy among uterine sarcomas. Gemcitabine was selected to be investigated together with fucoidan. Gemcitabine is commonly used anti-tumor drug belonging to the group of antimetabolites. It express antimetabolic effect by interruption of DNA synthesis. It is widely used as an option of standard approach for the systemic therapy in most of uterine sarcomas subtypes [3,11,12]. Furthermore, it is a comparator in many ongoing clinical trials enrolling uterine sarcoma patients [12]. Combinations of gemcitabine with novel agents including olaratumab, nivolumab, and pazopanib are under investigation in phases II–III of clinical trials [13].

This study was aimed to test if the combination of gemcitabine and fucoidan will have better therapeutic effect in uterine sarcomas and carcinosarcomas cell lines than the compounds applied alone and to assess the type of interactions between them in concomitant treatment.

2. Materials and Methods

2.1. Reagents

Gemcitabine (100 mg/mL, 0.38 mMol/mL) was purchased from Accord Healthcare (UK), and fucoidan (*Undaria pinnatifida*) was obtained from Sigma-Aldrich (St. Louis, MO, USA). Fucoidan was diluted in respective complete culture medium at a concentration of 10 mg/mL just before use.

The Roswell Park Memorial Institute 1640 (RPMI-1640), Eagle's Minimum Essential Medium (MEM), McCoy's 5a Medium Modified, fetal bovine serum (FBS), trypsin-EDTA were purchased from PAN-Biotech (Aidenbach, Germany), penicillin-streptomycin and MTT (3-(4,5-dimethylthiazol-2-yl)-2,5-diphenyltetrazolium bromide) were obtained from Sigma-Aldrich (St. Louis, MO, USA). The Cell Proliferation ELISA System assay kit was purchased from Roche (Molecular Biochemicals, Manneihm, Germany). PE Active Caspase-3 Apoptosis Kit and Propidium iodide utilizing the PI/RNase Staining Buffer were obtained from Becton Dickinson Biosciences (San Jose, CA, USA).

2.2. Cell Lines and Cultures

Carcinosarcoma cell lines (SK-UT-1, SK-UT1-B), uterine leiomyosarcoma cell line (MES-SA), and endometrial stromal sarcoma cell line (ESS-1) were obtained from the Laboratorio de Investigación Traslacional (MD Anderson Cancer Center, Madrid). The selection of cell lines was done in order to assess the differences in response to tested compounds among wide range of uterine sarcomas. Each cell line used for experiments is a model of particular tumor what makes obtained data more reproducible and comparable with other research.

The cells were cultured in MEM (SK-UT-1, SK-UT1-B), McCoy's 5a Medium Modified (MES-SA), RPMI-1640 (ESS-1), containing 10% (SK-UT-1, SK-UT-1B, MES-SA) or 20% (ESS-1) FBS and 1% penicillin-streptomycin at 37 °C in a humidified 5% CO_2 atmosphere. Cells from the 4th to 9th passage were used for all experiments.

Detailed characteristic of cell lines is presented in Table 1.

Table 1. The characteristics of used cell lines [14,15].

Cell line:	SKUT-1 *	SKUT-1B *	MES-SA	ESS-1
Organism:	Homo sapiens, human	Homo sapiens, human	Homo sapiens, human	Homo sapiens, human
Tissue:	uterus	uterus/endometrium	uterus	uterus
Culture properties:	adherent	adherent	adherent	adherent
Disease:	grade III, mesodermal tumor (mixed); consistent with leiomyosarcoma	grade III, mesodermal tumor (mixed); consistent with leiomyosarcoma	grade III, recurrent, uterine leiomyosarcoma [14]	endometrial stromal sarcoma
Age:	75 years	75 years	56 years	76 years
Gender:	female	female	female	female
Ethnicity:	Caucasian	Caucasian	Caucasian	Caucasian

* SKUT-1 and SKUT-1B cell lines were derived from the same patient from different sites of tumor SKUT-1 line is a model of sarcomatous part of the tumor (forms spindle cell sarcomas), SKUT-1B is a model of carcinomatous part of the tumor (forms well differentiated adenocarcinomas).

2.3. Cell Viability Assay

Cells were platted on 96-well microplates SK-UT-1, SK-UT-1B, MES-SA, and ESS-1 (3×10^4 cells/mL) and cells were incubated in the presence of gemcitabine (0.1–200 ng/mL), fucoidan (0.01–5 mg/mL) and mixtures of both compounds for 96 h. The maximum concentration of gemcitabine achieved in human serum was 26.79 ± 10.06 μg/mL, reported by Wang et al. so the concentrations used in our experiments were much lower comparing to ones available in vivo [16]. The safe concentrations of fucoidan in human serum are still under investigations and no conclusive data are available up to date. Afterwards, the cells were incubated for 3 h with the MTT. During the time MTT was metabolized by living cells to purple formazan crystals, which were later solubilized in SDS buffer (10% SDS in 0.01 N HCl) overnight. Separate experiments were performed in triplicate. The optical density of the product was measured at 570 nm with the use of an ELX-800 plate reader (Bio-Tek, Instruments, Winooski, VT, USA) and analyzed using Gen5 software (Bio-Tek, Instruments, Winooski, VT, USA).

2.4. Assessment of Apoptosis

Cell lines were treated with tested compounds and mixtures for 48 h, as a control cells without treatment were used. Afterwards cells were harvested, fixed and permeabilized using the Cytofix. All of experiments were performed according to the manufacturer's instructions of PE Active Caspase-3 Apoptosis Kit (BD Biosciences, San Jose, CA, USA). Labeled cells were analyzed by flow cytometer FACSCalibur (Becton Dickinson, Franklin Lakes, NJ, USA), operating with CellQuest software (BD Biosciences, San Jose, CA, USA) to quantitatively assess the caspase-3 activity. The methodology of apoptosis assessment were described in detail elsewhere [9].

2.5. Cell Cycle Analysis

Cell lines were treated with tested compound and mixture for 48 h (as a control we used cells without exposure to tested compounds) and then fixed in 80% ethanol at −20 °C for 24 h. The experiment was conducted by utilizing PI/RNase Staining Buffer (BD Biosciences, USA) according to the manufacturer's instructions. Cell cycle analysis was performed using flow cytometry (FACSCalibur (Becton Dickinson, USA). Acquisition rate was at least 60 events/sec in low acquisition mode and at least 10,000 events were measured. Methodology of cell cycle analysis was presented in detail previously [10].

2.6. Isobolographic and Statistical Analysis

In order to determine the inhibition rate of cell viability, measured by MTT assay, per dose of gemcitabine and fucoidan, log-probit linear regression analysis was performed according to method described by Litchfield and Wilcoxon [17,18]. Median inhibitory concentrations (IC_{50}) for gemcitabine and fucoidan in ESS-1, SK-UT-1, and SK-UT-1B cell lines were calculated according to method previously described [19]. Due to lack of cells response, the IC_{50} was not achievable for fucoidan in MES-SA cell line [10]. Parallelism between dose–response curves for gemcitabine and fucoidan in ESS-1, SK-UT-1, and SK-UT-1B cell lines was confirmed by log-probit method, as it was described in detail previously [18]. Next, isobolographic analysis of interactions between drugs for the combination of gemcitabine and fucoidan in tested cell lines were performed according to the method presented by i.a. Tallarida et al. [20]. The median additive inhibitory concentrations ($IC_{50\ add}$) for two-drug mixtures were theoretically calculated according to the method described elsewhere [20]. The calculated values were used for performing MTT tests on ESS-1, SK-UT-1, and SK-UT-1B cell lines—the assessment of experimentally derived $IC_{50\ mix}$ values for tested drug combinations in a fixed 1:1 ratio. The particular drug concentrations (gemcitabine and fucoidan) in the mixture were calculated by multiplying $IC_{50\ mix}$ values accordingly to proportions in additive mixture. Detailed description of isobolographic method was introduced by Tallarida, Grabovsky and Luszczki [20–22]. The results of MTT test were analyzed by one-way ANOVA test, Tukey's Multiple Comparison Post-test using GraphPad Prism 5.0 (GraphPad Softwere Inc., San Diego, CA, USA). The $p < 0.05$ was considered as statistically significant.

3. Results

3.1. Cell Viability Assay

Anti-proliferative effects of gemcitabine on tested cell lines is presented on Figure 1. Experimentally determined IC_{50} values for gemcitabine in SK-UT-1, SK-UT-1B, ESS-1, and MES-SA cell lines, were 31.173, 25.243, 13.875, and 72.482 ng/mL respectively.

As we previously reported fucoidan significantly affects SK-UT-1, SK-UT-1B, and ESS-1 cell lines, meanwhile MES-SA cells seem to be resistant for this agent. IC_{50} was 0.966, 3.348, and 0.848 mg/mL respectively, it was not possible to determine IC_{50} for fucoidan in MES-SA cell line due to insufficient response to treatment [9]. The IC_{50} values are summarized in Supplementary Table S1.

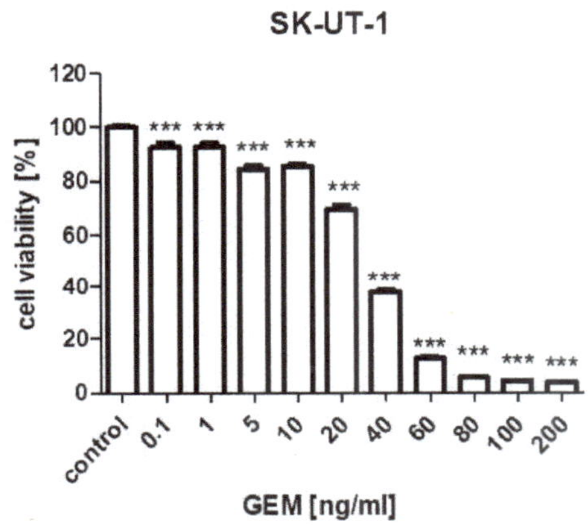

A

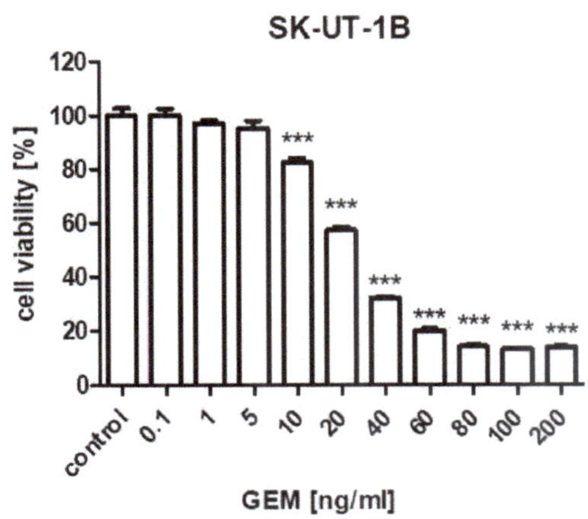

B

Figure 1. *Cont.*

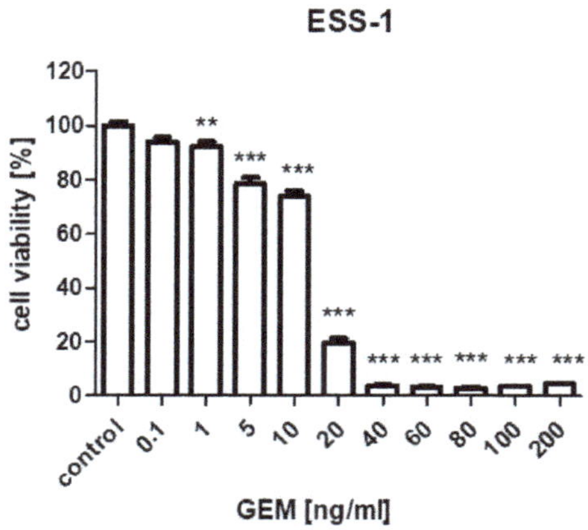

C

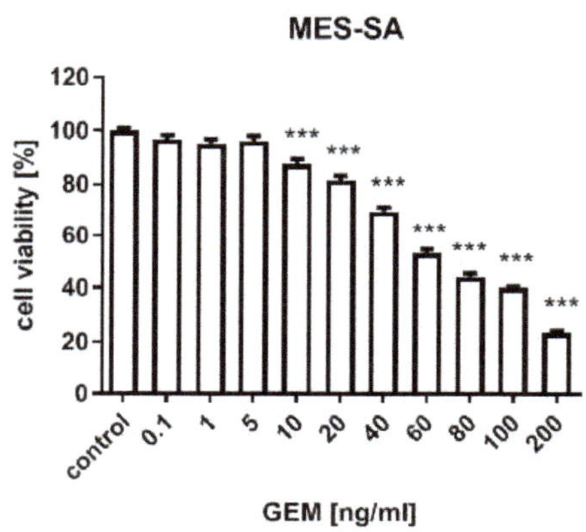

D

Figure 1. The influence of gemcitabine on the proliferation of carcinosarcoma cell lines (SK-UT-1 (**A**), SK-UT1-B (**B**)), endometrial stromal sarcoma cell line (ESS-1 (**C**)) and uterine leiomyosarcoma cell line (MES-SA (**D**)). The cells were treated with the gemcitabine at various concentrations for 96 h. (** $p < 0.01$, *** $p < 0.001$ were considered as statistically significant).

3.2. Isobolographic Anaysis

Additive effect of the combined treatment with gemcitabine and fucoidan was observed in ESS-1 and SK-UT-1 cell lines. Although the supra-additive (synergistic) effect was noticed in SK-UT-1B cell line.

The details of results obtained in isobolographic analysis are presented on Figures 2–4.

In Figures 2–4 the median inhibitory concentrations (IC_{50}) for gemcitabine (GEM) and fucoidan (FUK) are plotted on the X- and Y-axes, respectively. The solid lines on both axes reflect the S.E.M. for the IC_{50} values for the studied drugs, when administered alone. The lower and upper isoboles of additivity represent the curves connecting the IC_{50} values for GEM and FUK administered alone. The dotted line illustrates the fixed-ratio of 1:1 for the combination of GEM with FUK. The points A′ and A″ depict the theoretically calculated $IC_{50\,add}$ values for both, lower and upper isoboles of additivity. The point M reflects the experimentally-derived $IC_{50\,mix}$ value for total dose of the mixture expressed as proportions of GEM and FUK that produced a 50% anti-proliferative effect (50% isobole) in the cancer cell line (SK-UT-1, SK-UT-1B, and ESS-1, respectively for Figures 2–4) measured in vitro by the MTT assay. On the graph, the S.E.M. values are presented as horizontal and vertical error bars for every IC_{50} value. Type I isobolographic analysis of interactions are presented in Supplementary Table S2. The effect of combined fucoidan and gemcitabine on the proliferation of tested cell lines were presented of Figure 5.

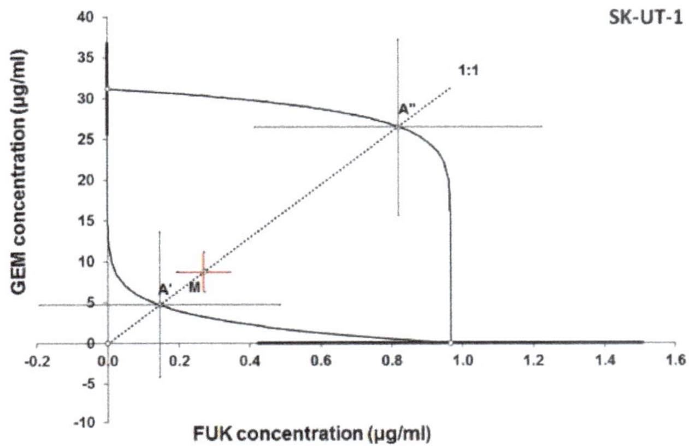

Figure 2. Isobologram showing interaction between gemcitabine (GEM) and fucoidan (FUK) with respect to their anti-proliferative effects in the cancer cell line (SK-UT-1) measured in vitro by the MTT assay. The experimentally-derived $IC_{50\ mix}$ value is placed within the area of additivity and indicates additive interaction between GEM and FUK in this cancer cell line.

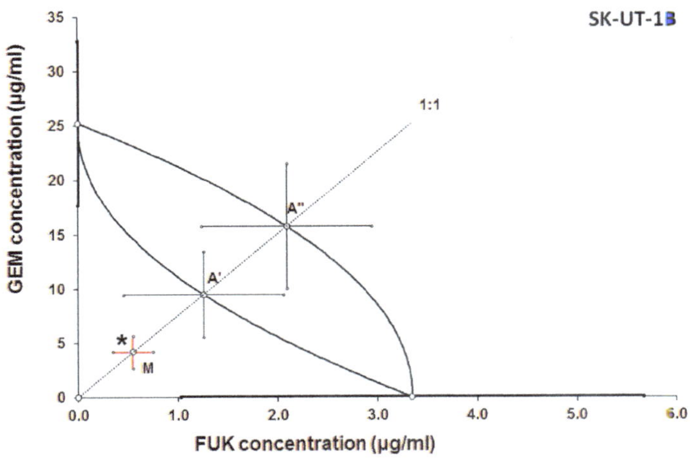

Figure 3. Isobologram showing interaction between gemcitabine (GEM) and fucoidan (FUK) with respect to their anti-proliferative effects in the cancer cell line (SK-UT-1B) measured in vitro by the MTT assay. Because the experimentally-derived $IC_{50\ mix}$ value is placed significantly below the point A′, the interaction between GEM and FUK for the cancer cell line SK-UT-1B is supra-additive (synergistic). * $p < 0.05$ vs. the respective $IC_{50\ add}$ values.

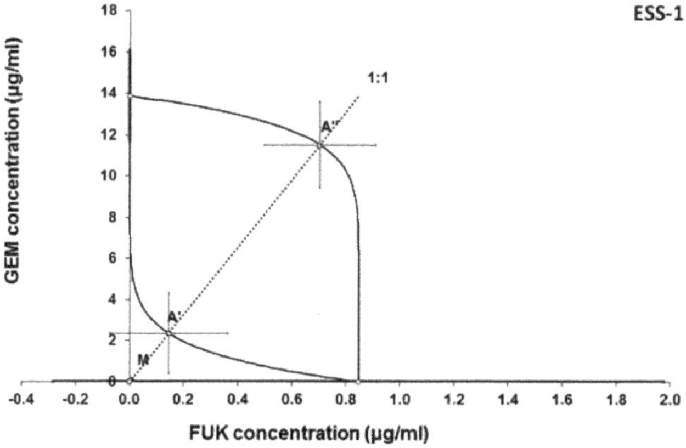

Figure 4. Isobologram showing interaction between gemcitabine (GEM) and fucoidan (FUK) with respect to their anti-proliferative effects in the cancer cell line (ESS-1) measured in vitro by the MTT assay. Although the experimentally-derived $IC_{50\ mix}$ value is placed below, but near to the point A', the interaction between GEM and FUK in this cancer cell line is additive.

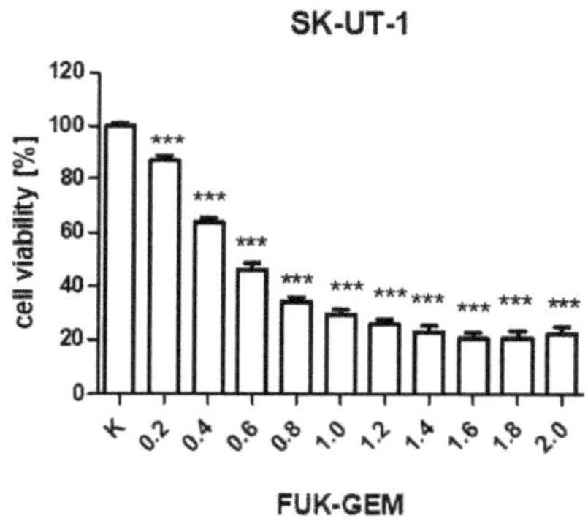

A

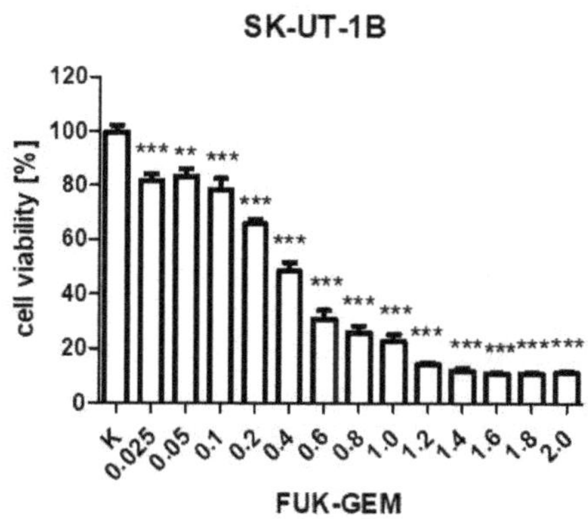

B

Figure 5. *Cont.*

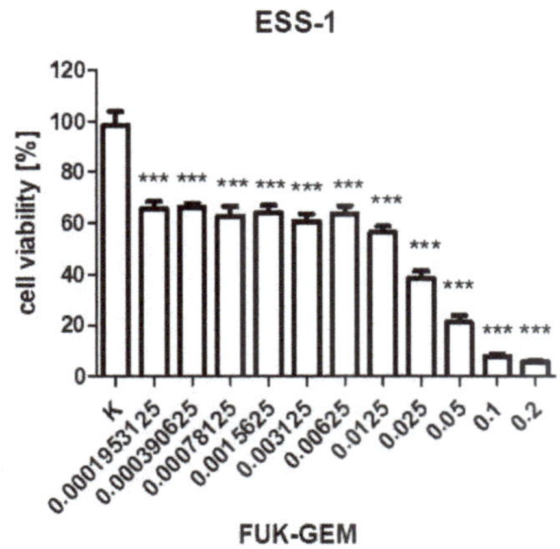

C

Figure 5. The influence of combined fucoidan and gemcitabine on the proliferation of carcinosarcoma cell lines (SK-UT-1 (**A**), SK-UT-1B (**B**)) and endometrial stromal sarcoma cell line (ESS-1 (**C**)). The cells were treated with fucoidan and gemcitabine at various concentrations for 96 h. (** $p < 0.01$, *** $p < 0.001$ were considered as statistically significant). The values on axis X represent the multiplicity of calculated IC_{50}. The combinations of gemcitabine and fucoidan were mixed 1:1 before added to cells.

3.3. *Assessment of Apoptosis*

The impact of gemcitabine and its combination with fucoidan on apoptosis, measured as a number of cells with activated caspase 9, is presented on Figure 6A–C. Gemcitabine was used in concentration of IC_{50}. Results obtained in isobolographic analysis were used to select appropriate concentrations of mixture for apoptosis assessment. For SK-UT-1 and SK-UT-1B concentrations of 0.5 IC50 for both agents were used. Due to very strong effect observed in ESS-1 cell line concentration 0.05 IC_{50} was selected for experiments.

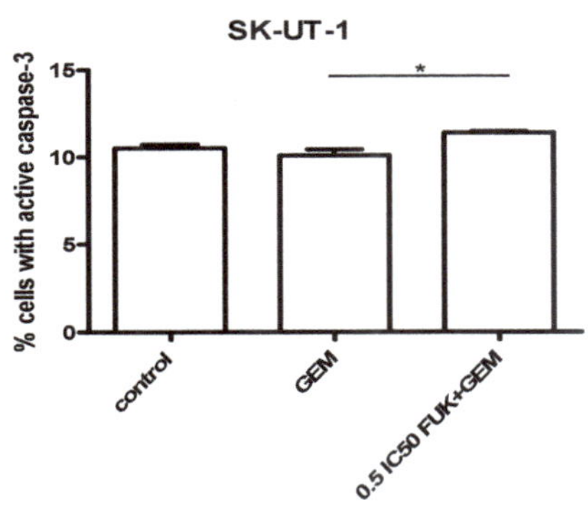

A

Figure 6. *Cont.*

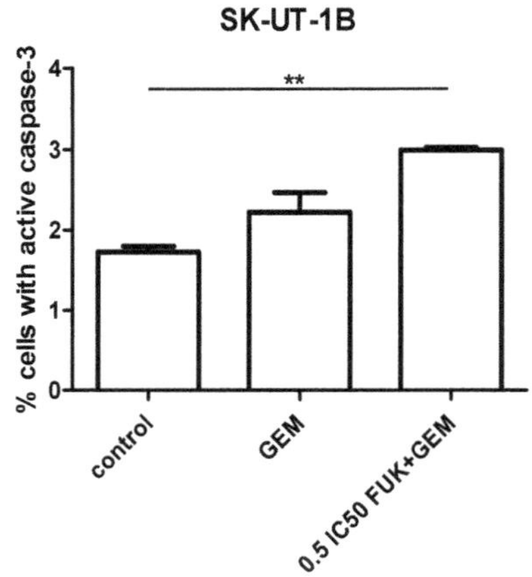

B

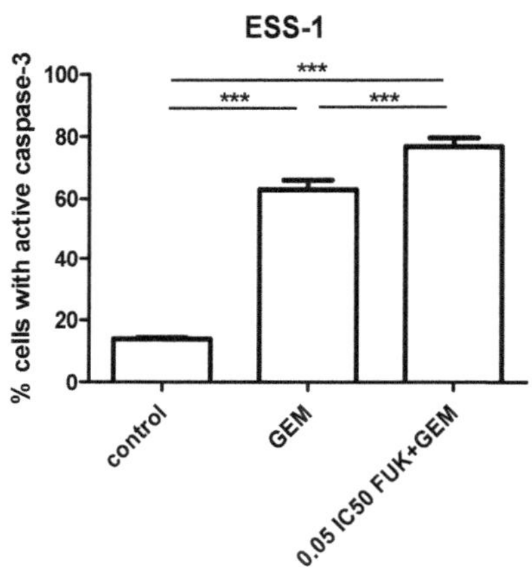

C

Figure 6. Effects of gemcitabine (in IC_{50} concentration) and mixture of fucoidan and gemcitabine (in concentrations of 0.5 IC_{50} for SK-UT-1 and SK-UT-1B and 0,05 IC50 for ESS-1) on caspase-3 activation in SK-UT-1 (**A**), SK-UT-1B (**B**), and ESS-1 (**C**) cells. Results are expressed as mean ± SD of three separate experiments (* $p < 0.05$, ** $p < 0.01$, *** $p < 0.001$ versus the control, one-way ANOVA test).

The strongest induction of apoptosis by both gemcitabine and mixture gemcitabine + fucoidan was observed in ESS-1 cell line. This effect was also observed in SK-UT-1B cells. Although the induction of apoptosis in SK-UT-1 cell line of both single agent and combination was very weak. Statistical significance was observed between single agent and combination but the differences comparing to control were not significant.

3.4. Cell Cycle

The results of cell cycle analysis of cells treated with gemcitabine and its mixture with fucoidan are presented in Figure 7. Statistical significance of differences between each phase of cell cycle in treatments and control are listed in Supplementary Table S3. Tested compounds were used in concentration as in apoptosis assessments.

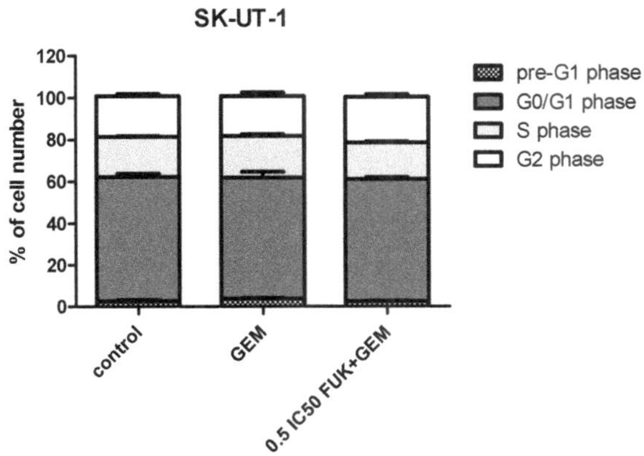

A

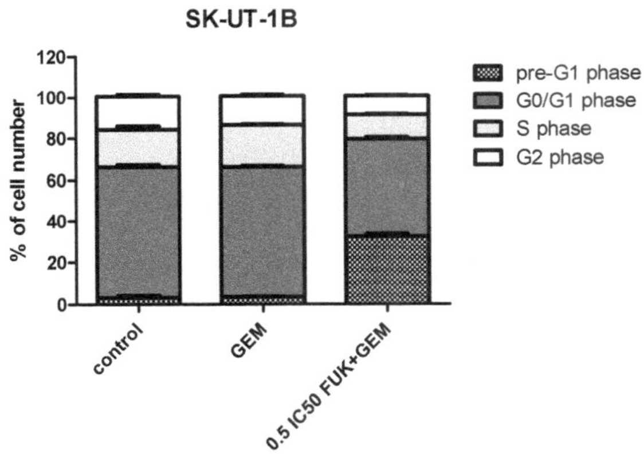

B

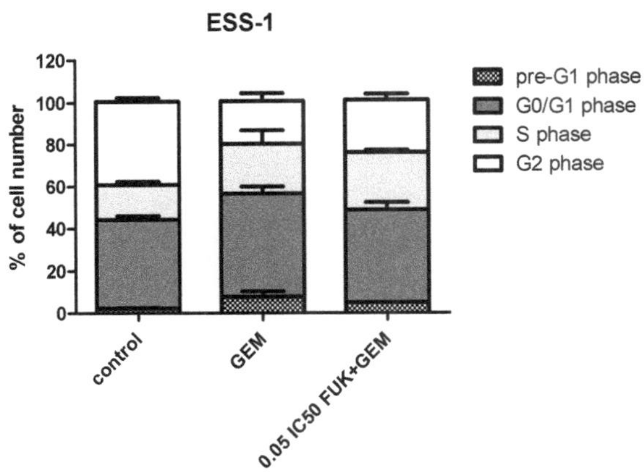

C

Figure 7. Effects of gemcitabine (in IC_{50} concentration) and mixture of fucoidan and gemcitabine (in concentrations of 0.5 IC_{50} for SK-UT-1 and SK-UT-1B and 0.05 IC_{50} for ESS-1) on cell cycle progression in SK-UT-1 (**A**), SK-UT1-B (**B**), and ESS-1 (**C**) cell lines. The results are presented as mean ± SD from three separate experiments.

The most significant impact of concomitant treatment with gemcitabine and fucoidan on cell cycle arrest (measured as percent of cells in phases pre-G1 and G0/G1) was detected in the SK-UT-1 B cell line (comparing to both control and gemcitabine treatment). In the ESS-1 cell line significant differences were noted between cells treated with gemcitabine and control, with no such differences between gemcitabine and mixture. The differences in number of cells in each phase of cell cycle in SK-UT-1 cells exposed to investigated agents were very small in spite of statistical significance observed in particular phases.

4. Discussion

This paper for our knowledge is the first report of interactions between fucoidan and gemcitabine in any model. The results obtained in our experiments show that both gemcitabine and fucoidan significantly affect cell viability in all tested cell lines (with exception of fucoidan in MES-SA cell line). The problem of MES-SA cell line resistance to fucoidan have been discussed previously [10]. In present study we observed also the worst response of MES-SA to gemcitabine among all of tested cell lines. Interestingly we used regular MES-SA line, but not its multidrug resistant variant MES-SA/Dx5. The mechanism of relative resistance of MES-SA cell line to fucoidan and gemcitabine is not known and requires further investigations. For our best knowledge the phenomenon of spontaneous MES-SA resistance to gemcitabine have not been published yet. Resistance to gemcitabine was artificially induced by genetic modifications (transfection) in gene coding deoxynucleoside kinase in MES-SA, although such modifications were not performed in the cell line we used [23]. Occurrence of some spontaneous mutations in deoxynucleoside kinase gene at least in part of cells could possibly explain our observation. The occurrence of serious side effects combined with limited activity makes systemic therapy of uterine sarcomas doubtful [13]. The most common unwanted results of chemotherapy are hematologic toxicities (leucopoenia, neutropenia, thrombocytopenia), fever, and concomitant infectious, that appear in the majority of treated patients [24]. Side effects of drugs are mostly dose-dependent, so it could be limited by the reduction of dosage, on the other hand, reduction of dose would decrease the therapeutic effect.

In multidrug regimens, in assumption, lower doses of multiple drugs may cause fewer side effects (or its lower intensity) without decreasing effectiveness or even increasing it [13]. But in clinical practice such approaches are commonly associated with higher rate of adverse events.

The rarity and heterogeneity of uterine sarcomas limit the possibilities of conducting clinical trials, which may be considered as a cause of the poor results of its treatment up to date. Even in preclinical studies number of research aimed to test new drugs combinations in this group of tumors is limited.

Coley et al. tested the combination of seliciclib (cyclin-dependent kinases (CDK)-inhibitor) and paclitaxel among selection of uterine sarcomas cell lines. Using isobolography, they observed synergism between both drugs. Interestingly, differences in activity depending on the sequence of each drug administration were reported. Although the study is interesting, seliciclib is not registered for the treatment of uterine sarcomas. Furthermore, paclitaxel in monotherapy is not recommended among this type of malignancies [25].

In the present study we propose a combination of standard chemotherapy with a natural compound that is known to be safe; its intake is not associated with serious adverse events [26]. Both substances were administered at the same time in order to avoid a "drug sequence effect" observed by Coley at al. that could influence the results. Isobolographic analysis showed additive interaction between gemcitabine and fucoidan in SK-UT-1 and ESS-1 cell lines and even supra-additive (synergy) in SK-UT-1B. No antagonism between tested substances was observed.

Cell viability was decreased equally or more than the sum of effects of single agents. If these observations are confirmed in animal models, clinical trials of fucoidan as an addition to treatment with gemcitabine may allow a decrease of its dose. Fucoidan is nowadays used as dietary supplement, and seaweeds containing it are widely used in Asian cuisine [26]. There are many researches aimed at checking its activity in various types of cancer. Some of them regard a combination of fucoidan and other agents (including cytostatics), and the results of most of them are promising. In our study we assessed the type of interactions between fucoidan and a cytostatic agent, using an isobolographic method, and calculated IC50 for the mixture. Significantly lower doses of therapeutic compounds may be associated with lower risk of adverse events. Furthermore, it may solve another problem that was widely discussed previously— the bioavailability of fucoidan. That is currently still under investigation both in animal models as well as in clinical trials—the results of these studies are awaited [27–29]. IC50 values for fucoidan obtained in combination with gemcitabine are much lower than 1 mg/mL: Such concentrations seem to be achievable in vivo. The doubtful concentrations were around 5 mg/mL. Although a few studies in this field were performed, precise determination of available serum concentrations among human are still under investigation [29,30].

Quite similar results were obtained by Mathew et al. who assessed IC50 for various types of fucoidan from 0.3 to 1.3 mg/mL (depending on plant that it was derived from). In this study they also performed analysis of fucoidans' impact on CYP450 and COMT (catechol-O-methyltransferase) pathways, concluding that it is limited, which might be considered as a confirmation of its safety [31].

Our findings are comparable with other studies assessing activity of fucoidan and cytotoxic drugs. Burney et al. assessed activity of 2 types of fucoidan in combination with paclitaxel and tamoxifen on mouse models of human breast or ovarian cancer. They concluded that interaction between tested compounds were additive or supra-additive (depending on combination and model used) [32].

Similar observations were taken by Zhang et al. who tested combinations of fucoidan with cisplatin, tamoxifen, and paclitaxel in breast cancer cell lines. They concluded that fucoidan "enhances the anti-cancer activity" of tested cytostatic drugs. Furthermore, they observed the induction of apoptosis and cell cycle arrest among cells treated with fucoidan what corresponds with our previous findings [9,33].

Observed additive and supra-additive interactions between various fucoidans and chemotherapeutic drugs require explanation. The mechanism of fucoidans' activity is not fully investigated. As it was mentioned above there are various cellular pathways that are affected by this compound. Its multipotential activity is, on the one hand, promising, and does not allow the overcoming of single pathway change, but on the other it is very difficult to understand completely. One of possible mechanism of synergistic interaction with chemotherapeutics is the ability to down-regulate expression of Bcl-xL and Mcl-1, known as anti-apoptotic proteins [34]. The addition of fucoidan to gemcitabine enhances its ability to induce apoptosis. This effect is especially noticeable in ESS-1 cell line where, even if mixture was added in concentration 0.05 of IC50, almost 80% of cells enter apoptosis and the difference between gemcitabine in monotherapy and the mixture reaches almost 20%. Weaker effect was observed in SK-UT-1B cells. In the SK-UT-1 cell line almost no impact of gemcitabine and fucoidan to apoptosis was noted, even if agents were used in concentration of 0.5 of IC50. Abudabbus et al. compared the ability of fucoidan in combination with cisplatin, doxorubicin and taxol, to induce apoptosis and to arrest the cell cycle among benign and malignant breast cells [35]. They reported a strong impact to cell cycle arrest and induction of apoptosis in malignant cells but no significant effect in normal cells. A comparable effect of fucoidan in monotherapy was described by Arumugam et al. in hepatoblastoma cell line [36]. So, our results confirm proapoptotic features of fucoidan in selected models. Interestingly, observations taken in experiments assessing apoptosis were confirmed also in cell cycle analysis. The strongest effect of combination of gemcitabine and fucoidan was observed in ESS-1 line. Moderate and weak changes in number of cells in particular cell cycle phases were observed in SK-UT-1 and SK-UT-1B cell lines respectively.

Gemcitabine is known to affect the cell cycle in various cell models and in vivo; this feature is the consequence of its mechanism of action by DNA damaging [37]. The process of gemcitabine-induced cell cycle arrest can by interrupted by activity of DNA repairing mechanisms such as Chk1 [38]. Such a mechanism might be responsible for the limited changes in cells distribution among cell cycle phases observed in SK-UT-1 and SK-UT-1B treated with gemcitabine.

Park et al. reported the effect of cell cycle arrest by fucoidan treatment in bladder cancer cells. They also revealed the its association with down-regulation of cyclin D1, cyclin E, and cyclin-dependent-kinases (Cdks) in a concentration-dependent manner, without any change in Cdk inhibitors (p21, p27) [39]. Interestingly Han et al. experimented on a colon cancer model and observed cell cycle arrest associated with an increased expression of p21 [40]. Different mechanisms of cell cycle arrest induced by fucoidan in different models, combined with potential activity of DNA repairing factors affecting activity of gemcitabine, can explain differences among tested cell lines, we observed. So far, no studies assessing the impact of combination of gemcitabine and fucoidan have been published.

The results and survey of literature indicate that testing multicompound regimens gives promising results and further research in the field could reveal a combination that could be beneficial for patients suffering from uterine sarcomas.

The study is not without limitations and should be considered carefully. The main limitation of the study is a cancer model that was used, 2D cell cultures, which do not allow investigation of the tumor microenvironment and may not fully reflect the response of tumor cells in vivo. Although promising, obtained results require clinical studies to be considered in practice.

The results obtained in our study indicate 3 potential directions for further investigations:

- Analysis of the mechanisms of the MES-SA cell line relative resistance to fucoidan and gemcitabine that may lead to identification of such mechanisms in tumors and make progress for overcoming them.
- Due to similarities in clinical course between carcinosarcomas and endometrial carcinomas the results obtained on SK-UT-1 and SK-UT-1B cell lines might be replicated on carcinomas cell lines and extend indications for future practical applications.
- Studies on 3D cell cultures or animal models have to be performed in order to confirm the activity of proposed combination on tissue models and to assess how the tumor microenvironment affects it.

5. Conclusions

Obtained data showed additive and supra-additive effect of fucoidan combined with gemcitabine in uterine endometrial stromal sarcoma (ESS-1) and carcinosarcoma (SK-UT-1, SK-UT-1B) cell lines, what confirms it has better or at least equal performance comparing to sum of effects of monotherapies.

The addition of fucoidan to gemcitabine enhances proapoptotic effect of gemcitabine in endometrial stromal sarcoma cells (ESS-1) but not in carcinosarcoma (SK-UT-1, SK-UT-1B) cell lines.

Gemcitabine in monotherapy do not induce cell cycle arrest in carcinosarcoma (SK-UT-1, SK-UT-1B) cell lines. Although the addition of fucoidan to gemcitabine induces it in model of carcinomatous part of carcinosarcoma (SK-UT-1B).

Differences in response to applied treatment among tested cell lines (in cell viability apoptosis and cell cycle distribution) can be explained by the multipotential and not-fully-investigated activity of fucoidan as well as differences in cellular mechanisms (such as DNA repairing) in selected models.

The relative resistance of uterine leiomyosarcoma cell line (MES-SA) to applied drugs combination justify searching for other therapeutic regimens to improve therapy efficacy.

Taking into consideration the disappointingly low effectiveness of systemic therapy among these types of cancer, a combination of gemcitabine and fucoidan seems to be a promising alternative, having the potential to increase effectiveness and safety of the treatment.

Supplementary Materials: The following are available online at http://www.mdpi.com/2072-6694/12/1/107/s1, Table S1: Anti-proliferative effects of gemcitabine (GEM) and fucoidan (FUK) administered singly in cancer cell lines ESS-1, SKUT-1, SKUT-1B and MES-SA, as measured in vitro by the MTT assay, Table S2: Type I isobolographic analysis of interactions (for non-parallel dose-response effects) between gemcitabine (GEM) and fucoidan (FUK) in three cancer cell lines ESS-1, SKUT-1 and SKUT-1B, as measured in vitro by the MTT assay, Table S3: Statistical significance of differences in percents of cells in each cell cycle phases among cells treated with gemcitabine or mixture (gemcitabine and fucoidan) and control (no treatment). The results were analyzed by one-way ANOVA test, Tukey's Multiple Comparison Post-test, $p < 0.05$ was considered as statistically significant. The results of experiment are presented on Figure 7.

Author Contributions: Conceptualization, M.B., W.B. and J.K.; methodology, K.O. and M.D.-G.; validation, A.W., M.D.-G.; formal analysis, J.K.; investigation, K.O., A.W. and M.D.-G.; resources, W.B. and G.M.-B.; data curation, J.Ł.; writing—original draft preparation, M.B.; writing—review and editing, K.O. and R.T.; supervision, J.K.; project administration, M.B.; funding acquisition, W.B., J.K. and R.T. All authors have read and agreed to the published version of the manuscript.

Funding: The research was founded by Medical University of Lublin (grants No. DS 120, DS 121 and MNmd129).

Acknowledgments: The research was performed in collaboration under umbrella of ENITEC (European Network of Individual Treatment in Endometrial Cancer), the research group established by ESGO (European Society of Gynaecological Oncology).

Conflicts of Interest: The authors declare no conflict of interest.

References

1. El-Khalfaoui, K.; du Bois, A.; Heitz, F. Current and future options in the management and treatment of uterine sarcoma. *Ther. Adv. Med. Oncol.* **2014**, *6*, 21–28. [CrossRef] [PubMed]
2. Benson, C.; Miah, A. Uterine sarcoma–current perspectives. *Int. J. Womens Health* **2017**, *9*, 597–606. [CrossRef] [PubMed]
3. National Comprehensive Cancer Network. Uterine Neoplasms (Version: 3.2019-February 11, 2019). Available online: https://www.nccn.org/professionals/physician_gls/pdf/uterine.pdf (accessed on 21 March 2019).
4. Ricci, S.; Giuntoli, R.L., II; Eisenhauer, E.; Lopez, M.A.; Krill, L.; Tanner, E.J., III; Gehrig, P.A.; Havrilesky, L.J.; Secord, A.A.; Levinson, K.; et al. Does adjuvant chemotherapy improve survival for women with early-stage uterine leiomyosarcoma? *Gynecol. Oncol.* **2013**, *131*, 629–633. [CrossRef] [PubMed]
5. Hensley, M.L.; Maki, R.; Venkatraman, E.; Geller, G.; Lovegren, M.; Aghajanian, C.; Sabbatini, P.; Tong, W.; Barakat, R.; Spriggs, D.R. Gemcitabine and docetaxel in patients with unresectable leiomyosarcoma: Results of a phase II trial. *J. Clin. Oncol.* **2002**, *20*, 2824–2831. [CrossRef]
6. Seddon, B.; Scurr, M.; Jones, R.L.; Wood, Z.; Propert-Lewis, C.; Fisher, C.; Flanagan, A.; Sunkersing, J.; A'Hern, R.; Whelan, J.; et al. A phase II trial to assess the activity of gemcitabine and docetaxel as first line chemotherapy treatment in patients with unresectable leiomyosarcoma. *Clin. Sarcoma Res.* **2015**, *5*, 13. [CrossRef]
7. Hensley, M.L.; Miller, A.; O'Malley, D.M.; Mannel, R.S.; Behbakht, K.; Bakkum-Gamez, J.N.; Michael, H. Randomized phase III trial of gemcitabine plus docetaxel plus bevacizumab or placebo as first-line treatment for metastatic uterine leiomyosarcoma: An NRG Oncology/Gynecologic Oncology Group study. *J. Clin. Oncol.* **2015**, *33*, 1180–1185. [CrossRef]
8. Van Weelden, G.; Bobiński, M.; Okła, K.; Van Weelden, W.J.; Romano, A.; Pijnenborg, J. Fucoidan Structure and Activity in Relation to Anti-Cancer Mechanisms. *Mar. Drugs* **2019**, *17*, 32. [CrossRef]
9. Bobiński, M.; Okła, K.; Bednarek, W.; Wawruszak, A.; Dmoszyńska-Graniczka, M.; Garcia-Sanz, P.; Wertel, I.; Kotarski, J. The Effect of Fucoidan, a Potential New, Natural, Anti-Neoplastic Agent on Uterine Sarcomas and Carcinosarcoma Cell Lines: ENITEC Collaborative Study. *Arch. Immunol. Ther. Exp.* **2019**, *67*, 125–131. [CrossRef]
10. Atashrazm, F.; Lowenthal, R.M.; Woods, G.M.; Holloway, A.F.; Karpiniec, S.S.; Dickinson, J.L. Fucoidan Suppresses the Growth of Human Acute Promyelocytic Leukemia Cells In Vitro and In Vivo. *J. Cell. Physiol.* **2016**, *231*, 688–697.
11. Brown, K.; Dixey, M.; Weymouth-Wilson, A.; Linclau, B. The synthesis of gemcitabine. *Carbohydr. Res.* **2014**, *387*, 59–73. [CrossRef]
12. Ducoulombier, A.; Cousin, S.; Kotecki, N.; Penel, N. Gemcitabine-based chemotherapy in sarcomas: A systematic review of published trials. *Crit. Rev. Oncol. Hematol.* **2016**, *98*, 73–80. [CrossRef]
13. Tantari, M.; Barra, F.; Di Domenico, S.; Ferraioli, D.; Vellone, V.G.; De Cian, F.; Ferrero, S. Current state of the art and emerging pharmacotherapy for uterine leiomyosarcomas. *Expert Opin. Pharmacother.* **2019**, *20*, 713–723. [CrossRef]
14. Harker, W.G.; MacKintosh, F.R.; Sikic, B.I. Development and characterization of a human sarcoma cell line, MES-SA, sensitive to multiple drugs. *Cancer Res.* **1983**, *43*, 4943–4950.
15. Cell Lines Characteristics Cards. Available online: https://www.atcc.org (accessed on 17 December 2019).

16. Wang, L.R.; Liu, J.; Huang, M.Z.; Xu, N. Comparison of pharmacokinetics, efficacy and toxicity profile of gemcitabine using two different administration regimens in Chinese patients with non-small-cell lung cancer. *J. Zhejiang Univ. Sci. B* **2007**, *8*, 307–313. [CrossRef]
17. Grabarska, A.; Łuszczki, J.J.; Nowosadzka, E.; Gumbarewicz, E.; Jeleniewicz, W.; Dmoszyńska-Graniczka, M.; Kowalczuk, K.; Kupisz, K.; Polberg, K.; Stepulak, A. Histone Deacetylase Inhibitor SAHA as Potential Targeted Therapy Agent for Larynx Cancer Cells. *J. Cancer* **2017**, *8*, 19–28. [CrossRef]
18. Litchfield, J.T., Jr.; Wilcoxon, F. A simplified method of evaluating dose-effect experiments. *J. Pharmacol. Exp. Ther.* **1949**, *96*, 99–113.
19. Wawruszak, A.; Luszczki, J.J.; Grabarska, A.; Gumbarewicz, E.; Dmoszynska-Graniczka, M.; Polberg, K.; Stepulak, A. Assessment of Interactions between Cisplatin and Two Histone Deacetylase Inhibitors in MCF7, T47D and MDA-MB-231 Human Breast Cancer Cell Lines-An Isobolographic Analysis. *PLoS ONE* **2015**, *10*, e0143013. [CrossRef]
20. Tallarida, R.J. An overview of drug combination analysis with isobolograms. *J. Pharmacol. Exp. Ther.* **2006**, *319*, 1–7. [CrossRef]
21. Luszczki, J.J. Isobolographic analysis of interaction between drugs with nonparallel dose-response relationship curves: A practical application. *Naunyn Schmiedebergs Arch. Pharmacol.* **2007**, *375*, 105–114. [CrossRef]
22. Grabovsky, Y.; Tallarida, R.J. Isobolographic analysis for combinations of a full and partial agonist: Curved isoboles. *J. Pharmacol. Exp. Ther.* **2004**, *310*, 981–986. [CrossRef]
23. Jordheim, L.P.; Galmarini, C.M.; Dumontet, C. Gemcitabine resistance due to deoxycytidine kinase deficiency can be reverted by fruitfly deoxynucleoside kinase, DmdNK, in human uterine sarcoma cells. *Cancer Chemother. Pharmacol.* **2006**, *58*, 547–554. [CrossRef] [PubMed]
24. Neoptolemos, J.P.; Palmer, D.H.; Ghaneh, P.; Psarelli, E.E.; Valle, J.W.; Halloran, C.M.; Faluyi, O.; O'Reilly, D.A.; Cunningham, D.; Wadsley, J.; et al. Comparison of adjuvant gemcitabine and capecitabine with gemcitabine monotherapy in patients with resected pancreatic cancer (ESPAC-4): A multicentre, open-label, randomised, phase 3 trial. *Lancet* **2017**, *389*, 1011–1024. [CrossRef]
25. Coley, H.M.; Shotton, C.F.; Kokkinos, M.I.; Thomas, H. The effects of the CDK inhibitor seliciclib alone or in combination with cisplatin in human uterine sarcoma cell lines. *Gynecol. Oncol.* **2007**, *105*, 462–469. [CrossRef] [PubMed]
26. Kim, K.J.; Lee, O.H.; Lee, H.H.; Lee, B.Y. A 4-week repeated oral dose toxicity study of fucoidan from the Sporophyll of Undaria pinnatifida in Sprague-Dawley rats. *Toxicology* **2010**, *267*, 154–158. [CrossRef] [PubMed]
27. Clinical trial: NCT03422055 Study of Tolerance, Biodistribution and Dosimetry of Fucoidan Radiolabeled by Technetium-99m (NANO-ATHERO). Available online: https://clinicaltrials.gov/ct2/show/NCT03422055 (accessed on 31 December 2019).
28. Hsu, H.Y.; Hwang, P.A. Clinical applications of fucoidan in translational medicine for adjuvant cancer therapy. *Clin. Transl. Med.* **2019**, *8*, 15. [CrossRef]
29. Zhang, W.; Sun, D.; Zhao, X.; Jin, W.; Wang, J.; Zhang, Q. Microanalysis and preliminary pharmacokinetic studies of a sulfated polysaccharide from Laminaria japonica. *Chin. J. Oceanol. Limnol.* **2016**, *34*, 177–185. [CrossRef]
30. Irhimeh, M.R.; Fitton, J.H.; Lowenthal, R.M.; Kongtawelert, P. A quantitative method to detect fucoidan in human plasma using a novel antibody. *Methods Find. Exp. Clin. Pharmacol.* **2005**, *27*, 705–710. [CrossRef]
31. Mathew, L.; Burney, M.; Gaikwad, A.; Nyshadham, P.; Nugent, E.K.; Gonzalez, A.; Smith, J.A. Preclinical Evaluation of Safety of Fucoidan Extracts from Undaria pinnatifida and Fucus vesiculosus for Use in Cancer Treatment. *Integr. Cancer Ther.* **2017**, *16*, 572–584. [CrossRef]
32. Burney, M.; Mathew, L.; Gaikwad, A.; Nugent, E.K.; Gonzalez, A.O.; Smith, J.A. Evaluation Fucoidan Extracts from Undaria pinnatifida and Fucus vesiculosus in Combination With Anticancer Drugs in Human Cancer Orthotopic Mouse Models. *Integr. Cancer Ther.* **2018**, *17*, 755–761. [CrossRef]
33. Zhang, Z.; Teruya, K.; Yoshida, T.; Eto, H.; Shirahata, S. Fucoidan extract enhances the anti-cancer activity of chemotherapeutic agents in MDA-MB-231 and MCF-7 breast cancer cells. *Mar. Drugs* **2013**, *11*, 81–98. [CrossRef]
34. Lee, E.F.; Harris, T.J.; Tran, S.; Evangelista, M.; Arulananda, S.; John, T.; Ramnac, C.; Hobbs, C.; Zhu, H.; Gunasingh, G.; et al. BCL-XL and MCL-1 are the key BCL-2 family proteins in melanoma cell survival. *Cell Death Dis.* **2019**, *10*, 342. [CrossRef] [PubMed]

35. Abudabbus, A.; Badmus, J.A.; Shalaweh, S.; Bauer, R.; Hiss, D. Effects of Fucoidan and Chemotherapeutic Agent Combinations on Malignant and Non-malignant Breast Cell Lines. *Curr. Pharm. Biotechnol.* **2017**, *18*, 748–757. [CrossRef]
36. Arumugam, P.; Arunkumar, K.; Sivakumar, L.; Murugan, M.; Murugan, K. Anticancer effect of fucoidan on cell proliferation, cell cycle progression, genetic damage and apoptotic cell death in HepG2 cancer cells. *Toxicol. Rep.* **2019**, *6*, 556–563.
37. da Silva, G.N.; de Castro Marcondes, J.P.; de Camargo, E.A.; da Silva Passos Junior, G.A.; Sakamoto-Hojo, E.T.; Salvadori, D.M.F. Cell cycle arrest and apoptosis in TP53 subtypes of bladder carcinoma cell lines treated with cisplatin and gemcitabine. *Exp. Biol. Med.* **2010**, *235*, 814. [CrossRef] [PubMed]
38. Montano, R.; Khan, N.; Hou, H.; Seigne, J.; Ernstoff, M.S.; Lewis, L.D.; Eastman, A. Cell cycle perturbation induced by gemcitabine in human tumor cells in cell culture, xenografts and bladder cancer patients: Implications for clinical trial designs combining gemcitabine with a Chk1 inhibitor. *Oncotarget* **2017**, *8*, 67754–67768. [CrossRef] [PubMed]
39. Park, H.Y.; Choi, I.W.; Kim, G.Y.; Kim, B.W.; Kim, W.J.; Choi, Y.H. Fucoidan induces G1 arrest of the cell cycle in EJ human bladder cancer cells through down-regulation of pRB phosphorylation. *Revista Brasileira De Farmacognosia* **2015**, *25*, 246–251. [CrossRef]
40. Han, Y.S.; Lee, J.H.; Lee, S.H. Antitumor Effects of Fucoidan on Human Colon Cancer Cells via Activation of Akt Signaling. *Biomol. Ther.* **2015**, *23*, 225–232. [CrossRef]

cancers

MDPI

Article

The α9 Nicotinic Acetylcholine Receptor Mediates Nicotine-Induced PD-L1 Expression and Regulates Melanoma Cell Proliferation and Migration

Hai Duong Nguyen [1,†], You-Cheng Liao [2], Yuan-Soon Ho [3,4,5,6,†], Li-Ching Chen [3,4,7], Hui-Wen Chang [5,†], Tzu-Chun Cheng [6], Donald Liu [8], Woan-Ruoh Lee [2,8,9], Shing-Chuan Shen [2,9,10], Chih-Hsiung Wu [11,12] and Shih-Hsin Tu [4,7,12,*]

1 International Master Program in Medicine, College of Medicine, Taipei Medical University, Taipei 110, Taiwan; m142106003@tmu.edu.tw

2 Graduate Institute of Medical Sciences, College of Medicine, Taipei Medical University, Taipei 110, Taiwan; d119108004@tmu.edu.tw (Y.-C.L.); cmbwrlee@tmu.edu.tw (W.-R.L.); scshen@tmu.edu.tw (S.-C.S.)

3 TMU Research Center of Cancer Translational Medicine, Taipei Medical University, Taipei 110, Taiwan; hoyuansn@tmu.edu.tw (Y.-S.H.); d117094003@tmu.edu.tw (L.-C.C.)

4 Taipei Cancer Center, Taipei Medical University, Taipei 110, Taiwan

5 Department of Medical Laboratory, Taipei Medical University Hospital, Taipei 110, Taiwan; g160090005@tmu.edu.tw

6 School of Medical Laboratory Science and Biotechnology, College of Medical Science and Technology, Taipei Medical University, Taipei 110, Taiwan; d119096007@tmu.edu.tw

7 Division of Breast Surgery, Department of Surgery, Taipei Medical University Hospital, Taipei 110, Taiwan

8 Department of Dermatology, Taipei Medical University Shuang Ho Hospital, New Taipei City 237, Taiwan; 16308@s.tmu.edu.tw

9 Department of Dermatology, School of Medicine, College of Medicine, Taipei Medical University, Taipei 110, Taiwan

10 International Master/PhD Program in Medicine, College of Medicine, Taipei Medical University, Taipei 101, Taiwan

11 Department of Surgery, EnChu Kong Hospital, New Taipei City 237, Taiwan; chwu@tmu.edu.tw

12 Department of Surgery, School of Medicine, College of Medicine, Taipei Medical University, Taipei 110, Taiwan

* Correspondence: drtu@h.tmu.edu.tw; Tel.: +886-2-2736-1661 (ext. 3420); Fax: +886-2-2739-3422

† These authors contributed equally to this work.

Received: 15 October 2019; Accepted: 2 December 2019; Published: 11 December 2019

Abstract: Cigarette smoking is associated with an increased risk of melanoma metastasis. Smokers show higher PD-L1 expression and better responses to PD-1/PD-L1 inhibitors than nonsmokers. Here, we investigate whether nicotine, a primary constituent of tobacco, induces PD-L1 expression and promotes melanoma cell proliferation and migration, which is mediated by the α9 nicotinic acetylcholine receptor (α9-nAChR). α9-nAChR overexpression in melanoma using melanoma cell lines, human melanoma tissues, and assessment of publicly available databases. α9-nAChR expression was significantly correlated with PD-L1 expression, clinical stage, lymph node status, and overall survival (OS). Overexpressing or knocking down α9-nAChR in melanoma cells up- or downregulated PD-L1 expression, respectively, and affected melanoma cell proliferation and migration. Nicotine-induced α9-nAChR activity promoted melanoma cell proliferation through stimulation of the α9-nAChR-mediated AKT and ERK signaling pathways. In addition, nicotine-induced α9-nAchR activity promoted melanoma cell migration via activation of epithelial-mesenchymal transition (EMT). Moreover, PD-L1 expression was upregulated in melanoma cells after nicotine treatment via the transcription factor STAT3 binding to the PD-L1 promoter. These results highlight that nicotine-induced α9-nAChR activity promotes melanoma cell proliferation, migration, and PD-L1 upregulation. This study may reveal important insights into the mechanisms underlying nicotine-induced melanoma growth and metastasis through α9-nAChR-mediated carcinogenic signals and PD-L1 expression.

Keywords: melanoma cells; nicotine; α9-nAChR; PD-L1; STAT3

1. Introduction

Melanoma is the most aggressive and lethal form of skin cancer, accounting for fewer than 5% of all malignant skin tumors, but 75% of all skin cancer deaths in the United States [1]. Melanomas are transformed, abnormal developments of melanocytes [2]. Unlike melanocytes, melanoma cells grow uncontrollably and create a thick mass with or without ulceration [3]. As melanomas become thicker and invade deeper, melanoma cells may spread into the dermal layer, nearby tissues and other parts of the body [4]. Melanoma is divided into various histopathological subtypes, including superficial spreading, nodular, lentigo maligna, acral lentiginous, desmoplastic, muscosal and amelanotic melanoma [5]. The different subtypes of melanoma can vary substantially in their molecular characterization and pathogenesis [5]. Melanoma is the most dangerous type of skin cancer because it is the type most likely to metastasize if not diagnosed in an early stage [6]. Programmed cell death protein 1 (PD-1) blockade is the first-line therapy for advanced melanoma patients [7,8]. The mechanism of melanoma tumorigenesis is not clear and may involve a combination of environment and genetic factors [9].

Nicotinic acetylcholine receptors (nAChRs) are common ligand-activated neurotransmitter receptors located throughout the body [10]. Several subtypes of nAChRs have been demonstrated to be closely correlated to the formation and progression of tumors [10]. α9-nAChR is one of the most recently discovered nAChR subtypes [11]. At first, α9-nAChR was found in the outer hair cells of the cochlea where it is involved in auditory functions [11,12]. Subsequently, many biological functions of α9-nAChR have been identified, including keratinocyte adhesion [13], inflammation [14], chronic pain [15,16], immune responses [17], endocrine activities [18], homeostasis of osteocytes and bone mass regulation [19], and breast epithelial cancer formation [20]. α9-nAChR expression levels in breast cancer patients are strongly correlated with staging and 5-year disease-specific survival rate, and highly expresses (7.84-fold higher than the level in normal tissue) in 186 (67.3%) of the 276 breast cancer paired samples [20]. α9-nAChR short interfering RNA in MDA-MB-231 cells inhibits cell proliferation in vitro and tumor growth in vivo [20]. The α9-nAChR expression is essential for mediating tumor metastasis through epithelial-mesenchymal transition (EMT) [21,22]. Expression of EMT-related gene α9-nAChR is increased in metastasis in triple-negative breast cancer [21,22]. Notably, Chun-Yu Lin et al. reported α9-nAChR subunit could potentially correlate with 38 cancer-related pathways in 15 cancer types and interact with 64 proteins to play important roles in biological functions related to human cancer development [23]. These studies suggest that α9-nAChR may initiate signal transduction pathways, thus inducing melanoma growth and metastasis during tumor progression.

Tobacco is a known cause of cancer mortality [24]. Tobacco smoke contains 7000 toxic and carcinogenic chemicals, and various classes, such as tobacco-specific nitrosamines, polycyclic aromatic hydrocarbon, and aldehydes, are capable of inducing DNA damage response that initiates turmorigenesis and enhances metastasis [25]. Exposure to tobacco smoke is an independent risk factor for cancer [25]. The risk and severity of cancer depend on the duration of exposure and the amount of tobacco smoking [26]. In melanoma, tobacco smoking supports metastasis of melanoma cells, leading to a significant decrease in 5-year disease-free and overall survival (OS) in smokers compared to nonsmokers [27]. The correlation between current smoking and nodal metastasis of primary cutaneous melanoma is direct and positive [28]. This correlation is independent of tumor thickness, ulceration, and other risk factors [28]. An epidemiological cohort study to assess the clinical significance of α9-nAChR expression in breast tumors of different stages and to correlate α9-nAChR expression with smoking history among Taiwanese women [20]. The results revealed the increased expression of α9-nAChR mRNA in tumor tissues from current smokers relative to those from passive smokers and nonsmokers (6.62-fold vs. 2.8-fold or 1.51-fold, respectively) [20]. Nicotine is an important and common component of tobacco responsible for addiction [29]. Many studies indicate that nicotine induces breast cancer growth and metastasis in vitro and in vivo through binding to and activating α9-nAChR [20,23,30,31]. Long-term exposure to extremely low doses of nicotine and 4-(methylnitrosamino)-1-(3-pyridyl)-1-butanone (NNK) induced nonmalignant breast epithelial cell transformation through activation of the α9-nAChR-mediated signaling pathways [32]. However, the mechanisms by which nicotine-induced α9-nAChR activity promotes melanoma growth and metastasis are not fully understood.

Programmed death-ligand 1 (PD-L1) is a type 1 transmembrane protein that is encoded by the CD274 gene in humans and is overexpressed in some kinds of cancers [33]. The overexpression of PD-L1 in tumors serves as an indicator of progression and a poor prognosis [34]. In contrast, PD-L1 expression is associated with an increased response to PD-1/PD-L1 inhibitor treatment [35]. PD-1 is an inhibitory receptor that is encoded by the PDCD1 gene and is located on the surface of all T cells [35]. The PD-1 receptor on activated T cells functions in the regulation of the immune system during various physiological responses, including autoimmune disease and cancer [36]. The binding of PD-1 to PD-L1 on the surface of cancer cells leads to an immunosuppressive effect based on the interaction between phosphatases (SHP-1 or SHP-2) and an immunoreceptor tyrosine-based switch motif (ITSM) [36]. PD-L1 not only has roles in the suppression of the immune system but also has distinct tumor-intrinsic roles in growth, metastasis, and resistance to therapy [37]. Tobacco smoking-related carcinogenesis affects cancer risk by increasing the somatic mutation load [38], thereby creating neoantigens which are strongly correlated with decreased progression-free survival in cancers [39]. In contrast, neoantigens are able to stimulate immune responses and pave the way for efficacious PD-1/PD-L1 immunotherapy [40]. Several recent studies have shown positive correlations between smoking and PD-L1 expression or PD-1/PD-L1 immune checkpoint inhibitor efficacy [41]. Some clinical trials have reported that the smoking status influences OS in anti-PD-1/PD-L1-treated lung cancer patients [42]. Tobacco-smoking patients show higher PD-L1 expression [43] and better treatment responses to anti-PD-1/PD-L1 immunotherapy than patients with lung cancer who have never smoked [41]. The overall response rate to anti-PD-1/PD-L1 antibodies is lower in patients who have never smoked than in former/current smokers [41]. Recently, Wang et al. found that tobacco smoke induced PD-L1 expression in lung epithelial cells and the main tobacco carcinogens benzo(a)pyrene (BaP), dibenz[a, h]anthracence (DbA), and benzo[g, h, I]perylene (BzP) upregulated PD-L1 expression [44]. The above results suggest the possibility that nicotine may induce a mechanism that drives expression of PD-L1 in melanoma cells through α9-nAChR.

In this study, we demonstrated that α9-nAChR expression was upregulated in melanoma and associated with PD-L1 expression at the mRNA and protein levels. Overexpression of α9-nAChR in melanoma cells upregulated PD-L1 expression and enhanced melanoma cell proliferation and migration. In contrast, knocking down α9-nAChR in melanoma cells reduced PD-L1 expression and inhibited melanoma cell proliferation and migration. Nicotine-induced α9-nAChR activity upregulated PD-L1 expression and promoted melanoma cell proliferation and migration. Our results indicate novel functions for α9-nAChR in melanoma cell proliferation, migration, and PD-L1 regulation.

2. Results

2.1. α9-nAChR Is Overexpressed in Melanoma

To evaluate the expression of nAChR subunits in three melanoma cell lines (A375, A2058, and MDA-MB 435) and a primary melanocyte cell line (HEMn-LP), we performed reverse transcriptase polymerase chain reaction (RT-PCR) to detect the mRNA levels of all alpha nAChR subunits. We found overexpression of α2-10 nAChR subunits in the A375, A2058, and MDA-MB 435 melanoma cells compared to the HEMn-LP melanocytes (Figure 1A). Quantification of the α1-10 nAChR mRNA levels in the A375, A2058, and MDA-MB 435 melanoma cells was performed based on intensities, and we found that α9-nAChR mRNA expression was more prominent than that of the other subunits of alpha nAChRs (* $p < 0.05$) (Figure 1B). α9-nAChR expression was detected in the three melanoma cell lines (A375, A2058 and MDA-MB 435) and primary melanocyte cell line (HEMn-LP) by RT-PCR (Figure 1A) and western blotting (Figure 1D and Figure S1).

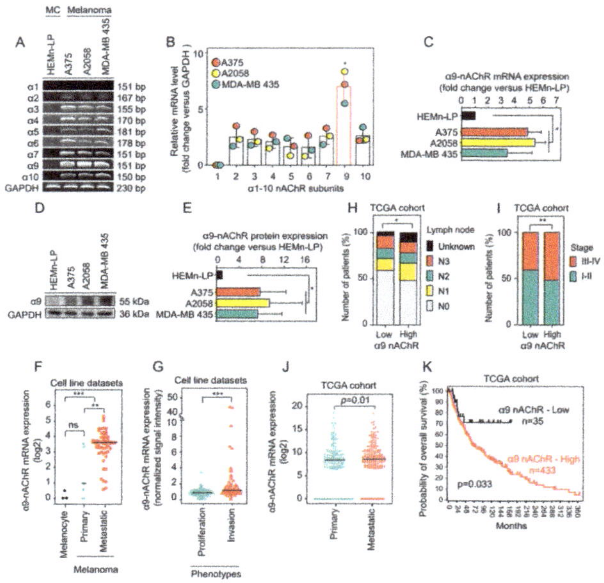

Figure 1. α9-nAChR expression levels and their correlations with clinicopathological parameters in multiple melanoma databases. (**A**) Detection of nAChR subunits in the primary epidermal melanocyte cell line HEMn-LP and the melanoma cell lines A375, A2058, and MDA-MB 435 by RT-PCR. (**B**) Relative mRNA expression of α1-10 nAChR subunits in the A375, A2058, and MDA-MB 435 melanoma cell lines. (**C**) Relative α9-nAChR mRNA expression in the HEMn-LP, A375, A2058, and MDA-MB 435 cell lines. (**D**,**E**) Determination of α9-nAChR mRNA levels using western blotting and statistical analysis of α9-nAChR protein levels. (**F**) The mRNA expression of α9-nAChR in two datasets from the public R2 MegaSampler platform (http://hgserver1.amc.nl/cgi-bin/r2/main.cgi) comprising melanocyte cell lines ($n = 3$) and primary ($n = 5$), and metastatic ($n = 58$) melanoma cell lines. (**G**) Screening of melanoma cell line datasets (http://www.jurno.ch/php/genehunter.html) for the mRNA expression of α9-nAChR. These cell lines were further subdivided into proliferative ($n = 101$) and invasive ($n = 90$) phenotypes. (**H**) α9-nAChR gene expression level in the TCGA-SKCM cohort ($n = 472$) downloaded from the UCSC Xena browser (https://xenabrowser.net/heatmap/). Melanoma patients were further divided into two groups based on the mean value of α9-nAChR mRNA expression, low α9-nAChR expression ($n = 169$) and high α9-nAChR expression ($n = 291$). Bar plots show the proportions of five subcategories of lymph node status in the high and low α9-nAChR level groups. (**I**) The frequencies of stages of I/II and III/IV in the high and low α9-nAChR level groups of the TCGA-SKCM cohort. (**J**) The differences in α9-nAChR expression between primary ($n = 211$) and metastatic ($n = 201$) groups. The result for the TCGA-SKCM cohort was processed using the UCSC Xena browser. (**K**) Kaplan–Meier analysis for melanoma patients based on the result from the public R2: Kaplan Meier Scanner software (https://hgserver1.amc.nl) showing a borderline difference between the groups with high (red, 433 samples) and low (black, 35 samples) α9-nAChR expression levels in the TCGA-SKCM cohort with the optimal cut-off value. (**C**,**E**) Results are shown as mean ± standard deviation (SD) of three individual experiments. *** $p < 0.001$, Student's t-test. (**F**,**G**,**J**) The data were analyzed by the Mann-Whitney test. The median of α9-nAChR expression in each group is shown by a horizontal line. *ns*, not significant; ** $p < 0.01$; *** $p < 0.001$. (**H**,**I**) The two groups' qualitative data were compared using the χ^2 test; * $p < 0.05$, ** $p < 0.01$.

Statistical analysis found that the α9-nAChR mRNA (Figure 1C) and protein levels (Figure 1E) were obviously elevated in the three melanoma cells compared to the HEMn-LP melanocytes (* $p < 0.05$).

Melanoma cell line datasets from the public R2 MegaSampler platform (http://hgserver1.amc.nl/cgi-bin/r2/main.cgi) were evaluated. We found that α9-nAChR mRNA expression in melanoma cell lines was significantly higher than that in melanocyte cell lines (*** $p < 0.001$) (Figure 1F). In addition, α9-nAChR mRNA expression in metastatic melanoma cell lines was higher than that in primary melanoma cell lines (** $p < 0.01$) (Figure 1F).

Melanoma cell lines stratified into either a proliferative or an invasive phenotype using the melanoma cell line datasets from HOPP Database (http://www.jurmo.ch/hopp/hopp_mpse.php) were defined by a specific gene expression pattern [45]. We analyzed α9-nAChR mRNA levels and found that they were significantly upregulated in the melanoma cells ($n = 176$) with the invasive phenotype ($n = 90$) compared to those with the proliferative phenotype ($n = 101$) (*** $p < 0.001$) (Figure 1G).

We examined α9-nAChR expression of human skin cutaneous melanoma (SKCM) using the data obtained from The Cancer Genome Atlas (TCGA) from the University of California Santa Cruz (UCSC) Xena browser (https://xenabrowser.net/). The samples were divided into primary and metastatic groups according to the TNM classification for malignant melanoma staging. We found that the metastatic group had higher α9-nAChR mRNA levels than the primary group (* $p = 0.01$) (Figure 1J). Moreover, Kaplan-Meier analysis based on the result from R2: Kaplan Meier Scanner software (https://hgserver1.amc.nl) to analyze the OS of TCGA-SKMC cohort stratified according to α9-nAChR mRNA expression with an optimal cut-off value. TCGA-SKCM cohort divided into high α9-nAChR mRNA expression (433 samples) and low α9-nAChR mRNA expression (35 samples) groups (Figure 1K). 0.9 is the ratio of high α9-nAChR mRNA expression samples (433 samples) to the total amount of samples (468 samples). The results showed that the patients with high α9-nAChR expression were shorter OS than patients with low α9-nAChR mRNA expression (* $p = 0.033$) (Figure 1K).

We evaluated the immunohistochemistry (IHC) staining of α9-nAChR in tissue microarray specimens using the IHC scoring system, and the scoring system was determined as no staining (0), weak (1+), moderate (2+), and strong (3+) based on α9-nAChR intensity (Figure 2A). Histological analysis of α9-nAChR expression in melanoma tissues (Mel, $n = 176$) and normal skin tissues (NS, $n = 16$) indicated that there were significantly upregulated in melanoma tissues compared with normal skin tissues (*** $p < 0.001$) (Figure 2C,D). Taken together, these results suggested that α9-nAChR is overexpressed in melanoma and correlates with melanoma stage and phenotype.

2.2. *Correlation of α9-nAChR Levels with Clinicopathological Features in Melanoma*

To investigate the relationships between α9-nAChR and clinicopathological parameters in melanoma, we downloaded α9-nAChR gene expression data for the TCGA-SKMC cohort. We divided the patients into two groups according to the mean level of α9-nAChR. The first group expressed low levels of α9-nAChR, and the second group expressed high levels of α9-nAChR. Table 1 summarizes the associations of α9-nAChR with melanoma clinicopathological parameters (age, sex, ulceration, Breslow depth, Clark level, tumor size, lymph node status, distant metastasis status, and stage of disease). High α9-nAChR mRNA expression was significantly associated with sex (* $p = 0.04$) (Table 1), and α9-nAChR mRNA expression was significantly higher in female groups than that in male group (* $p < 0.05$) (Figure S2); however, no strong relationships were found for age, ulceration, Breslow depth or Clark level.

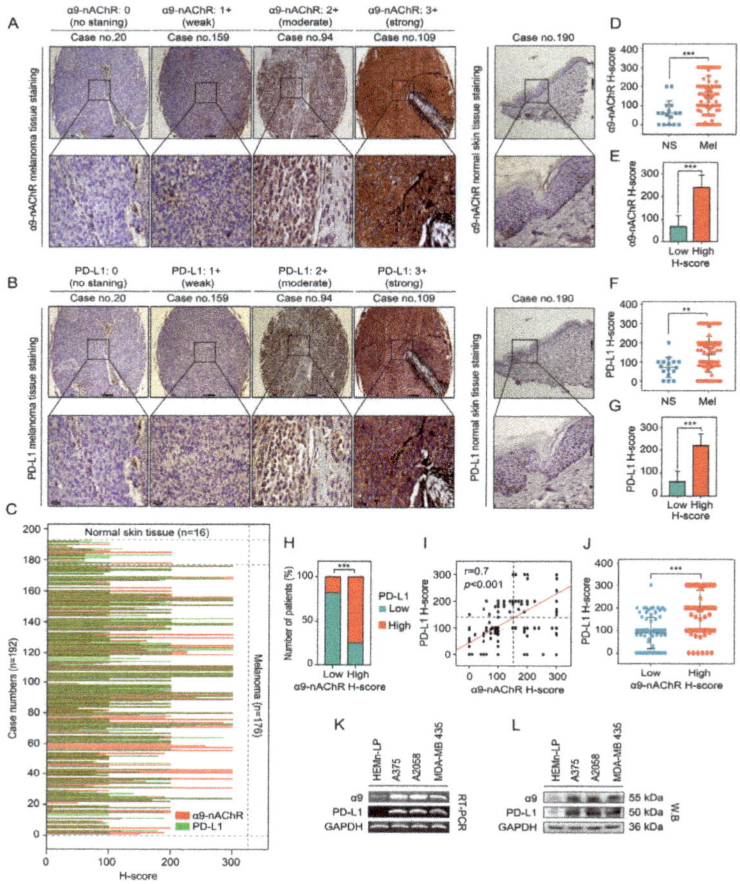

Figure 2. Correlations between α9-nAChR and PD-L1 in melanoma. (**A**,**B**) Representative images for α9-nAChR and PD-L1 immunohistochemistry (IHC) scoring system in the melanoma tissues and normal skin tissues. The scoring system was determined as no staining (0), weak (1+), moderate (2+), and strong (3+) based on α9-nAChR and PD-L1 intensity. (**C**) The α9-nAChR (red lines) and PD-L1 (green lines) expression profiles of melanoma tissues (n = 176) and normal skin tissues (n = 16) were detected by IHC staining based on H-score value. (**D**,**F**) Histological analysis of α9-nAChR and PD-L1 expression in melanoma tissues (Mel, n = 176) compared with normal skin tissues (NS, n = 16). (**E**,**G**) Histological analysis of α9-nAChR and PD-L1 expression in the tissue microarrays (n = 192) based on H-score value. The tissues were divided into two groups based on the mean value of α9-nAChR or PD-L1 H-score, low α9-nAChR (n = 97) or high α9-nAChR (n = 95) and low PD-L1 (n = 104) or high PD-L1 (n = 88). (**H**) The association between α9-nAChR and PD-L1 expression in the tissue microarrays (n = 192) as categorical variables. The χ^2 test was employed to assess the correlation between a9-nAChR and PD-L1 expression in samples. *** $p < 0.001$. (**I**) Correlation between α9-nAChR and PD-L1 expression in the melanoma tissue microarrays (n = 192). Pearson's rank correlation measured the strength of the association between α9-nAChR and PD-L1 expression. *** $p < 0.001$. (**J**) PD-L1 expression levels in melanoma tissues with high α9-nAChR (n = 89) or low α9-nAChR (n = 87) expression. The melanoma tissues were divided into low or high α9-nAChR based on the mean value of α9-nAChR. (**K**) α9-nAChR and PD-L1 mRNA levels detected by PT-PCR. (**L**) α9-nAChR and PD-L1 protein levels detected by western blotting. The data are presented as the mean ± SD, ** $p < 0.01$, *** $p < 0.001$, Student's t-test.

Table 1. Correlations between α9-nAChR mRNA expression and clinicopathological parameters of melanoma patients determined using the TCGA-SKCM cohort (n = 472).

Parameters		CHRNA9 Low mRNA n (%)	CHRNA9 High mRNA n (%)	*p*-Value
Age (year)	≥40	147 (32%)	257 (55%)	0.33
	<40	18 (4%)	42 (9%)	
Sex	Female	114 (24%)	179 (38%)	0.04 *
	Male	53 (11%)	126 (27%)	
Ulceration	Yes	61 (19%)	106 (34%)	0.08
	No	40 (13%)	107 (34%)	
Breslow depth (mm)	≥2	76 (21%)	145 (40%)	0.8
	<2	46 (13%)	93 (26%)	
Clark level	I-III	32 (10%)	69 (21%)	0.24
	IV-V	85 (26%)	136 (42%)	
Tumor size	T1	14 (3%)	27 (7%)	0.87
	T2	27 (7%)	52 (13%)	
	T3	32 (8%)	59 (14%)	
	T4	60 (15%)	93 (23%)	
	Unknown	15 (4%)	32 (8%)	
Lymph node status	N0	94 (21%)	141 (31%)	0.035 *
	N1	20 (4%)	54 (12%)	
	N2	17 (4%)	32 (7%)	
	N3	22 (5%)	34 (8%)	
	Unknown	6 (1%)	30 (7%)	
Distant metastasis status	M0	148 (33%)	270 (61%)	0.39
	M1	10 (2%)	15 (3%)	
Stage	I-II	94 (25%)	123 (30%)	0.009 **
	III-IV	64 (15%)	131 (32%)	

Clinicopathological parameters were assessed using the χ^2 test analysis. * $p < 0.05$, ** $p < 0.01$. We divided the patients into two groups according to the mean level of α9-nAChR mRNA expression. The first group expressed low levels of α9-nAChR, and the second group expressed high levels of α9-nAChR.

We next explored the correlations between α9-nAChR mRNA expression and tumor staging and observed that high α9-nAChR expression was significantly associated with lymph node metastasis (* $p = 0.035$) (Figure 1H and Table 1) and clinical stage (** $p = 0.009$) (Figure 1I and Table 1); however, no significant associations were found for tumor size or distant metastasis status in the TCGA-SKCM cohort. From these results, we suggest that α9-nAChR correlates with sex, lymph node metastasis status and clinical stage.

2.3. Correlation Between α9-nAChR and PD-L1 in Melanoma

We investigate whether the correlation between α9-nAChR and PD-L1 in melanoma. We analyzed the expression of α9-nAChR and PD-L1 in the melanoma tissues ($n = 176$) and normal skin tissues ($n = 16$) using the IHC staining (Figure 2A–C). The expression levels of α9-nAChR and PD-L1 were significantly increased in melanoma tissues compared with skin normal tissues (Figure 2D,F). The patient tissues were divided into two groups based on the mean value of the α9-nAChR H-score, low α9-nAChR H-score ($n = 97$) and high α9-nAChR H-score ($n = 95$) or PD-L1 H-score, low PD-L1 H-score ($n = 104$) and high PD-L1 H-score ($n = 88$). 14 cases of the normal skin tissues ($n = 14/16$) had a low α9-nAChR H-score (Figure 2C,E) and 15 cases of the normal skin tissues ($n = 15/16$) had a low PD-L1 H-score (Figure 2C,G). The chi-square (χ^2) test was applied to assess the correlation between α9-nAChR and PD-L1 mRNA expression in samples as categorical variables (Table S1). We found that α9-nAChR expression was significantly associated with PD-L1 expression in the tissue microarrays (*** $p < 0.001$) (Figure 2H). Subsequently, we identified a strong correlation between α9-nAChR and PD-L1 expression levels in the tissue microarray data (r = 0.7, *** $p < 0.001$) (Figure 2I).

To further evaluate the correlation between α9-nAChR and PD-L1 mRNA expression, we analyzed the melanoma cell line datasets ($n = 176$) and TCGA-SKCM cohort ($n = 472$). We found that the mRNA expression of α9-nAChR was significantly associated with PD-L1 expression in the melanoma cell line datasets (*** $p < 0.001$) (Figure S3A,B) and TCGA-SKCM cohort (** $p < 0.01$) (Figure S4A,B). Furthermore, we identified a moderate correlation melanoma cell line datasets ($r = 0.22$, ** $p = 0.003$) (Figure S3C) and weak correlation between α9-nAChR and PD-L1 mRNA levels in the TCGA-SKCM cohort ($r = 0.115$, * $p = 0.012$) (Figure S4C). Melanoma is classified into various histopathological subtypes, and distinct subtypes can vary substantially in their molecular characterization and pathogenesis, which TCGA-SKCM cohort has not mentioned. Therefore, the distinct subtypes of melanoma might have different correlations between α9-nAChR and PD-L1 mRNA levels led to a weak correlation in TCGA-SKCM cohort (r = 0.115, * $p = 0.012$) (Figure 2H).

To gain more insight about the correlation of α9-nAChR and PD-L1 expression, RT-PCR and western blotting were performed in HEMn-LP, A375, A2058, and MDA-MB 435 cells using α9-nAChR and PD-L1-specific primers and antibodies, respectively. GAPDH was used as a loading control. α9-nAChR and PD-L1 expression were prominently higher in the A375, A2058, and MDA-MB 435 melanoma cells than in the HEMn-LP melanocytes at the mRNA (Figure 2K) and protein levels (Figure 2L and Figure S7). α9-nAChR expression was coupled with PD-L1 expression in the HEMn-LP, A375, A2058, and MDA-MB 435 cells at the mRNA (Figure 2K) and protein levels (Figure 2L and Figure S7). These experiments were repeated three times per cell line. Band intensities were measured using ImageJ software, and the amounts of α9-nAChR and PD-L1 at the mRNA and protein levels in the A375, A2058, and MDA-MB 435 melanoma cells were normalized to the α9-nAChR and PD-L1 mRNA and protein levels in the HEMn-LP cells. We found strong correlations between α9-nAChR and PD-L1 at the mRNA ($r = 0.96$, *** $p < 0.001$) (Figure S5A) and protein levels ($r = 0.66$, * $p = 0.02$) (Figure S5B) in the HEMn-LP, A375, A2058, and MDA-MB 435 cells.

On the other hand, we compared PD-L1 expression levels between the low α9-nAChR (melanoma tissue microarrays, $n = 87$; TCGA-SKCM cohort, $n = 167$; melanoma cell line datasets, $n = 156$) and high α9-nAChR (melanoma tissue microarrays, $n = 89$; TCGA-SKCM cohort, $n = 305$; melanoma cell line datasets, $n = 20$) expression groups and found that PD-L1 expression levels were significantly higher in the samples with high α9-nAChR expression than in the samples with low α9-nAChR expression (** $p < 0.01$, *** $p < 0.001$) (Figure 2J, Figures S3D and S4D). According to the results, we suggest that α9-nAChR expression correlates with the PD-L1 expression in melanoma.

2.4. α9-nAChR Induces PD-L1 Expression and Regulates in Proliferation and Migration

The results for the melanoma databases, cell lines, and tissue microarrays indicated that α9-nAChR was highly associated with tumor growth, metastasis, and PD-L1 expression in melanoma patients and with cell proliferation, migration, and PD-L1 expression in melanoma cell lines. To evaluate the roles of α9-nAChR in the proliferation, migration and PD-L1 upregulation of melanoma cells, we generated stable α9-nAChR-siRNA-expressing A2058 cells, α9-nAChR-overexpressing A2058 cells and α9-nAChR-overexpressing MDA-MB 435 cells by transfection with α9-nAChR-small interfering RNA (siRNA) or pcDNA3.1-α9-nAChR overexpression vectors. Stable scrambled siRNA-A2058 cells, pcDNA3.1-A2058 cells, and pcDNA3.1-MDA-MB 435 cells were used as control groups. The protein levels of α9-nAChR expression in the cells with α9-nAChR knockdown or overexpression were compared to the protein levels in the control cells (Figures 3A and 4A, Figures S8 and S10). α9-nAChR overexpression significantly increased cell proliferation in the A2058 (Figure 3B) and MDA-MB 435 cells (Figure 3C) after five days incubation by a cell count assay (*** $p < 0.001$). These results were confirmed by a soft-agar colony assay at 21 days (*ns*, not significant; * $p < 0.05$; *** $p < 0.001$) (Figure 3D,E). In contrast, compared to stable scrambled siRNA-A2058 cells, stable α9-nAChR-siRNA-expressing A2058 cells substantially decreased the numbers of cells and colonies after 5 and 21 days, respectively (*** $p < 0.001$) (Figure 4C,D). Together, these data demonstrate that α9-nAChR regulates melanoma cell proliferation in vitro.

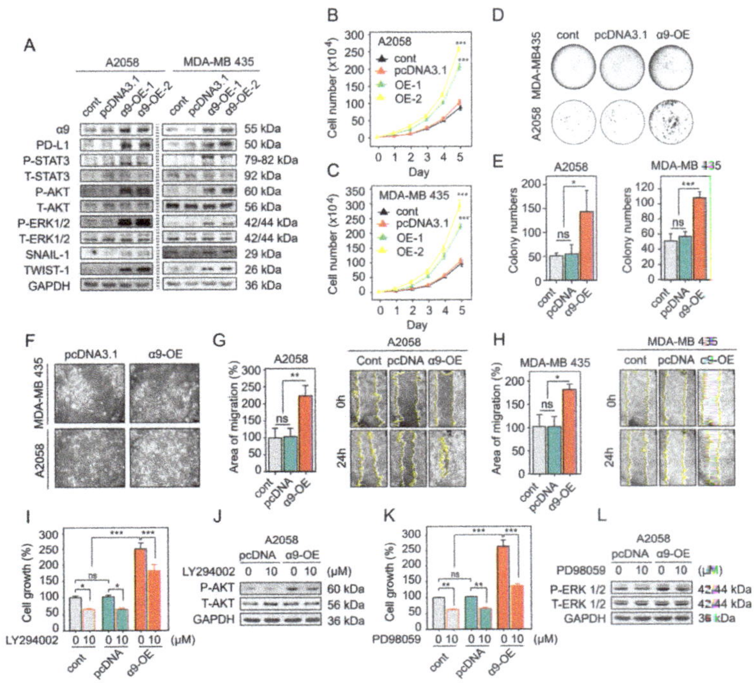

Figure 3. α9-nAChR overexpression induces PD-L1 upregulation and promotes melanoma cell proliferation and migration. (**A**) Overexpression of α9-nAChR activated the phosphorylation of AKT, ERK, and STAT3 and induced PD-L1, Snail-1, and Twist-1 protein levels in A2058 and MDA-MB-435 cells, which were evaluated by western blotting. (**B**,**C**) α9-nAChR overexpression significantly increased the proliferation of A2058 and MDA-MB435 cells compared to control cells, as determined by a cell count assay. (**D**) Soft-agar growth assays were performed with A2058 and MDA-MB-435 cells transfected with pcDNA3.1 or pcDNA3.1-α9-nAChR overexpression vectors. (**E**) Colony numbers were quantified and presented in the histogram. (**F**) Cell morphology of A2058 and MDA-MD-435 cells transfected with pcDNA3.1 or pcDNA3.1-α9-nAChR overexpression vector. (**G**,**H**) Representative micrographs and statistical analysis demonstrated that α9-nAChR overexpression increased the migration areas of A2058 and MDA-MB 435 cells at 24 h after scratching. (**I**,**K**) Percentages of cell proliferation for the A2058 cells, pcDNA3.1-A2058 cells, and α9-nAChR-overexpressing A2058 cells were treated with or without the PI3K inhibitor LY294002 (10 μm) or MEK inhibitor PD98059 (10 μm) for 48 h determined by an MTT assay. (**J**,**L**) Western blotting to detect the protein levels of P-AKT and P-ERK in pcDNA3.1-A2058 cells and α9-nAChR-overexpressing A2058 cells with or without LY294002 or PD98059 treatment. The data are presented as the mean ± SD of three independent experiments. * $p < 0.05$, ** $p < 0.01$, *** $p < 0.001$, Student's t-test.

We performed a scratch-wound healing assay to examine whether α9-nAChR affects the migration of melanoma cells. After scratching, cells along the wound-edge migrated into the empty space continuously so that the migration area, as indicated by comparing the distance between the opposite cell edges at 0 h and 24 h, became wider over time. We compared stable α9-nAChR-overexpressing A2058 cells to stable pcDNA3.1-A2058 cells and showed apparent increases in the migration areas at 24 h after scratching (*ns*, not significant; ** $p < 0.01$) (Figure 3G). For stable α9-nAChR-overexpressing MDA-MB 435 cells, the migration areas were also significantly larger than those of pcDNA3.1-MDA-MB 435 cells at 24 h after scratching (*ns*, not significant; * $p < 0.05$) (Figure 3H). In contrast, compared to corresponding stable scrambled siRNA-A2058 cells, stable α9-nAChR-siRNA-expressing A2058 cells exhibited substantially reduced migration areas at 24 h after scratching (Figure 4E, right panel).

Statistical analysis demonstrated that the migration areas of the stable α9-nAChR-siRNA-expressing A2058 cells were significantly narrower than those of the stable scrambled siRNA-A2058 cells at 24 h after scratching (* $p < 0.05$) (Figure 4E, left panel). Taken together, our results demonstrate that α9-nAChR positively regulates melanoma cell migration.

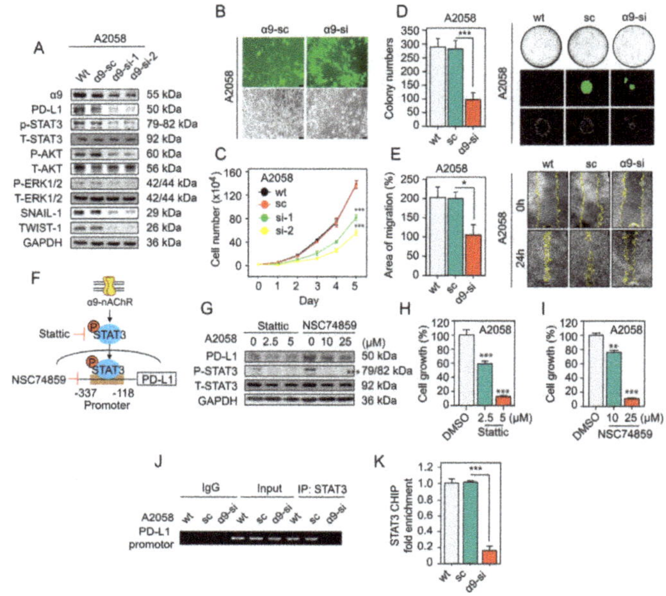

Figure 4. Knocking down α9-nAChR expression inhibits melanoma cell proliferation and migration and affects the protein level of PD-L1. (**A**) Western blotting results showing the protein level changes in α9-nAChR, PD-L1, P-STAT3, P-AKT, P-ERK, Snail-1 and Twist-1 after knocking down α9-nAChR expression in A2058 cells. GAPDH served as a control. (**B**) Fluorescence micrographs showing GFP expression (bottom panels) of α9-nAChR-siRNA-expressing or scrambled siRNA-A2058 cells at 48 h after pSUPER-α9-nAChR-si or pSUPER-α9-nAChR-scramble vector transfection, and corresponding phase-contrast microscopy images showing cell morphology (top panels). (**C**) Cell proliferation of α9-nAChR-siRNA-expressing A2058 cells monitored by a cell count assay. (**D**) Cell proliferation of α9-siRNA-expressing A2058 cells assessed by a soft-agar growth assay. The colonies formed in the low density seeding assay were counted after 21 days (left panels). The amounts, sizes, and fluorescence microscopy images of colonies are shown (right panels). (**E**) Migratory capacity measured at 24 h using a wound-healing assay with A2058 cells stably expressing α9-nAChR-siRNA (si), scrambled siRNA (sc) or wild-type (wt) (left panels). Quantification of the migratory areas (right panels) were measured using ImageJ software. (**F**) The STAT3 binding site of the PD-L1 promoter and targets of STAT3 phosphorylation inhibitor Stattic and STAT3 DNA-binding inhibitor NSC74859. (**G**) Effects of treatment with the inhibitors Stattic and NSC74858 on P-STAT3, and PD-L1 protein levels in A2058 cells evaluated by western blotting. GAPDH served as a control. (**H**,**I**) Inhibitors Stattic and NSC74859 inhibited A2058 cell proliferation. (**J**,**K**) ChIP assay using STAT3-precipitated DNA samples from wild-type, scrambled siRNA, and α9-nAChR-siRNA-expressing A2058 cells. The DNA fragments were subjected to RT-PCR and real-time PCR (qPCR) using PD-L1 promoter primer pairs as shown in Table S2. In RT-PCR, a rabbit IgG antibody was used as a negative control, and total genomic DNA was used as a positive control. In qPCR, relative fold enrichments of PD-L1 promoter region were calculated after normalization to GAPDH. Each of the immunoprecipitations was replicated three times, and each sample was quantified at least in triplicate. The data are presented as the mean ± SD of independent experiments. * $p < 0.05$, ** $p < 0.01$, *** $p < 0.001$, Student's t-test.

α9-nAChR expression correlated with PD-L1 expression in melanoma (Figure 2 and Figures S3–S5). To further understand the relationship between α9-nAChR and PD-L1 in melanoma cells, we performed western blotting to assess PD-L1 protein levels in stable α9-nAChR-knockdown and α9-nAChR-overexpressing cells. We found that the protein level of PD-L1 was significantly upregulated in stable α9-nAChR-overexpressing A2058 and MDA-MB 435 cells compared to the corresponding stable pcDNA3.1-expressing A2058 and MDA-MB 435 cells (Figure 3A and Figure S8). On the other hand, the protein level of PD-L1 was substantially downregulated in the stable α9-nAChR-siRNA-expressing A2058 cells compared to the stable scrambled siRNA-A2058 cells (Figure 4A and Figure S10). According to these results, we suggest that α9-nAChR induces melanoma cell proliferation, migration, and PD-L1 upregulation.

2.5. α9-nAChR Regulates Melanoma Cell Proliferation via the AKT and ERK Signaling Pathways

To investigate the mechanisms by which α9-nAChR regulates proliferation, western blot analysis was performed with melanoma cells to detect the protein levels of AKT and ERK, which are well-documented signaling molecules associated with tumor growth [2]. We found that α9-nAChR affected the phosphorylated forms of AKT and ERK1/2 in α9-nAChR-siRNA-expressing A2058 cells, α9-nAChR-overexpressing A2058 cells and α9-nAChR-overexpressing MDA-MB 435 cells (Figures 3A and 4A, Figures S8 and S10). As shown in Figure 3A and Figure S8, overexpression of α9-nAChR in A2058 cells and MDA-MB 435 cells substantially promoted the activation of the AKT and ERK signaling pathways. In contrast, α9-nAChR depletion in A2058 cells significantly decreased AKT and ERK activation (Figure 4A and Figure S10).

To elucidate the role of α9-nAChR in promoting melanoma cell proliferation via the AKT and ERK signaling pathways, we examined the proliferative ability of stable α9-nAChR-overexpressing A2058 cells treated with the PIK3 inhibitor LY 294002 (10 μm) or MEK inhibitor PD 98059 (10 μm) for 48 h and compared the proliferation of these cells to that of stable pcDNA3.1-A2058 cells treated with the inhibitors. Western blotting was performed to evaluate the inhibitory control of PIK3 inhibitor LY 294002 or MEK inhibitor PD 98059 on the phosphorylation of AKT and ERK (Figure 3J,L and Figure S9). We found that an inhibitory effect on the proliferation of the A2058 cells when α9-nAChR overexpression was combined with the PI3K or MEK inhibitor (*** $p < 0.001$) (Figure 3I,K). Together, these data suggest that α9-nAChR plays a role in the proliferation of melanoma cells through the activation of the AKT and ERK signaling pathways.

2.6. α9-nAChR Promotes Melanoma Cell Migration through EMT

The fact that α9-nAChR overexpression/knockdown greatly affected melanoma cell movement strongly suggested a major function for α9-nAChR in controlling melanoma cell migration. To elucidate the underlying mechanism, we observed morphological changes in cells with α9-nAChR overexpression or knockdown compared to cells transfected with control vectors. The results demonstrated that α9-nAChR overexpression in A2058 and MDA-MB 435 cells induced mesenchymal-like melanoma cells with loss of cell-cell adhesions junctions (Figure 3F). Conversely, α9-nAChR knockdown in A2058 cells resulted in increased cell-cell adhesions junctions and epithelial-like melanoma cells (Figure 4B). The molecular changes in the levels of EMT markers (Twist-1 and Snail-1) which are pivotal in controlling cancer cell migration via cell transformation [46], also indicated the occurrence of EMT in cells with α9-nAChR overexpression/knockdown. Moreover, it is known that MEK and PI3K cascades promote EMT by upregulating Twist-1 and Snail-1 expression [46]. The western blot analysis clearly showed that compared to stable scrambled siRNA-A2058 cells, stable α9-nAChR-siRNA-expressing A2058 cells inhibited the activation of AKT and EKR phosphorylation and significantly downregulated the expression of the EMT markers Twist-1 and Snail-1 (Figure 4A and Figure S10). In contrast, compared to pcDNA3.1-A2058 cells and pcDNA3.1-MDA-MB 435 cells, α9-nAChR overexpressing A2058 cells and α9-nAChR overexpressing MDA-MB 435 cells exhibited activated MEK and PI3K signaling pathways and substantially increased expression of EMT markers, including Twist-1 and

Snail-1, at the protein level (Figure 3A and Figure S8). Based on these results, we suggest that α9-nAChR regulates melanoma cell migration through EMT.

2.7. α9-nAChR Regulates PD-L1 Expression via the STAT3 Signaling Pathway

We investigate whether the STAT3 signaling pathway regulates PD-L1 expression. We treated A2058 cells with the STAT3 phosphorylation inhibitor Stattic (2.5 μm or 5 μm) and STAT3 DNA-binding inhibitor NSC74859 (10 μm or 25 μm) to inhibit STAT3 activation (Figure 4F). Western blotting and MTT assay were performed to evaluate the PD-L1 expression and cell growth, respectively. Stattic and NSC74859 significantly inhibited STAT3 activation, leading to the suppression of PD-L1 expression (Figure 4G and Figure S11) and cell growth (Figure 4H,I) in A2058 cells. From this evidence, we conclude that STAT3 signaling pathway regulates cell growth and PD-L1 expression in melanoma cells.

We examined the role of the α9-nAChR-mediated STAT3 signaling pathway in regulating PD-L1 expression. We found that α9-nAChR overexpression in A2058 and MDA-MB 435 cells substantially increased phosphorylated STAT3 expression (Figure 3A and Figure S8). In contrast, knocking down α9-nAChR in A2058 cells significantly decreased phosphorylated STAT3 expression (Figure 4A and Figure S10).

Several recent findings have mentioned that -337 to -118 fragment of the PD-L1 core promoter is enriched with consensus binding sites for STAT3, which is known to regulate PD-L1 expression transcriptionally (Figure 4F) [47,48]. To investigate the mechanism involving the transcription factor STAT3 in α9-nAChR-mediated PD-L1 expression, we performed a chromatin immunoprecipitation (ChIP) assay with an anti-STAT3 rabbit polyclonal antibody and primer pairs specific for the PD-L1 gene promoter to compare stable α9-nAChR-siRNA-expressing A2058 cells with stable scrambled siRNA-A2058 cells. The expression of PD-L1 was evaluated by RT-PCR and qPCR. We found that the transcription factor STAT3 directly bound to the promoter of PD-LP at -337 to -118 region (Figure 4J). Moreover, knocking down α9-nAChR expression decreased the PD-L1 level by reducing STAT3 binding to the PD-L1 promoter (*** $p < 0.001$) (Figure 4K). Thus, we suggest that PD-L1 expression is directly regulated by the STAT3 downstream signaling transduction pathway activated by α9-nAChR.

2.8. Nicotine-Induced α9-nAChR Activity Upregulates PD-L1 Expression and Promotes Melanoma Cell Proliferation and Migration

We examined the effects of nicotine-induced α9-nAChR activity on cell proliferation in the HEMn-LP melanocytes and A375, A2058, and MDA-MB 435 melanoma cells using nicotine at final concentrations of 0.001–1 μm to treat the cells for 24–72 h. As shown in Figure 5A,B and Figure S12, we found that nicotine-induced α9-nAChR activity significantly promoted cell proliferation in a concentration and time-dependent manner in the three melanoma cell lines (A375, A2058, and MDA-MB 435), with maximal proliferation occurring with the 1 μm dose (ns, not significant; * $p < 0.05$; ** $p < 0.01$; *** $p < 0.001$). However, we found no significant effect of nicotine-induced α9-nAChR activity on HEMn-LP melanocyte growth (ns, not significant) (Figure 5A, Figures S6 and S18). Compared with that in stable scrambled siRNA-A2058 cells, α9-nAChR-siRNA-expressing A2058 cells showed an inhibition of nicotine-stimulated proliferation (Figure 6K,L and Figure S17). These results suggest that nicotine-induced α9-nAChR activity promotes melanoma cell proliferation.

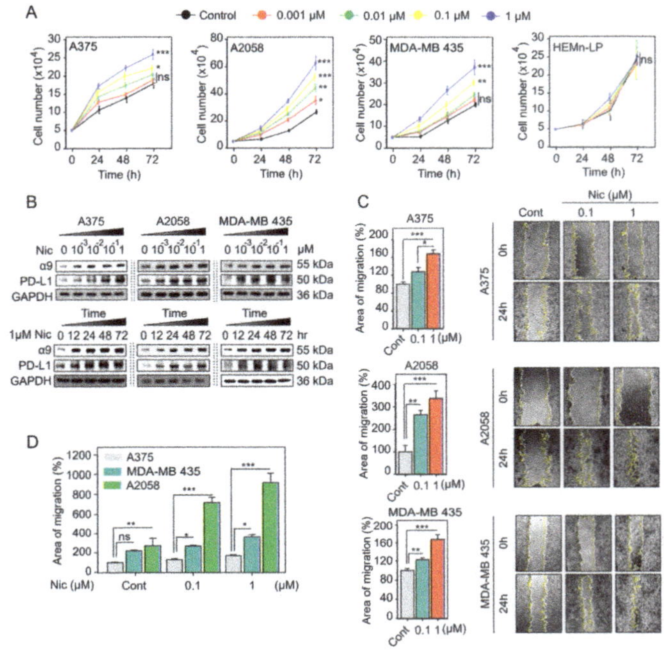

Figure 5. Nicotine-induced α9-nAChR activity promotes melanoma cell proliferation, migration, and PD-L1 upregulation. (**A**) Nicotine induced melanoma cell proliferation. Cell count assays were performed with the A375, A2058, and MDA-MB 435 melanoma cells and HEMn-LP melanocytes, and the cells were treated with nicotine at the final concentrations of 0.001–1 μm for 24–72 h. (**B**) Melanoma cells were exposed to nicotine at the indicated concentrations for indicated time periods. α9-nAChR and PD-L1 protein levels were determined by western blotting. (**C**) Nicotine promoted melanoma cell migration. Cell migration capabilities were evaluated using a wound-healing assay. Cells were scratched using 10 or 100 μL tips and treated with nicotine at the indicated concentrations for 24 h. (**D**) Comparing migration area among the A375, A2058, and MDA-MB 435 melanoma cells. The data are presented as the mean ± SD of three independent experiments. ns, not significant; * $p < 0.05$; ** $p < 0.01$; *** $p < 0.001$; Student's t-test.

To examine the effects of nicotine on migration in the A375, A2058, and MDA-MB 435 melanoma cells, we performed a scratch-wound healing assay. After scratching, the cells were treated with nicotine at a final concentration of 0.1 or 1 μm for 24 h. By comparing migration areas at 0 h and 24 h, we found that the migration areas were significantly larger in the nicotine treatment groups than in the groups without nicotine treatment after 24 h (* $p < 0.05$, ** $p < 0.01$, *** $p < 0.001$) (Figure 5C). When comparing migration area among the three cell lines, the migration areas of the A2058 cells were the largest (*ns*, not significant; * $p < 0.05$; ** $p < 0.01$; *** $p < 0.001$) (Figure 5D). Moreover, exposure of all three cell lines to nicotine induced increases in Twist-1 and Snail-1 protein levels (Figure 6B and Figure S14). These results suggest that exposure to nicotine induces EMT and promotes cell migration through α9-nAChR in the A375, A2058, and MDA-MB 435 melanoma cells. To investigate nicotine-induced α9-nAChR and PD-L1 upregulation, we performed western blotting to detect α9-nAChR and PD-L1 protein levels in the A375, A2058 and MDA-MB 435 melanoma cells treated with nicotine at final concentrations of 0.001–1 μm for 24 h or at a concentration of 1 μm for 12–72 h. We found that nicotine induced both α9-nAChR and PD-L1 upregulation in a concentration- and time-dependent manner (Figure 5B and Figure S12). However, in the HEMn-LP melanocytes, nicotine-induced α9-nAChR activity did not upregulate PD-L1 expression (Figures S6 and S18).

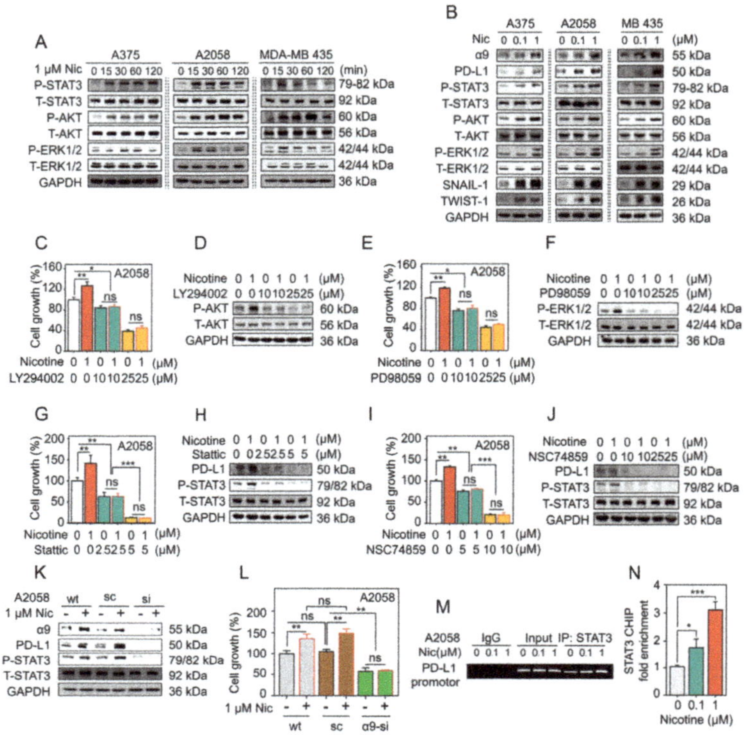

Figure 6. α9-nAChR mediates nicotine-induced PD-L1 expression and regulates melanoma cell proliferation. (**A**) Nicotine increased STAT3, AKT, and ERK phosphorylation in a time-dependent manner in the A375, A2058, and MDA-MB-435 cells, as assessed by western blotting. (**B**) Nicotine induced the phosphorylation of STAT3, AKT, and ERK and upregulated the protein levels of α9-nAChR, PD-L1, Snail-1 and Twist-1 in a dose-dependent manner. The results were determined by western blotting. (**C–J**) A2058 cells were pretreated with the inhibitor LY294002 (10 μm or 25 μm), PD98059 (10 μm or 25 μm), Stattic (2.5 μm or 5 μm) or NSC74859 (10 μm or 25 μm) for 3-24 h and subsequently treated with or without nicotine (1 μm) for additional 24-48 h. Western blot analysis was used to detect protein levels. Cell proliferation was assessed by an MTT assay. (**K**,**L**) Wild-type, scrambled siRNA, and α9-nAChR-si-expressing A2058 cells were treated with or without 1 μm nicotine. Western blot analysis was used to assess protein levels. Cell growth was assessed by an MTT assay. (**M**,**N**) After A2058 cells were exposed to 0.1 μm or 1 μm nicotine for 24 h, a ChIP assay was performed using an anti-STAT3 antibody. The protein-chromatin immunoprecipitates were subjected to RT-PCR and qPCR. In RT-PCR, a rabbit IgG antibody was used as a negative control, and total genomic DNA was used as a positive control. In qPCR, relative enrichments of PD-L1 promoter region were calculated after normalization to GAPDH. Each of the immunoprecipitations was replicated three times, and each sample was quantified at least in triplicate. The data are presented as the mean ± SD of three independent experiments. ns, not significant; * $p < 0.05$; ** $p < 0.01$; *** $p < 0.001$; Student's t-test.

Wild-type, scrambled siRNA, and α9-nAChR-siRNA-expressing A2058 cells were treated with or without 1 μm nicotine. After 24 h, we compared α9-nAChR-siRNA-expressing A2058 cells treated with 1 μm nicotine to those not treat with 1 μm nicotine. We found that knocking down α9-nAChR led to an inhibition of nicotine-induced PD-L1 protein levels through α9-nAChR-mediated STAT3 signaling pathway (Figure 6K and Figure S17). Together, these results suggest that α9-nAChR mediates nicotine-induced PD-L1 expression.

2.9. Nicotine-Induced α9-nAChR Activity Significantly Increases Melanoma Cell Proliferation via the AKT and ERK Signaling Pathways

To examine the mechanisms by which the AKT and ERK signaling pathways induced nicotine-stimulated α9-nAChR activity promote proliferation in the A375, A2058, and MDA-MB 435 melanoma cells, we treated cells with 1 μm nicotine at the indicated time points (Figure 6A and Figure S13). Nicotine increased AKT and ERK phosphorylation in a time-dependent manner in the A375, A2058 and MDA-MB 435 melanoma cells, as assessed by immunoblotting with anti–phospho-AKT and anti–phospho-ERK antibodies (Figure 6A and Figure S13). To evaluate concentration-dependent responses to nicotine, we treated cells with nicotine at final concentrations of 0.1–1 μm for 24 h. We found that nicotine increased the phosphorylation of AKT and ERK level in the three cell lines in a concentration-dependent manner (Figure 6B and Figure S14).

To examine the potential roles of PI3K and MEK inhibitors in the prevention of nicotine-induced melanoma cell proliferation, we pretreated A2058 cells with LY 294002 (10 μm and 25 μm) or PD 98059 (10 μm and 25 μm) for 3 h and then treated the cells with or without 1 μm nicotine for 24–48 h. As shown in Figure 6C–F and Figure S15, LY294002 and PD98059 significantly inhibited the activation of the AKT and ERK signaling pathways, leading to the inhibition of nicotine-induced melanoma cell proliferation. Moreover, knocking down α9-nAChR expression led to an inhibition of the nicotine-induced proliferation mediated by the AKT and ERK signaling pathways (Figure 6L). Together, these data indicate that nicotine-induced α9-nAChR activity significantly increases melanoma cell proliferation via the AKT and ERK signaling pathways.

2.10. Nicotine-Induced α9-nAChR Activity Upregulates PD-L1 Expression via STAT3 Signaling Pathway

The transcription factor STAT3 is required for regulating PD-L1 expression. To examine whether nicotine-induced α9-nAChR activity activated the transcription factor STAT3 in response to PD-L1 upregulation, we performed western blotting and ChIP assays. Nicotine induced STAT3 phosphorylation and upregulated PD-L1 protein levels in a time- and dose-dependent manner (Figures 5B and 6A,B, Figures S12–S14)). Pretreatment of A2058 cells with the STAT3 phosphorylation inhibitor Stattic (2.5 μm or 5 μm) or STAT3 DNA-binding inhibitor NSC74859 (10 μm or 25 μm) to block the activation of the STAT3 signaling pathway was performed for 24 h prior to a coincubation with or without 1 μm nicotine stimulation. As shown in Figure 6H,J and Figure S16, the inhibitors Stattic and NSC74859 significantly inhibited the stimulation of the nicotine-activated STAT3 signaling pathway that leads to PD-L1 upregulation. Moreover, the inhibitors Stattic and NSC74859 prevented the nicotine-induced melanoma cell proliferation (Figure 6G,I). Knocking down α9-nAChR expression led to an inhibition of the nicotine-induced PD-L1 protein level through the STAT3 signaling pathway (Figure 6K and Figure S17). We added 0.1 or 1 μm nicotine to A2058 cells for 24 h, and then performed a ChIP assay with an anti-STAT3 antibody and primer pairs specific for the PD-L1 gene promoter. The expression of PD-L1 was evaluated by RT-PCR and qPCR. We found that nicotine activated the transcription factor STAT3, which directly bound to the promoter of PD-L1 in -337 to -118 region (Figure 6M). Moreover, nicotine upregulated PD-L1 levels by increasing STAT3 binding to the PD-L1 promoter in a dose-dependent manner (* $p < 0.05$, *** $p < 0.001$) (Figure 6N). Together, these results indicate that nicotine-induced α9-nAChR activity activates the STAT3 signaling pathway that influences melanoma cell proliferation and upregulates PD-L1 expression, and the STAT3 inhibitors Stattic and NSC74859 prevent nicotine-induced PD-L1 expression and proliferation.

3. Discussion

Smoking is the leading cause of premature death and causes so many types of cancers [49]. Increasing evidence indicates that nicotine, which is one of the harshest substances in tobacco smoke, contributes to carcinogenesis in various cancer types, affects cancer progression and results in a poor prognosis in patients [30]. Nicotine binds to nAChRs and induces various downstream signaling pathways involved in cancer promotion, progression, and malignancy [30]. Recently, various epidemiological studies reported that tobacco smoking significantly increases metastasis in melanoma [27,28]. Moreover, a previous study indicated that α9-nAChR had important roles in promoting breast cancer growth and metastasis in vitro and in vivo [20]. In our results, we showed, for the first time, the associations of α9-nAChR expression with clinicopathological features of melanoma patients, including clinical staging, lymph node status and OS (Figure 1 and Table 1). α9-nAChR expression was significantly higher in melanoma cells than in melanocytes based on four cell lines (A375, A2058, and MDA-MB 435 melanoma cell lines and HEMN-LP melanocyte cell line) and melanoma cell line datasets (Figure 1). α9-nAChR mRNA level was higher in melanoma cells with the invasive phenotype than in those with the proliferative phenotype (Figure 1). In the TCGA-SKCM cohort, the metastatic patient group had higher α9-nAChR mRNA level than the primary patient group. In vitro, overexpression of α9-nAChR via transfection substantially increased melanoma cell proliferation and migration (Figure 3). By contrast, the suppression of α9-nAChR expression significantly inhibited melanoma cell growth and migration (Figure 4). Taken together, these findings suggest that α9-nAChR expression is not only relevant to melanoma growth and metastasis but also an unfavorable prognostic factor in melanoma patients.

α9-nAChR mRNA expression significantly associated with sex (* $p = 0.04$) (Table 1), and female had a higher α9-nAChR mRNA expression as compared to male (* $p < 0.05$) (Figure S2). 17β-estradiol upregulated α9-nAChR expression via the phosphorylation of estrogen receptor-alpha (ERα)-mediated the binding of AP1 and transcription factors to the α9-nAChR gene promoter, and –induced α9-nAChR protein production in breast cancer cells [31]. This study may provide us the clues to explain the mechanism of female sex hormone-induced α9-nAChR expression.

Numerous findings indicate that PD-L1 is present in tumor cells [50]. Tumor-expressed PD-L1 binds to PD-1-expressed on T cells, leading to the alterations of tumor immunopathogenesis which helps tumor cells evade the immune response [50]. Furthermore, the roles for tumor-intrinsic-PD-L1 in the regulation of EMT, cancer stemness, tumor development, metastasis, and resistance to therapy have recently been discovered [37]. PD-L1 overexpression, knockdown or knockout in glioblastoma multiforme was investigated, and the results demonstrated that PD-L1 promoted cancer cell growth and metastasis in vitro and in vivo [51]. In our study, the expression of PD-L1 is overexpressed in melanoma cells and tissues compared with melanocytes and normal skin tissues, respectively (Figure 2); and the positive correlation between α9-nAChR and PD-L1 expression by using melanoma cell lines, tissue microarrays, and public databases (Figure 2 and Figures S3–S5). In particular, α9-nAChR knockdown or overexpression induced changes in PD-L1 expression (Figures 3 and 4). We suggest that α9-nAChR promotes melanoma cell proliferation and migration through the regulation of PD-L1.

The transcription factor STAT3 plays a key role in the regulation of PD-L1 expression in various cancer types [52]. In non-Hodgkin lymphoma, STAT3 binds to the PD-L1 gene promoter, resulting in upregulation of PD-L1 expression in vitro and in vivo [53]. STAT3 suppression decreases PD-L1 expression in ALK-positive anaplastic large cell lymphoma (ALCL) [54] and KRAS-mutant non-small cell lung cancer (NSCLC) cells [55]. In our study, STAT3 was activated and regulated by α9-nAChR (Figures 3 and 4). Moreover, we found that STAT3 bound to the PD-L1 promoter in melanoma cells, and knocking down α9-nAChR expression inhibited STAT3 binding to the PD-L1 promoter, leading to PD-L1 downregulation (Figure 4). In contrast, the stimulation of α9-nAChR expression with nicotine increases STAT3 binding to PD-L1 promoter, leading to PD-L1 upregulation (Figure 6).

α9-nAChR synergizes mainly with epidermal growth factor (EGF) receptor, insulin-like growth factor 1 (IGF-1) receptor, vascular endothelial growth factor (VEGF) receptor in lung cancer cells [56], and estrogen receptor α (ERα) in breast cancer cells [31]. Moreover, EGFR has been shown to regulate the expression of PD-L1 through the activation of the IL6/JAK/STAT3 pathway in NSCLC [57]. 17β-Estradiol-induced PD-L1 expression through ERα-mediated AKT signaling [58]. We suggest that not only is STAT3 activated through α9-nAChR, but the interaction of α9-nAChR with other receptors also regulates PD-L1 expression.

PD-L1 expression can be inducible or constitutive and may change over time in response to different stimuli such as EGF, cytokines, interferon-gamma (IFNγ), 17β-estradiol, BaP, DbA, and BzP [44,58,59]. In our study, we found that nicotine remarkably increased PD-L1 protein expression in a dose- and time-dependent manner through α9-nAChR (Figure 5), suggesting that nicotine-induced α9-nAChR activity can regulate anti-cancer immunity. In addition, nicotine-induced α9-nAChR stimulated cancer cell proliferation and migration to drive cancer development and progression (Figure 5). Our study suggested that a new role for nicotine in suppressing anticancer immunity through α9-nAChR-induced PD-L1 expression in melanoma.

The ERK, AKT and STAT3 signaling pathways have been previously described to regulate melanoma cell proliferation [2,60]. In this study, we found MEK, AKT and STAT3 inhibitors interfered with melanoma cell proliferation and attenuated nicotine-induced melanoma cell proliferation (Figures 3, 4 and 6). α9-nAChR regulated melanoma cell proliferation (Figures 3 and 4). Stimulation of α9-nAChR with nicotine or overexpression of α9-nAChR promoted the activation of the AKT, ERK and STAT3 signaling pathways (Figures 3, 5 and 6). In contrast, knocking down α9-nAChR expression inhibited the AKT, ERK and STAT3 signaling pathways (Figures 4 and 6). These observations, along with the rest of the results in this study, suggest an important role for the α9-nAChR-mediated AKT, ERK and STAT3 signaling pathways in regulating melanoma cell proliferation.

In a previous study, Chernyavsky et al. found that stimulation of α9-nAChR initiated cell migration in epidermal keratinocytes through activation of the phosphorylation of intercellular junction proteins (desmoglein 3 and β-catenin), and focal adhesion proteins (paxillin and focal adhesion kinase (FAK)), thereby separating cell-cell adhesion complexes; in contrast, knocking down α9-nAChR expression inhibited the phosphorylation of adhesion and cytoskeletal proteins [61]. Moreover, α9-nAChR regulated the shape and adhesion of keratinocytes through the activation of several kinases, including EGFR, Scr, PLC, and PKC, and GTPases (Rho and Rac) [61]. Recently, Chun et al. found that α9-nAChR in breast cancer cells interacted with EGFR and EEBB2, and these genes were coexpressed with numerous downregulated genes, such as nectin-3 (NECTIN3) in the adherens junction pathway [23]. In our present study, we also investigated the participation of α9-nAChR in the regulation of melanoma cell migration (Figures 3 and 4), and stimulation of α9-nAChR by nicotine significantly increased melanoma cell migration (Figure 5). Furthermore, we found that the protein levels of the EMT markers Twist-1 and Snail-1 and morphological changes were affected by the overexpression or suppression of α9-nAChR (Figures 3 and 4). These observations suggest that the α9-nAChR network is significantly associated with migration in melanoma cells.

Snail and Twist-1 transcription factors contribute to repression of the epithelial phenotype and activation of the mesenchymal phenotype [62]. Snail-1 and Twist-1 EMT markers are not enough for proving EMT program, but their expression is activated early in EMT, and they thus have central roles in metastatic cancer [62]. In many previous studied, researchers mentioned that AKT and EKR signaling pathways promote EMT markers, including Snail-1 and Twist-1 [62] and cigarette smoking-induced AKT and ERK signaling pathways promotes the EMT program in cancers [63]. Nicotine/ α9-nAChR activates AKT, ERK pathways and EMT markers in melanoma cells (Figures 3, 4 and 6). We suggest that nicotine/ α9-nAChR may induce the EMT program through AKT or ERK pathways, or both.

In this study, the expression of α9-nAChR was significantly associated with metastasis and OS. α9-nAChR had roles in inducing PD-L1 expression and regulating melanoma cell proliferation and migration. Notably, nicotine-induced α9-nAChR activity promoted melanoma cell proliferation through stimulation of the α9-nAChR-mediated AKT, ERK and STAT3 signaling pathways. In addition, nicotine-induced α9-nAchR activity promoted melanoma cell migration by activating the EMT pathway. PD-L1 expression was upregulated in melanoma cells after nicotine treatment via α9-nAchR-mediated activation of the binding of the transcription factor STAT3 to the PD-L1 promoter.

4. Materials and Methods

4.1. Cell Culture

The human primary neonatal foreskin melanocyte cell line HEMn-LP from a lightly pigmented donor was cultured at 37 °C and 5% CO2 in Medium 254 supplemented with human melanocyte growth supplement (HMGS). Melanocyte cell line, medium, and supplement listed above were purchased from Invitrogen (Carlsbad, CA, USA). The human melanoma cell line MDA-MB 435 was purchased from the American Tissue Cell Culture Collection (ATCC, Manassas, VA, USA), and the melanoma cell lines A2058 and A375 were obtained from Dr. Shing-Chuan Shen (Taipei Medical University, Taipei, Taiwan). All melanoma cell lines were maintained at 37 °C and 5% CO2 in Dulbecco's modified Eagle's medium (DMEM, Gibco, Grand Island, NY, USA) supplemented with 10% fetal bovine serum (FBS, Gibco) and 1% penicillin-streptomycin-neomycin (PSN) antibiotic mixture (Gibco).

4.2. Bioinformatics Analysis

4.2.1. Melanoma Cell Line Dataset Analysis

Cell line datasets from the public R2 MegaSampler platform (https://hgserver1.amc.nl/cgi-bin/r2/main.cgi) were used to screen α9-nAChR mRNA expression in melanocyte, primary melanoma, and metastatic melanoma groups. We compared the α9-nAChR mRNA expression among the groups in the cell line datasets. Data from six different datasets in the melanoma cell phenotype-specific gene expression database (HOPP Database) (http://www.jurmo.ch/php/ genehunter.html) were used to assess α9-nAChR mRNA expression and PD-L1 expression in melanoma cell lines [45]. The melanoma cell lines from these datasets were divided into proliferative and invasive phenotypes. We conducted a statistical comparison of α9-nAChR mRNA expression values between the two phenotype groups using the nonparametric Mann-Whitney U-test. In addition, α9-nAChR and PD-L1 expression was used to divide the melanoma cell lines into high and low expression groups based on the mean level of α9-nAChR or PD-L1 expression. The χ^2 test was used to analyze the correlation between α9-nAChR (low or high) and PD-L1 (low or high) expression. Pearson and Spearman's rank correlation analysis was used to evaluate the strength of the association between α9-nAChR and PD-L1 expression. A p-value <0.05 was considered statistically significant.

4.2.2. TCGA-SKCM Cohort Analysis

We downloaded gene expression data from the cancer genomic browser of UCSC (https://xenabrowser.net/). We selected gene expression of α9-nAChR and PD-L1, and then clinicopathological parameters (sex, age, ulceration, Breslow depth, Clerk level, tumor size, lymph node status, distant metastasis status, stage) were selected from the phenotype search option. All patient data were downloaded. Furthermore, OS time and events were also obtained. α9-nAChR and PD-L1 mRNA expression were used to divide the patients into high and low expression groups according to the mean level of α9-nAChR or PD-L1 mRNA expression. The χ^2 test was used to analyze the correlations between α9-nAChR (low or high) mRNA expression, PD-L1 (low or high) mRNA expression, and clinicopathological parameters. Pearson and Spearman's rank correlation analysis was used to evaluate the strength of the association between α9-nAChR and PD-L1 mRNA expression. A p-value <0.05 was considered statistically significant.

4.3. Kaplan-Meier Analysis

Kaplan-Meier analysis based on the result from publicly available R2: Kaplan Meier Scanner software (https://hgserver1.amc.nl), selected the TCGA-SKCM cohort, then selected the optimal cut-off value to separate α9-nAChR mRNA expression into high or low group, and finally selected OS. A p-value <0.05 was considered statistically significant.

4.4. Nicotine Treatment

Cells were seeded in 60 mm dishes or 6-well plates at 5×10^5 cells per dish or 5×10^4 cells per well, respectively. Nicotine was purchased from Sigma-Aldrich (St. Louis, MO, USA, Catalog #54-11-5). The final concentrations were 0.001, 0.001, 0.01, 0.1 and 1 μm/mL, and cells were treated for the indicated time periods. The cells were harvested and analyzed by a cell count assay and western blotting at the indicated time points.

4.5. Inhibitor Treatment

Cells were seeded at 5×10^5 cells per dish in 60 mm dishes or 5×10^3 cells per well in 96-well plates. These cells were pretreated with 10–25 μm LY 294002 (#1130), 10–25 μm PD 98059 (#1231), 2.5–5 μm Stattic (#2798) or 5–10 μm NSC74859 (#4655) for 3–24 h and then cultured in the presence or absence of nicotine (1 μm) for 24–48 h. All inhibitors were purchased from TOCRIS Bioscience (Minneapolis, MN, USA). The cells were harvested to assess cell proliferation and viability as well as protein expression, which was analyzed by western blotting.

4.6. Cell Proliferation and Viability Assays

4.6.1. Cell Counting Assay

Cell numbers were determined by a cell counting assay. Cells were seeded in a 6-well plate at a density of 5×10^4 cells per well and then incubated in medium for the indicated times. The cells were collected and suspended in a growth medium. After staining with trypan blue, the cell number was counted with a hemocytometer. This assay was repeated four times with duplicate samples.

4.6.2. MTT Assay

Cell growth and viability were determined using a 3-(4,5-dimethylthiazol-2-yl)-2,5-diphenyltetrazolium (MTT) assay. Cells were seeded in triplicate in 96-well plates at a density of 5000 cells per well. Cells were treated with a compound at the indicated concentration for the indicated time periods. The medium was changed every two to three days. The optical density at 570 nm was measured using the SynergyTM HT Multi-Detection Microplate Reader (BioTek Instruments, Winooski, VT, USA). The assay was repeated three times with duplicate samples.

4.6.3. Soft-Agar Growth Assay

Anchorage-independent cell growth was analyzed using a soft-agar colony formation assay. Cells were seeded in a 0.4% top agarose II (Amresco, Solon, OH, USA) solution in growth medium over a base layer of 0.9% agarose II in 6-well plates at a density of 1×10^4 cells/mL. Soft-agar cultures were maintained at 37 °C for an additional 21 days and observed for the appearance of the colonies using the Leica DMI 4000 B Microscope Imaging System (Leica Microsystems, Wetzlar, Germany). The assay was repeated four times with duplicate samples. Images were acquired using a BioSpectrumTM 500 Imaging system, and colonies were counted under a light microscope.

4.7. Wound healing Scratch Assay

Cell migration was evaluated by a scratch-wound healing assay. In total, 5×10^5 cells per well were seeded in 12-well plates. After the cells grew to approximately 90% confluence, the cell monolayers were wounded by manually scraping the monolayers with a 10 μL or 100 μL sterile pipette tips. Then cells were washed three times with 1× PBS and fresh medium was added. Migrated cells were observed and captured by using the Leica DMI 4000 B Microscope Imaging System. This assay was performed at 0 h and 24 h after wounding. Wound closure was quantified as a percentage (versus the control set to 100%) that indicated the relative distance between the two scratch edges (scratch wound) using ImageJ software.

4.8. Construction of Vectors

4.8.1. α9-nAChR siRNA

The sequences of the α9-nAChR-scramble and α9-nAChR-siRNA primers are shown in Table S2. α9-nAChR-siRNA was constructed using two independent siRNAs. The scrambled siRNA sequence was used as a control. After BLAST analysis to verify the absence of significant sequence homology with other genes, the selected sequences were inserted into the *p*SUPER vector and digested with bglII and HindIII to generate the *p*SUPER-α9-nAChR-siRNA and *p*SUPER- α9-nAChR-scramble vectors.

4.8.2. α9-nAChR Overexpression

The sequences of the α9-specific overexpression primers are shown in Table S2. After BLAST analysis to verify the absence of significant sequence homology with other human genes, the selected sequences were inserted into the pcDNA3.1 vector (Invitrogen) and digested with bglII and HindIII to generate pcDNA3.1-α9-nAChR overexpression vectors. The identity of the construct was confirmed by DNA sequence analysis.

4.9. Generation of Stable Cell Lines

For stable transfections, 5×10^6 cells were washed twice with 1× PBS and mixed with 10 μg of vectors (*p*SUPER-α9-nAChR-siRNA, *p*SUPER-α9-nAChR-scramble, pcDNA3.1 or pcDNA3.1-α9 nAChR overexpression vector). The mixtures were transfected using the pipette-type microporator MP-100 electroporation (Invitrogen). After 48 h, the selection medium supplemented with G418 concentration of 5 mg/mL (A2058 cells) or 2 mg/mL (MDA-MB 435 cells) was added to the transfected cells. Every 2–3 days, the selection medium was replaced with new selection medium. After 30 days, the clones were isolated and evaluated for knockdown or overexpression at protein levels, and these clones were compared with scramble clones by western blotting.

4.10. RNA Isolation and RT-PCR Analysis

Total RNA was isolated from human cell lines using TRIzol (Invitrogen) according to the manufacturer's protocol. The specific primers were synthesized as shown in Table S2. The PCR amplicons were analyzed on 1.2% agarose gels (Amresco) stained with ethidium bromide. Images were acquired by an INFINITY-a digital imaging system (Vilber Lourmet, France), and the intensities of the bands were quantified using ImageJ software. The relative mRNA levels were normalized against the GAPDH mRNA levels.

4.11. Protein Extraction, Western Blotting, and Antibody

To determine protein expression, we harvested cellular proteins for western blotting as previously described [20]. Cells were washed three times with ice-cold 1× PBS and lysed on ice in a cell lysis buffer (50 mM Tris-HCl, pH 8.0; 120 mM NaCl2; 0.5% NP-40; 100 mM sodium fluoride; and 200 M sodium orthovanadate) supplemented with protease inhibitors. Proteins (50 μg per lane) were separated by gel electrophoresis on 10% SDS-PAGE gels, transferred to nitrocellulose membrane by electroblotting. The membranes were probed with appropriate primary antibodies against α9-nAChR (Thermo Fisher Scientific, Rockford, IL, USA, #PA5-46826), PD-L1 (Genetex, Irvine, CA, USA, #GTX104763; Thermo Fisher Scientific, #14-5983-82; Santa Cruz Biotechnology, Santa Cruz, CA, USA, #sc-293425), total-AKT (Genetex, #GTX121937), phospho-AKT (Cell Signaling Technology, Beverly, MA, USA, #9271), total-ERK1/2 (Genetex, #GTX113094), phospho-ERK1/2 (Santa Cruz Biotechnology, #sc-7383), total-STAT3 (Santa Cruz Biotechnology, #sc-8059), phospho-STAT3 (Cell Signaling Technology, #9145), Snail-1 (Cell signaling Technology, #6615), Twist-1 (Santa Cruz Biotechnology, #sc-81417) and GAPDH (Genetex, #GTX627408) followed by incubation with an HRP-conjugated secondary antibody.

Band intensity was quantified using ImageJ software. The relative protein levels were normalized against the GAPDH protein levels.

4.12. ChIP Assay

ChIP assays of the cultured cells were performed as described previously [31]. In brief, cells were cross-linked with a final concentration of 1% formaldehyde at 25 °C for 15 min and quenched with 0.125 M glycine for 5 min. Chromatin was sheared by sonication to average lengths of 500 bp. Chromatin was immunoprecipitated using 10 μg of anti-STAT3 monoclonal antibody (mAb) and protein A agarose beads at 4 °C overnight. DNA was recovered from the protein A complex using an elution buffer (10 mmol/L Tris, 10 mmol/L EDTA, 1% SDS, 150 mmol/L NaCl, and 5 mmol/L DTT, pH 8.0) and purified using a Macherey-Nagel PCR Clean-up kit. The DNA fragments were subjected to RT-PCR and qPCR using PD-L1 promoter primer pairs as shown in Supplemental Table S2. Nonimmunoprecipitated lysates (input) and immunoprecipitates obtained with an anti-rabbit IgG (Santa Cruz Biotechnology, #sc-2007) served as positive and negative controls, respectively. Relative enrichments of PD-L1 promoter region were calculated after normalization to GAPDH. Each of the immunoprecipitations was replicated three times, and each sample was quantified at least in triplicate.

4.13. IHC Staining and Scoring System

4.13.1. IHC Staining

Tissue microarrays used in the present study were purchased from US Biomax, Inc. (Rockville, MD, USA). The tissue microarrays have 192 specimens, including 112 specimens of malignant melanoma, 64 specimens of metastatic malignant melanoma, eight specimens of adjacent normal skin tissue, and eight specimens of skin normal tissue. Each core of the tissue microarrays represents a specimen. US Biomax, Inc. also provided patient information concerning age, gender, clinical stage, TNM, and pathological diagnosis. Standard immunohistochemical procedures were implemented as follows: Paraffin sections were baked at 60 °C for 1h, de-paraffinized in xylene, rehydrated through graded ethanol, quenched for endogenous peroxidase activity in 3% hydrogen peroxide for 10 min, and processed for antigen retrieval by high pressure cooking in citrate antigen retrieval solution (pH = 6.0) for about 30 min. Sections was blocked with normal serum at room temperature for 30 min. Then, the samples on the slides were incubated at 37 °C for 2 h with rabbit polyclonal antibodies against α9-nAchR (1:100, Thermo Fisher Scientific, #PA5-77511) and mouse monoclonal antibodies against PD-L1 (1:100, Santa Cruz Biotechnology, #sc-293425) in a moist chamber. The slides were subsequently incubated with goat anti-Mouse/Rabbit IgG VisUCyte HRP Polymer Antibody (VC002, R&D Systems, Minneapolis, MN, USA) for 1 h at room temperature. Immunostaining was performed using the Liquid DAB-Substrate Chromogen System (K3468, Dako, Glostrup, Denmark), which resulted in a brown-colored precipitate at the antigen site. Subsequently, sections were counterstained with hematoxylin and the procedure of dehydration was implemented. Finally, covers slips were applied in a non-aqueous mounting medium.

4.13.2. Immunohistochemistry Score

The staining scores were evaluated independently by two investigators, who were blinded to the information on the microarrays. For IHC analysis, the results were analyzed using Image-Pro Plus 6.0 software. The expression of each protein expression was evaluated by intensity scoring on scale of "0" to "3+" (0: no staining, +: weak staining, 2+: moderate staining, 3+: high staining) and proportion (0%, 1–24%, 25–50%, >50%). The final immunoreactive score was determined by multiplying the intensity score by the proportion of positive cells.

4.14. Statistical Analysis

Data were analyzed using SPSS version 22.0 (SPSS Inc., Chicago, IL, USA) and SigmaPlot graphing software (San Jose, CA, USA). Continuous variables are presented as the mean ± standard deviation and were analyzed by two-tailed unpaired Student's t-test for two groups. Nonmormally distributed data are expressed as the median and were assessed by the nonparametric Mann-Whitney U-test. The χ^2 test analysis was used to study the associations between α9-nAChR (low or high) and PD-L1 (low or high) expression or different clinicopathological features. Pearson and Spearman's rank correlation analysis was used for the association between α9-nAChR and PD-L1 expression levels. Kaplan-Meier analysis was used to plot survival curves, which were compared by the log-rank test. A p-value <0.05 (two-sided) was considered statistically significant.

5. Conclusions

This study may reveal important insights into the mechanisms underlying nicotine-induced melanoma cell proliferation and migration mediated through α9-nAChR-initiated carcinogenic signaling and PD-L1 expression. α9-nAChR could serve as a therapeutic target to decrease the growth and metastasis in melanoma (Figure 7).

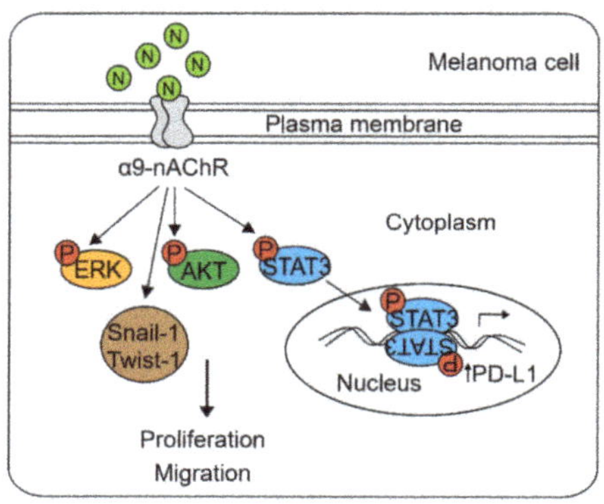

Figure 7. A schematic molecular model of nicotine-induced α9-nAChR activity in melanoma cells. α9-nAChR mediates nicotine-induced PD-L1 expression and regulates melanoma cell proliferation and migration. Nicotine-induced α9-nAChR activity upregulates Snail-1 and Twist-1 expression and activates the phosphorylation of AKT, ERK, and STAT3.

Supplementary Materials: The following material is available online at http://www.mdpi.com/2072-6694/11/12/1991/s1, Table S1: The chi-square test was employed to assess the correlation between a9-nChAR and PD-L1 expression in the tissue microarrays (n = 192) as categorical variables, Table S2: RT-PCR primers, Figure S1: Uncropped scans with size marker indications of western blots shown in Figure 1D, Figure S2: α9-nAChR mRNA level in female (n = 179) and male (n = 293) groups, Figure S3: Correlation between α9-nAChR and PD-L1 mRNA expression in the melanoma cell line datasets, Figure S4: Correlation between α9-nAChR and PD-L1 mRNA expression in the TCGA-SKCM cohort, Figure S5: Correlations between α9-nAChR and PD-L1 in melanoma cell lines, Figure S6: Nicotine-induced α9-nAChR activity did not influence PD-L1 protein levels in the HEMn-LP melanocytes, and densitometry readings/intensity ratios, Figure S7: Uncropped scans with size marker indications of western blots shown in Figure 2L, and densitometry readings/intensity ratios, Figure S8: Uncropped scans with size marker indications of western blots shown in Figure 3A, and densitometry readings/intensity ratios, Figure S9: Uncropped scans with size marker indications of western blots shown in Figure 3J,L, and densitometry readings/intensity ratios, Figure S10: Uncropped scans with size marker indications of western blots shown in Figure 4A, and densitometry readings/intensity ratios, Figure S11: Uncropped scans with size marker indications of western blots shown in Figure 4G, and densitometry readings/intensity ratios, Figure S12: Uncropped scans with size marker indications of western blots shown in Figure 5B, and densitometry readings/intensity ratios, Figure S13: Uncropped scans with size marker indications of western blots shown in Figure 6A, and densitometry readings/intensity ratios, Figure S14: Uncropped scans with size marker indications of western blots shown in Figure 6B, and densitometry readings/intensity ratios, Figure S15: Uncropped scans with size marker indications of western blots shown in Figure 6D,F, and densitometry readings/intensity ratios, Figure S16: Uncropped scans with size marker indications of western blots shown in Figure 6H,J, and densitometry readings/intensity ratios, Figure S17: Uncropped scans with size marker indications of western blots shown in Figure 6K, and densitometry readings/intensity ratios, Figure S18: Uncropped scans with size marker indications of western blots shown in Figure S6, and densitometry readings/intensity ratios.

Author Contributions: Conceptualization, H.D.N.; Data curation, H.D.N.; Formal analysis, H.D.N. and Y.-S.H.; Investigation, H.D.N., Y.-S.H., H.-W.C. and S.-H.T.; Methodology, H.D.N., Y.-C.L., L.-C.C. and T.-C.C.; Resources, Y.-S.H., H.-W.C., L.-C.C., D.L., W.-R.L., S.-C.S., C.-H.W. and S.-H.T.; Software, H.D.N; Validation, H.D.N.; Visualization, H.D.N.; Writing-original draft, H.D.N.; Writing-review & editing, H.D.N., Y.-S.H. and S.-H.T; Supervision, Y.-S.H.; Project administration, Y.-S.H., H.-W.C. and S.-H.T.; Funding acquisition, Y.-S.H., C.-H.W. and S.-H.T.

Funding: This study was supported by the Health and Welfare surcharge of tobacco products grant (MOHW108-TDU-B-212-124014), by the "TMU Research Center of Cancer Translational Medicine" from The Featured Areas Research Center Program within the framework of the Higher Education Sprout Project by the Ministry of Education (MOE) in Taiwan. These grants support the animal model experiments and resource and helping for clinical related information interpretation. The Ministry of Science and Technology, Taiwan (MOST108-2320-B-038-033-MY3 and MOST108-2320-B-038-002 awarded to Dr. Ho, MOST 106-2320-B-038-061-MY3 awarded to Dr. Chen, MOST106-2314-B-038-053-MY3 awarded to Dr. Tu and MOST104-2314-B-038-059-MY3 and NSC 101-2314-B-038-014-MY3 awarded to Dr. Wu.). This grant supports the human resource and most of the consumable materials.

Acknowledgments: We thank Shing-Chuan Shen, Donald Liu, and Woan-Ruoh Lee for the International Clinical Training Program (ICTP) in the Department of Dermatology, Taipei Medical University Shuang Ho Hospital; we thank Shing-Chuan Shen for the cell lines; we thank You-Cheng Liao, Li-Ching Chen and Tzu-Chun Cheng for administrative and technical support; we thank Yuan-Soon Ho for revising the manuscript. This study was supported by research grants of the Health and Welfare Surcharge on Tobacco Products, the "TMU Research Center of Cancer Translational Medicine" from The Featured Areas Research Center Program within the framework of the Higher Education Sprout Project by the Ministry of Education, and the Ministry of Science and Technology, Taiwan.

Conflicts of Interest: The authors declared that they have no conflicts of interest.

References

1. Gershenwald, J.E.; Guy, G.P. Stemming the Rising Incidence of Melanoma: Calling Prevention to Action. *J. Natl. Cancer Inst.* **2016**, *108*, djv381. [CrossRef] [PubMed]
2. Domingues, B.; Lopes, J.; Soares, P.; Populo, H. Melanoma treatment in review. *ImmunoTargets Ther.* **2018**, *7*, 35–49. [CrossRef] [PubMed]
3. Dummer, R.; Hauschild, A.; Lindenblatt, N.; Pentheroudakis, G.; Keilholz, U.; on behalf of the ESMO Guidelines Committee. Cutaneous melanoma: ESMO clinical practice guidelines for diagnosis, treatment and follow-up. *Ann. Oncol.* **2015**, *26*, 126–132. [CrossRef] [PubMed]
4. Kaufman, H.L.; Margolin, K.; Sullivan, R. Management of Metastatic Melanoma in 2018. *JAMA Oncol.* **2018**, *4*, 857–858. [CrossRef] [PubMed]
5. Wolff, K.; Goldsmith, L.A.; Katz, S.I.; Gilchrest, B.A.; Paller, A.S.; Leffell, D.J. Melanoma. Fitzpatrick's dermatology in general medicine, 2 volumes. *Transplantation* **2008**, *85*, 654.
6. Rosko, A.J.; Vankoevering, K.K.; McLean, S.A.; Johnson, T.M.; Moyer, J.S. Contemporary management of early-stage melanoma: A systematic review. *JAMA Facial Plast. Surg.* **2017**, *19*, 232–238. [CrossRef] [PubMed]
7. Swetter, S.M.; Tsao, H.; Bichakjian, C.K.; Curiel-Lewandrowski, C.; Elder, D.E.; Gershenwald, J.E.; Guild, V.; Grant-Kels, J.M.; Halpern, A.C.; Johnson, T.M.; et al. Guidelines of care for the management of primary cutaneous melanoma. *J. Am. Acad. Dermatol.* **2019**, *80*, 208–250. [CrossRef]
8. Martins, F.; Sofiya, L.; Sykiotis, G.P.; Lamine, F.; Maillard, M.; Fraga, M.; Shabafrouz, K.; Ribi, C.; Cairoli, A.; Guex-Crosier, Y.; et al. Adverse effects of immune-checkpoint inhibitors: epidemiology, management and surveillance. *Nat. Rev. Clin. Oncol.* **2019**, *16*, 563–580. [CrossRef]
9. Bressac-de-Paillerets, B.; Avril, M.-F.; Chompret, A.; Demenais, F. Genetic and environmental factors in cutaneous malignant melanoma. *Biochimie* **2002**, *84*, 67–74. [CrossRef]
10. Wu, C.H.; Lee, C.H.; Ho, Y.S. Nicotinic acetylcholine receptor-based blockade: Applications of molecular targets for cancer therapy. *Clin. Cancer Res.* **2011**, *17*, 3533–3541. [CrossRef]
11. Elgoyhen, A.B.; Johnson, D.S.; Boulter, J.; Vetter, D.E.; Heinemann, S. α9: An acetylcholine receptor with novel pharmacological properties expressed in rat cochlear hair cells. *Cell* **1994**, *79*, 705–715. [CrossRef]
12. Glowatzki, E.; Fuchs, P.A. Cholinergic synaptic inhibition of inner hair cells in the neonatal mammalian cochlea. *Science* **2000**, *288*, 2366–2368. [CrossRef] [PubMed]

13. Nguyen, V.T.; Ndoye, A.; Grando, S.A. Novel human alpha9 acetylcholine receptor regulating keratinocyte adhesion is targeted by Pemphigus vulgaris autoimmunity. *Am. J. Pathol.* **2000**, *157*, 1377–1391. [CrossRef]
14. Liu, Q.; Whiteaker, P.; Morley, B.J.; Shi, F.-D.; Lukas, R.J. Distinctive Roles for α7*- and α9*-Nicotinic Acetylcholine Receptors in Inflammatory and Autoimmune Responses in the Murine Experimental Autoimmune Encephalomyelitis Model of Multiple Sclerosis. *Front. Cell. Neurosci.* **2017**, *11*, 287–300. [CrossRef] [PubMed]
15. Romero, H.K.; Christensen, S.B.; Di Cesare Mannelli, L.; Gajewiak, J.; Ramachandra, R.; Elmslie, K.S.; Vetter, D.E.; Ghelardini, C.; Iadonato, S.P.; Mercado, J.L.; et al. Inhibition of α9α10 nicotinic acetylcholine receptors prevents chemotherapy-induced neuropathic pain. *Proc. Natl. Acad. Sci. USA* **2017**, *114*, 1825–1832. [CrossRef] [PubMed]
16. Bufalo, A.; Cesario, A.; Salinaro, G.; Fini, M.; Russo, P. Alpha9Alpha10 Nicotinic Acetylcholine Receptors as Target for the Treatment of Chronic Pain. *Curr. Pharm. Des.* **2014**, *20*, 6042–6047. [CrossRef] [PubMed]
17. Luebke, A.E.; Makoni, P.D.; Guth, S.M.; Lysakowski, A. Alpha-9 nicotinic acetylcholine receptor immunoreactivity in the rodent vestibular labyrinth. *J. Comp. Neurol.* **2005**, *492*, 323–333. [CrossRef]
18. Colomer, C.; Olivos-Oré, L.A.; Vincent, A.; McIntosh, J.M.; Artalejo, A.R.; Guérineau, N.C. Functional characterization of α9-containing cholinergic nicotinic receptors in the rat adrenal medulla: Implication in stress-induced functional plasticity. *J. Neurosci.* **2010**, *30*, 6732–6742. [CrossRef]
19. Baumann, L.; Kauschke, V.; Vikman, A.; Dürselen, L.; Krasteva-Christ, G.; Kampschulte, M.; Heiss, C.; Yee, K.T.; Vetter, D.E.; Lips, K.S. Deletion of nicotinic acetylcholine receptor alpha9 in mice resulted in altered bone structure. *Bone* **2019**, *120*, 285–296. [CrossRef]
20. Lee, C.H.; Huang, C.S.; Chen, C.S.; Tu, S.H.; Wang, Y.J.; Chang, Y.J.; Tam, K.W.; Wei, P.L.; Cheng, T.C.; Chu, J.S.; et al. Overexpression and activation of the α9-nicotinic receptor during tumorigenesis in human breast epithelial cells. *J. Natl. Cancer Inst.* **2010**, *102*, 1322–1335. [CrossRef]
21. Huang, L.C.; Lin, C.L.; Qiu, J.Z.; Lin, C.Y.; Hsu, K.W.; Tam, K.W.; Lee, J.Y.; Yang, J.M.; Lee, C.H. Nicotinic acetylcholine receptor subtype alpha-9 mediates triple-negative breast cancers based on a spontaneous pulmonary metastasis mouse model. *Front. Cell. Neurosci.* **2017**, *11*, 336–350. [CrossRef] [PubMed]
22. Lee, K.L.; Kuo, Y.C.; Ho, Y.S.; Huang, Y.H. Triple-negative breast cancer: Current understanding and future therapeutic breakthrough targeting cancer stemness. *Cancers* **2019**, *11*, 1334. [CrossRef] [PubMed]
23. Lin, C.-Y.; Lee, C.-H.; Chuang, Y.-H.; Lee, J.-Y.; Chiu, Y.-Y.; Wu Lee, Y.-H.; Jong, Y.-J.; Hwang, J.-K.; Huang, S.-H.; Chen, L.-C.; et al. Membrane protein-regulated networks across human cancers. *Nat. Commun.* **2019**, *10*, 3131–3148. [CrossRef] [PubMed]
24. Balogh, E.P.; Dresler, C.; Fleury, M.E.; Gritz, E.R.; Kean, T.J.; Myers, M.L.; Nass, S.J.; Nevidjon, B.; Toll, B.A.; Warren, G.W. Reducing tobacco-related cancer incidence and mortality: Summary of an institute of medicine workshop. *Oncologist* **2014**, *19*, 21–31. [CrossRef] [PubMed]
25. Rodgman, A.; Perfetti, T.A. The chemical components of tobacco and tobacco smoke. In *The Chemical Components of Tobacco and Tobacco Smoke*; CRC Press: Boca Raton, FL, USA, 2016; pp. 1–2332.
26. Schane, R.E.; Ling, P.M.; Glantz, S.A. Health effects of light and intermittent smoking: A review. *Circulation* **2010**, *121*, 1518–1522. [CrossRef]
27. Shaw, H.M.; Milton, G.W. Smoking and the development of metastases from malignant melanoma. *Int. J. Cancer* **1981**, *28*, 153–156. [CrossRef]
28. Jones, M.S.; Jones, P.C.; Stern, S.L.; Elashoff, D.; Hoon, D.S.B.; Thompson, J.; Mozzillo, N.; Nieweg, O.E.; Noyes, D.; Hoekstra, H.J.; et al. The Impact of Smoking on Sentinel Node Metastasis of Primary Cutaneous Melanoma. *Ann. Surg. Oncol.* **2017**, *24*, 2089–2094. [CrossRef]
29. Harris, C.C. Tobacco smoking, E-cigarettes, and nicotine harm. *Proc. Natl. Acad. Sci. USA* **2018**, *115*, 1406–1407. [CrossRef]
30. Grando, S.A. Connections of nicotine to cancer. *Nat. Rev. Cancer* **2014**, *14*, 419–429. [CrossRef]
31. Lee, C.-H.; Chang, Y.-C.; Chen, C.-S.; Tu, S.-H.; Wang, Y.-J.; Chen, L.-C.; Chang, Y.-J.; Wei, P.-L.; Chang, H.-W.; Chang, C.-H.; et al. Crosstalk between nicotine and estrogen-induced estrogen receptor activation induces α9-nicotinic acetylcholine receptor expression in human breast cancer cells. *Breast Cancer Res. Treat.* **2011**, *129*, 331–345. [CrossRef]

32. Fararjeh, A.-F.S.; Tu, S.-H.; Chen, L.-C.; Cheng, T.-C.; Liu, Y.-R.; Chang, H.-L.; Chang, H.-W.; Huang, C.-C.; Wang, H.-C.R.; Hwang-Verslues, W.W.; et al. Long-term exposure to extremely low-dose of nicotine and 4-(methylnitrosamino)-1-(3-pyridyl)-1-butanone (NNK) induce non-malignant breast epithelial cell transformation through activation of the a9-nicotinic acetylcholine receptor-mediated signaling pathway. *Environ. Toxicol.* **2019**, *34*, 73–82. [PubMed]
33. Salmaninejad, A.; Valilou, S.F.; Shabgah, A.G.; Aslani, S.; Alimardani, M.; Pasdar, A.; Sahebkar, A. PD-1/PD-L1 pathway: Basic biology and role in cancer immunotherapy. *J. Cell. Physiol.* **2019**, *234*, 16824–16837. [CrossRef] [PubMed]
34. Wang, X.; Teng, F.; Kong, L.; Yu, J. PD-L1 expression in human cancers and its association with clinical outcomes. *Onco. Targets. Ther.* **2016**, *9*, 5023–5039. [PubMed]
35. Sharpe, A.H.; Lafleur, M.W.; Muroyama, Y.; Drake, C.G. Therapy Inhibitors of the PD-1 Pathway in Tumor. *J. Immunol.* **2018**, *200*, 375–383.
36. Arasanz, H.; Gato-Cañas, M.; Zuazo, M.; Ibañez-Vea, M.; Breckpot, K.; Kochan, G.; Escors, D. PD1 signal transduction pathways in T cells. *Oncotarget* **2017**, *8*, 51936–51945. [CrossRef]
37. Dong, P.; Xiong, Y.; Yue, J.; Hanley, S.J.B.; Watari, H. Tumor-intrinsic PD-L1 signaling in cancer initiation, development and treatment: Beyond immune evasion. *Front. Oncol.* **2018**, *8*, 386–394. [CrossRef]
38. Alexandrov, L.B.; Ju, Y.S.; Haase, K.; Van Loo, P.; Martincorena, I.; Nik-Zainal, S.; Totoki, Y.; Fujimoto, A.; Nakagawa, H.; Shibata, T.; et al. Mutational signatures associated with tobacco smoking in human cancer. *Science* **2016**, *354*, 618–622. [CrossRef]
39. Lu, Y.; Robbins, P.F. Seminars in Immunology Cancer immunotherapy targeting neoantigens. *Semin. Immunol.* **2016**, *28*, 22–27. [CrossRef]
40. Yi, M.; Qin, S.; Zhao, W.; Yu, S.; Chu, Q.; Wu, K. The role of neoantigen in immune checkpoint blockade therapy. *Exp. Hematol. Oncol.* **2018**, *7*, 28–39. [CrossRef]
41. Li, B.; Huang, X.; Fu, L. Impact of smoking on efficacy of PD-1/PD-l1 inhibitors in non-small cell lung cancer patients: A meta-analysis. *Onco. Targets. Ther.* **2018**, *11*, 3691–3696. [CrossRef]
42. Cao, L.; Wang, X.; Li, S.; Zhi, Q.; Wang, Y.; Wang, L.; Li, K.; Jiang, R. PD-L1 is a prognostic biomarker in resected NSCLC patients with moderate/high smoking history and elevated serum SCCA level. *J. Cancer* **2017**, *8*, 3251–3260. [CrossRef]
43. Calles, A.; Liao, X.; Sholl, L.M.; Rodig, S.J.; Freeman, G.J.; Butaney, M.; Lydon, C.; Dahlberg, S.E.; Hodi, F.S.; Oxnard, G.R.; et al. Expression of PD-1 and Its Ligands, PD-L1 and PD-L2, in Smokers and Never Smokers with KRAS-Mutant Lung Cancer. *J. Thorac. Oncol.* **2015**, *10*, 1726–1735. [CrossRef] [PubMed]
44. Wang, G.Z.; Zhang, L.; Zhao, X.C.; Gao, S.H.; Qu, L.W.; Yu, H.; Fang, W.F.; Zhou, Y.C.; Liang, F.; Zhang, C.; et al. The Aryl hydrocarbon receptor mediates tobacco-induced PD-L1 expression and is associated with response to immunotherapy. *Nat. Commun.* **2019**, *10*, 1125–1138. [CrossRef] [PubMed]
45. Widmer, D.S.; Cheng, P.F.; Eichhoff, O.M.; Belloni, B.C.; Zipser, M.C.; Schlegel, N.C.; Javelaud, D.; Mauviel, A.; Dummer, R.; Hoek, K.S. Systematic classification of melanoma cells by phenotype-specific gene expression mapping. *Pigment Cell Melanoma Res.* **2012**, *25*, 343–353. [CrossRef] [PubMed]
46. Lamouille, S.; Xu, J.; Derynck, R. Molecular mechanisms of epithelial-mesenchymal transition. *Nat. Rev. Mol. Cell Biol.* **2014**, *15*, 178–196. [CrossRef] [PubMed]
47. Shen, X.; Zhang, L.; Li, J.; Li, Y.; Wang, Y.; Xu, Z.X. Recent Findings in the Regulation of Programmed Death Ligand 1 Expression. *Front. Immunol.* **2019**, *10*, 1337–1362. [CrossRef]
48. Wang, W.B.; Yen, M.L.; Liu, K.J.; Hsu, P.J.; Lin, M.H.; Chen, P.M.; Sudhir, P.R.; Chen, C.H.; Chen, C.H.; Sytwu, H.K.; et al. Interleukin-25 Mediates Transcriptional Control of PD-L1 via STAT3 in Multipotent Human Mesenchymal Stromal Cells (hMSCs) to Suppress Th17 Responses. *Stem Cell Rep.* **2015**, *5*, 392–404. [CrossRef]
49. Bergen, A.W.; Caporaso, N. Cigarette smoking. *J. Natl. Cancer Inst.* **1999**, *91*, 1365–1375. [CrossRef]
50. Wang, Y.; Wang, H.; Yao, H.; Li, C.; Fang, J.Y.; Xu, J. Regulation of PD-L1: Emerging routes for targeting tumor immune evasion. *Front. Pharmacol.* **2018**, *9*, 536–549. [CrossRef]
51. Qiu, X.Y.; Hu, D.X.; Chen, W.-Q.; Chen, R.Q.; Qian, S.R.; Li, C.Y.; Li, Y.J.; Xiong, X.X.; Liu, D.; Pan, F.; et al. PD-L1 confers glioblastoma multiforme malignancy via Ras binding and Ras/Erk/EMT activation. *Biochim. Biophys. Acta Mol. Basis Dis.* **2018**, *1864*, 1754–1769. [CrossRef]

52. Zerdes, I.; Matikas, A.; Bergh, J.; Rassidakis, G.Z.; Foukakis, T. Genetic, transcriptional and post-translational regulation of the programmed death protein ligand 1 in cancer: biology and clinical correlations. *Oncogene* **2018**, *37*, 4639–4661. [CrossRef] [PubMed]
53. Song, T.L.; Nairismägi, M.-L.; Laurensia, Y.; Lim, J.-Q.; Tan, J.; Li, Z.-M.; Pang, W.-L.; Kizhakeyil, A.; Wijaya, G.-C.; Huang, D.-C.; et al. Oncogenic activation of the STAT3 pathway drives PD-L1 expression in natural killer/T-cell lymphoma. *Blood* **2018**, *132*, 1146–1158. [CrossRef] [PubMed]
54. Zhang, J.-P.; Song, Z.; Wang, H.-B.; Lang, L.; Yang, Y.-Z.; Xiao, W.; Webster, D.E.; Wei, W.; Barta, S.K.; Kadin, M.E.; et al. A novel model of controlling PD-L1 expression in ALK + anaplastic large cell lymphoma revealed by CRISPR screening. *Blood* **2019**, *134*, 171–185. [CrossRef] [PubMed]
55. Chen, N.; Fang, W.; Zhan, J.; Hong, S.; Tang, Y.; Kang, S.; Zhang, Y.; He, X.; Zhou, T.; Qin, T. Upregulation of PD-L1 by EGFR activation mediates the immune escape in EGFR-driven NSCLC: implication for optional immune targeted therapy for NSCLC patients with EGFR mutation. *J. Thorac. Oncol.* **2015**, *10*, 910–923. [CrossRef] [PubMed]
56. Chernyavsky, A.I.; Shchepotin, I.B.; Galitovkiy, V.; Grando, S.A. Mechanisms of tumor-promoting activities of nicotine in lung cancer: Synergistic effects of cell membrane and mitochondrial nicotinic acetylcholine receptors. *BMC Cancer* **2015**, *15*, 152–164. [CrossRef]
57. Zhang, N.; Zeng, Y.; Du, W.; Zhu, J.; Shen, D.; Liu, Z.; Huang, J.-A. The EGFR pathway is involved in the regulation of PD-L1 expression via the IL-6/JAK/STAT3 signaling pathway in EGFR-mutated non-small cell lung cancer. *Int. J. Oncol.* **2016**, *49*, 1360–1368. [CrossRef]
58. Yang, L.; Huang, F.; Mei, J.; Wang, X.; Zhang, Q.; Wang, H.; Xi, M.; You, Z. Posttranscriptional Control of PD-L1 Expression by 17β-Estradiol via PI3K/Akt Signaling Pathway in ERα-Positive Cancer Cell Lines. *Int. J. Gynecol. Cancer* **2017**, *27*, 196–205. [CrossRef]
59. García-Aranda, M.; Redondo, M. Targeting protein kinases to enhance the response to anti-PD-1/PD-l1 immunotherapy. *Int. J. Mol. Sci.* **2019**, *20*, 2296. [CrossRef]
60. Kortylewski, M.; Jove, R.; Yu, H. Targeting STAT3 affects melanoma on multiple fronts. *Cancer Metastasis Rev.* **2005**, *24*, 315–327. [CrossRef]
61. I Chernyavsky, A.; Arredondo, J.; Vetter, D.; Grando, S. Central role of α9 acetylcholine receptor in coordinating keratinocyte adhesion and motility at the initiation of epithelialization. *Exp. Cell Res.* **2007**, *313*, 3542–3555. [CrossRef]
62. Xu, R.; Won, J.Y.; Kim, C.H.; Kim, D.E.; Yim, H. Roles of the Phosphorylation of Transcriptional Factors in Epithelial-Mesenchymal Transition. *J. Oncol.* **2019**, *2019*, 1–11. [CrossRef] [PubMed]
63. Vu, T.; Jin, L.; Datta, P. Effect of Cigarette Smoking on Epithelial to Mesenchymal Transition (EMT) in Lung Cancer. *J. Clin. Med.* **2016**, *5*, 44. [CrossRef] [PubMed]

cancers

MDPI

Article

Streptomyces sp. MUM256: A Source for Apoptosis Inducing and Cell Cycle-Arresting Bioactive Compounds against Colon Cancer Cells

Loh Teng-Hern Tan [1,2], Chim-Kei Chan [3], Kok-Gan Chan [4,5,*], Priyia Pusparajah [6], Tahir Mehmood Khan [7], Hooi-Leng Ser [1,2], Learn-Han Lee [1,7,8,9,*] and Bey-Hing Goh [7,8,10,11,*]

1 Novel Bacteria and Drug Discovery (NBDD) Research Group, Microbiome and Bioresource Research Strength, Jeffrey Cheah School of Medicine and Health Sciences, Monash University Malaysia, Bandar Sunway 47500, Selangor Darul Ehsan, Malaysia; tenghern@gmail.com or loh.teng.hern@monash.edu (L.T.-H.T.); hooileng_ser@y7mail.com (H.-L.S.)

2 Institute of Biomedical and Pharmaceutical Sciences, Guangdong University of Technology, Guangzhou 510006, China

3 de Duve Institute, Université catholique de Louvain, Avenue Hippocrate 74, 1200 Brussels, Belgium; chimkei@gmail.com

4 International Genome Centre, Jiangsu University, Zhenjiang 212013, China

5 Division of Genetics and Molecular Biology, Institute of Biological Sciences, Faculty of Science, University of Malaya, Kuala Lumpur 50603, Malaysia

6 Medical Health and Translational Research Group, Jeffrey Cheah School of Medicine and Health Sciences, Monash University Malaysia, Bandar Sunway 47500, Malaysia; priyia.pusparajah@monash.edu

7 Institute of Pharmaceutical Science, University of Veterinary and Animal Science Lahore, Punjab 54000, Pakistan; tahir.khan@uvas.edu.pk

8 Health and Well-Being Cluster, Global Asia in the 21st Century (GA21) Platform, Monash University Malaysia, Bandar Sunway 47500, Malaysia

9 Key Laboratory of Microbial Metabolism, Joint International Research Laboratory of Metabolic and Developmental Sciences, School of Life Sciences and Biotechnology, Shanghai Jiao Tong University, Shanghai 200240, China

10 Biofunctional Molecule Exploratory (BMEX) Research Group, School of Pharmacy, Monash University Malaysia, Bandar Sunway 47500, Selangor Darul Ehsan, Malaysia

11 College of Pharmaceutical Sciences, Zhejiang University, 866 Yuhangtang Road, Hangzhou 310058, China

* Correspondence: kokgan@um.edu.my (K.-G.C.); lee.learn.han@monash.edu or leelearnhan@yahoo.com (L.-H.L.); goh.bey.hing@monash.edu (B.-H.G.)

Received: 23 October 2019; Accepted: 25 October 2019; Published: 6 November 2019

Abstract: New and effective anticancer compounds are much needed as the incidence of cancer continues to rise. Microorganisms from a variety of environments are promising sources of new drugs; *Streptomyces* sp. MUM256, which was isolated from mangrove soil in Malaysia as part of our ongoing efforts to study mangrove resources, was shown to produce bioactive metabolites with chemopreventive potential. This present study is a continuation of our previous efforts and aimed to investigate the underlying mechanisms of the ethyl acetate fraction of MUM256 crude extract (MUM256 EA) in inhibiting the proliferation of HCT116 cells. Our data showed that MUM256 EA reduced proliferation of HCT116 cells via induction of cell-cycle arrest. Molecular studies revealed that MUM256 EA regulated the expression level of several important cell-cycle regulatory proteins. The results also demonstrated that MUM256 EA induced apoptosis in HCT116 cells mediated through the intrinsic pathway. Gas chromatography-mass spectrometry (GC-MS) analysis detected several chemical compounds present in MUM256 EA, including cyclic dipeptides which previous literature has reported to demonstrate various pharmacological properties. The cyclic dipeptides were further shown to inhibit HCT116 cells while exerting little to no toxicity on normal colon cells in this study. Taken together, the findings of this project highlight the important role of exploring the mangrove microorganisms as a bioresource which hold tremendous promise for the development of chemopreventive drugs against colorectal cancer.

Keywords: *Streptomyces*; mangrove; anti-proliferative; apoptosis; colon cancer

1. Introduction

Colorectal cancer (CRC) constitutes the third most common cancer diagnosed and the fourth leading cause of cancer mortality globally [1]. According to Arnold et al. [2], CRC incidence and mortality is correlated well to human development levels. The study described that the rate of CRC cases is increasing rapidly in many low-income and middle-income countries while stable or decreasing trends are seen in highly developed countries [2]. Treatment of CRC usually involves surgical removal alone when detected early whereas late stage of advanced or metastasized CRC requires radiotherapy or chemotherapy [3]. However, chemotherapy suffers from significant limitations including insufficient selectivity for tumor cells, leading to many adverse side effects [4,5]. Although targeted therapies have been regarded as the most successful treatments of cancer for the past few decades, genetic mutations greatly affect the efficacy of various targeted drug therapies and leading to mutation-driven resistance of targeted therapy [6–8]. Thus, there is an urgent need to search for alternative chemotherapeutic/chemoprevention agents which may overcome the limitations of chemotherapy, and natural products are an excellent resource to explore [9].

Natural products have provided the inspiration for the development of alternative agents for the prevention or therapy of cancer, and using natural products to reduce cancer prevalence has received substantial attention [10–14]. The natural products of interest can be found in fruits, vegetables, herbs and medicinal plants [15–19]. Aside from plant-based natural products, microorganisms are also reported to be a useful source of bioactive compounds, including chemotherapeutic agents [20–23]. For instance, actinomycin was discovered as the first microbial metabolite isolated from the bacterium *Streptomyces antibioticus* in 1940 [24] to be used in cancer therapy. Since then, many more microbial metabolites with antitumor properties were discovered including anthracyclines, bleomycin, mitosanes, mithramycin, pentostatin and calicheamicins [25]. Currently, there is evidence demonstrating that the mangrove derived microbial metabolites could be the next bioresources for potential cancer therapeutic agents [26–29]. Thus, we explored the potential of *Streptomyces* isolated from Malaysian mangrove soil with a focus on its ability to produce metabolites exhibiting chemopreventive activity.

This work represents part of an ongoing project to discover anticancer compounds from mangrove resources, and our screening of the various isolated *Streptomyces* strains led to the discovery of *Streptomyces* sp. MUM256 which possesses the potential to produce active metabolites that induced cell-cycle arrest and apoptosis. In the earlier study [30], we demonstrated that the methanol extract of *Streptomyces* sp. MUM256 exhibited antioxidant and cytotoxic properties. The present study is a continuation of this work aiming to investigate the underlying mechanisms of the cytotoxic and antiproliferative effects of the ethyl acetate fraction of *Streptomyces* sp. MUM256 crude extract (MUM256 EA) against the HCT116 cell line. We demonstrated that the MUM256 EA induced cell-cycle arrest by downregulating several important cell-cycle regulatory proteins and induced apoptosis via interactions with the intrinsic pathway in colon cancer cells (Figure 1). Thus, we believe these results provide new insight into the development of mangrove-derived *Streptomyces* metabolites against CRC.

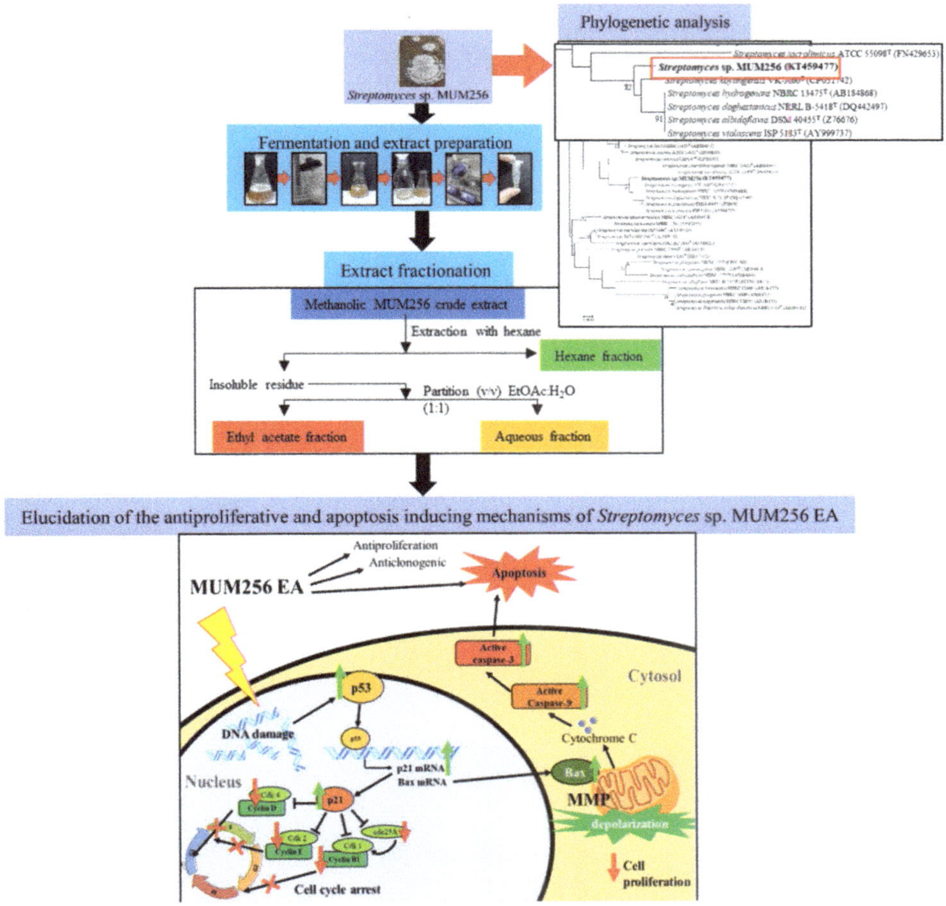

Figure 1. The summarized flow chart of this study. The figure illustrates the fermentation, crude extract extraction, fractionation and elucidated mechanisms of MUM256 EA in cell-cycle arrest and apoptosis induction.

2. Results

2.1. Phylogenetic Analysis of Streptomyces sp. MUM256

Given that the publicly available database for 16S rRNA gene sequence, such as Ezbiocloud, is regularly updated by adding new bacteria species with validly published names, a new phylogenetic tree was constructed for strain MUM256 based on its 16S rRNA gene sequence (GenBank accession number KT459477) (Figure 2). Based on the blast result of the Ezbiocloud database, the 16S rRNA gene sequence of strain MUM256 demonstrated highest similarity to *S. hydrogenans* NBRC13475^T (99.70%), *S. daghestanicus* NRRL B-5418^T (99.70%), *S. albidoflavus* DSM40455^T (99.70%), *S. violascens* ISP5183^T (99.70%) followed by *S. koyangensis* VK-A60^T (99.48%). According to Figure 2, the 16S rRNA sequence of strain MUM256 formed a distinct clade with strains *S. koyangensis* VK-A60^T, *S. hydrogenans* NBRC13475^T, *S. daghestanicus* NRRL B-5418^T, *S. albidoflavus* DSM40455^T and *S. violascens* ISP5183^T at bootstrap value of 82%, showing relatively high confidence level of the association (Figure 2).

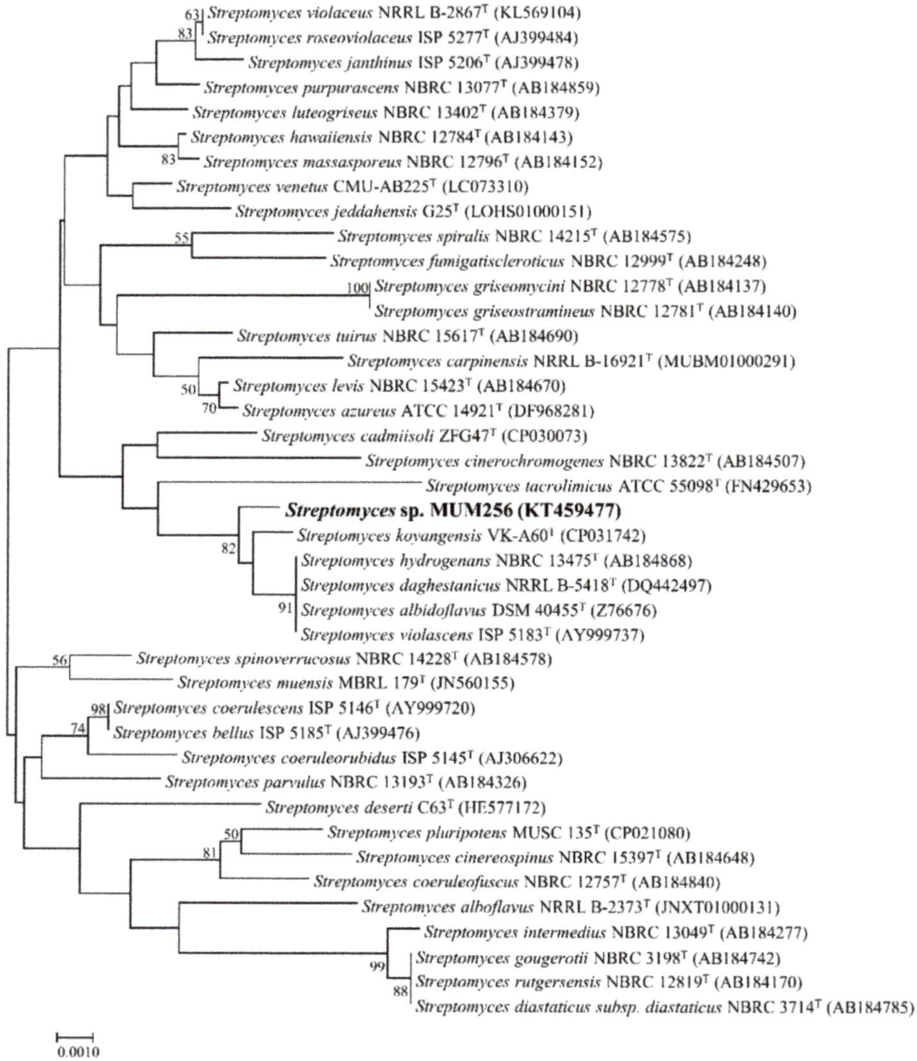

Figure 2. Neighbour-joining phylogenetic tree based on 16S rRNA gene sequence of strain MUM256 (1343bp). The tree illustrates the relationship between strain MUM256 and closely related strains. Numbers at nodes indicate percentages of 1000 bootstrap re-samplings. Bar, 0.001 substitutions per site.

2.2. To Examine the Cytotoxic Effect of Streptomyces sp. MUM256 Fractions against Colon Cancer Cell HCT116

Three different fractions were obtained from the methanolic MUM256 extract after being subjected to sequential fractionation with three types of solvents, namely hexane, ethyl acetate and water. Figure 3a demonstrates the cell viability of HCT116 after exposure to MUM256 extract and the respective fractions for 72 h. The ethyl acetate fraction of MUM256 extract was shown to exhibit the highest cytotoxicity towards HCT116 among the fractions tested, followed by the hexane fraction and the aqueous fraction as the least toxic against HCT116 cells. The toxicity of MUM256 EA was also evaluated on a normal colon cell line CCD-18Co. The MUM256 EA exhibits significantly lesser toxicity towards a normal colon cell (CCD-18Co) at all the concentrations tested in this study (Figure 3b).

The IC_{50} of MUM256 EA towards CCD-18Co was measured at 215 µg/mL which is 1.72 higher than its cytotoxicity towards colon cancer cell (HCT116) with IC_{50} of 88.44 µg/mL. This result demonstrates that the MUM256 EA displays a slight preferential cytotoxicity against HCT116 colon cancer cells over a CCD-18Co normal colon cell.

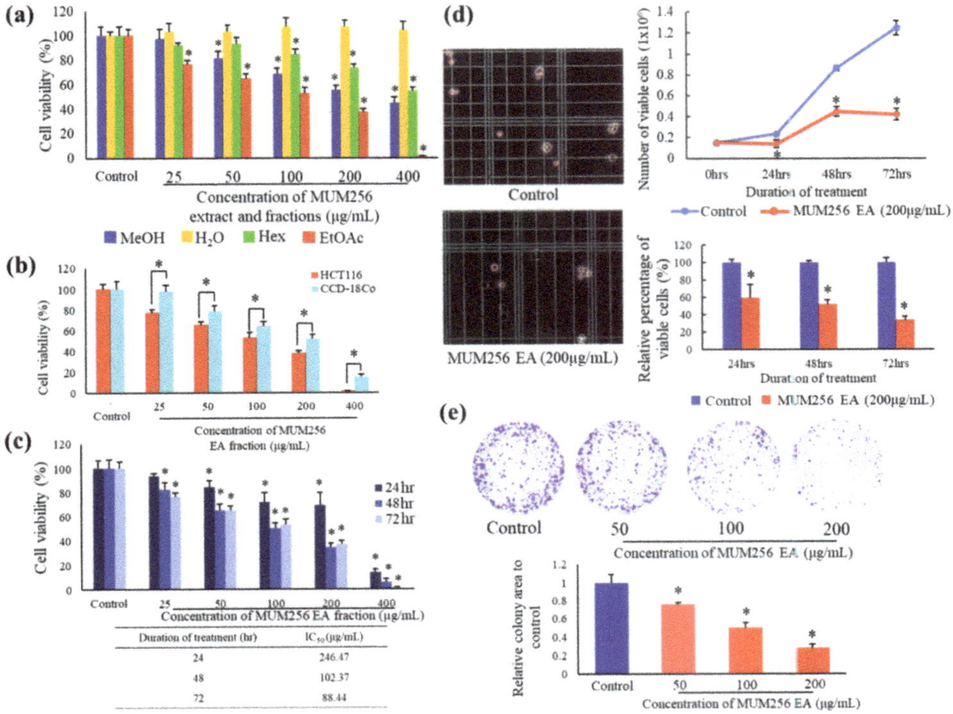

Figure 3. Cytotoxic and antiproliferative properties of MUM256 EA against HCT116 cells. (**a**) Cytotoxic effect of MUM256 crude extract (MeOH: methanol) and 3 different fractions (H_2O: water; Hex: hexane; EtOAc: ethyl acetate) against HCT116 cells at a series of concentrations after 72 h treatment, ($n = 4$, * $p < 0.05$). (**b**) The toxicity of MUM256 EA towards CCD-18Co as compared to its cytotoxicity against HCT116 after 72 h exposure, ($n = 4$, * $p < 0.05$). (**c**) Cytotoxic effect of MUM256 EA at a series of concentrations against HCT116 cells after 24, 48 and 72 h of exposure, ($n = 4$, * $p < 0.05$). (**d**) Line graph shows the inhibitory effect of MUM256 EA fraction on HCT116 cells. Bar graph illustrates the relative percentage of number of viable cells after exposure to MUM256 EA fraction after normalized to 100% for number of viable cells exposed to vehicle control (0.5% dimethyl sulfoxide (DMSO)) ($n = 3$), * $p < 0.05$ indicates significant difference between the untreated and the treated cells. (**e**) Images of colony formation after 7 days of control and MUM256 EA pre-treated cells in 6 well-plate. Crystal violet staining was used for visualization and quantification of colonies. The area of colonies formed are shown as bar graph with the mean relative mean colony area to control ± standard deviation (SD) ($n = 3$, * $p < 0.05$).

2.3. *MUM256 EA Suppresses Cell Viability and Proliferation in HCT116 Cells*

To further examine the cytotoxic effects of MUM256 EA using MTT assay, the HCT116 cells were treated with increasing concentrations of MUM256 EA for 24, 48 and 72 h. As shown in Figure 3c, when treated for 24, 48 or 72 h in the concentration range of 25 to 400 µg/mL, MUM256 EA reduced the cell viability of HCT116 cells; MUM256 EA demonstrated cytotoxic effect against HCT116 cells in a significant dose-dependent manner at both 48 and 72 h timepoints. At 400 µg/mL, the cell viability of

HCT116 was reduced by MUM256 EA in a time-dependent manner. The concentrations of MUM256 EA required to produce a 50% reduction in cell viability (IC_{50}) were determined by a regression analysis; the value of the half maximal inhibitory concentration (IC_{50}) decreases according to the time of treatment (IC_{50}-24 h: 246.47 µg/mL > IC_{50}-48 h: 102.37 µg/mL > IC_{50}-72 h: 88.44 µg/mL).

Trypan blue dye exclusion assay showed that the growth of HCT116 cells were reduced upon MUM256 EA treatment as demonstrated by the lower number of viable HCT116 cells in the wells treated with MUM256 EA when compared to the wells with untreated cells (Figure 3d). According to Figure 3d, a significantly lower number of viable cells were detected following exposure to 200 µg/mL MUM256 EA: 59.7%, 51.7% and 33.6% at 24, 48 and 72 h, respectively, when compared to control.

To further evaluate MUM256 EA's inhibitory effects on cancer cell proliferation, we performed colony formation/clonogenic survival assay on HCT116 cells upon exposure to MUM256 EA treatment, as a test reflecting long-term effect/toxicity [31]. The assay clearly demonstrated that clone formation of HCT116 cells was reduced in a dose-dependent manner (Figure 3e). Treatment with MUM256 EA resulted in a reduction in both numbers of colonies formed as well as the size of the colonies-colony numbers for treated samples were significantly reduced and appeared smaller than those of the control. Therefore, these results confirm that MUM256 EA is cytotoxic and anti-proliferative towards HCT116 cells.

2.4. Morphological Changes Induced by MUM256 EA

The morphological examination revealed that the treatment of MUM256 EA visibly altered the cell morphology of the viable HCT116 cells. In Figure 4a, most of the untreated HCT116 cells (control) appeared as normal angular and spindle shapes, but most of the cells lost these features after treatment with MUM256 EA. For example, it was observed that the treatment caused rounding and detachment of HCT116 cells. In addition, reduced number of cells and cell shrinkage with lesser cytoplasm mass were also observed (indicated by arrows) in Figure 4a. These abnormal morphological changes of the cells upon exposure to MUM256 EA has provided some insight on its cytotoxic effect towards the HCT116 cells.

Furthermore, flow cytometry was also utilized to examine the morphological changes of the cells after treatment with MUM256 EA. The results showed a decrease in forward light scatter (Figure 4b); the median channel fluorescence was lowered in the treated cells with increasing dose of treatment. Meanwhile, the side light scatter was increased in the treated cells significantly with a 1.48-fold increase in the median channel fluorescence compared to the control (Figure 4b). The reduced forward light scatter indicates cell shrinkage, whereas the increased side scatter may be caused by chromatin condensation, nuclear fragmentation and crosslinking of cytoplasmic proteins by activated transglutamine, resulting in the cell becoming more reflective and refractive [32].

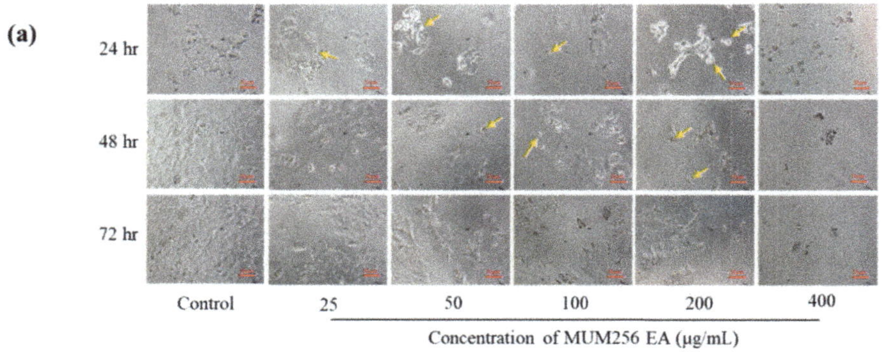

Figure 4. *Cont.*

(b)

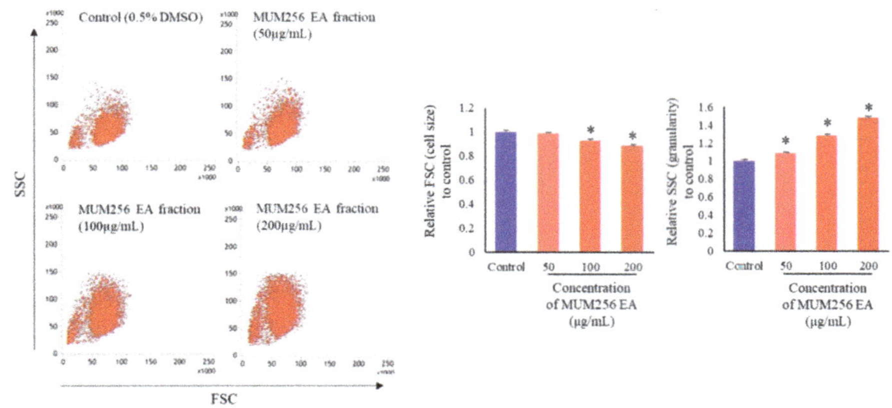

Figure 4. Morphological changes of HCT116 cells after exposure to MUM256 EA treatment. (**a**) Phase contrast microscopic evaluation of HCT116 cells under ×10 magnification. Decreased number of HCT116 cells were observed across the 3 timepoints with increasing concentration of MUM256 EA treatment. Furthermore, arrows indicate the rounded cells due to cytoplasmic shrinkage and detached cells which was increased in number with increasing concentration of MUM256 EA treatment. (**b**) Representative forward scatter/side scatter (FSC/SSC) dot plot of HCT116 cells after exposure to MUM256 EA at different concentrations. The x-axis shows a decreasing FSC in HCT116 cells treated with increasing concentrations of MUM256 EA. The FSC represents the diffracted light and it is related to cell size. The y-axis shows an increasing SSC in HCT116 treated with increasing concentrations of MUM256 EA. The SSC represent the reflected and refracted light and is related to cell granularity and complexity. The bar graphs show the relative median FSC and SSC fluorescence signals of treated cell population as compared to control ± SD (n = 3, * $p < 0.05$).

2.5. Cell Cycle Arrest Effect of MUM256 EA in HCT116 Cells

In Figure 5a, exposure of HCT116 to 200 μg/mL of MUM256 EA resulted in a statistically significant increase in the percentage of cells in G_0/G_1 phase from 44.7% (untreated cells) to 67.4%, with a concomitant decrease in percentage of cells in the G_2 phase from 32.68% (untreated cells) to 14.31%. However, by 48 h, cells started accumulating in the G_2 phase from 13.74% to 22.13% and followed by accumulation from 11.14% to 22.39% by 72 h (Figure 5a). These results suggested that MUM256 EA hampers cell-cycle progression by arresting the cells in G_1 and G_2/M phase, subsequently lead to inhibition of cell proliferation.

Expressions of Cell-Cycle Regulatory Proteins

After treatment with MUM256 EA for 24, 48 and 72 h, both reverse-transcription polymerase chain reaction (RT-PCR) and Western blotting analysis showed the expression levels of $p21^{Waf1/Cip1}$ were significantly elevated as compared to untreated HCT116 cells (Figure 5b,c). As one of the transcriptional targets of p53, the expressions of $p21^{Waf1/Cip1}$ could be regulated by p53 [33–35]. Thus, the changes of p53 in response to MUM256 EA in HCT116 was also investigated using intracellular flow cytometry. The p53 protein expression was shown to be increased upon treatment with MUM256 EA for 24 h (Figure 8c). This result supports that p53 activation plays a role in MUM256 EA-induced p21 up-regulation in HCT116 cells.

Besides that, this study also demonstrated that the MUM256 EA treatment downregulated several important cell cycle regulator genes, including CDK2, CDK4 and cdc25A phosphatase (Figure 5b). Western blot analysis also showed that MUM256 EA induced reduction in cyclin B1 protein levels in the HCT116 cells after 24, 48 and 72 h of exposure (Figure 5c).

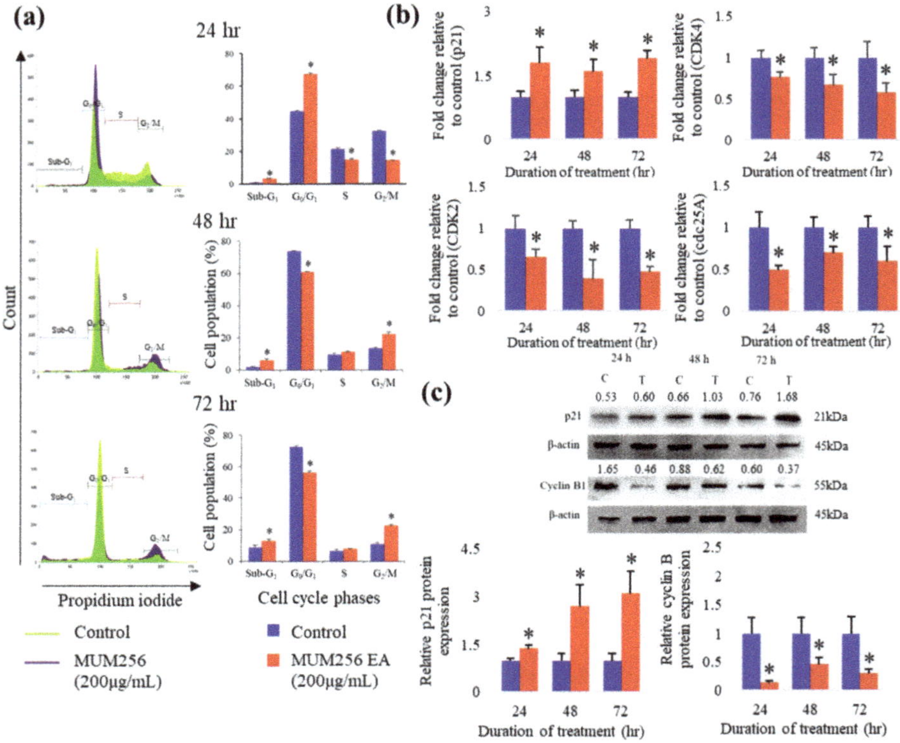

Figure 5. Cell-cycle arrest at G1 and G2 phase induced by MUM256 EA through modulation of cell cycle regulatory proteins. (**a**) Effect of MUM256 EA on cell cycle distribution. HCT116 cells were treated with 200 µg/mL of MUM256 EA for indicated durations (24, 48 and 72 h). Cells were harvested and fixed with ice-cold 70% ethanol overnight. Before flow cytometric analysis, the cells were incubated in phosphate-buffered saline (PBS) containing 25 µg/mL PI, 0.1% Triton-X-100 and 100 µg/mL RNase A for 30 min in the dark at room temperature. The cellular DNA content was determined by flow cytometry. Representative of histograms showing X-axis, DNA contents; y-axis, cell population (%). Each bar represents at least three individual experiments, (* $p < 0.05$). (**b**) Real-time polymerase chain reaction (PCR) analysis of p21, CDK2, CDK4 and cdc25A expressions in HCT116 cells treated with 200µg/mL MUM256 EA. The fold change of the gene was normalized against 18S rRNA expression using the formula $2^{-\Delta\Delta CT}$. Bar graph represents the comparison between treatment and control by *t*-test, * $p < 0.05$ ($n = 3$). (**c**) The levels of p21 and cyclin B1 proteins were determined by Western blotting. HCT116 cells were treated with DMSO and MUM256 EA at 200 µg/mL for indicated durations and the expression of p21 and cyclin B1 were determined by Western blotting with specific antibody. One representative Western blot of three is presented (Figure S1). Each has the expression of β-actin as the internal control. Protein expression was quantified by the densitometry analysis using Image J and normalized against β-actin expression. The values above each band indicate the densitometry ratio of each protein of interest normalized against respective β-actin expression. Bar graph represents the comparison between treatment and control by *t*-test, * $p < 0.05$ ($n = 3$).

2.6. Nuclear Condensation and DNA Fragmentation

Differences were observed in the nuclei of treated and untreated HCT116 after staining with Hoechst 33258 (Figure 6a). The Hoechst 33258 dye stains morphologically normal nuclei dimly blue, whereas treated cells demonstrate smaller nuclei with brilliant blue staining due to nuclear condensation (yellow arrows). This result demonstrated that MUM256 EA induces morphological changes with characteristics of apoptotic cell death. This result is also consistent with the detection of a significantly increased percentage of subG1 cell populations as an index of the apoptotic DNA fragmentation.

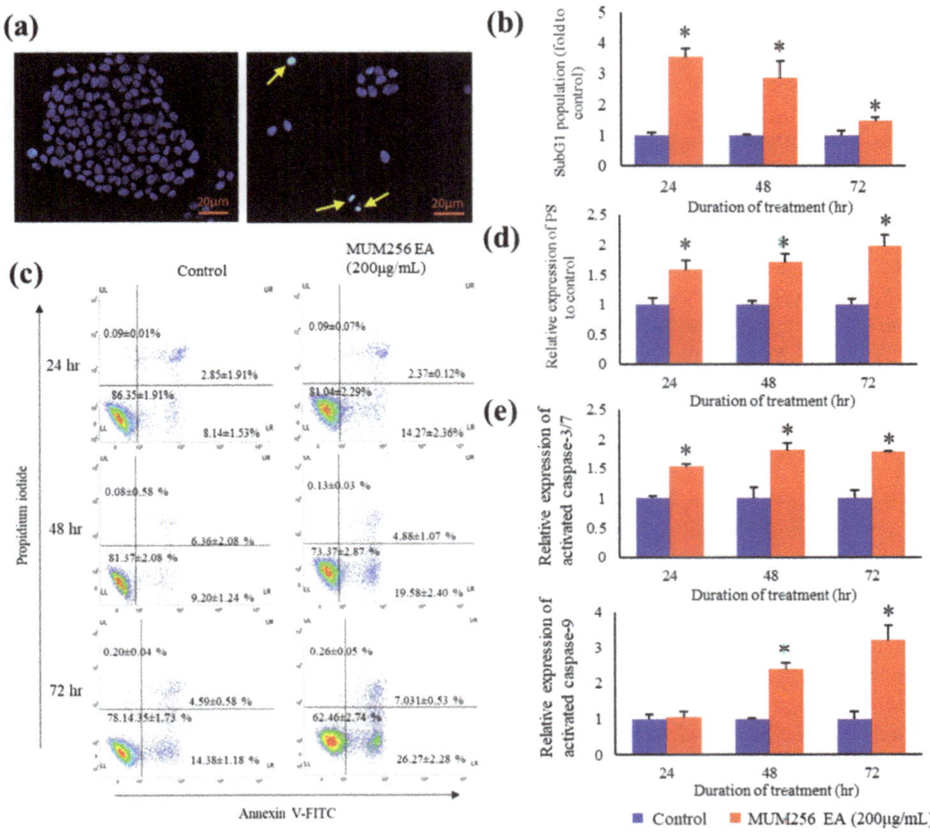

Figure 6. Apoptosis evaluation of MUM256 EA on HCT116 cells. (**a**) Nuclear morphology of HCT116 stained by Hoechst 33342 dye. Comparison of the nuclear morphological features of untreated and treated HCT116 under fluorescence microscope. Arrow indicates the abnormal nuclear morphological features (nuclear shrinkage or condensation shown by the bright-blue fluorescent and nuclear fragmentation) resulted from the cytotoxic effect of MUM256 EA fraction, indicating sign of apoptosis. (**b**) Bar graph represents the mean percentage of subG1 population ± S.D. ($n = 3$, * $p < 0.05$). (c) The apoptosis-inducing effect of MUM256 EA fraction was examined by flow cytometric analysis using double staining of Annexin V-FITC and PI on HCT116 cells for 24, 48 and 72 h. Quantification of early apoptosis (LR) (Annexin V positive/PI negative) and late apoptosis (UR) (Annexin V positive/PI positive) are shown by the representative four quadrants plot. The bar graph represents the mean percentage of population ± S.D., ($n = 3$, * $p < 0.05$). (**d**) Both the early apoptotic portion and the late apoptotic portion were included as the total apoptotic cells upon treatment for 24, 48 and 72 h. The bar graph represents the mean percentage of apoptotic cells ± S.D, ($n = 3$, * $p < 0.05$). (**e**) Flow cytometry analysis of caspase-3/7 and 9 activities in HCT116 cells using a non-cytotoxic fluorescent labelled inhibitors of caspase-3/7 and 9. HCT116 cells were treated with 200 µg/mL MUM256 EA fraction for various times as indicated, or with DMSO as a control, then stained with FAM-DEVD-FMK or FAM-LEHD-FMK which recognizes and binds to active caspase-3/7 or 9, respectively, before analysed by flow cytometry (BD FacVerse). The bar graph represents the median of fluorescence intensity (MFI) of active caspase-3/7 or 9 straining in treated cells normalized to control ± SD. ($n = 3$, * $p < 0.05$).

MUM256 EA treatment increased the percentage of cells in the subG_1 phase after treatment for 24, 48 and 72 h, respectively, as compared to control (Figure 6b). Intranucleosomal DNA fragmentation is a major hallmark of apoptosis [36,37]. PI is a fluorescent dye that binds stoichiometrically to nucleic acids [38]; and as fluorescence emission is proportional to DNA content of the cells, it allows the analysis of a cell population's replication state. Hence, cells undergoing apoptosis can be identified from the DNA content histograms as those with lower fluorescent signal due to the low molecular weight of DNA fragments present in these cells compared to the G1 cells [37]. These data suggest that the reduced cell proliferation by MUM256 EA could be contributed by the induction of apoptosis.

2.7. Exposure of Phosphatidylserine

In this study, additional evidence for the occurrence of apoptosis was obtained by double staining of the cells with propidium iodide (PI) and Annexin V-FITC. Annexin V is a protein that binds with high affinity to phosphatidylserine (PS), which is translocated from the inner to the outer membrane leaflet during early apoptosis. Meanwhile, the healthy cells have phosphatidylserines on the inner leaflet of the plasma membrane. This technique is then combined with DNA binding dye (propidium iodide) which is impenetrable to intact membrane, hence allowing for the determination of the apoptotic state of a cell. As shown in Figure 6c, the percentage of the early apoptotic cells (PS positive and PI negative cells) increased in response to treatment with MUM256 EA for 24, 48 and 72 h. Overall, the percentage of apoptotic cells (PS positive cells) were increased in HCT116 cells treated with MUM256 EA for 24, 48 and 72 h (Figure 6d).

2.8. Caspase Activation

To investigate the effect of MUM256 EA in activation of caspases, flow cytometric analysis was performed using non-cytotoxic fluorescent labelled inhibitors, FAM-DEVD-FMK and FAM-LEHD-FMK that bind irreversibly to cysteine of the active center of caspase-3/7 or caspase-9 in the cells, respectively, hence allowing accurate quantification of the caspase activity in the cells. The results showed that the expression of active caspase-3/7 in HCT116 cells were significantly increased upon treatment with MUM256 EA for 24, 48 and 72 h as demonstrated by the increased median fluorescence intensity (MFI) (Figure 6e). Similarly, the flow-cytometric analysis demonstrated that MUM256 EA induced an increasing trend of caspase-9 activation in HCT116 cells (Figure 6e). Thus, MUM256 EA-induced cell death was accompanied by an increase in caspase activities, which then stimulated the molecular cascade of apoptosis. To further validate the apoptosis inducing effect of MUM256 EA, pretreatment of cells with either z-VAD-fmk, a broad-spectrum caspase inhibitor, or z-DEVD-fmk, a caspase-3 inhibitor, diminished the cell death induced by MUM256 EA (Figure 7). These data suggest that the MUM256 EA-induced apoptosis involves a caspase-dependent pathway in HCT116 cells.

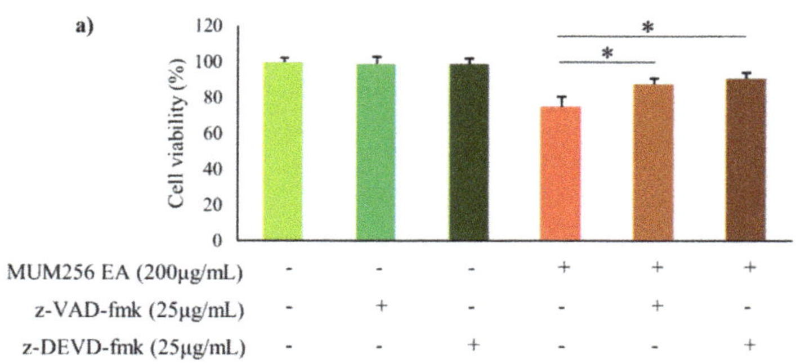

Figure 7. *Cont.*

b)

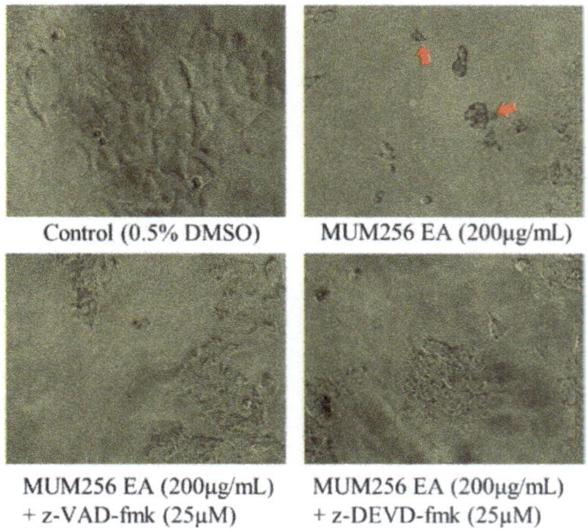

Figure 7. Effect of caspase inhibitors on MUM256 EA induced apoptosis in HCT116 cells. (**a**) Cells were pretreated with either z-VAD-fmk (25 µg/mL) or z-DEVD-fmk (25 µg/mL) for 1 h, then treated as described (Control—0.5% DMSO, with or without—200 µg/mL MUM256 EA) for 24 h. Cell viability was measured by MTT (3-(4,5-dimethylthiazol-2-yl)-2,5-diphenyltetrazolium bromide) assay, (n = 4, * $p < 0.05$). (**b**) Morphological changes of HCT116 cells upon indicated treatments were observed by phase-contrast microscopy (magnification ×20). Arrows indicate the treated cells with apoptotic morphology.

2.9. Mitochondrial Membrane Potential (MMP)

To determine whether the caspase-dependent apoptosis in HCT116 cells was mediated through the mitochondrial apoptotic pathway in response to MUM256 EA treatment, the effect of MUM256 EA on the mitochondrial membrane permeability was investigated. JC-1 dye was used to assess the mitochondrial transmembrane potential of HCT116 cells in response to the treatment with MUM256 EA. JC-1 dye exhibits variable characteristics when bound to the membrane of apoptotic cells versus non-apoptotic cells. In non-apoptotic cells, JC-1 dye appears in aggregate form, emitting orange fluorescence at 590 nm when bound on polarized mitochondrial membrane with high membrane potential; but exists in monomeric form that emits green fluorescence at 527 nm in cells with disrupted mitochondrial potential [39]. The results of flow cytometry showed that treatment of HCT116 cells with MUM256 EA resulted in the increase of green-fluorescence-positive cells as well as the decrease in the JC-1 orange/green ratio in treated cells, suggesting depolarization of MMP (Figure 8a). This result suggested that the MUM256 EA induced cell apoptosis by causing the collapse of mitochondrial transmembrane potential in HCT116 cells.

2.10. Bax Pro-Apoptotic Protein Expressions

Given that mitochondrial transmembrane permeabilization represents a process of point-of-no-return, it is highly regulated, largely by members of the BCL-2 protein family [40]. Thus, we investigated the effect of MUM256 EA on the expression levels of Bax—which is a member of the BCL-2 family—in HCT116 cells. As shown in Figure 8b, RT-PCR analysis showed that treatment of HCT116 cells with MUM256 EA led to significantly increased expression of Bax at 24, 48 and 72 h. In line with the RNA expression result, the intracellular flow cytometry analysis showed the level of proapoptotic protein Bax in HCT116 cells was increased significantly upon treatment with MUM256 EA for 24, 48 and 72 h. This result suggested that the overall increase in Bax expressions could be the contributor to the mitochondrial mediated apoptosis induced by MUM256 EA in HCT116 cells.

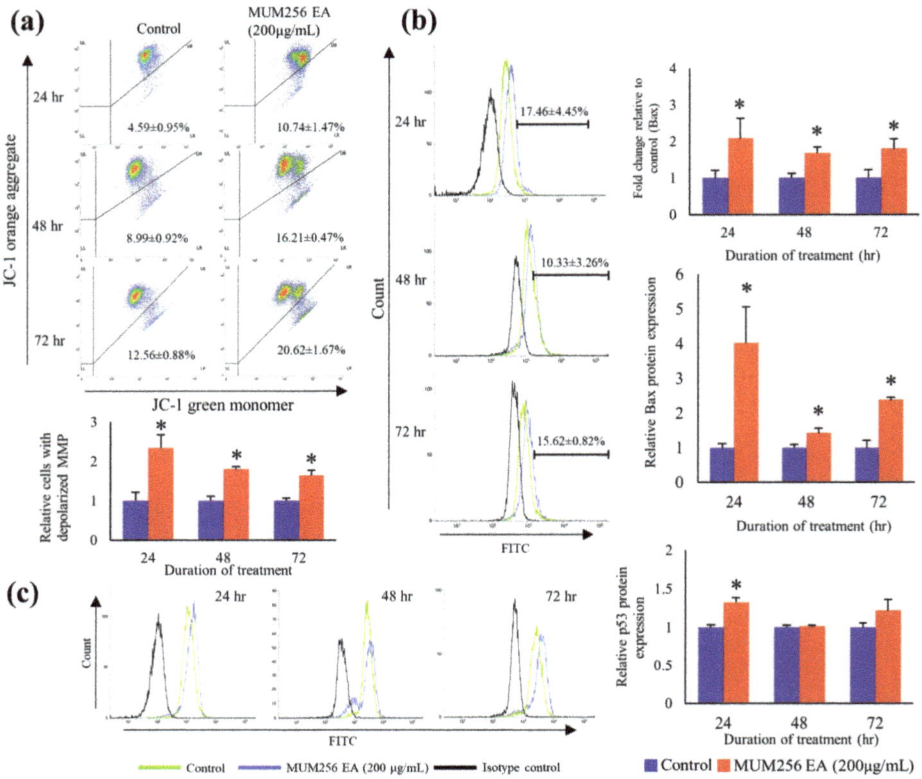

Figure 8. Effect of MUM256 EA on the mitochondrial membrane potential, proapoptotic protein (Bax) and p53 status in HCT116 cells. (**a**) HCT116 cells were treated with MUM256 EA for indicated durations and then were stained with JC-1 dye to detect the change of MMP by flow cytometry. Dot plots are representative of three independent experiments. Bar graph shows the ratio of JC-1 orange/green normalized to control, decrease in the ratio indicating MMP depolarization upon treatment, * $p < 0.05$ ($n = 3$). (**b**) Real-time PCR analysis of Bax expression in HCT116 cells treated with 200 µg/mL MUM256 EA fraction. The fold change of the gene was normalized against porphobilinogen deaminase (PBGD) expression using the formula $2^{-\Delta\Delta CT}$. Flow cytometry analysis of Bax protein expression in HCT116 cells using intracellular immunofluorescence staining method. Representative overlay of histograms showing Bax-associated fluorescence intensity after exposure to MUM256 EA fraction at 200 µg/mL for 24, 48 and 72 h. Bar graph represents the relative expression levels of Bax protein in treated HCT116 cells compared to control. Values are mean percentage of cells with expressed Bax ± SD. of three independent experiments. Asterisks indicate a significant difference compared to control (* $p < 0.05$). (**c**) Flow cytometry analysis of p53 protein expression in HCT116 cells using intracellular immunofluorescence staining method. Representative overlay of histograms showing p53-associated fluorescence intensity after exposure to MUM256 EA fraction at 200 µg/mL for 24, 48 and 72 h. Relative expression levels of p53 protein in HCT116 cells. Values are median fluorescence intensity ± SD of three independent experiments. Asterisks indicate a significant difference compared to control (* $p < 0.05$).

2.11. Detection of Bioactive Constituents in Streptomyces sp. MUM256 EA Extract Using Gas Chromatography-Mass Spectrometry

To further elucidate the potential chemical compounds that may have contributed to the antioxidant and cytotoxic properties, gas chromatography-mass spectrometry (GC-MS) was used to detect the chemical compounds present in MUM256 EA. GC-MS has been widely used as the analytical tool for molecular detection and identification in drug discovery. Numerous studies also utilized GC-MS to profile the bioactive compounds present in the secondary metabolites of *Streptomyces* bacteria [41–43]. The results of GC-MS analysis revealed that MUM256 EA contains several groups of chemical compounds, including the pyrrole, pyrazine and cyclic dipeptides. The identification of the chemical compounds was performed by comparing their mass spectra to standard mass spectra available in the database of NIST 05 Spectral library. Table 1 tabulates the retention time, molecular weight and molecular formula of the chemical compounds. Figure 9a,b depicts the GC chromatogram and chemical structures detected in MUM256 EA, respectively.

Table 1. Chemical constituents detected by GC-MS analysis.

No.	Constituents	Retention Time	Molecular Formula	Chemical Group	Molecular Weight	Similarity (%)
1	5-Pyrrolidino-2-pyrrolidone	54.839	$C_8H_{14}N_2O$	Heterocyclic, pyrrolidine	154	93.8
2	5-Isopropylidene-3,3-dimethyl-dihydrofuran-2-one	58.047	$C_9H_{14}O_2$	Heterocyclic, cyclic ether	154	98.2
3	Pyrrolo[1,2-a]pyrazine-1,4-dione, hexahydro-3-(2-methylpropyl)-	58.654	$C_{11}H_{18}N_2O_2$	Cyclic dipeptides	210	91.6
4	4(1H)-Pyrimidinone, 6-amino-2-methyl-5-nitroso-	58.957	$C_5H_6N_4O_2$	Heterocyclic, pyridine	154	95.9
5	Pyrrolidine-2-carboxamide, 1-benzyloxycarbonyl-N-(4-tolyl)-	70.314	$C_{20}H_{22}N_2O_3$	Heterocyclic, pyrrolidine	338	96.6
6	Pyrrolo[1,2-a]pyrazine-1,4-dione, hexahydro-3-(phenylmethyl)-	71.593	$C_{14}H_{16}N_2O_2$	Cyclic dipeptides	244	99.0

(a)

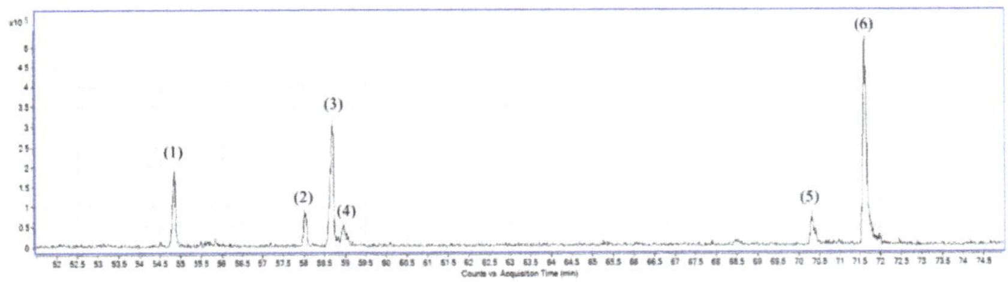

(b)

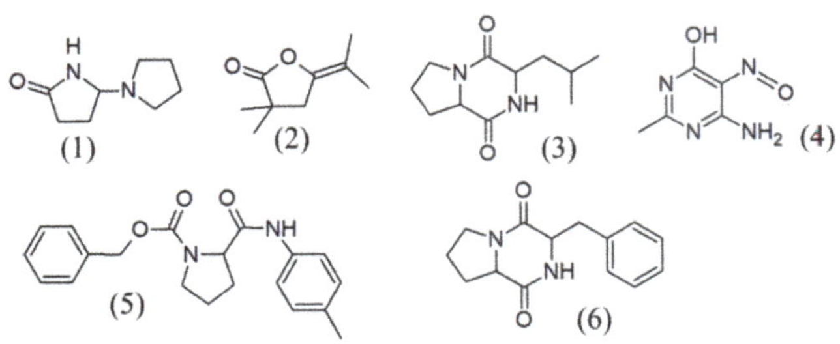

Figure 9. Gas chromatography-mass spectrometry (GC-MS) analysis of MUM256 EA. (**a**) The GC chromatogram of MUM256 EA. (**b**) The chemical structures of the detected constituents present in MUM256 EA. The identification of the chemical compounds was performed by comparing their mass spectra to standard mass spectra available in the database of NIST 05 Spectral library.

2.12. The Cytotoxic Effects of the Two Main Bioactive Constituents Detected by Gas Chromatography-Mass Spectrometry (GC-MS) against HCT116 and CCD-18Co Cells

To examine the cytotoxic properties of the chemical compounds in MUM256 EA, pure Compounds (**3**) and (**6**) were obtained and tested against HCT116 colon cancer cells and the CCD-18Co normal colon cells. Based on the results, Compound (**6**) showed stronger cytotoxic effect against HCT116 cells as compared to Compound (**3**) after 72 h exposure (Figure 10a). Interestingly, both compounds were not toxic towards the CCD-18Co cells with no significant reduction in cell viability at all the concentrations tested after 72 h of exposure (Figure 10b), except Compound (**3**) reduced the cell viability of CCD-18Co to 86% at 400 μg/mL. Both compounds also showed cytotoxic effect against another colon cancer (HT-29 cells) (Figure S2). To investigate potential combination effects between Compounds (**3**) and (**6**), HCT116 cells were treated with Compounds (**3**) and (**6**) at a ratio of 3:5 based on peak height shown in the GC-MS total ion chromatogram. However, the results did not indicate any significant combined effect of both compounds against HCT116 (Figure 10c).

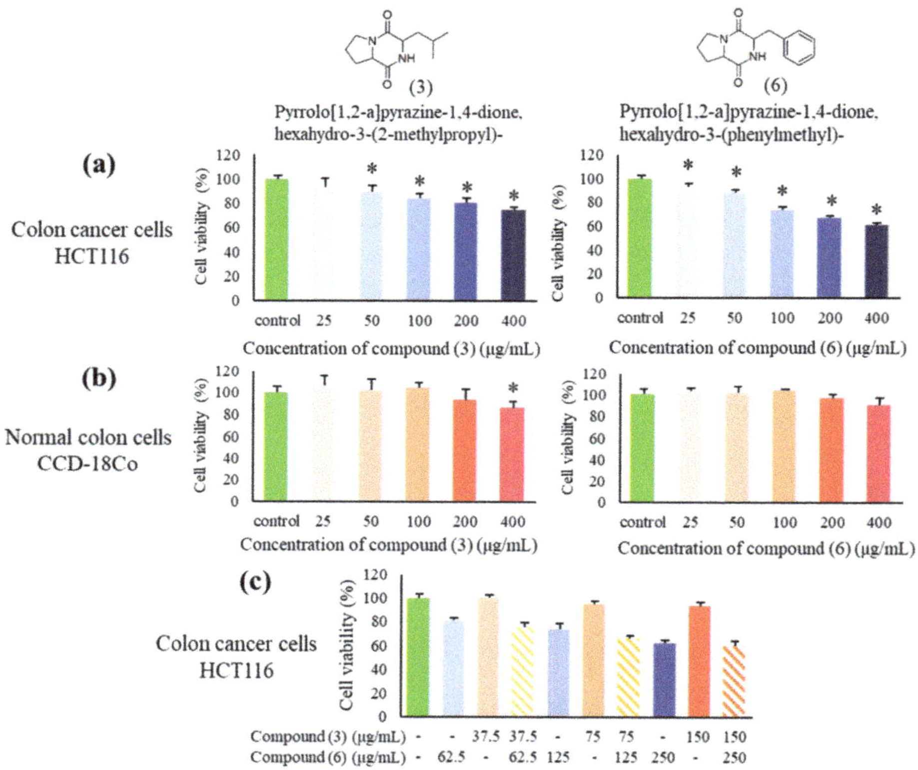

Figure 10. Cytotoxic effects of the two main Compounds (**3**) and (**6**) on HCT116 and CCD-18Co cells. (**a**) Cytotoxic effect of pyrrolo [1,2-a] pyrazine-1,4-dione, hexahydro-3-(2-methylpropyl)- (**3**) and pyrrolo [1,2-a] pyrazine-1,4-dione, hexahydro-3-(phenylmethyl)- (**6**) against HCT116 cells at a series of concentrations after 72 h treatment, ($n = 5$, * $p < 0.05$). (**b**) The toxicity of pyrrolo [1,2-a] pyrazine-1,4-dione, hexahydro-3-(2-methylpropyl)- (**3**) and pyrrolo [1,2-a]pyrazine-1,4-dione, hexahydro-3-(phenylmethyl)- (**6**) towards CCD-18Co after 72 h exposure, ($n = 5$, * $p < 0.05$). (**c**) The combined effect of Compounds (**3**) and (**6**) on HCT116 at respective concentrations after 72 h treatment, ($n = 5$, * $p < 0.5$).

3. Discussion

Actinobacteria, including the genus *Streptomyces*, from unexplored ecosystems, such as mangrove environments, represent a prolific source of bioactive compounds which can contribute immensely to humans in various biotechnological and pharmaceutical applications [44–48]. The capability of these mangrove-derived *Streptomyces* to produce unique secondary metabolites with interesting bioactivities has been linked to the highly dynamic mangrove ecosystem that exerts significant influence on bacterial metabolic and physiological adaptations [26,49]. In the present study, *Streptomyces* sp. MUM256 was demonstrated to produce bioactive compounds—present in the ethyl acetate fraction of crude extract—which exhibited promising antiproliferative activity against colon cancer cells.

In this study, MUM256 EA demonstrated the strongest cytotoxicity effects on HCT116 cells as examined by the 3-(4,5-dimethylthiazol-2-yl)-2,5-diphenyl tetrazolium bromide (MTT) assay which measures the mitochondrial activity in viable cells [50]. Based on the morphological changes assessed by phase-contrast microscopy and flow cytometry, MUM256 EA is suggested to exert its cytotoxic effect by causing the reduced viability of HCT116 through mechanisms associated to apoptotic cell death. This is because among the predominant features of apoptotic cells were their diminished size, condensation of chromatin, segregation of nucleoli, nuclear fragmentation and formation of apoptotic bodies [51,52], features which we observed in treated cells. However, the decreased forward light-scatter signal is not a specific marker of apoptosis since mechanically broken and fragmented cells, isolated cell nuclei, and necrotic cells also have reduced light-scatter properties. Therefore, a more specific assay should be accompanied with these observations to determine the mode of cell death.

Carcinogenesis is a result of cell-cycle disorganization, leading to uncontrolled cellular proliferation [53]. Regulatory proteins like the cyclins, cyclin-dependent kinases (CDK) and their substrate proteins, CDK inhibitors and tumor-suppressor genes are major proteins that control the cell-cycle progression [54]. In tumors, these cell cycle regulators are altered, resulting in failure to control the correct entry and progression through the cell cycles; thus agents that suppress proliferation of cancer cells by regulating these cell cycle regulators have therapeutic value [55]. In the present study, we demonstrated that MUM256 EA reduced the proliferation of colon cancer cells by arresting cells at G1 and G2 phases through the modulation of cell-cycle regulatory proteins including p21, cyclin B1, CDK2, CDK4 and cdc25A phosphatase.

CDK inhibitory protein, p21, is an essential signaling molecule in regulating the cell-cycle progression by interacting directly with cyclin/CDK complexes. It inhibits CDK activities, which in turn results in decreased cell proliferation. Hence, the expression of CDK inhibitory protein p21 has been exploited in the development of chemotherapeutic drugs that disrupt carcinogenesis by impeding the cell cycle in cancer cells [56]. Consistent with this notion, the increased p21 expression level in response to MUM256 EA treatment was accompanied by the decreased expression levels of CDK2 and CDK4 in HCT116 cells. CDK4 associates with cyclin D to regulate the cell-cycle progression in the G1 phase; the cyclin D/CDK4 senses cell mitogenic signals and prompts the cells to initiate DNA synthesis [55,57]. Similarly, CDK2 associates with cyclins E/A to form complexes which activate the downstream targets, including Rb and E2F transcription factor, hence facilitating the transition of G1 to S phase and DNA synthesis [58]. Thus, the accumulation of HCT116 cells in G1 phase in response to MUM256 EA could be due to the downregulation of CDK2 and CDK4 which are crucial in controlling G1/S transition and G1 progression.

The cdc25A is found to be involved in both early G1/S and late G2/M transitions. In the regulation of S phase entry, it activates CDK2 through its dephosphorylating action [59]. Meanwhile, overexpression of cdc25A was also found to mediate mitosis through the regulation of G2/M transition by activating the cyclin B1/CDK1 complexes [60,61]. Thus, we examined the expression levels of cdc25A and cyclin B1 in HCT116 cells upon treatment with MUM256 EA. The expression of cdc25A phosphatase was reduced in HCT116 cells upon treatment with MUM256 EA. Furthermore, a previous study demonstrated that p21 plays an important role in mediating the degradation of cyclin B1 in response to DNA damage and is necessary for the maintenance of G2 cell-cycle arrest in HCT116 cells [62]. Thus, the upregulation of p21 and downregulation of cdc25A in response to MUM256 EA may be involved in the downregulation of cyclin B1 which lead to G2/M cell-cycle arrest in HCT116 cells.

Dysregulation of the apoptotic mechanism has been associated with carcinogenesis. Therefore, any defects along the apoptotic pathway can lead to unchecked cell proliferation, development and progression of cancer and cancer resistance to drug therapies. Thus, apoptosis induction in cancer cells by modulating the pro-apoptotic proteins involved in the apoptotic pathways is an effective strategy for cancer treatment. We demonstrated that MUM256 EA induced cell death in HCT116 via apoptosis as shown by the increased number of Annexin-V positive cells. Furthermore, the use of fluorescent labelled inhibitors, FAM-DEVD-FMK and FAM-LEHD-FMK and the use of caspase inhibitors (benzoylcarbonyl-valyl-alanyl-aspartyl-[O-methyl]-fluoromethylketone (z-VAD-fmk) and benzoylcarbonyl-aspartyl-glutamyl-valyl-aspartyl-fluoromethylketone (z-DEVD-fmk)) further confirmed the role of caspases in MUM256 EA-induced apoptosis. Activated initiator caspases cleave and activate effector caspases, which are involved in cleavage of multiple cellular protein substrates, resulting in triggering of apoptosis [40].

Our data showed that MUM256 EA treatment resulted in increased HCT116 cells with depolarised MMP as compared to the untreated of HCT116 cells. The loss of mitochondrial membrane potential is one of the hallmarks of apoptosis [63]. This result suggests that MUM256 EA-induced cell apoptosis is mediated by mitochondrial-dependent pathways in HCT116 cells. Nevertheless, the occurrence of mitochondrial depolarization also could be a consequence of apoptosis induced by MUM256 EA.

The apoptotic mitochondrial events are tightly regulated by the members of Bcl-2 family of proteins [64]. As a proapoptotic protein, the active Bax functions to permeabilise the mitochondrial outer membrane through the formation of oligomers which induce pore formation at the mitochondrial outer membrane [65]. It is also well established that Bax accumulation causes the release of cytochrome c from mitochondria, leading to the activation of caspase cascade [66,67]. The present findings indicate that MUM256 EA treatment elevated both RNA and protein expression of Bax in HCT116 cells which confirmed the involvement of intrinsic pathway of apoptosis in HCT116 cells.

Literature shows that p53 mediated G_1 and G_2 cell cycle arrest could also be accomplished through transcriptional activation of its downstream targets p21, GADD45α, 14-3-3ε [68]. Our results also support this notion that p53 activation may be involved in the cell cycle arrest in HCT116 cells mediated by MUM256 EA-induced p21 up-regulation. Furthermore, p53 plays an important role in apoptosis via induction of pro-apoptotic Bax protein in response to cellular stress responses. Thus, the increased p53 protein expression suggests that p53 activation may also play an important in the MUM256 EA-induced apoptosis in HCT116 cells. Overall, these data suggest that MUM256 EA triggers sequential activation of p53/Bax/caspase-9/caspase-3/7 cascade to induce the mitochondrial-mediated apoptotic cell death program in HCT116 cells.

According to GC-MS analysis, several groups of chemical compounds such as pyrrole, pyrazine and cyclic dipeptides were detected in the MUM256 EA. The chemical compounds detected in this study also have been evidenced previously in the microbial fermentation broth/extracts, including those isolated from Actinobacteria and the genus *Streptomyces*. For example, 5-pyrrolidino-2-pyrrolidone (**1**) [43], pyrrolo [1,2-a] pyrazine-1,4-dione, hexahydro-3-(2-methylpropyl)- (**3**) [69,70], pyrrolo [1,2-a] pyrazine-1,4-dione, hexahydro-3-(phenylmethyl)- (**6**) [70].

Both pyrrolo [1,2-a] pyrazine-1,4-dione, hexahydro-3-(2-methylpropyl)- (**3**) and pyrrolo [1,2-a] pyrazine-1,4-dione, hexahydro-3-(phenylmethyl)- (**6**) were detected and they belong to the chemical group of proline-containing cyclic dipeptide or 2,5-diketopiperazines (DKP). They appear to be the two major constituents present in MUM256 EA (Figure 9a). DKP consists of a group of the simplest peptide derivatives that are found ubiquitously in nature [71]. Numerous studies have also pointed out the detection of these peptides in the fermentation culture of microorganisms [72–74]. Previous studies also reported that these cyclic dipeptide compounds possess antioxidant and cytotoxic properties [75–78]. A number of previous studies also elucidated the underlying mechanism of action of pyrrolo [1,2a] pyrazine-1,4-dione, hexahydro-3-(phenylmethyl)- (**6**) in killing cancer cells [79–81]. To sum up the results of GC-MS analysis, several chemical compounds were detected which have been noted to have pharmacological properties by previous studies, suggesting that the anti-colon cancer properties of MUM256 EA could be contributed by these chemical compounds. In this study, both pyrrolo [1,2-a] pyrazine-1,4-dione, hexahydro-3-(2-methylpropyl)- (**3**) and pyrrolo [1,2a] pyrazine-1,4-dione, hexahydro-3-(phenylmethyl)- (**6**) were demonstrated to inhibit the growth of HCT116 cells, indicating that these compounds may have been responsible for the anti-colon cancer properties of MUM256 EA.

However, the efficacy of these compounds and their combined effect was not as strong as MUM256 EA against HCT116 colon cells, and this may be due to possible interactions between Compounds (**1**), (**2**), (**4**) and (**5**) in MUM256 EA with the cancer cells. Thus, future study should further identify and validate the potential lead compound as well as elucidate the mechanisms underlying the cell cycle arrest and apoptosis induction. Nonetheless, this study has shed light on the *Streptomyces* sp. strain MUM256 as a promising source for bioactive compounds with anti-colon cancer activities.

4. Materials and Methods

4.1. Phylogenetic Analysis of Strain MUM256

The 16S rRNA gene of strain MUM256 (GenBank accession number KT459477) was subjected to multiple alignment analysis using CLUSTAL-X software with representative sequences of related type strains in genus *Streptomyces* [82]. The reference sequences were obtained from the GenBank/European Molecular Biology Laboratory (EMBL)/DNA Data Bank of Japan (DDBJ) databases on 20th June 2019. The phylogenetic tree was reconstructed according to neighbor-joining algorithm with the use of MEGA version 6.0 software. Kimura's two-parameter model was selected for computing the evolutionary distances of neighbor-joining algorithm [83]. Bootstrap analysis based on 1000 resamplings method of Felsentein was performed to evaluate the topology of the constructed phylogenetic tree [84]. The sequence similarities were calculated based on Ezbiocloud server (http://www.ezbiocloud.net/) [85].

4.2. Fermentation and Preparation of Extracts for Bioactivities Screening

A 14-day culture of *Streptomyces* sp. MUM256 in TSB was used as the inoculum for fermentation process using Han's fermentation media 1 (HFM1) as the fermentative medium [49,86]. Fermentation was carried out in an Erlenmeyer flask containing 200 mL of HFM1 for 7–10 days at 28 °C with shaking speed at 200 rpm. The resulting HFM1 medium was subjected to centrifugation at 12,000× *g* for 15 min. The supernatant was filtered and freeze-dried before proceeding to extraction with methanol. After 72 h of extraction with methanol, the supernatant was filtered and collected. The residue was re-extracted twice under the same conditions. A rotary vacuum evaporator was used to remove the methanol from the crude extract at 40 °C, and DMSO was used to dissolve the crude extract prior to assay.

4.3. Fractionation of Methanolic Crude Extract of Streptomyces sp. MUM256

The MUM256 crude extract was subjected to sequential fractionation using selected solvents namely hexane, ethyl acetate and water. The solid methanolic MUM256 extract was suspended and macerated with hexane solvent. The hexane fraction was obtained after filtration through filter paper. The residue was partitioned between equal volumes of ethyl acetate and distilled water. Both the hexane and ethyl acetate fractions were concentrated using a rotary evaporator while the aqueous fraction was subjected to freeze drying. The dried fractions were suspended in DMSO before further analysis.

4.4. Cell Culture

Both human colon cancer (HCT116, HT-29) and normal colon (CCD-18Co) cell lines were purchased from American Type Culture Collection (ATCC, Virginia, VA, USA). The cells were cultured in RPMI (Roswell Park Memorial Institute)-1640 (Gibco, Life Technologies, Gaithersburg, MD, USA) supplemented with 10% fetal bovine serum and 1× antibiotic-antimycotic (Gibco) at 37 °C in a humidified incubator containing 5% CO_2 and 95% air. The cultures were viewed using an inverted microscope to assess the degree of confluency and to ensure they were free from bacterial and fungal contamination.

4.5. Cell Viability Test Using MTT (3-(4,5-Dimethylthiazol-2-yl)-2,5-Diphenyltetrazolium Bromide) Assay

The cytotoxic effect of the extract/fractions on HCT116, HT-29 and CCD-18Co cell lines were determined using 3-(4,5-dimethylthiazol-2-yl)-2,5-diphenyltetrazolium bromide (MTT) assay [87]. The cells (5×10^3 cells/well) were seeded into 96-well plates and left to adhere overnight in the incubator. *Streptomyces* extract/fraction was added into each well in a series of concentrations (25 to 400 μg/mL) in the presence or absence of apoptosis inhibitors (z-VAD-fmk or z-DEVD-fmk). Pyrrolo [1,2-a] pyrazine-1,4-dione, hexahydro-3-(2-methylpropyl)- (Henan Tianfu Chemical Co., Ltd., Zhengzhou, China) and pyrrolo [1,2-a] pyrazine-1,4-dione, hexahydro-3-(phenylmethyl)- (Chemieliva Pharmaceutical Co., Ltd., Chongqing, China) were added into each well in a series of concentrations. DMSO (0.5% (v/v)) was used as the solvent and the negative control in all the experiments. The treatment of cells was conducted according to the indicated durations. MTT assay was performed with the addition of 20 μL of 5 mg/mL of MTT (Sigma, Saint Louis, MO, USA) into each well and incubated at 37 °C in a humid atmosphere with 5% CO_2, 95% air for 4 h. The formazan crystals were dissolved in DMSO (100 μL) after the removal of the medium. The absorbance of dissolved formazan solution was measured spectrophotometrically at 570 nm (with 650 nm as reference wavelength) using a microplate reader.

4.6. *Trypan Blue Exclusion Assay*

Trypan blue was used to stain and differentiate live and dead cells. Under phase contrast microscopy, the live cells appear colorless and bright while the dead cells stained blue after exposure to trypan blue. The live cells were counted using a haemocytometer.

4.7. *Clonogenic Survival Assay*

HCT116 cells were seeded in 6-well plates at 1500 cells/well, and the indicated concentrations of MUM256 EA were added the following day. After 24 h of treatment, the media containing the MUM256 EA was aspirated and rinsed with phosphate-buffered saline (PBS). Fresh culture media was added, and the cells were left for 1 week to form colonies [88]. The culture media was replaced with fresh culture media every 2 days. Colonies were fixed in ice-cold methanol and subsequently stained with 0.01% crystal violet in dH_2O for 10 min. Excess stain was rinsed twice with dH_2O and allowed to air dry. Images of each well were taken, and the colony area was quantitated using the 'colony area' plugin of ImageJ software [89].

4.8. *Phase Contrast Microscopy*

The effect of MUM256 EA on the HCT116 cells was examined morphologically with an inverted light microscope.

4.9. *Flow Cytometry*

Flow cytometric analysis was performed on a BD FACSVerse™ flow cytometer (BD Bioscience, San Jose, CA, USA). For phenotypic characterization of cells, the cells were harvested and washed twice with PBS after treatment. The aliquots of cell suspension were analysed according to cell size (as defined by FSC value) and their granularity (as defined by SSC value) using the FSC/SSC dot plot diagram. A total of 10,000 events were acquired for each sample and presented on the FSC/SSC dot plot.

4.10. *Cell-Cycle Analysis Using Propidium Iodide (PI)*

DNA content and cell cycle distribution were evaluated using propidium iodide (PI) staining [90]. After exposure to MUM256 EA for a designated period, the cells were harvested and washed with PBS before being fixed with 70% ice-cold ethanol at −20 °C overnight. Fixed cells were washed twice and stained in a buffer containing 25 μg/mL PI, 0.1% Triton-X-100 and 100 μg/mL RNase A for 30 min in a dark at room temperature. The PI-stained cells were acquired up to 10,000 events and analysed using BD FACSVerse™ flow cytometer (BD Bioscience, San Jose, CA, USA).

4.11. *Quantification of Gene Expression Using Reverse-Transcription Quantitative Polymerase Chain Reaction (PCR)*

To investigate the antiproliferative and apoptotic effects of the *Streptomyces* sp. MUM256 EA, the expression of genes associated with apoptosis and cell cycle regulation were examined using qPCR. After treatment, the total RNA from cancer cells was extracted using TRIZOL reagent. The RNA concentration was determined by ultraviolet (UV) spectrophotometry. Reverse transcription of the extracted RNA was performed using a commercial cDNA reverse transcription kit. Quantitative real-time PCR was conducted in a reaction mixture containing cDNA, specific primers and SYBR Green qPCR master mix. The following primer pair sequences were used: CDK2, 5′-GGTCCTCCACCGAGACCTTAA-3′ (forward) and 5′-CAGGGACTCCAAAAGCTCTGG-3′ (reverse); CDK4, 5′-CAGTGTACAAGGCCCGT GATC-3′ (forward) and 5′-ACGAACTGTGCTGATGGGAAG-3′ (reverse); cdc25A, 5′-CCCCAAAGG AACCATTGAGA-3′ (forward) and 5′-CTGATGTTTCCCAGCAACTG-3′ (reverse); $p21^{WAF1/CIP1}$, 5′-GGACAGCAGAGGAAGACCATGT-3′ (forward) and 5′-GCCGTTTTCGACCCTGAGA-3′ (reverse); 18S rRNA, 5′-GGCCCTGTAATTGGAATGAGTC-3′ (forward) and 5′-CCAAGATCCAACTACGA GCTT-3′ (reverse); Bax, 5′-GTCGCCCTTTTCTACTTTGCCAG-3′ (forward) and 5′-TCCAGCCCAACA GCCGCTCC-3′ (reverse); PBGD, 5′-ACCATCGGAGCCATCTGCAAG-3′ (forward) and 5′-CCCACC ACACTCTTCTCTGGCA-3′ (reverse). Primers specificity was confirmed by peak melt curve before use. The amplification was then performed as described by Goh, et al. [91]. The RNA expression level of the genes of interest was then analysed and normalized to the house keeping genes (18SrRNA and PBGD) expression levels.

4.12. Western Blot Analysis

The cells were harvested and lysed in cold radioimmunoprecipitation assay (RIPA) buffer containing protease and phosphatase inhibitors after treatment with MUM256 EA for designated periods. The total content of protein was quantified using bicinchoninic acid (BCA) assay. Twenty-five micrograms of total protein were loaded into 10% sodium dodecyl sulphate-polyacrylamide gel electrophoresis (SDS-PAGE) and transferred onto a nitrocellulose membrane. The membrane was blocked with skim milk for 1 h prior to incubation with primary antibodies (p21, cyclin B1 and β-actin, Cell Signaling Technology, Denver, MA, USA) overnight at 4 °C. After substantial washing, the membrane was reacted with anti-rabbit/mouse immunoglobulin G-horseradish peroxidase-labelled secondary antibodies for 1 h at room temperature. The membrane was stained with enhanced chemiluminescence (ECL) detection kit prior to detection using gel documentation system. Image J was used to analyse and quantify the band intensities.

4.13. Hoechst 33342 Nuclear Staining

Hoechst 33342 nuclear stain was used to detect the apoptotic cells by observing the apoptotic nuclear morphology using fluorescence microscopy. After treatment, the cells were harvested in PBS and then exposed to 0.05 g/L Hoechst 33342 dye in PBS for 30 min at room temperature. The samples were observed and analysed under fluorescence microscope [90].

4.14. Externalization of Phosphatidylserine Detection by Flow Cytometry

The cells were harvested after treatment with MUM256 EA. A cell suspension at a density of 1×10^6 /mL was prepared and subjected to centrifugation to obtain the cell pellet. The cell pellet was washed with pre-cooled PBS prior to the staining with Annexin V-FITC and propidium iodide in the 1X binding buffer [90]. After incubation at room temperature for 15 minutes in the dark, cells were analyzed by flow cytometry. A total of 10,000 events were collected and analysed using BD FACSVerse™ flow cytometer (BD Bioscience, San Jose, CA, USA).

4.15. Measurement of Caspases-3/7 and -9 Activities

Using Carboxyfluorescein FLICA Apoptosis Detection Kit (Immunochemistry technologies, LLC), the caspases-3/7 and -9 activities were measured in HCT116 cells after treatment by flow cytometer. The treated cells were harvested and incubated with 1x FAM-DEVD-FMK or FAM-LEHD-FMK reagent for 1 h at 37 °C. The stained cells were washed twice with washing solution before subjected to 10,000 events acquisition and analysis by using FACSVerse™ flow cytometer (BD Bioscience, San Jose, CA, USA).

4.16. Mitochondrial Membrane Potential Analysis

JC-1 (5,5′,6,6′-tetrachloro-1,1′,3,3′-tetrathylbenzimidazolcarbocyanine iodide) dye from BD™ MitoScreen kit ((BD Bioscience, San Jose, CA, USA) was used to evaluate the status of mitochondrial membrane potential of the cells. After exposure to *Streptomyces* sp. MUM256 EA for a designated period, cells were harvested and washed twice before being subjected to 10,000 events acquisition and analysis by using BD FACSVerse™ flow cytometer (BD Bioscience, San Jose, CA, USA).

4.17. Measurement of p53 and Bax Proteins by Flow Cytometry

The expression level of p53 and Bax proteins was determined by immunofluorescence staining using flow cytometry. After the treatment with MUM256 EA, cells were washed twice with PBS, fixed and subsequently permeabilized using Cytofix/Cytoperm™ kit (BD Biosciences, San Jose, CA, USA). The cells were resuspended in 500 µL of fixation/permeabilization solution and incubated at 4 °C for 20 min. The cells were washed twice with Perm/Wash™ buffer and incubated for another 15 min in 1 mL of Perm/Wash™ buffer. For the detection of Bax or p53 proteins, the fixed and permeabilized cells were incubated with 100 µL of Perm/Wash™ buffer containing corresponding antibodies. Briefly, the indirect antibodies staining was conducted by incubating the cells with mouse anti-human monoclonal antibody (p53, Bax) (Santa Cruz, TX, USA) at 4 °C in the dark for 30 min. After the washing procedure, the cells were further incubated with fluorescein isothiocyanate (FITC)-conjugated goat anti-mouse IgG_1 (Santa Cruz, USA) at 4 °C in the dark for 30 min. Finally, the cells were then washed with Perm/Wash™ buffer before being subjected to 10,000 events acquisition and analysis by using FACSVerse™ flow cytometer (BD Bioscience). FITC-conjugated normal mouse IgG_1 (Santa Cruz, USA) was used as the isotype control to differentiate non-specific background signal.

4.18. Chemical Profiling of MUM256 EA Using GC-MS

Gas chromatography-mass spectrometry (GC-MS) analysis was performed to profile the constituents present in MUM256 EA [92]. The analysis was conducted using Agilent technologies 6980N (GC) equipped with 5979 Mass Selective Detector (MS), HP-5MS (5% phenyl methyl siloxane) capillary column of dimensions 30.0 m × 250 µm × 0.25 µm and helium was used as the carrier gas at 1 mL/ min. For the initial 10 min, the column was operated at 40 °C, followed by an increase of 3 °C/min to 250 °C and was kept isothermal for 5 min. The MS was operating at 70 eV. Identification was performed by comparing the mass spectral data of the detected constituents in MUM256 EA with those available in the database NIST 05 Spectral library.

4.19. Statistical Analysis

All the experiments were performed at least in triplicate. The results were expressed as mean ± standard deviation (SD) and analysed using SPSS statistical analysis software. One-way analysis of variance (ANOVA) and Student's *t*-test were performed to determine if there was a significant difference between the treated and untreated groups. A difference was considered statistically significant when $p \leq 0.05$.

5. Conclusions

In summary, these findings show that MUM256 EA has an inhibitory effect on the proliferating and colony-forming abilities of HCT116 cells. Our work demonstrates that MUM256 EA induced G1 and G2/M cell-cycle arrest in HCT116 cells possibly linked to upregulation of p21 and p53 and downregulation of cyclin B1, CDK2, CDK4 and cdc25A phosphatase. Morphological changes observed suggest the occurrence of apoptosis in HCT116 cells upon treatment with MUM256 EA. The results demonstrated that MUM256 EA induced apoptosis in HCT116 cells mediated through intrinsic pathway with activation of p53, Bax, caspases-9 and -3/7. GC-MS analysis of MUM256 EA detected the presence of cyclic dipeptides which were demonstrated to have various pharmacological properties from previous literature suggesting that these compounds may have contributed to the cytotoxic properties of MUM256 EA. Furthermore, the proline-based cyclic dipeptides detected in MUM256 EA inhibited the growth HCT116 cells while showing little to no toxicity towards CCD-18Co normal colon cells. Taken together, these findings highlight the important role of exploring the microorganisms from the mangrove as a bioresource, especially for proline-based cyclic dipeptides. Further characterization of the lead compounds in this fraction may yield bioactive agents which hold tremendous promise for the development of chemopreventive drugs, and *in vivo* investigation of MUM256 EA is underway.

Supplementary Materials: The following are available online at http://www.mdpi.com/2072-6694/11/11/1742/s1: Figure S1: The Western blot of protein of interest (p21 and cyclin B) and respective β-actin expressions, Figure S2: Cytotoxic effects of the two main compounds (3) and (6) on HT-29 colon cancer cells.

Author Contributions: The writing was performed by L.T.-H.T., P.P., L.-H.L. and B.-H.G. While C.-K.C., T.M.K., K.-G.C., H.-L.S., L.-H.L. and B.-H.G. provided technical support, vital guidance and insight to the work. The project was conceptualized by L.T.-H.-L. and B.-H.G.

Funding: This research is financially supported by the MOSTI eScience Fund (02-02-10-SF0215), University grants from the University of Malaya (FRGS grant no. FP022-2018A), External Industry Grants from Biotek Abadi Sdn Bhd (vote no. GBA-81811A), Monash Global Asia in the 21st Century (GA21) research grant (GA-HW-19-L01 & GA-HW-19-S01) and Fundamental Research Grant Scheme (FRGS/1/2019/WAB09/MUSM/02/1 & FRGS/1/2019/SKK08/MUSM/02/7).

Acknowledgments: This work was inspired by Monash PhD Research Training Module entitled "Bioprospective of Microbes with Biopharmaceutical Potential with Bioinformatics and Drug Discovery Platforms". We would like to thank Prof. Chow Sek Chuen (from School of Science, Monash University Malaysia) for providing us with the caspase inhibitors z-VAD-fmk and z-DEVD-fmk.

Conflicts of Interest: The authors declare no conflict of interest.

References

1. Ferlay, J.; Soerjomataram, I.; Dikshit, R.; Eser, S.; Mathers, C.; Rebelo, M.; Parkin, D.M.; Forman, D.; Bray, F. Cancer incidence and mortality worldwide: Sources, methods and major patterns in GLOBOCAN 2012. *Int. J. Cancer* **2015**, *136*, E359–E386. [CrossRef] [PubMed]
2. Arnold, M.; Sierra, M.S.; Laversanne, M.; Soerjomataram, I.; Jemal, A.; Bray, F. Global patterns and trends in colorectal cancer incidence and mortality. *Gut* **2017**, *66*, 683–691. [CrossRef] [PubMed]
3. Lee, M.-T.G.; Chiu, C.-C.; Wang, C.-C.; Chang, C.-N.; Lee, S.-H.; Lee, M.; Hsu, T.-C.; Lee, C.-C. Trends and Outcomes of Surgical Treatment for Colorectal Cancer between 2004 and 2012-an Analysis using National Inpatient Database. *Sci. Rep.* **2017**, *7*, 2006. [CrossRef] [PubMed]
4. Chatterjee, K.; Zhang, J.; Honbo, N.; Karliner, J.S. Doxorubicin cardiomyopathy. *Cardiology* **2010**, *115*, 155–162. [CrossRef] [PubMed]
5. Steele, T.A. Chemotherapy-induced immunosuppression and reconstitution of immune function. *Leuk. Res.* **2002**, *26*, 411–414. [CrossRef]
6. Jin, J.; Wu, X.; Yin, J.; Li, M.; Shen, J.; Li, J.; Zhao, Y.; Zhao, Q.; Wu, J.; Wen, Q. Identification of genetic mutations in cancer: Challenge and opportunity in the new era of targeted therapy. *Front. Oncol.* **2019**, *9*, 263. [CrossRef] [PubMed]
7. Khoo, X.-H.; Paterson, I.C.; Goh, B.-H.; Lee, W.-L. Cisplatin-Resistance in Oral Squamous Cell Carcinoma: Regulation by Tumor Cell-Derived Extracellular Vesicles. *Cancers* **2019**, *11*, 1166. [CrossRef] [PubMed]
8. Eng, S.-K.; Loh, T.H.T.; Goh, B.-H.; Lee, W.-L. KRAS as Potential Target in Colorectal Cancer Therapy. In *Natural Bio-active Compounds*; Springer: New York, NY, USA, 2019; pp. 389–424.
9. Chan, C.K.; Tang, L.Y.; Goh, B.H.; Kadir, H.A. Targeting apoptosis via inactivation of PI3K/Akt/mTOR signaling pathway involving NF-κB by geraniin in HT-29 human colorectal adenocarcinoma cells. *Prog. Drug Discov. Biomed. Sci.* **2019**, *2*, a0000030.
10. Pan, P.; Skaer, C.; Yu, J.; Zhao, H.; Ren, H.; Oshima, K.; Wang, L.S. Berries and other natural products in the pancreatic cancer chemoprevention in human clinical trials. *J. Berry Res.* **2017**, *7*, 147–161. [CrossRef]
11. Kotecha, R.; Takami, A.; Espinoza, J.L. Dietary phytochemicals and cancer chemoprevention: A review of the clinical evidence. *Oncotarget* **2016**, *7*, 52517–52529. [CrossRef]

12. Tan, L.T.H.; Low, L.E.; Tang, S.Y.; Yap, W.H.; Chuah, L.H.; Chan, C.K.; Lee, L.H.; Goh, B.H. A reliable and affordable 3D tumor spheroid model for natural product drug discovery: A case study of curcumin. *Prog. Drug Discov. Biom. Sci.* **2019**, *2*, a0000017.
13. Tan, L.T.-H.; Chan, K.-G.; Pusparajah, P.; Lee, W.-L.; Chuah, L.-H.; Khan, T.M.; Lee, L.-H.; Goh, B.-H. Targeting membrane lipid a potential cancer cure? *Front. Pharmacol.* **2017**, *8*, 12. [CrossRef] [PubMed]
14. Low, L.E.; Tan, L.T.-H.; Goh, B.-H.; Tey, B.T.; Ong, B.H.; Tang, S.Y. Magnetic cellulose nanocrystal stabilized Pickering emulsions for enhanced bioactive release and human colon cancer therapy. *Int. J. Biol. Macromol.* **2019**, *127*, 76–84. [CrossRef] [PubMed]
15. Tan, L.T.H.; Lee, L.H.; Yin, W.F.; Chan, C.K.; Abdul Kadir, H.; Chan, K.G.; Goh, B.H. Traditional Uses, Phytochemistry, and Bioactivities of Cananga odorata (Ylang-Ylang). *Evid. Based Complement. Alternat. Med.* **2015**, *2015*, 896314. [CrossRef] [PubMed]
16. Tang, C.; Hoo, P.C.; Tan, L.T.; Pusparajah, P.; Khan, T.M.; Lee, L.H.; Goh, B.H.; Chan, K.G. Golden Needle Mushroom: A Culinary Medicine with Evidenced-Based Biological Activities and Health Promoting Properties. *Front. Pharmacol.* **2016**, *7*, 474. [CrossRef]
17. Tan, H.-L.; Chan, K.-G.; Pusparajah, P.; Lee, L.-H.; Goh, B.-H. Gynura procumbens: An overview of the biological activities. *Front. Pharmacol.* **2016**, *7*, 52. [CrossRef]
18. Goh, J.X.H.; Tan, L.T.-H.; Goh, J.K.; Chan, K.G.; Pusparajah, P.; Lee, L.-H.; Goh, B.-H. Nobiletin and derivatives: Functional compounds from citrus fruit peel for colon cancer chemoprevention. *Cancers* **2019**, *11*, 867. [CrossRef]
19. Tay, K.-C.; Tan, L.T.-H.; Chan, C.K.; Hong, S.L.; Chan, K.-G.; Yap, W.H.; Pusparajah, P.; Lee, L.-H.; Goh, B.-H. Formononetin: A review of its anticancer potentials and mechanisms. *Front. Pharmacol.* **2019**, *10*, 820. [CrossRef]
20. Zhang, J.-Y.; Tao, L.-Y.; Liang, Y.-J.; Yan, Y.-Y.; Dai, C.-L.; Xia, X.-K.; She, Z.-G.; Lin, Y.-C.; Fu, L.-W. Secalonic acid D induced leukemia cell apoptosis and cell cycle arrest of G1 with involvement of GSK-3β/β-catenin/c-Myc pathway. *Cell Cycle* **2009**, *8*, 2444–2450. [CrossRef]
21. Lee, L.-H.; Chan, K.-G.; Stach, J.; Wellington, E.M.; Goh, B.-H. The search for biological active agent (s) from actinobacteria. *Front. Microbiol.* **2018**, *9*, 824. [CrossRef]
22. Azman, A.-S.; Othman, I.; Fang, C.-M.; Chan, K.-G.; Goh, B.-H.; Lee, L.-H. Antibacterial, anticancer and neuroprotective activities of rare Actinobacteria from mangrove forest soils. *Indian J. Microbiol.* **2017**, *57*, 177–187. [CrossRef] [PubMed]
23. Lee, L.-H.; Azman, A.-S.; Zainal, N.; Eng, S.-K.; Ab Mutalib, N.-S.; Yin, W.-F.; Chan, K.-G. Microbacterium mangrovi sp. nov., an amylolytic actinobacterium isolated from mangrove forest soil. *Int. J. Syst. Evol. Microbiol.* **2014**, *64*, 3513–3519. [CrossRef] [PubMed]
24. Pugh, L.H.; Katz, E.; Waksman, S.A. Antibiotic and cytostatic properties of the actinomycins. *J. Bacteriol.* **1956**, *72*, 660.
25. Katz, L.; Baltz, R.H. Natural product discovery: Past, present, and future. *J. Ind. Microbiol. Biotechnol.* **2016**, *43*, 155–176. [CrossRef] [PubMed]
26. Ser, H.L.; Tan, L.T.; Law, J.W.; Chan, K.G.; Duangjai, A.; Saokaew, S.; Pusparajah, P.; Ab Mutalib, N.S.; Khan, T.M.; Goh, B.H.; et al. Focused Review: Cytotoxic and Antioxidant Potentials of Mangrove-Derived Streptomyces. *Front. Microbiol.* **2017**, *8*, 2065. [CrossRef] [PubMed]
27. Law, J.W.; Ser, H.L.; Duangjai, A.; Saokaew, S.; Bukhari, S.I.; Khan, T.M.; Ab Mutalib, N.S.; Chan, K.G.; Goh, B.H.; Lee, L.H. Streptomyces colonosanans sp. nov., A Novel Actinobacterium Isolated from Malaysia Mangrove Soil Exhibiting Antioxidative Activity and Cytotoxic Potential against Human Colon Cancer Cell Lines. *Front. Microbiol.* **2017**, *8*, 877. [CrossRef]
28. Ser, H.-L.; Palanisamy, U.D.; Yin, W.-F.; Chan, K.-G.; Goh, B.-H.; Lee, L.-H. Streptomyces malaysiense sp. nov.: A novel Malaysian mangrove soil actinobacterium with antioxidative activity and cytotoxic potential against human cancer cell lines. *Sci. Rep.* **2016**, *6*, 24247. [CrossRef]
29. Tan, L.T.-H.; Chan, K.-G.; Pusparajah, P.; Yin, W.-F.; Khan, T.M.; Lee, L.-H.; Goh, B.-H. Mangrove derived Streptomyces sp. MUM265 as a potential source of antioxidant and anticolon-cancer agents. *BMC Microbiol.* **2019**, *19*, 38. [CrossRef]
30. Tan, L.T.H.; Ser, H.L.; Yin, W.F.; Chan, K.G.; Lee, L.H.; Goh, B.H. Investigation of Antioxidative and Anticancer Potentials of Streptomyces sp. MUM256 Isolated from Malaysia Mangrove Soil. *Front. Microbiol.* **2015**, *6*, 1316. [CrossRef]

31. Munshi, A.; Hobbs, M.; Meyn, R.E. Clonogenic cell survival assay. In *Chemosensitivity*; Springer: New York, NY, USA, 2005; pp. 21–28.
32. Wlodkowic, D.; Skommer, J.; Darzynkiewicz, Z. Cytometry of apoptosis. Historical perspective and new advances. *Exp. Oncol.* **2012**, *34*, 255–262.
33. Taylor, W.R.; Stark, G.R. Regulation of the G2/M transition by p53. *Oncogene* **2001**, *20*, 1803–1815. [CrossRef] [PubMed]
34. Zhang, X.; Song, X.; Yin, S.; Zhao, C.; Fan, L.; Hu, H. p21 induction plays a dual role in anti-cancer activity of ursolic acid. *Exp. Biol. Med. (Maywood)* **2016**, *241*, 501–508. [CrossRef] [PubMed]
35. Waldman, T.; Kinzler, K.W.; Vogelstein, B. p21 is necessary for the p53-mediated G1 arrest in human cancer cells. *Cancer Res.* **1995**, *55*, 5187–5190. [PubMed]
36. Wyllie, A.H. Glucocorticoid-induced thymocyte apoptosis is associated with endogenous endonuclease activation. *Nature* **1980**, *284*, 555–556. [CrossRef] [PubMed]
37. Henry, C.M.; Hollville, E.; Martin, S.J. Measuring apoptosis by microscopy and flow cytometry. *Methods* **2013**, *61*, 90–97. [CrossRef] [PubMed]
38. Riccardi, C.; Nicoletti, I. Analysis of apoptosis by propidium iodide staining and flow cytometry. *Nat. Protoc.* **2006**, *1*, 1458–1461. [CrossRef] [PubMed]
39. Darzynkiewicz, Z.; Bedner, E. Analysis of apoptotic cells by flow and laser scanning cytometry. *Methods Enzymol.* **1999**, *322*, 18–39.
40. Green, D.; Kroemer, G. The central executioners of apoptosis: Caspases or mitochondria? *Trends Cell Biol.* **1998**, *8*, 267–271. [CrossRef]
41. Elleuch, L.; Shaaban, M.; Smaoui, S.; Mellouli, L.; Karray-Rebai, I.; Fourati-Ben Fguira, L.; Shaaban, K.A.; Laatsch, H. Bioactive secondary metabolites from a new terrestrial Streptomyces sp. TN262. *Appl. Biochem. Biotechnol.* **2010**, *162*, 579–593. [CrossRef]
42. Kavitha, A.; Savithri, H.S. Biological Significance of Marine Actinobacteria of East Coast of Andhra Pradesh, India. *Front. Microbiol.* **2017**, *8*, 1201. [CrossRef]
43. Zothanpuia, A.K.P.; Chandra, P.; Leo, V.V.; Mishra, V.K.; Kumar, B.; Singh, B.P. Production of Potent Antimicrobial Compounds from Streptomyces cyaneofuscatus Associated with Fresh Water Sediment. *Front. Microbiol.* **2017**, *8*, 68. [CrossRef] [PubMed]
44. Tan, L.T.-H.; Chan, K.-G.; Lee, L.-H.; Goh, B.-H. Streptomyces bacteria as potential probiotics in aquaculture. *Front. Microbiol.* **2016**, *7*, 79. [CrossRef] [PubMed]
45. Tan, L.T.H.; Mahendra, C.K.; Yow, Y.Y.; Chan, K.G.; Khan, T.M.; Lee, L.H.; Goh, B.H. *Streptomyces* sp. MUM273b: A mangrove-derived potential source for antioxidant and UVB radiation protectants. *MicrobiologyOpen* **2019**, *8*, e859. [CrossRef] [PubMed]
46. Ser, H.-L.; Ab Mutalib, N.-S.; Yin, W.-F.; Goh, B.-H.; Lee, L.-H.; Chan, K.-G. Genome sequence of Streptomyces antioxidans MUSC 164T isolated from mangrove forest. *Prog. Microbes Mol. Biol.* **2018**, *1*, a0000001.
47. Lee, L.-H.; Zainal, N.; Azman, A.-S.; Ab Mutalib, N.-S.; Hong, K.; Chan, K.-G. Mumia flava gen. nov., sp. nov., an actinobacterium of the family Nocardioidaceae. *Int. J. Syst. Evol. Microbiol.* **2014**, *64*, 1461–1467. [CrossRef]
48. Lee, L.-H.; Cheah, Y.-K.; Sidik, S.M.; Xie, Q.-Y.; Tang, Y.-L.; Lin, H.-P.; Ab Mutalib, N.-S.; Hong, K. Barrientosiimonas humi gen. nov., sp. nov., an actinobacterium of the family Dermacoccaceae. *Int. J. Syst. Evol. Microbiol.* **2013**, *63*, 241–248. [CrossRef]
49. Hong, K.; Gao, A.H.; Xie, Q.Y.; Gao, H.; Zhuang, L.; Lin, H.P.; Yu, H.P.; Li, J.; Yao, X.S.; Goodfellow, M.; et al. Actinomycetes for marine drug discovery isolated from mangrove soils and plants in China. *Mar. Drugs* **2009**, *7*, 24–44. [CrossRef]
50. Twentyman, P.R.; Luscombe, M. A study of some variables in a tetrazolium dye (MTT) based assay for cell growth and chemosensitivity. *Br. J. Cancer* **1987**, *56*, 279. [CrossRef]
51. Kerr, J.F.; Wyllie, A.H.; Currie, A.R. Apoptosis: A basic biological phenomenon with wideranging implications in tissue kinetics. *Br. J. Cancer* **1972**, *26*, 239. [CrossRef]
52. Gerschenson, L.E.; Rotello, R.J. Apoptosis: A different type of cell death. *FASEB J.* **1992**, *6*, 2450–2455. [CrossRef]
53. Reddy, L.; Odhav, B.; Bhoola, K.D. Natural products for cancer prevention: A global perspective. *Pharmacol. Ther.* **2003**, *99*, 1–13. [CrossRef]

54. Grana, X.; Reddy, E.P. Cell cycle control in mammalian cells: Role of cyclins, cyclin dependent kinases (CDKs), growth suppressor genes and cyclin-dependent kinase inhibitors (CKIs). *Oncogene* **1995**, *11*, 211–219. [PubMed]
55. Malumbres, M.; Barbacid, M. Cell cycle, CDKs and cancer: A changing paradigm. *Nat. Rev. Cancer* **2009**, *9*, 153–166. [CrossRef] [PubMed]
56. Karimian, A.; Ahmadi, Y.; Yousefi, B. Multiple functions of p21 in cell cycle, apoptosis and transcriptional regulation after DNA damage. *DNA Repair (Amst)* **2016**, *42*, 63–71. [CrossRef] [PubMed]
57. Malumbres, M. Cyclin-dependent kinases. *Genome Biol.* **2014**, *15*, 122. [CrossRef]
58. Chohan, T.A.; Qian, H.; Pan, Y.; Chen, J.-Z. Cyclin-dependent kinase-2 as a target for cancer therapy: Progress in the development of CDK2 inhibitors as anti-cancer agents. *Curr. Med. Chem.* **2015**, *22*, 237–263. [CrossRef]
59. Busino, L.; Chiesa, M.; Draetta, G.F.; Donzelli, M. Cdc25A phosphatase: Combinatorial phosphorylation, ubiquitylation and proteolysis. *Oncogene* **2004**, *23*, 2050–2056. [CrossRef]
60. Chen, M.-S.; Ryan, C.E.; Piwnica-Worms, H. Chk1 kinase negatively regulates mitotic function of Cdc25A phosphatase through 14-3-3 binding. *Mol. Cell. Biol.* **2003**, *23*, 7488–7497. [CrossRef]
61. Mailand, N.; Podtelejnikov, A.V.; Groth, A.; Mann, M.; Bartek, J.; Lukas, J. Regulation of G2/M events by Cdc25A through phosphorylation-dependent modulation of its stability. *EMBO J.* **2002**, *21*, 5911–5920. [CrossRef]
62. Gillis, L.D.; Leidal, A.M.; Hill, R.; Lee, P.W. p21Cip1/WAF1 mediates cyclin B1 degradation in response to DNA damage. *Cell Cycle* **2009**, *8*, 253–256. [CrossRef]
63. Fesik, S.W. Promoting apoptosis as a strategy for cancer drug discovery. *Nat. Rev. Cancer* **2005**, *5*, 876. [CrossRef] [PubMed]
64. Kirkin, V.; Joos, S.; Zornig, M. The role of Bcl-2 family members in tumorigenesis. *Biochim. Biophys. Acta* **2004**, *1644*, 229–249. [CrossRef] [PubMed]
65. Er, E.; Oliver, L.; Cartron, P.-F.; Juin, P.; Manon, S.; Vallette, F.M. Mitochondria as the target of the pro-apoptotic protein Bax. *Biochim. Biophys. Acta (BBA)-Bioenerg.* **2006**, *1757*, 1301–1311. [CrossRef] [PubMed]
66. Kroemer, G.; Reed, J.C. Mitochondrial control of cell death. *Nat. Med.* **2000**, *6*, 513–519. [CrossRef] [PubMed]
67. Lin, M.; Tang, S.; Zhang, C.; Chen, H.; Huang, W.; Liu, Y.; Zhang, J. Euphorbia factor L2 induces apoptosis in A549 cells through the mitochondrial pathway. *Acta Pharm. Sin. B* **2017**, *7*, 59–64. [CrossRef] [PubMed]
68. Johnson, D.G.; Walker, C.L. Cyclins and cell cycle checkpoints. *Annu. Rev. Pharmacol. Toxicol.* **1999**, *39*, 295–312. [CrossRef] [PubMed]
69. Rhee, K.H. Isolation and characterization of Streptomyces sp KH-614 producing anti-VRE (vancomycin-resistant enterococci) antibiotics. *J. Gen. Appl. Microbiol.* **2002**, *48*, 321–327. [CrossRef] [PubMed]
70. Macherla, V.R.; Liu, J.; Bellows, C.; Teisan, S.; Nicholson, B.; Lam, K.S.; Potts, B.C. Glaciapyrroles A, B, and C, pyrrolosesquiterpenes from a Streptomyces sp. isolated from an Alaskan marine sediment. *J. Nat. Prod.* **2005**, *68*, 780–783. [CrossRef]
71. Prasad, C. Bioactive cyclic dipeptides. *Peptides* **1995**, *16*, 151–164. [CrossRef]
72. Vazquez-Rivera, D.; Gonzalez, O.; Guzman-Rodriguez, J.; Diaz-Perez, A.L.; Ochoa-Zarzosa, A.; Lopez-Bucio, J.; Meza-Carmen, V.; Campos-Garcia, J. Cytotoxicity of cyclodipeptides from Pseudomonas aeruginosa PAO1 leads to apoptosis in human cancer cell lines. *BioMed Res. Int.* **2015**, *2015*, 197608. [CrossRef]
73. Ser, H.-L.; Yin, W.-F.; Chan, K.-G.; Khan, T.M.; Goh, B.-H.; Lee, L.-H. Antioxidant and cytotoxic potentials of Streptomyces gilvigriseus MUSC 26T isolated from mangrove soil in Malaysia. *Prog. Microbes Mol. Biol.* **2018**, *1*, a0000002.
74. Würth, R.; Barbieri, F.; Florio, T. New molecules and old drugs as emerging approaches to selectively target human glioblastoma cancer stem cells. *BioMed Res. Int.* **2014**, *2014*, 126586. [CrossRef] [PubMed]
75. Ser, H.-L.; Palanisamy, U.D.; Yin, W.-F.; Malek, A.; Nurestri, S.; Chan, K.-G.; Goh, B.-H.; Lee, L.-H. Presence of antioxidative agent, Pyrrolo [1, 2-a] pyrazine-1, 4-dione, hexahydro-in newly isolated *Streptomyces mangrovisoli* sp. nov. *Front. Microbiol.* **2015**, *6*, 854. [CrossRef] [PubMed]
76. Law, J.W.-F.; Ser, H.-L.; Ab Mutalib, N.-S.; Saokaew, S.; Duangjai, A.; Khan, T.M.; Chan, K.-G.; Goh, B.-H.; Lee, L.-H. Streptomyces monashensis sp. nov., a novel mangrove soil actinobacterium from East Malaysia with antioxidative potential. *Sci. Rep.* **2019**, *9*, 3056. [CrossRef]
77. Tan, L.T.-H.; Chan, K.-G.; Chan, C.K.; Khan, T.M.; Lee, L.-H.; Goh, B.-H. Antioxidative potential of a *Streptomyces* sp. MUM292 isolated from mangrove soil. *BioMed Res. Int.* **2018**, *2018*, 4823126. [CrossRef]

78. Tan, L.T.-H.; Chan, K.-G.; Khan, T.M.; Bukhari, S.I.; Saokaew, S.; Duangjai, A.; Pusparajah, P.; Lee, L.-H.; Goh, B.-H. *Streptomyces* sp. MUM212 as a source of antioxidants with radical scavenging and metal chelating properties. *Front. Pharmacol.* **2017**, *8*, 276. [CrossRef]
79. Brauns, S.C.; Dealtry, G.; Milne, P.; Naude, R.; Van de Venter, M. Caspase-3 activation and induction of PARP cleavage by cyclic dipeptide cyclo(Phe-Pro) in HT-29 cells. *Anticancer Res.* **2005**, *25*, 4197–4202.
80. Brauns, S.C.; Milne, P.; Naude, R.; Van de Venter, M. Selected cyclic dipeptides inhibit cancer cell growth and induce apoptosis in HT-29 colon cancer cells. *Anticancer Res.* **2004**, *24*, 1713–1719.
81. Rhee, K.-H. Inhibition of DNA topoisomerase I by cyclo (L-prolyl-L-phenylalanyl) isolated from Streptomyces sp. AMLK-335. *J. Microbiol. Biotechnol.* **2002**, *12*, 1013–1016.
82. Thompson, J.D.; Gibson, T.J.; Higgins, D.G. Multiple sequence alignment using ClustalW and ClustalX. *Curr. Protoc. Bioinform.* **2003**, *1*, 2–3. [CrossRef]
83. Kumar, S.; Tamura, K.; Nei, M. MEGA: Molecular evolutionary genetics analysis software for microcomputers. *Bioinformatics* **1994**, *10*, 189–191. [CrossRef] [PubMed]
84. Felsenstein, J. Confidence limits on phylogenies: An approach using the bootstrap. *Evolution* **1985**, *39*, 783–791. [CrossRef] [PubMed]
85. Yoon, S.-H.; Ha, S.-M.; Kwon, S.; Lim, J.; Kim, Y.; Seo, H.; Chun, J. Introducing EzBioCloud: A taxonomically united database of 16S rRNA gene sequences and whole-genome assemblies. *Int. J. Syst. Evol. Microbiol.* **2017**, *67*, 1613. [PubMed]
86. Lee, L.H.; Cheah, Y.K.; Mohd Sidik, S.; Ab Mutalib, N.S.; Tang, Y.L.; Lin, H.P.; Hong, K. Molecular characterization of Antarctic actinobacteria and screening for antimicrobial metabolite production. *World J. Microbiol. Biotechnol.* **2012**, *28*, 2125–2137. [CrossRef] [PubMed]
87. Chan, C.K.; Tan, L.T.; Andy, S.N.; Kamarudin, M.N.A.; Goh, B.H.; Kadir, H.A. Anti-neuroinflammatory Activity of *Elephantopus scaber* L. via Activation of Nrf2/HO-1 Signaling and Inhibition of p38 MAPK Pathway in LPS-Induced Microglia BV-2 Cells. *Front. Pharmacol.* **2017**, *8*, 397. [CrossRef] [PubMed]
88. Tessmann, J.W.; Buss, J.; Begnini, K.R.; Berneira, L.M.; Paula, F.R.; de Pereira, C.M.P.; Collares, T.; Seixas, F.K. Antitumor potential of 1-thiocarbamoyl-3,5-diaryl-4,5-dihydro-1H-pyrazoles in human bladder cancer cells. *Biomed. Pharmacother.* **2017**, *94*, 37–46. [CrossRef]
89. Guzman, C.; Bagga, M.; Kaur, A.; Westermarck, J.; Abankwa, D. ColonyArea: An ImageJ plugin to automatically quantify colony formation in clonogenic assays. *PLoS ONE* **2014**, *9*, e92444. [CrossRef]
90. Goh, B.H.; Kadir, H.A. In vitro cytotoxic potential of Swietenia macrophylla King seeds against human carcinoma cell lines. *J. Med. Plants Res.* **2011**, *5*, 1395–1404.
91. Goh, B.H.; Chan, C.K.; Kamarudin, M.N.A.; Abdul Kadir, H. Swietenia macrophylla King induces mitochondrial-mediated apoptosis through p53 upregulation in HCT116 colorectal carcinoma cells. *J. Ethnopharmacol.* **2014**, *153*, 375–385. [CrossRef]
92. Supriady, H.; Kamarudin, M.N.A.; Chan, C.K.; Goh, B.H.; Kadir, H.A. SMEAF attenuates the production of pro-inflammatory mediators through the inactivation of Akt-dependent NF-κB, p38 and ERK1/2 pathways in LPS-stimulated BV-2 microglial cells. *J. Funct. Foods* **2015**, *17*, 434–448. [CrossRef]

cancers

MDPI

Article

Manoalide Preferentially Provides Antiproliferation of Oral Cancer Cells by Oxidative Stress-Mediated Apoptosis and DNA Damage

Hui-Ru Wang [1], Jen-Yang Tang [2,3], Yen-Yun Wang [4,5,6], Ammad Ahmad Farooqi [7], Ching-Yu Yen [8], Shyng-Shiou F. Yuan [4,6,9], Hurng-Wern Huang [1,*] and Hsueh-Wei Chang [4,6,10,11,12,*]

1 Institute of Biomedical Science, National Sun Yat-sen University, Kaohsiung 80424, Taiwan
2 Department of Radiation Oncology, Faculty of Medicine, College of Medicine, Kaohsiung Medical University, Kaohsiung 80708, Taiwan
3 Department of Radiation Oncology, Kaohsiung Medical University Hospital, Kaohsiung 80708, Taiwan
4 Cancer Center, Kaohsiung Medical University Hospital, Kaohsiung Medical University, Kaohsiung 80708, Taiwan
5 School of Dentistry, College of Dental Medicine, Kaohsiung Medical University, Kaohsiung 80708, Taiwan
6 Center for Cancer Research, Kaohsiung Medical University, Kaohsiung 80708, Taiwan
7 Department of Molecular Oncology, Institute of Biomedical and Genetic Engineering (IBGE), Islamabad 54000, Pakistan
8 Department of Oral and Maxillofacial Surgery Chi-Mei Medical Center, Tainan 71004, Taiwan
9 Translational Research Center, Kaohsiung Medical University Hospital, Kaohsiung 80708, Taiwan
10 Department of Medical Research, Kaohsiung Medical University Hospital, Kaohsiung 80708, Taiwan
11 Institute of Medical Science and Technology, National Sun Yat-sen University, Kaohsiung 80424, Taiwan
12 Department of Biomedical Science and Environmental Biology, Kaohsiung Medical University, Kaohsiung 80708, Taiwan
* Correspondence: sting@mail.nsysu.edu.tw (H.-W.H.); changhw@kmu.edu.tw (H.-W.C.); Tel.: +886-7-525-2000 (ext. 5814) (H.-W.H.); +886-7-312-1101 (ext. 2691) (H.-W.C.); Fax: +886-7-525-0197 (H.-W.H.); +886-7-312-5339 (H.-W.C.)

Received: 15 May 2019; Accepted: 2 September 2019; Published: 4 September 2019

Abstract: Marine sponge-derived manoalide has a potent anti-inflammatory effect, but its potential application as an anti-cancer drug has not yet been extensively investigated. The purpose of this study is to evaluate the antiproliferative effects of manoalide on oral cancer cells. MTS assay at 24 h showed that manoalide inhibited the proliferation of six types of oral cancer cell lines (SCC9, HSC3, OC2, OECM-1, Ca9-22, and CAL 27) but did not affect the proliferation of normal oral cell line (human gingival fibroblasts (HGF-1)). Manoalide also inhibits the ATP production from 3D sphere formation of Ca9-22 and CAL 27 cells. Mechanically, manoalide induces subG1 accumulation in oral cancer cells. Manoalide also induces more annexin V expression in oral cancer Ca9-22 and CAL 27 cells than that of HGF-1 cells. Manoalide induces activation of caspase 3 (Cas 3), which is a hallmark of apoptosis in oral cancer cells, Ca9-22 and CAL 27. Inhibitors of Cas 8 and Cas 9 suppress manoalide-induced Cas 3 activation. Manoalide induces higher reactive oxygen species (ROS) productions in Ca9-22 and CAL 27 cells than in HGF-1 cells. This oxidative stress induction by manoalide is further supported by mitochondrial superoxide (MitoSOX) production and mitochondrial membrane potential (MitoMP) destruction in oral cancer cells. Subsequently, manoalide-induced oxidative stress leads to DNA damages, such as γH2AX and 8-oxo-2′-deoxyguanosine (8-oxodG), in oral cancer cells. Effects, such as enhanced antiproliferation, apoptosis, oxidative stress, and DNA damage, in manoalide-treated oral cancer cells were suppressed by inhibitors of oxidative stress or apoptosis, or both, such as *N*-acetylcysteine (NAC) and Z-VAD-FMK (Z-VAD). Moreover, mitochondria-targeted superoxide inhibitor MitoTEMPO suppresses manoalide-induced MitoSOX generation and γH2AX/8-oxodG DNA damages. This study validates the preferential antiproliferation effect of manoalide and explores the oxidative stress-dependent mechanisms in anti-oral cancer treatment.

Keywords: marine sponge; natural product; anticancer drug; oral cancer inhibition

1. Introduction

Oral cancer is one of the high incidence cancers worldwide [1], especially in Southeast Asia and Taiwan. Betel quid chewing, smoking, and alcohol consumption are high risk factors for oral cancer [2]. Oral cancer causes serious morbidity and mortality [3]. Current therapies for oral cancer patients include surgery or chemoradiation, or both. However, chemoradiation commonly shows severe side effects in oral cancer patients [4]. Therefore, continuous drug screening and development for oral cancer therapy remains a challenge.

Marine natural products provide an abundant resource for development of anti-cancer agents [5–8]. Besides corals, marine sponges provide diverse sources of natural products from the ocean. Marine sponges are marine resources with a wide range of bioactive compounds and secondary metabolites with potential therapeutic effects [9–11]. Bioactive compounds of marine sponges and their microbial consortia are known for their anticancer, anti-inflammatory, antiviral, and antibiotic effects [11,12].

In 1980, manoalide, an antibiotic sesterterpenoid isolated from the marine sponge *Luffariella variabilis*, was discovered [13]. In 1999, manoalide was reported to function as an analgesic and anti-inflammatory agent [14]. This anti-inflammatory effect may be caused by the inhibition of phospholipase A2 (PLA2) by manoalide [14]. Moreover, manoalide also functions as inhibitors for phospholipase C (PLC) [15,16] and calcium channels [17]. Manoalide reached Phase II (antipsoriatic) clinical trial, although it was discontinued due to formulation problems [18].

In addition to anti-inflammatory effects, the anti-cancer effects of manoalide have not been extensively studied. For example, manoalide showed a cytotoxic effect against murine lymphoma LI210 and human epidermoid carcinoma KB cells [19]. However, the anticancer effect against oral cancer cells was not studied as yet.

Natural products, such as marine sponges, commonly showed antioxidant properties [20,21]. Some marine sponge-derived natural products showed both cytotoxic and antioxidant activities [22–24]. Manoalide inhibits superoxide production in colon cancer cells (HT29-D4) [25], suggesting that manoalide may have an antioxidant potential. Interestingly, antioxidants possess double-edge sword activities to regulate cellular reactive oxygen species (ROS). For example, antioxidants at physiological concentrations may decrease ROS and benefit cell health but induce ROS that damage cells at high concentrations [26]. Hence, the ROS modulating effect of manoalide on oral cancer cells warrants further investigation. Furthermore, drugs with inducing ROS generation ability may preferentially kill cancer cells but show little damage to normal cells [27]. Whether manoalide causes a preferential killing to oral cancer cells needs further to be examined.

In this study, we hypothesized that manoalide may preferentially inhibit the proliferation of oral cancer cells. To examine this hypothesis, the preferential antiproliferation effect of manoalide on oral cancer cells was studied by analyzing cell survival, cell cycle, apoptosis, oxidative stress, and DNA damage.

2. Results

2.1. *Cell Viability of Manoalide-Treated Oral Cancer and Normal Oral Cells with or Without Pretreatments of NAC or Z-VAD*

Cell viability was determined by mitochondrial enzyme activity-based MTS assay. Figure 1A shows that manoalide dose-responsively decreases the viability (%) of oral cancer cells (CAL 27, Ca9-22, HSC3, OECM-1, SCC9, and OC-2), but it only slightly decreases oral normal cells (human gingival fibroblasts (HGF-1)), i.e., their IC_{50} values of manoalide are 7.8, 9.1, 14.9, 17.4, and 18.5 μM at 24 h MTS assay. Among the oral cancer cells, Ca9-22 and CAL 27 cells belong to different oral locations (gingival and tongue) and show higher cytotoxicity upon manoalide treatment. Accordingly, Ca9-22 and CAL 27 cells were selected for the following assays to investigate the detailed mechanisms of anti-oral cancer cells by manoalide. Figure 1B shows that 48 and 72 h treatments of manoalide dose-responsively decrease the viability (%) of oral cancer cells, but it only slightly decreases oral normal cells (HGF-1), i.e., the IC_{50} values of manoalide-treated oral cancer Ca9-22 and CAL 27 cells are 5.3 versus 14.0 μM and 3.1 versus 7.5 μM at 48 and 72 h MTS assay, respectively. Furthermore, the photo images of 3D sphere formation pattern of oral cancer cells are provided (Figure S1A). Its cell viability needs to be determined by ATP detection. As shown in Figure 1C, the ATP-detected 3D sphere formation ability of oral cancer cells (Ca9-22 and CAL 27) was decreased by manoalide treatment.

To address the role of oxidative stress and apoptosis in cell viability, the ROS scavenger *N*-acetylcysteine (NAC) [28,29] and apoptosis inhibitor Z-VAD-FMK (Z-VAD) [30] were used. The cell morphologies were abnormal in manoalide-treated oral cancer (Ca9-22 and CAL 27) cells, especially at higher concentrations (Figure S1B). However, these manoalide-induced abnormal changes on morphologies were recovered by NAC pretreatment and partly recovered by Z-VAD pretreatment (Figure S1B). Moreover, manoalide-suppressed cell viabilities in oral cancer cells were completely inhibited by a NAC pretreatment and partly inhibited by a Z-VAD pretreatment (Figure 1D).

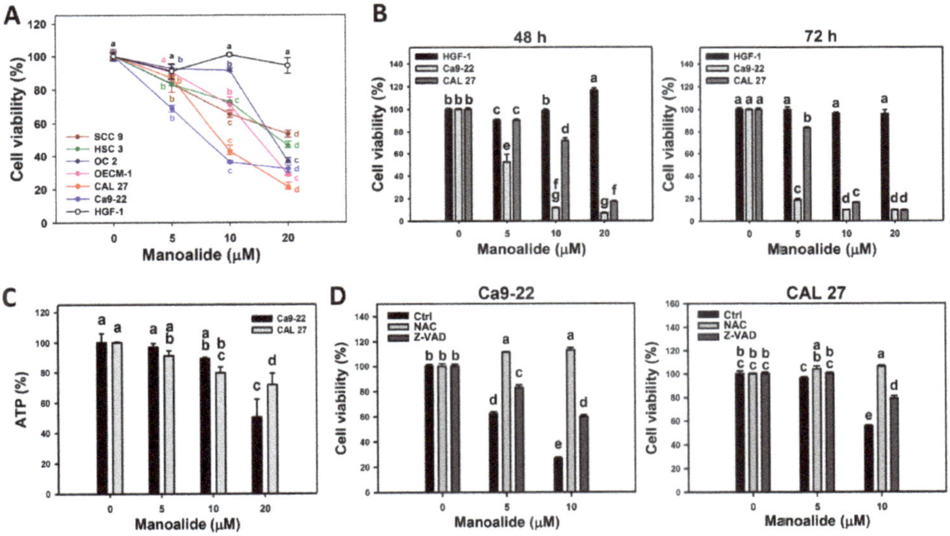

Figure 1. Cell viabilities of oral cancer cells after manoalide treatment and its *N*-acetylcysteine (NAC)/apoptosis inhibitor Z-VAD-FMK (Z-VAD) effects. Cells were treated with 0, 5, 10, and 20 μM of manoalide. All treatments have the same concentration of DMSO. (**A**) MTS assay-based cell viabilities for 24 h. Oral cancer (CAL 27, Ca9-22, OECM-1, OC-2, HSC 3, and SCC9) cells and oral normal (HGF-1) cells were included. (**B**) MTS assay-based cell viabilities for 48 and 72 h for oral cancer (CAL 27 and Ca9-22) and oral normal (HGF-1) cells. (**C**) Statistical of 3D spheroid formation for manoalide-treated oral cancer (Ca9-22 and CAL 27) cells for 72 h. (**D**) NAC and Z-VAD effects on MTS viability of manoalide-treated oral cancer cells. Pretreatment conditions were 8 mM, 1 h for NAC and 100 μM, 2 h for Z-VAD. Following pretreatment or not, oral cancer (Ca9-22 and CAL 27) cells were post-incubated with 5 and 10 μM manoalide for 24 h. Data, means ± SDs (n = 3). Data were analyzed by one-way ANOVA with Tukey HSD Post Hoc Test. Data showing the same small lettersrepresent nonsignificant differences whereas data showing no overlapping same small letters are significant difference (p < 0.05–0.001).

2.2. Cell Cycle Changes of Manoalide-Treated Oral Cancer Cells with or Without Pretreatments of NAC or Z-VAD

7-Aminoactinomycin D (7AAD) is a DNA staining dye for measuring the different cell cycle phases. Figure S2A shows the pattern changes of cell cycle progression for oral cancer cells (Ca9-22 and CAL 27) after manoalide treatment. The subG1 and > 4N populations appear at 10 and 20 μM of manoalide for Ca9-22 cells and at 20 μM for CAL 27 cells. Figure 2A shows that the subG1 populations are increased after manoalide treatment.

To address the role of oxidative stress and apoptosis in cell cycle distribution, the NAC and Z-VAD were used. Figure S2B shows the effect of NAC and Z-VAD pretreatments on pattern of cell cycle progression for manoalide-treated oral cancer cells and shows cell cycle disturbances (subG1 and > 4N populations). Figure 2B shows these manoalide-induced subG1 accumulations were recovered by NAC pretreatment and partly recovered by Z-VAD pretreatment.

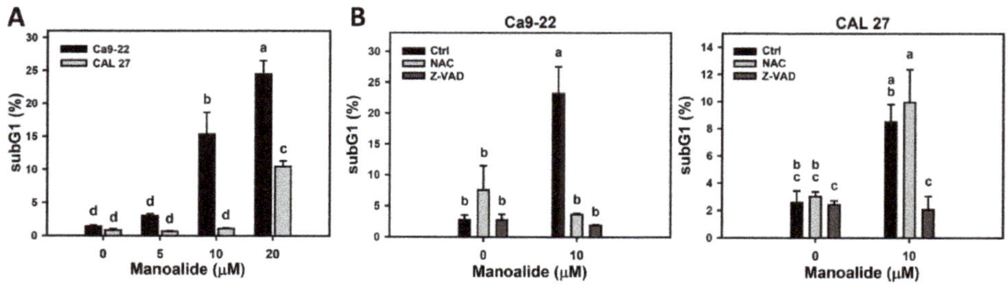

Figure 2. Cell cycle changes of manoalide-treated oral cancer (Ca9-22 and CAL 27) cells. (**A**) Statistical results of the subG1 (%) for manoalide-treated oral cancer cells in Figure S2A. Cells were treated with 0, 5, 10, and 20 μM of manoalide for 24 h. (**B**) Statistical result of the subG1 (%) for NAC, Z-VAD, and/or manoalide-treated oral cancer cells in Figure S2B. Cells were pretreated with 8 mM, 1 h for NAC or 100 μM and 2 h for Z-VAD, and they were then post-incubated with 10 μM of manoalide for 24 h. Data, means ± SDs ($n = 3$). Data were analyzed by one-way ANOVA with Tukey HSD Post Hoc Test. Data showing no overlapping same small letters represent significant difference ($p < 0.05$–0.001).

2.3. Apoptosis of Manoalide-Treated Oral Cancer Cells with or Without Pretreatments of NAC or Z-VAD

Apoptosis was detected by the annexin V/7AAD method. Figure S3A shows that the populations of oral cancer (Ca9-22 and CAL 27) cells shift from annexin V (−)/7ADD (−) to annexin V (+)/7ADD (−) at 5 μM of manoalide and further shift to annexin V (+)/7ADD (+) at 10 and 20 μM. In contrast, normal oral cells (HGF-1) show only a slight shift to apoptosis region. Therefore, cell populations of oral cancer cells shift from alive, early apoptosis, to late apoptosis when the concentrations of manoalide increase. Figure 3A shows that manoalide mainly induces early apoptosis at 5 μM, moderately induces late apoptosis at 10 μM, and mainly induces late apoptosis at 20 μM in oral cancer cells. However, manoalide-treated HGF-1 cells induce little apoptosis, which is undetectable at 5 and 10 μM and is less than 15% for early apoptosis at 20 μM.

The involvement of oxidative stress in the apoptosis for manoalide-treated oral cancer cells were further examined (Figure S3B and Figure 3B). Figure S3B shows that the populations of manoalide-induced late apoptosis shift to early apoptosis or living status by NAC or Z-VAD pretreatment. Figure 3B shows that NAC or Z-VAD pretreatments decrease the manoalide-induced apoptosis for both oral cancer cells, Ca9-22 and CAL 27.

To further validate that apoptosis is induced by manoalide in oral cancer cells, western blotting analysis was performed. Figure 3C shows that manoalide induces overexpression of cleaved forms of caspase 3 (c-Cas 3) in both oral cancer cells, Ca9-22 and CAL 27. Figure S4A showing the procaspase 3 and c-Cas 3 patterns also supports this finding. Moreover, this manoalide-induced c-Cas 3 is suppressed by NAC or Z-VAD pretreatments in both oral cancer cells, Ca9-22 and CAL 27 (Figure 3D). All the raw data for western blotting are provided (Figures S4B and S5).

To identify the initiator caspase responsible for the activation of Cas-3, the involvements of Cas 8 and Cas 9 were examined by using its inhibitors [31]. Figure S6A shows that the immunofluorescence of c-Cas 8- and c-Cas 9-staining of manoalide-treated oral cancer cells is higher than the control. Figure S6B shows that the populations of manoalide-induced c-Cas 3 shift to a low level by pretreatments of Cas 8 or Cas 9 inhibitors. Figure 3E shows that Cas 8 or Cas 9 inhibitor pretreatments decrease the manoalide-induced c-Cas 3 activation for both oral cancer cells, Ca9-22 and CAL 27, suggesting that initiator caspase, such as both Cas 8 and Cas 9, are responsible for the c-Cas-3 activation in manoalide-treated oral cancer cells.

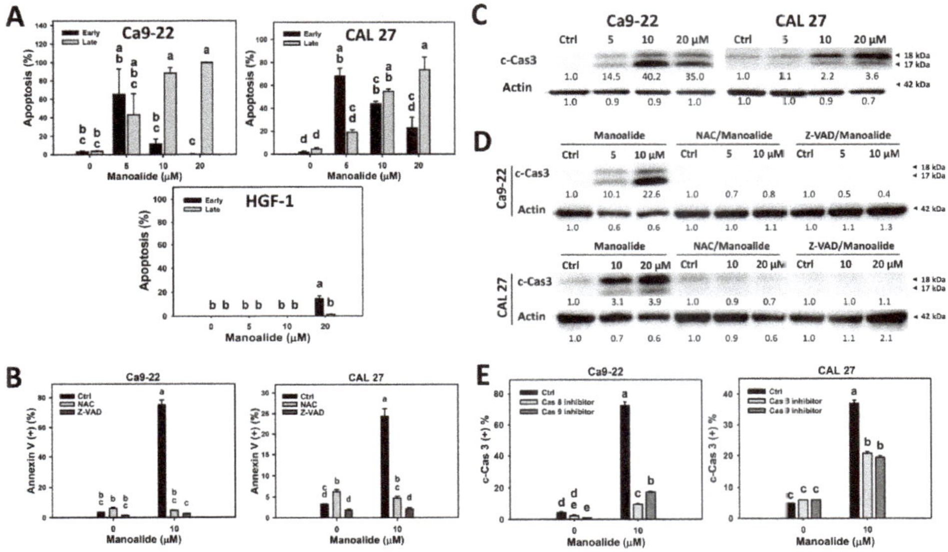

Figure 3. Apoptosis changes in manoalide-treated oral cancer (Ca9-22 and CAL 27) cells and normal oral (HGF-1) cells. (**A**) Statistical results of the annexin V/7AAD method in manoalide-treated oral cancer cells and normal oral (HGF-1) cells in Figure S3A. Cells were treated with different concentrations of manoalide for 24 h. Early and late apoptosis were, respectively, counted by the populations in the annexin V (+)/7AAD (−) and annexin V (+)/7AAD (+) regions, i.e., Q3 and Q2. (**B**) Statistics results of annexin V/7AAD method in NAC, Z-VAD, and/or manoalide-treated oral cells in Figure S3B. Cells were pretreated with NAC (8 mM, 1 h) or Z-VAD (100 µM, 2 h), and posttreated with manoalide (10 µM, 24 h). Apoptosis was represented by the sum of early and late apoptosis, i.e., annexin V (+)/7AAD (+ or −). (**C**) Western blotting for detecting apoptosis in manoalide-treated oral cancer cells. (**D**) Western blotting for detecting apoptosis in NAC, Z-VAD, and/or manoalide-treated oral cells. Cleaved forms caspase 3 (c-Cas 3) were used to detect apoptosis. Actin was the internal control. (**E**) Statistical results of c-Cas 3 positive levels in Cas 8 inhibitor, Cas 9 inhibitor, and/or manoalide-treated oral cells in Figure S6. Cells were pretreated with Cas 8 inhibitor Z-IETD-FMK (100 µM, 2 h) or Cas 9 inhibitor Z-LEHD-FMK (100 µM, 2 h), and posttreated with manoalide (10 µM, 24 h). Data were analyzed by one-way ANOVA with Tukey HSD Post Hoc Test. Data, means ± SDs ($n = 3$). Data showing no overlapping same small letters represent significant difference ($p < 0.05$–0.001).

2.4. ROS Production of Manoalide-Treated Oral Cancer and Normal Oral Cells

2′,7′-dichlorodihydrofluorescein diacetate (DCFH-DA) can react with ROS to generate products for flow cytometry detection [32]. Figure S7A shows the ROS patterns of manoalide-treated oral cancer (Ca9-22 and CAL 27) and normal oral (HGF-1) cells. Figure 4A shows that the ROS productions of

Ca9-22 and CAL 27 cells are dramatically induced when the concentrations of manoalide increase. In contrast, the ROS productions of HGF-1 cells stay unchanged at 5%, at less than 10 μM and slightly increased to 25% at 20 μM.

To address the role of oxidative stress and apoptosis in manoalide-induced ROS production, pretreatments of NAC and Z-VAD were used, and its ROS patterns were shown in Figure S7B. Figure 4B shows that the manoalide-induced ROS productions are inhibited by NAC and Z-VAD pretreatments for both oral cancer cells, Ca9-22 and CAL 27.

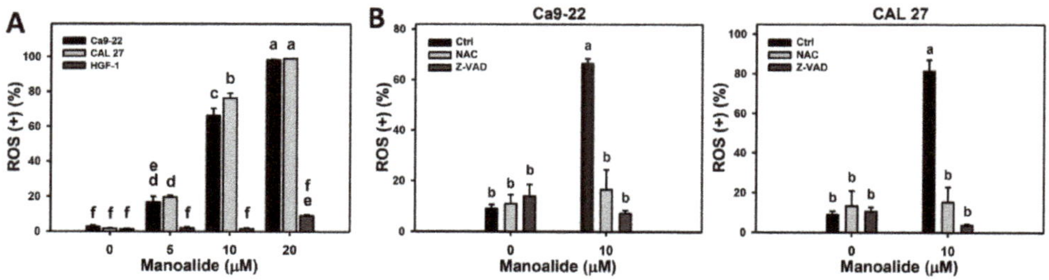

Figure 4. ROS changes in manoalide-treated oral cancer (Ca9-22 and CAL 27) and normal oral (HGF-1) cells. Cells were treated with different concentrations of manoalide for 10 min. (**A**). Statistical results of ROS (+) (%) for manoalide-treated oral cancer and oral normal cells in Figure S7A. (**B**) Statistical results in NAC, Z-VAD, and/or manoalide-treated oral cells in Figure S7B. Cells were pretreated with NAC (8 mM, 1 h) or Z-VAD (100 μM, 2 h), and posttreated with manoalide (10 μM, 10 min). Data were analyzed by one-way ANOVA with Tukey HSD Post Hoc Test. Data, means ± SDs ($n = 3$). Data showing no overlapping same small letters represent significant differences ($p < 0.05$–0.001).

2.5. Mitochondrial Superoxide (MitoSOX) Production of Manoalide-Treated Oral Cancer Cells with or Without Pretreatment of MitoTEMPO

MitoSOX™ Red can react with intra-mitochondrial superoxide to generate products for flow cytometry detection [33]. Figure S8A shows the populations of oral cancer (Ca9-22 and CAL 27) cells shift to MitoSOX (+) region when the concentrations of manoalide increase. Figure 5A shows that the MitoSOX production of Ca9-22 and CAL 27 cells dramatically increase. Figure S8B shows the positive control treatment (betulinic acid; BA) [34] for MitoSOX patterns in oral cancer. Figure 5B shows that the manoalide induces more MitoSOX productions than that of BA in Ca9-22 and CAL 27 cells.

Moreover, the involvement of MitoSOX in manoalide-treated oral cancer cells was further validated using MitoSOX inhibitor (MitoTEMPO). Figure S8C shows the MitoSOX patterns of MitoTEMPO pretreatment effects against manoalide-treated oral cancer cells. Figure 5C shows that MitoTEMPO pretreatment decreases the manoalide-induced MitoSOX production for both oral cancer cells, Ca9-22 and CAL 27.

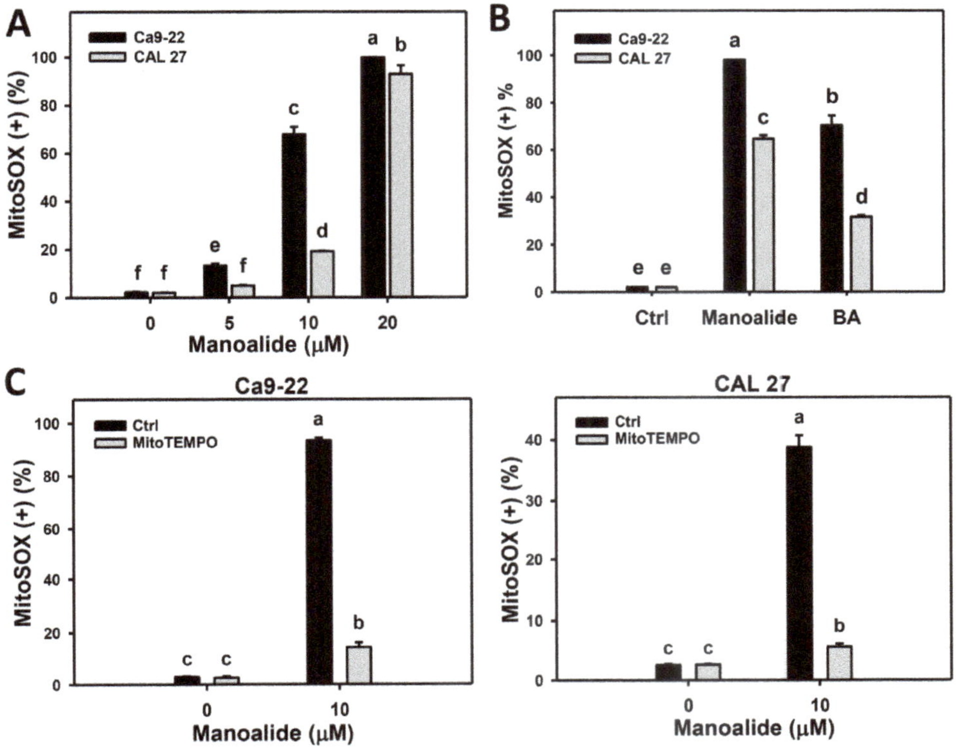

Figure 5. Change of mitochondrial superoxide (MitoSOX) production in manoalide-treated oral cancer (Ca9-22 and CAL 27) cells. (**A**) Statistical results of MitoSOX (+) (%) for manoalide-treated oral cancer cells in Figure S8A. Cells were treated with different concentrations of manoalide for 24 h. (**B**) Statistical results of positive control of MitoSOX (+) (%) for oral cancer cells in Figure S8B. Cells were treated with betulinic acid (BA; 25 μM, 24 h) as the positive control treatment for comparison to manoalide (10 μM, 24 h). (**C**) Statistical results of MitoSOX (+) (%) in MitoSOX inhibitor (MitoTEMPO) and/or manoalide-treated oral cells in Figure S8C. Cells were pretreated with MitoTEMPO (20 μM, 1 h) and posttreated with manoalide (10 μM, 24 h). Data were analyzed by one-way ANOVA with Tukey HSD Post Hoc Test. Data, means ± SDs ($n = 3$). Data showing no overlapping same small letters represent significant difference ($p < 0.05$–0.001).

2.6. Membrane Potential (MitoMP) of Manoalide-Treated Oral Cancer Cells with or Without Pretreatments of NAC or Z-VAD

JC-1 aggregate (red fluorescent) form can concentrate at mitochondria, and JC-1 monomer form (green fluorescent) escape from mitochondria, reflecting the intact and depolarized MitoMP, respectively [35]. These fluorescent signals were detected by flow cytometry. Figure S9A shows that the populations of oral cancer (Ca9-22 and CAL 27) cells shift from JC-1 aggregates MitoMP (+) to JC-1 monomers MitoMP (−) region when the concentrations of manoalide increase. Figure 6A shows that the JC-1 monomers MitoMP (−) population is dose-responsively increased in manoalide-treated oral cancer cells. Figure S9B shows the positive control treatment (betulinic acid; BA) for MitoMP patterns in oral cancer. Figure 6B shows that the manoalide induces more JC-1 monomers MitoMP (−) productions than that of BA in Ca9-22 and CAL 27 cells.

To address the role of oxidative stress and apoptosis in manoalide-suppressed MitoMP, pretreatments of NAC and Z-VAD were used and its MitoMP patterns are shown in Figure S9C.

Figure 6C shows that the manoalide-induced JC-1 monomer generations are inhibited by NAC and Z-VAD pretreatments for both oral cancer cells, Ca9-22 and CAL 27.

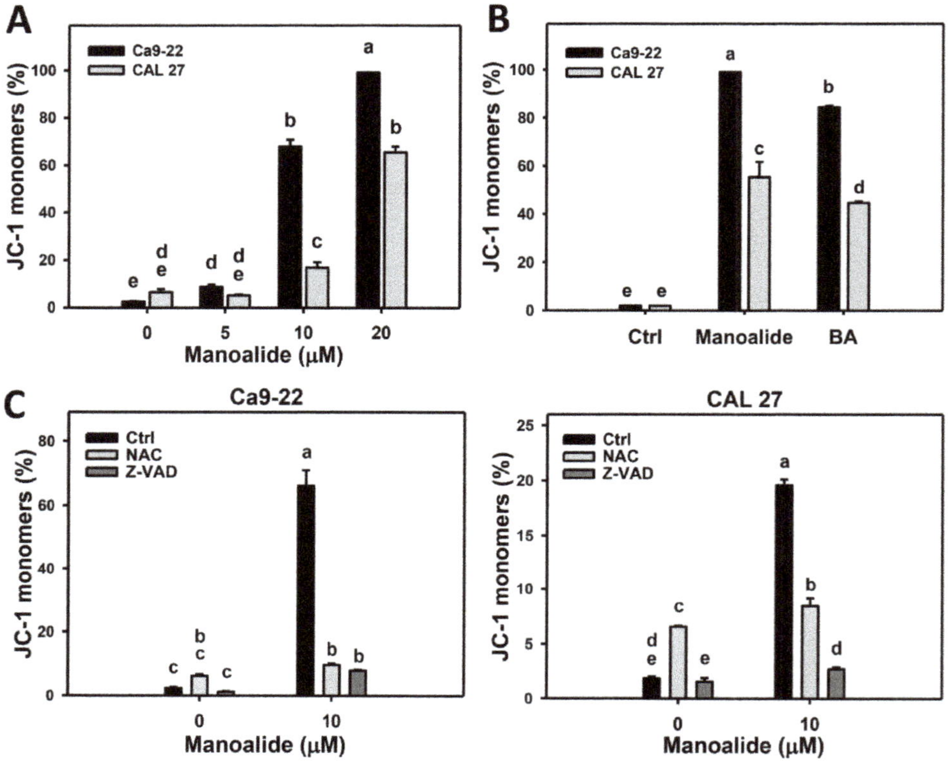

Figure 6. Change of membrane potential (MitoMP) in manoalide-treated oral cancer (Ca9-22 and CAL 27) cells. (**A**) Statistical results of JC-1 monomers (%) for manoalide-treated oral cancer cells in Figure S9A. Cells were treated with different concentrations of manoalide for 24 h. High JC-1 monomers (%) indicates low MitoMP, i.e., the MitoMP depolarization. (**B**) Statistical results of positive control of low MitoMP for oral cancer cells in Figure S9B. Cells were treated with betulinic acid (BA; 25 μM, 24 h) as the positive control treatment for comparison to manoalide (10 μM, 24 h). (**C**) Statistical results of JC-1 monomers (%) for NAC, Z-VAD, and/or manoalide-treated oral cells in Figure S9C. Cells were pretreated with NAC (8 mM, 1 h) or Z-VAD (100 μM, 2 h) and posttreated with manoalide (10 μM, 24 h). Data, means ± SDs ($n = 3$). Data were analyzed by one-way ANOVA with Tukey HSD Post Hoc Test. Data showing no overlapping same small letters represent significant difference ($p < 0.05$–0.001).

2.7. Flow Cytometry-Based DNA Damage Changes of Manoalide-Treated Oral Cancer Cells with or Without Pretreatments of NAC or Z-VAD

γH2AX is known as a DNA double strand break marker [36]. Figure S10A shows the populations of oral cancer (Ca9-22 and CAL 27) cells shift to γH2AX (+) region when the concentration of manoalide increases. Figure 7A shows that the γH2AX (+) population is dose-responsively increased in manoalide-treated oral cancer cells. To address the role of oxidative stress and apoptosis in manoalide-induced γH2AX, pretreatments of NAC and Z-VAD were used, and its γH2AX patterns are shown in Figure S10B. Figure 7B shows that the manoalide-induced γH2AX (+) (%) are inhibited by NAC and Z-VAD pretreatments for both oral cancer cells, Ca9-22 and CAL 27. Moreover, the involvement of MitoSOX for manoalide-induced γH2AX in oral cancer cells was further examined using MitoSOX inhibitor (MitoTEMPO). Figure S10C shows the γH2AX patterns of MitoTEMPO pretreatment effects against manoalide-treated oral cancer cells. Figure 7C shows that MitoTEMPO pretreatment decreases the manoalide-induced γH2AX (+) (%) for both oral cancer cells, Ca9-22 and CAL 27.

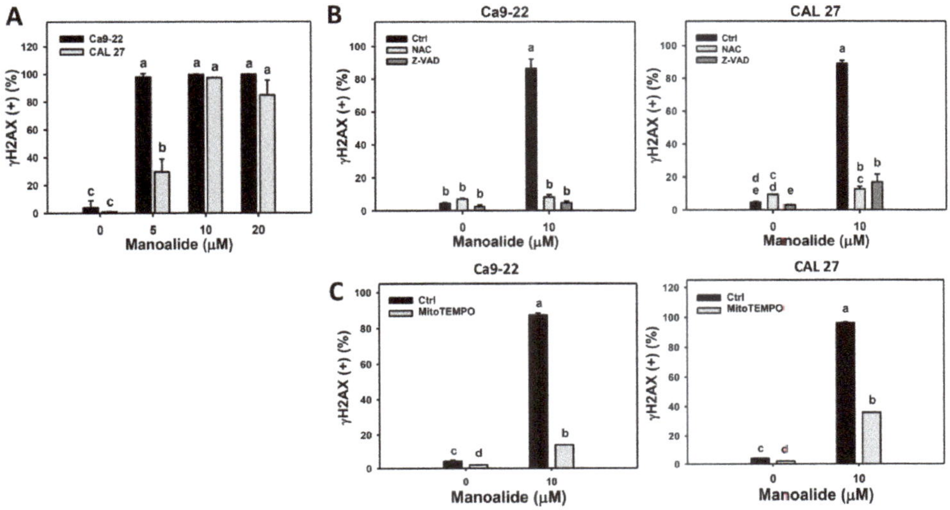

Figure 7. Change of γH2AX DNA damage in manoalide-treated oral cancer (Ca9-22 and CAL 27) cells. Cells were treated with the indicated concentrations of manoalide for 24 h. (**A**) Statistical results of γH2AX (+) (%) for manoalide-treated oral cancer cells in Figure S10A. (**B**) Statistical results of γH2AX (+) (%) in NAC, Z-VAD, and/or manoalide-treated oral cancer cells in Figure S10B. Cells were pretreated with 8 mM, 1 h for NAC or 100 μM, 2 h for Z-VAD, and then post-incubated with 10 μM of manoalide for 24 h. (**C**) Statistical results of γH2AX (+) (%) in MitoSOX inhibitor (MitoTEMPO) and/or manoalide-treated oral cancer cells in Figure S10C. Cells were pretreated with MitoTEMPO (20 μM, 1 h) and posttreated with manoalide (10 μM, 24 h). Data were analyzed by one-way ANOVA with Tukey HSD Post Hoc Test. Data, means ± SDs ($n = 3$). Data showing no overlapping same small letters represent significant difference ($p < 0.05$–0.001).

8-Oxo-2′-deoxyguanosine (8-oxodG) is one of the typical types of oxidative DNA damage [37]. Figure S11A shows the populations of oral cancer (Ca9-22 and CAL 27) cells shift to the 8-oxodG (+) region when the concentrations of manoalide increase, while the populations of 8-oxodG (+) in normal oral (HGF-1) cells are few. Figure 8A shows that the 8-oxodG (+) population is dose-responsively increased in manoalide-treated oral cancer cells; however, 8-oxodG (+) was rarely appeared in normal oral (HGF-1) cells. To address the role of oxidative stress and apoptosis in manoalide-induced 8-oxodG, pretreatments of NAC and Z-VAD were used and its 8-oxodG patterns are shown in Figure S11B. Figure 8B shows that the manoalide-induced 8-oxodG (+) (%) is inhibited by NAC and Z-VAD pretreatments for both oral cancer cells, Ca9-22 and CAL 27. Moreover, the involvement of MitoSOX for manoalide-induced 8-oxodG in oral cancer cells was further examined using MitoSOX inhibitor (MitoTEMPO). Figure S11C shows the 8-oxodG patterns of MitoTEMPO pretreatment effects against manoalide-treated oral cancer cells. Figure 8C shows that MitoTEMPO pretreatment decreases the manoalide-induced 8-oxodG (+) (%) for both oral cancer cells, Ca9-22 and CAL 27.

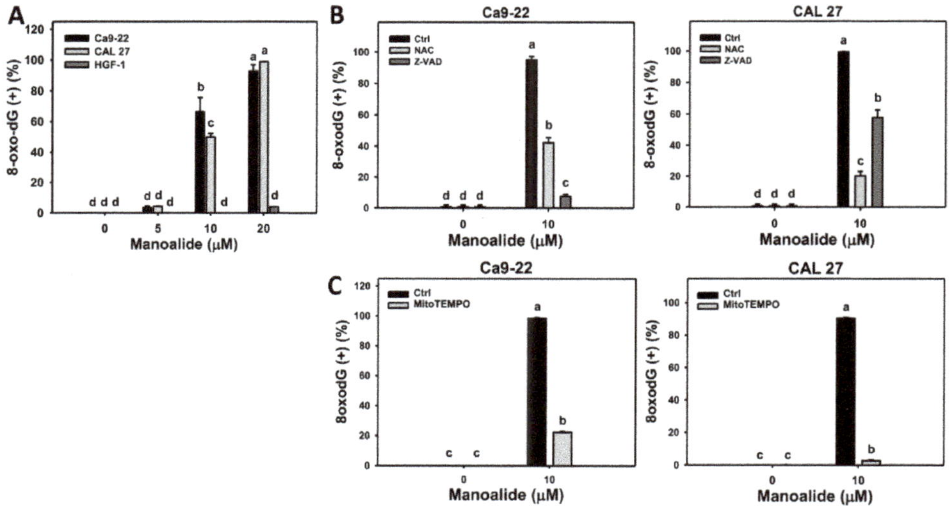

Figure 8. Change of 8-oxodG DNA damage in manoalide-treated oral cancer (Ca9-22 and CAL 27) and normal oral (HGF-1) cells. Cells were treated with the indicated concentrations of manoalide for 24 h. (**A**) Statistical results of 8-oxodG (+) (%) for manoalide-treated oral cancer cells in Figure S11A. (**B**) Statistical results of 8-oxodG (+) (%) in NAC, Z-VAD, and/or manoalide-treated oral cancer cells in Figure S11B. Cells were pretreated with 8 mM, 1 h for NAC or 100 μM, 2 h for Z-VAD, and then post-incubated with 10 μM of manoalide for 24 h. (**C**) Statistical results of 8-oxodG (+) (%) in MitoSOX inhibitor (MitoTEMPO) and/or manoalide-treated oral cancer cells in Figure S11C. Cells were pretreated with MitoTEMPO (20 μM, 1 h) and posttreated with manoalide (10 μM, 24 h). Data were analyzed by one-way ANOVA with Tukey HSD Post Hoc Test. Data, means ± SDs (n = 3). Data showing no overlapping same small letters represent significant difference (p < 0.05–0.001).

3. Discussion

The hypothesis that manoalide may preferentially inhibit the proliferation of oral cancer cells was validated in this study. In the following, we compare the manoalide sensitivity in different cancer cells and discuss the role of oxidative stress in preferential killing, apoptosis, and DNA damage in oral cancer cells.

3.1. Manoalide Sensitivity in Different Cancer Cells

Manoalide showed a cytotoxic effect against human epidermoid carcinoma KB cells (IC_{50} = 0.725 μM) [19] without providing information for treatment time and methods. Moreover, KB cells were reported as misidentified; these were HeLa cells, rather than oral cancer cells [38]. In the current study, oral cancer Ca9-22, CAL 27, OECM1, OC2, and HSC3 cells, respectively, show the IC_{50} values of manoalide with 7.8, 9.1, 14.9, 17.4, and 18.5 μM at 24 h MTS assay. It is noted that the drug safety for manoalide in normal cell lines was firstly demonstrated in normal oral cells (HGF-1), which remained healthy below 25 μM in a 24 h MTS assay. These results suggest that manoalide provides a preferentially killing effect to oral cancer cells and shows little damage to normal oral cells.

3.2. Manoalide Induced Oxidative Stress Contributes to Preferential Killing Against Oral Cancer Cells

Both 2,2-azinobis (3-ethyl-benzothiazoline-6-sulfonic acid) (ABTS) and hydroxyl scavenging activities of manoalide showed IC_{50} values for 14.3 and 18 μM (Figure S12), suggesting that manoalide causes the antioxidant abilities. Recently, antioxidants were reported to display dual concentration effects [26,39], i.e., low and high concentrations, respectively, decrease and increase intracellular ROS levels. Similarly, we found that manoalide differentially induced ROS production between oral cancer cells and normal oral cells, i.e., manoalide induces higher ROS production in oral cancer cells than in normal oral cells (Figure 4). Moreover, manoalide also induces other types of oxidative stresses, such as MitoSOX production and MitoMP depolarization (Figures 5 and 6). Using NAC or MitoTEMPO pretreatments, oxidative stresses, such as ROS generation (Figure 4B), MitoSOX production (Figure 5C), and MitoMP depolarization (Figure 6C), as well as antiproliferation, were suppressed (Figure 1D). Therefore, the preferential killing effect of manoalide is oxidative stress-dependent in oral cancer cells. This finding also supports the rationale that ROS-modulating drugs provided preferential killing effects against several types of cancer cells [27,40,41].

Manoalide displays a preferential killing against oral cancer cells with little damage to normal oral cells. Similarly, betulinic acid (BA) selectively inhibits proliferation against a number of cancer cells but not on normal cells (peripheral blood lymphoblast) [42,43], and therefore BA was chosen as a positive control. Like manoalide, BA induces ROS generation, apoptosis, and proliferation, and these effects are suppressed by NAC treatment [34]. As shown in Figures 5B and 6B, manoalide induces more MitoSOX generation and MitoMP depletion (JC-1 monomers generation) than that of BA, suggesting that manoalide is an effective oxidative stress inducer compared to BA in oral cancer cells. The anticancer effect of BA was independent on p53 mutant or wild types [42,43]. In the current study, all oral cancer cell lines harboring mutant p53 [44,45] and the role of p53 status warrants detailed investigation in future.

3.3. *Manoalide Induced Oxidative stress Contributes to Apoptosis and DNA Damage Against Oral Cancer Cells*

ROS-modulating drugs commonly induce apoptosis [29,40,41,46–49]. This is indicated by NAC and Z-VAD as apoptosis inhibitors. In our study, both show the suppressing effect on manoalide-induces subG1 accumulation and apoptosis (Figures 2B and 3B), suggesting that oxidative stress plays a vital role in manoalide-induced apoptosis. Additionally, we found that Cas 8 and Cas 9 inhibitors suppressed the manoalide-induced c-Cas 3 activation using flow cytometry (Figure 3E). Accordingly, the role of extrinsic and intrinsic apoptosis may be involved in manoalide-induced apoptosis, and it warrants detailed investigation in future.

It is noted that apoptosis inhibitor Z-VAD cannot completely recover the manoalide-induced antiproliferation against oral cancer cells. Half and one-third of cell viabilities for Ca9-22 and CAL 27 cells were unable to be recovered (Figure 1D). These results suggest that apoptosis cannot completely attribute to antiproliferation effects of manoalide-treated oral cancer cells. Detailed studies of the involvement of other non-apoptosis mechanisms after manoalide treatment are warranted.

Moreover, oxidative stress is a high risk factor for inducing DNA damage [41,47]. Consistently, DNA double strand breaks (γH2AX) and oxidative DNA damage (8-oxodG) were induced in oral cancer cells upon manoalide exposure. Both γH2AX and 8-oxodG levels in manoalide-treated oral cancer cells were suppressed by NAC pretreatments (Figures 7B and 8B). DNA damage also has a potential to induce apoptosis [50]. Accordingly, oxidative stress may induce DNA damage and lead to apoptosis. It is noted that the manoalide-induced γH2AX/8-oxodG expressions (Figures 7B and 8B) and ROS production (Figure 4B) were also suppressed by Z-VAD pretreatments, suggesting that apoptosis may crosstalk to DNA damage in addition to oxidative stress.

It was reported that superoxide anion, such as MitoSOX, cannot cross the mitochondrial membrane [51]. However, we found that manoalide-induced MitoSOX generation and γH2AX/8-oxodG expressions (Figures 7C and 8C) were suppressed by MitoTEMPO pretreatment. Since MitoTEMPO is the mitochondria-targeted superoxide inhibitor [52], the role of MitoSOX on DNA damage is explored by MitoTEMPO pretreatment. Our finding suggests that MitoSOX may directly or indirectly induce DNA double strand breaks and oxidative DNA damage. Accordingly, the validation and mechanism of MitoSOX-induced DNA damage warrants detailed investigation in future. Therefore, both intracellular ROS and mitochondrial superoxide (MitoSOX) may contribute to the manoalide-induced DNA damage in oral cancer cells.

The possible mechanism for preferential killing of manoalide against oral cancer cells but less damage to normal oral cells is discussed as follows. In HGF-1 cells, the ROS production is few (Figure 4A), leads to fewer annexin V-detected apoptosis (Figure 3A) and 8-oxodG DNA damage (Figure 8A) than that of oral cancer cells, and causes the oral cancer cell death but keeps normal oral cells alive.

3.4. Potential Target Molecules of Manoalide

Manoalide is known as an irreversible inhibitor for PLC [16] and PLA2 [53], as well as calcium channel blockers (CCBs) [17]. PLC inhibitors, such as U73122, were reported to induce apoptosis of human umbilical vein endothelial cells (HUVEC) [54]. PLA2 inhibitors, such as quercetin, were also summarized to inhibit inflammation and cancer proliferation [55]. CCBs, such as verapamil and diltiazem, have been reviewed for antiproliferation against several types of cancer cells in vitro and in vivo [56]. Accordingly, PLC, PLA2, and CCBs are the potential targets for manoalide. It warrants detailed investigation to explore the role of these potential targets between oral cancer and normal oral cells in future.

4. Materials and Methods

4.1. Cell and Drug Information

All human oral cancer cell lines (Ca9-22, CAL 27, HSC-3, OC-2, and SCC-9) and a normal oral cell line (HGF-1) were used from Health Science Research Resources Bank (HSRRB) (Osaka, Japan) and American Type Culture Collection (ATCC; Manassas, VA, USA) except for OECM1 [57], a generous gift from Dr. Wan-Chi Tsai (Kaohsiung Medical University, Taiwan). Cells were cultured in 5% CO_2 at 37 °C with humidity and maintained by regular formula (Gibco, Grand Island, NY, USA) with 10% fetal bovine serum as previously described [49].

Manoalide (CAYMAN CHEMICAL, Ann Arbor, MI, USA) was dissolved in dimethyl sulfoxide (DMSO) for treatment. A ROS scavenger *N*-acetylcysteine (NAC) [58] (Sigma-Aldrich; St. Louis, MO, USA) was dissolved in double distilled water. The mitochondrial superoxide inhibitor MitoTEMPO [59] (Cayman Chemical, Ann Arbor, MI, USA), panapoptosis inhibitor Z-VAD-FMK [60], Cas 8 inhibitor Z-IETD-FMK, and Cas 9 inhibitor Z-LEHD-FMK (Selleckchem.com; Houston, TX, USA) was dissolved in DMSO. All experiments contain the same concentration of DMSO.

4.2. Cell Viability Assay

After drug treatment, the mitochondrial activity-based cell viability was determined by MTS assay (CellTiter 96 Aqueous One Solution, Promega, Madison, WI, USA) at 24 h [61], and the 3D microtissue spheroids viability of oral cancer cells was measured by the CellTiter-Glo® 3D Cell Viability Assay (Promega, Madison, WI, USA) coupling with ATP level detection at 72 h [62].

4.3. Cell Cycle Assay

7AAD (Biotium, Inc., Hayward, CA, USA), a DNA dye, was applied to cell cycle analysis [63]. Briefly, drugs-treated cells were stained with 7AAD (1 μg/mL, 37 °C, 30 min). Finally, the cell cycle change was analyzed by Accuri™ C6 flow cytometry (Becton-Dickinson, Mansfield, MA, USA).

4.4. Annexin V/7AAD Assay for Apoptosis

Annexin V (Strong Biotech Corporation, Taipei, Taiwan) coupled with 7AAD was used for apoptosis analysis [64]. Briefly, drugs-treated cells were incubated with the mixture of annexin V-fluorescein isothiocyanate (FITC) (10 μg/mL) and 7AAD (1 μg/mL) at 37 °C for 30 min. Finally, the apoptosis expression was analyzed by Accuri™ C6 flow cytometry.

4.5. Western Blotting and c-Cas 3-Based Flow Cytometry for Apoptosis

Detailed steps of western blotting were previously described [60]. Briefly, the primary apoptosis antibodies (diluted 1:1000) including cleaved caspase-3 (c-Cas 3) rabbit mAb (Cell Signaling Technology, Inc., Danvers, MA, USA) were used. The internal control primary antibody (diluted 1:5000) was mAb-β-actin (Sigma-Aldrich, St. Louis, MO, USA). Following secondary antibody treatment, these protein signals were detected using enhanced chemiluminescence (ECL) substrate (WesternBright™ ECL HRP, Advansta, Menlo Park, CA, USA).

For c-Cas 3-based flow cytometry, cells were fixed with 70% ethanol, washed, and incubated with 1 μg/mL of c-Cas 3 (Asp175) rabbit mAb (Cell Signaling Technology) at 4 °C for overnight. After washing, cells were incubated with a secondary polyclonal antibody conjugated with Alexa Fluor 488 (ThermoFisher Scientific, San Jose, CA, USA) at room temperature for 1 h. Finally, the c-Cas 3 expression was analyzed by Accuri™ C6 flow cytometry. Cas 8 inhibitor Z-IETD-FMK (100 μM, 2 h) or Cas 9 inhibitor Z-LEHD-FMK (100 μM, 2 h) were applied to examine the involvement of Cas 8 and Cas 9 in apoptosis.

4.6. ROS Assay

DCFH-DA (Sigma-Aldrich; St. Louis, MO, USA) was used for ROS detecting dye [32]. Briefly, drugs-treated cells were incubated with DCFH-DA reagent (10 μM, 37 °C, 30 min). Finally, the ROS level was analyzed by Accuri™ C6 flow cytometry.

4.7. MitoSOX Assay

MitoSOX™ Red (Molecular Probes, Invitrogen, Eugene, OR, USA) was used as mitochondrial superoxide detecting dye [33]. Briefly, drugs-treated cells were incubated with MitoSOX reagent [48,49] (5 μM, 37 °C, 30 min). Finally, the MitoSOX level was analyzed by Accuri™ C6 flow cytometry.

4.8. MitoMP Assay

JC-1 (Merckmillipore) was used to detect mitochondrial membrane potential (MitoMP). JC-1 aggregate form generated red fluorescence indicating the normal function for MitoMP [35]. In contrast, JC-1 monomer form generated green fluorescence, indicating the dysfunction for MitoMP. Therefore, green fluorescent signals were counted as the decrease of MitoMP. Briefly, drugs-treated cells were treated with JC-1 (0.1 mM, 37 °C, 30 min). Finally, the MitoMP level was analyzed by Accuri™ C6 flow cytometry.

4.9. γH2AX Assay

DNA double strand break marker (γH2AX) was detected by antibody-based flow cytometry [40]. Briefly, drugs-treated cells were incubated with mouse primary antibody p-Histone H2A.X (Ser 139) (Santa Cruz Biotechnology, Santa Cruz, CA, USA) (1:100 dilution, 4 °C, 1 h) and washed for incubation with the secondary antibody-labeled with Alexa Fluor 488 (Cell Signaling Technology) (1:10000 dilution, 4 °C, 1 h). Finally, the γH2AX level was analyzed by Accuri™ C6 flow cytometry.

4.10. 8-oxodG Assay

8-oxodG was detected by antibody-based flow cytometry using a fluorometric OxyDNA assay kit (#500095; EMD Millipore, Darmstadt, Germany) [65,66]. Briefly, drugs-treated cells were incubated in antibody-labeled with FITC (10× dilution, 4 °C, 1 h). Finally, the 8-oxodG level was analyzed by Accuri™ C6 flow cytometry.

4.11. Statistical Analysis

Using JMP® 12 software, the significance of multiple comparisons between different treatments were analyzed by one-way ANOVA with Tukey HSD Post Hoc Test. Data showing no overlapping same small letters represent significant difference.

5. Conclusions

This study confirmed the hypothesis that manoalide may preferentially inhibit the proliferation of oral cancer cells. We found several types of results supporting that oxidative stresses were induced by manoalide. The oxidative stresses, such as intracellular ROS and MitoSOX/MitoMP, were also involved in manoalide-induced apoptosis and DNA damages in oral cancer cells. Finally, these mechanisms may contribute to preferentially inhibit the proliferation of oral cancer cells (Figure 9). Taken together, this study firstly shows that manoalide preferentially kills oral cancer cells without cytotoxic side effects to normal oral cells.

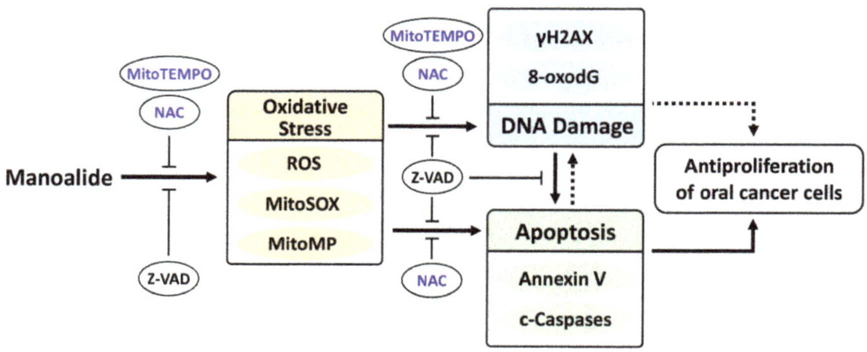

Figure 9. Expected mechanism of manoalide-induced preferential antiproliferation on oral cancer cells. Manoalide may preferentially kill oral cancer cells by inducing oxidative stress, such as ROS and MitoSOX productions, as well as MitoMP depolarization. These oxidative stresses may induce DNA double strand break damage and oxidative DNA 8-oxodG DNA damage. DNA damage may induce apoptosis and inhibit proliferation [67]. Finally, these oxidative stress and DNA damage changes cooperate to induce apoptosis [50] and lead to preferential antiproliferation of oral cancer cells. Additionally, apoptosis may also regulate DNA damage and oxidative stress. Solid lines (arrow and T) indicate the activating and inhibiting results from this study, whereas dashed lines indicate the mechanism is supported by literature.

Supplementary Materials: The following are available online at http://www.mdpi.com/2072-6694/11/9/1303/s1, Figure S1: 3D spheroid formation and cell morphology of oral cancer cells after manoalide treatment and its NAC/Z-VAD effects, Figure S2: Cell cycle changes of manoalide-treated oral cancer (Ca9-22 and CAL 27) cells, Figure S3: Apoptosis changes in manoalide-treated oral cancer (Ca9-22 and CAL 27) cells and normal oral (HGF-1) cells, Figure S4: Western blotting for procaspase 3 and cleaved caspase 3 (c-Cas 3) in manoalide-treated oral cancer (Ca9-22 and CAL 27) cells, Figure S5: Raw data for western blotting of cleaved caspase 3 (c-Cas 3) with or without NAC and Z-VAD pretreatment in manoalide-treated oral cancer (Ca9-22 and CAL 27) cells, Figure S6: Fluorescence staining of cleaved caspases 8 and 9 (c-Cas 8/9) and the effects of Cas 8/9 inhibitors on c-Cas 3-based flow cytometry, Figure S7: ROS changes in manoalide-treated oral cancer (Ca9-22 and CAL 27) and normal oral (HGF-1) cells, Figure S8: Change of MitoSOX production in manoalide-treated oral cancer (Ca9-22 and CAL 27) cells, Figure S9: Change of MitoMP in manoalide-treated oral cancer (Ca9-22 and CAL 27) cells, Figure S10: Change of γH2AX DNA damage in manoalide-treated oral cancer (Ca9-22 and CAL 27) cells, Figure S11: Change of 8-oxodG DNA damage in manoalide-treated oral cancer (Ca9-22 and CAL 27) cells, and Figure S12: ABTS and hydroxyl scavenging assays for manoalide.

Author Contributions: H.-R.W. and H.-W.C. contributed to original draft preparation. H.-R.W. performed the cytotoxicity assays, flow cytometry, and western blotting. J.-Y.T., Y.-Y.W., A.A.F., C.-Y.Y., and S.-S.F.Y. contributed to methodology, data analysis, and statistics. H.-W.H. and H.-W.C. designed the whole experiment and improved the manuscript.

Funding: This work was partly supported by funds of the Ministry of Science and Technology (MOST 108-2320-B-037-015-MY3, MOST 108-2314-B-037-020, MOST 108-2314-B-384-002), the National Sun Yat-sen University-KMU Joint Research Project (#NSYSUKMU 108-P001), the Chimei-KMU jointed project (108CM-KMU-11), the Kaohsiung Medical University Hospital (KMUH107-7R74), the Kaohsiung Medical University Research Center (KMU-TC108A04), and the Health and welfare surcharge of tobacco products, the Ministry of Health and Welfare, Taiwan, Republic of China (MOHW 108-TDU-B-212-124016).

Acknowledgments: The authors thank Hans-Uwe Dahms for editing the manuscript, the Center for Research Resources and Development of Kaohsiung Medical University for providing the service of LSRII system, and Sheng-Chieh Wang for ABTS/hydroxyl scavenging testing.

Conflicts of Interest: The authors declare no conflict of interest. The funders had no role in the design of the study; in the collection, analyses, or interpretation of data; in the writing of the manuscript, or in the decision to publish the results.

References

1. Montero, P.H.; Patel, S.G. Cancer of the oral cavity. *Surg. Oncol. Clin. N. Am.* **2015**, *24*, 491–508. [CrossRef] [PubMed]
2. Ko, Y.C.; Huang, Y.L.; Lee, C.H.; Chen, M.J.; Lin, L.M.; Tsai, C.C. Betel quid chewing, cigarette smoking and alcohol consumption related to oral cancer in Taiwan. *J. Oral Pathol. Med.* **1995**, *24*, 450–453. [CrossRef] [PubMed]
3. Petersen, P.E. Oral cancer prevention and control—The approach of the World Health Organization. *Oral Oncol.* **2009**, *45*, 454–460. [CrossRef] [PubMed]
4. Silverman, S., Jr. Oral cancer: Complications of therapy. *Oral Surg. Oral Med. Oral Pathol. Oral Radiol. Endod.* **1999**, *88*, 122–126. [CrossRef]
5. Singh, R.; Sharma, M.; Joshi, P.; Rawat, D.S. Clinical status of anti-cancer agents derived from marine sources. *Anticancer Agents Med. Chem.* **2008**, *8*, 603–617. [CrossRef] [PubMed]
6. Sithranga Boopathy, N.; Kathiresan, K. Anticancer drugs from marine flora: An overview. *J. Oncol.* **2010**, *2010*, 214186. [CrossRef]
7. Farooqi, A.A.; Fayyaz, S.; Hou, M.F.; Li, K.T.; Tang, J.Y.; Chang, H.W. Reactive oxygen species and autophagy modulation in non-marine drugs and marine drugs. *Mar. Drugs* **2014**, *12*, 5408–5424. [CrossRef]
8. Lee, M.G.; Liu, Y.C.; Lee, Y.L.; El-Shazly, M.; Lai, K.H.; Shih, S.P.; Ke, S.C.; Hong, M.C.; Du, Y.C.; Yang, J.C.; et al. Heteronemin, a marine sesterterpenoid-type metabolite, induces apoptosis in prostate LNcap cells via oxidative and ER stress combined with the inhibition of topoisomerase II and Hsp90. *Mar. Drugs* **2018**, *16*, 204. [CrossRef]
9. Mehbub, M.F.; Lei, J.; Franco, C.; Zhang, W. Marine sponge derived natural products between 2001 and 2010: Trends and opportunities for discovery of bioactives. *Mar. Drugs* **2014**, *12*, 4539–4577. [CrossRef]
10. Mehbub, M.F.; Perkins, M.V.; Zhang, W.; Franco, C.M.M. New marine natural products from sponges (Porifera) of the order Dictyoceratida (2001 to 2012); A promising source for drug discovery, exploration and future prospects. *Biotechnol. Adv.* **2016**, *34*, 473–491. [CrossRef]
11. Calcabrini, C.; Catanzaro, E.; Bishayee, A.; Turrini, E.; Fimognari, C. Marine sponge natural products with anticancer potential: An updated review. *Mar. Drugs* **2017**, *15*, 310. [CrossRef] [PubMed]
12. Perdicaris, S.; Vlachogianni, T.; Valavanidis, A. Bioactive natural substances from marine sponges: New developments and prospects for future pharmaceuticals. *Nat. Prod. Chem. Res.* **2013**, *1*, 114. [CrossRef]
13. de Silva, E.D.; Scheuer, P.J. Manoalide, an antibiotic sesterterpenoid from the marine sponge *Luffariella variabilis* (polejaeff). *Tetrahedron Lett.* **1980**, *21*, 1611–1614. [CrossRef]
14. Soriente, A.; De Rosa, M.M.; Scettri, A.; Sodano, G.; Terencio, M.C.; Paya, M.; Alcaraz, M.J. Manoalide. *Curr. Med. Chem.* **1999**, *6*, 415–431. [PubMed]
15. Muallem, S.; Loessberg, P.; Sachs, G.; Wheeler, L.A. Agonist-sensitive and -insensitive intracellular Ca^{2+} pools. Separate Ca^{2+}-releasing mechanisms revealed by manoalide and benzohydroquinone. *Biochem. J.* **1991**, *279*, 367–375. [CrossRef] [PubMed]
16. Bennett, C.F.; Mong, S.; Wu, H.L.; Clark, M.A.; Wheeler, L.; Crooke, S.T. Inhibition of phosphoinositide-specific phospholipase C by manoalide. *Mol. Pharmacol.* **1987**, *32*, 587–593. [PubMed]
17. Wheeler, L.A.; Sachs, G.; De Vries, G.; Goodrum, D.; Woldemussie, E.; Muallem, S. Manoalide, a natural sesterterpenoid that inhibits calcium channels. *J. Biol. Chem.* **1987**, *262*, 6531–6538. [PubMed]

18. Kijjoa, A.; Sawangwong, P. Drugs and cosmetics from the sea. *Mar. Drugs* **2004**, *2*, 73–82. [CrossRef]
19. Kobayashi, J.; Zeng, C.M.; Ishibashi, M.; Sasaki, T. Luffariolides F and G, new manoalide derivatives from the Okinawan marine sponge *Luffariella* sp. *J. Nat. Prod.* **1993**, *56*, 436–439. [CrossRef]
20. Thambidurai, Y.; D, S.; SKM, H.; T, A.K. Free radical scavenging activity of marine sponges collected from Kovalam, Chennai. *Asian J. Pharm. Clin. Res.* **2017**, *10*, 321. [CrossRef]
21. Utkina, N.K. Antioxidant activity of zyzzyanones and makaluvamines from the marine sponge *Zyzzya fuliginosa*. *Nat. Prod. Commun.* **2013**, *8*, 1551–1552. [CrossRef] [PubMed]
22. Chairman, K.; Singh, A.J.A.R.; Alagumuthu, G. Cytotoxic and antioxidant activity of selected marine sponges. *Asian Pac. J. Trop. Dis.* **2012**, *2*, 234–238. [CrossRef]
23. Abdillah, S.; Nurhayati, A.P.D.; Nurhatika, S.; Setiawan, E.; Heffen, W.L. Cytotoxic and antioxidant activities of marine sponge diversity at Pecaron Bay Pasir Putih Situbondo East Java, Indonesia. *J. Pharm. Res.* **2013**, *6*, 685–689. [CrossRef]
24. Rivera, A.P.; Uy, M.M. In vitro antioxidant and cytotoxic activities of some marine sponges collected off Misamis Oriental Coast, Philippines. *E-J. Chem.* **2012**, *9*, 354–358. [CrossRef]
25. de Carvalho, D.D.; Sadok, A.; Bourgarel-Rey, V.; Gattacceca, F.; Penel, C.; Lehmann, M.; Kovacic, H. Nox1 downstream of 12-lipoxygenase controls cell proliferation but not cell spreading of colon cancer cells. *Int. J. Cancer* **2008**, *122*, 1757–1764. [CrossRef] [PubMed]
26. Bouayed, J.; Bohn, T. Exogenous antioxidants—Double-edged swords in cellular redox state: Health beneficial effects at physiologic doses versus deleterious effects at high doses. *Oxid. Med. Cell Longev.* **2010**, *3*, 228–237. [CrossRef] [PubMed]
27. Suzuki-Karasaki, Y.; Suzuki-Karasaki, M.; Uchida, M.; Ochiai, T. Depolarization controls TRAIL-sensitization and tumor-selective killing of cancer cells: Crosstalk with ROS. *Front. Oncol.* **2014**, *4*, 128. [CrossRef] [PubMed]
28. Hseu, Y.C.; Lee, M.S.; Wu, C.R.; Cho, H.J.; Lin, K.Y.; Lai, G.H.; Wang, S.Y.; Kuo, Y.H.; Kumar, K.J.; Yang, H.L. The chalcone flavokawain B induces G2/M cell-cycle arrest and apoptosis in human oral carcinoma HSC-3 cells through the intracellular ROS generation and downregulation of the Akt/p38 MAPK signaling pathway. *J. Agric. Food Chem.* **2012**, *60*, 2385–2397. [CrossRef] [PubMed]
29. Shih, H.C.; El-Shazly, M.; Juan, Y.S.; Chang, C.Y.; Su, J.H.; Chen, Y.C.; Shih, S.P.; Chen, H.M.; Wu, Y.C.; Lu, M.C. Cracking the cytotoxicity code: Apoptotic induction of 10-acetylirciformonin B is mediated through ROS generation and mitochondrial dysfunction. *Mar. Drugs* **2014**, *12*, 3072–3090. [CrossRef]
30. Tang, J.Y.; Wu, C.Y.; Shu, C.W.; Wang, S.C.; Chang, M.Y.; Chang, H.W. A novel sulfonyl chromen-4-ones (CHW09) preferentially kills oral cancer cells showing apoptosis, oxidative stress, and DNA damage. *Environ. Toxicol.* **2018**, *33*, 1195–1203. [CrossRef]
31. Nakagawa, Y.; Takahashi, A.; Kajihara, A.; Yamakawa, N.; Imai, Y.; Ota, I.; Okamoto, N.; Mori, E.; Noda, T.; Furusawa, Y.; et al. Depression of p53-independent Akt survival signals in human oral cancer cells bearing mutated p53 gene after exposure to high-LET radiation. *Biochem. Biophys. Res. Commun.* **2012**, *423*, 654–660. [CrossRef] [PubMed]
32. Yeh, C.C.; Yang, J.I.; Lee, J.C.; Tseng, C.N.; Chan, Y.C.; Hseu, Y.C.; Tang, J.Y.; Chuang, L.Y.; Huang, H.W.; Chang, F.R.; et al. Anti-proliferative effect of methanolic extract of *Gracilaria tenuistipitata* on oral cancer cells involves apoptosis, DNA damage, and oxidative stress. *BMC Complement. Altern. Med.* **2012**, *12*, 142. [CrossRef] [PubMed]
33. Mukhopadhyay, P.; Rajesh, M.; Yoshihiro, K.; Hasko, G.; Pacher, P. Simple quantitative detection of mitochondrial superoxide production in live cells. *Biochem. Biophys. Res. Commun.* **2007**, *358*, 203–208. [CrossRef] [PubMed]
34. Shen, H.; Liu, L.; Yang, Y.; Xun, W.; Wei, K.; Zeng, G. Betulinic acid inhibits cell proliferation in human oral squamous cell carcinoma via modulating ROS-regulated p53 signaling. *Oncol. Res.* **2017**, *25*, 1141–1152. [CrossRef] [PubMed]
35. Jin, C.Y.; Park, C.; Lee, J.H.; Chung, K.T.; Kwon, T.K.; Kim, G.Y.; Choi, B.T.; Choi, Y.H. Naringenin-induced apoptosis is attenuated by Bcl-2 but restored by the small molecule Bcl-2 inhibitor, HA 14-1, in human leukemia U937 cells. *Toxicol. In Vitro* **2009**, *23*, 259–265. [CrossRef] [PubMed]
36. Kuo, L.J.; Yang, L.X. Gamma-H2AX—A novel biomarker for DNA double-strand breaks. *In Vivo* **2008**, *22*, 305–309. [PubMed]

37. Roszkowski, K.; Jozwicki, W.; Blaszczyk, P.; Mucha-Malecka, A.; Siomek, A. Oxidative damage DNA: 8-oxoGua and 8-oxodG as molecular markers of cancer. *Med. Sci. Monit.* **2011**, *17*, CR329–CR333. [CrossRef]
38. Vaughan, L.; Glanzel, W.; Korch, C.; Capes-Davis, A. Widespread use of misidentified cell line KB (HeLa): Incorrect attribution and its impact revealed through mining the scientific literature. *Cancer Res.* **2017**, *77*, 2784–2788. [CrossRef]
39. Yen, C.Y.; Hou, M.F.; Yang, Z.W.; Tang, J.Y.; Li, K.T.; Huang, H.W.; Huang, Y.H.; Lee, S.Y.; Fu, T.F.; Hsieh, C.Y.; et al. Concentration effects of grape seed extracts in anti-oral cancer cells involving differential apoptosis, oxidative stress, and DNA damage. *BMC Complement. Altern. Med.* **2015**, *15*, 94. [CrossRef]
40. Chiu, C.C.; Haung, J.W.; Chang, F.R.; Huang, K.J.; Huang, H.M.; Huang, H.W.; Chou, C.K.; Wu, Y.C.; Chang, H.W. Golden berry-derived 4beta-hydroxywithanolide E for selectively killing oral cancer cells by generating ROS, DNA damage, and apoptotic pathways. *PLoS ONE* **2013**, *8*, e64739. [CrossRef]
41. Chang, H.W.; Li, R.N.; Wang, H.R.; Liu, J.R.; Tang, J.Y.; Huang, H.W.; Chan, Y.H.; Yen, C.Y. Withaferin A induces oxidative stress-mediated apoptosis and DNA damage in oral cancer cells. *Front. Physiol.* **2017**, *8*, 634. [CrossRef] [PubMed]
42. Zuco, V.; Supino, R.; Righetti, S.C.; Cleris, L.; Marchesi, E.; Gambacorti-Passerini, C.; Formelli, F. Selective cytotoxicity of betulinic acid on tumor cell lines, but not on normal cells. *Cancer Lett.* **2002**, *175*, 17–25. [CrossRef]
43. Shankar, E.; Zhang, A.; Franco, D.; Gupta, S. Betulinic acid-mediated apoptosis in human prostate cancer cells involves p53 and nuclear factor-kappa B (NF-kappaB) pathways. *Molecules* **2017**, *22*, 264. [CrossRef]
44. Lin, S.C.; Liu, C.J.; Chiu, C.P.; Chang, S.M.; Lu, S.Y.; Chen, Y.J. Establishment of OC3 oral carcinoma cell line and identification of NF-kappa B activation responses to areca nut extract. *J. Oral Pathol. Med.* **2004**, *33*, 79–86. [CrossRef] [PubMed]
45. Hamroun, D.; Kato, S.; Ishioka, C.; Claustres, M.; Beroud, C.; Soussi, T. The UMD TP53 database and website: Update and revisions. *Hum. Mutat.* **2006**, *27*, 14–20. [CrossRef] [PubMed]
46. Huang, C.H.; Huang, Z.W.; Ho, F.M.; Chan, W.H. Berberine impairs embryonic development in vitro and in vivo through oxidative stress-mediated apoptotic processes. *Environ. Toxicol.* **2018**, *33*, 280–294. [CrossRef] [PubMed]
47. Hung, J.H.; Chen, C.Y.; Omar, H.A.; Huang, K.Y.; Tsao, C.C.; Chiu, C.C.; Chen, Y.L.; Chen, P.H.; Teng, Y.N. Reactive oxygen species mediate Terbufos-induced apoptosis in mouse testicular cell lines via the modulation of cell cycle and pro-apoptotic proteins. *Environ. Toxicol.* **2016**, *31*, 1888–1898. [CrossRef] [PubMed]
48. Chang, Y.T.; Huang, C.Y.; Tang, J.Y.; Liaw, C.C.; Li, R.N.; Liu, J.R.; Sheu, J.H.; Chang, H.W. Reactive oxygen species mediate soft corals-derived sinuleptolide-induced antiproliferation and DNA damage in oral cancer cells. *Onco Targets Ther.* **2017**, *10*, 3289–3297. [CrossRef]
49. Chang, Y.T.; Huang, C.Y.; Li, K.T.; Li, R.N.; Liaw, C.C.; Wu, S.H.; Liu, J.R.; Sheu, J.H.; Chang, H.W. Sinuleptolide inhibits proliferation of oral cancer Ca9-22 cells involving apoptosis, oxidative stress, and DNA damage. *Arch. Oral Biol.* **2016**, *66*, 147–154. [CrossRef]
50. Norbury, C.J.; Zhivotovsky, B. DNA damage-induced apoptosis. *Oncogene* **2004**, *23*, 2797–2808. [CrossRef]
51. Inoue, M.; Sato, E.F.; Nishikawa, M.; Park, A.M.; Kira, Y.; Imada, I.; Utsumi, K. Mitochondrial generation of reactive oxygen species and its role in aerobic life. *Curr. Med. Chem.* **2003**, *10*, 2495–2505. [CrossRef] [PubMed]
52. Liang, H.L.; Sedlic, F.; Bosnjak, Z.; Nilakantan, V. SOD1 and MitoTEMPO partially prevent mitochondrial permeability transition pore opening, necrosis, and mitochondrial apoptosis after ATP depletion recovery. *Free Radic. Biol. Med.* **2010**, *49*, 1550–1560. [CrossRef] [PubMed]
53. Lombardo, D.; Dennis, E.A. Cobra venom phospholipase A2 inhibition by manoalide. A novel type of phospholipase inhibitor. *J. Biol. Chem.* **1985**, *260*, 7234–7240. [PubMed]
54. Miao, J.Y.; Kaji, K.; Hayashi, H.; Araki, S. Inhibitors of phospholipase promote apoptosis of human endothelial cells. *J. Biochem.* **1997**, *121*, 612–618. [CrossRef] [PubMed]
55. Yarla, N.; Satyakumar, K.; Srinivasu, D.; DSVGK K, A.G.; Dharmapuri, G.; Putta, G.; Jagarlapoodi, S.; Bheeram, V.; Sadu, S.; Duddukuri, G. Phospholipase A2: A potential therapeutic target in inflammation and cancer (In silico, In vitro, In vivo and clinical approach). *J. Cancer Sci. Ther.* **2015**, *7*, 8.
56. Mason, R.P. Calcium channel blockers, apoptosis and cancer: Is there a biologic relationship? *J. Am. Coll. Cardiol.* **1999**, *34*, 1857–1866. [CrossRef]

57. Yang, C.Y.; Meng, C.L. Regulation of PG synthase by EGF and PDGF in human oral, breast, stomach, and fibrosarcoma cancer cell lines. *J. Dent. Res.* **1994**, *73*, 1407–1415. [CrossRef] [PubMed]
58. Huang, C.H.; Yeh, J.M.; Chan, W.H. Hazardous impacts of silver nanoparticles on mouse oocyte maturation and fertilization and fetal development through induction of apoptotic processes. *Environ. Toxicol.* **2018**, *33*, 1039–1049. [CrossRef]
59. Wang, T.S.; Lin, C.P.; Chen, Y.P.; Chao, M.R.; Li, C.C.; Liu, K.L. CYP450-mediated mitochondrial ROS production involved in arecoline N-oxide-induced oxidative damage in liver cell lines. *Environ. Toxicol.* **2018**, *33*, 1029–1038. [CrossRef]
60. Chen, C.Y.; Yen, C.Y.; Wang, H.R.; Yang, H.P.; Tang, J.Y.; Huang, H.W.; Hsu, S.H.; Chang, H.W. Tenuifolide B from *Cinnamomum tenuifolium* stem selectively inhibits proliferation of oral cancer cells via apoptosis, ROS generation, mitochondrial depolarization, and DNA damage. *Toxins (Basel)* **2016**, *8*, 319. [CrossRef]
61. Yeh, C.C.; Tseng, C.N.; Yang, J.I.; Huang, H.W.; Fang, Y.; Tang, J.Y.; Chang, F.R.; Chang, H.W. Antiproliferation and induction of apoptosis in Ca9-22 oral cancer cells by ethanolic extract of *Gracilaria tenuistipitata*. *Molecules* **2012**, *17*, 10916–10927. [CrossRef] [PubMed]
62. Liu, P.F.; Chang, H.W.; Cheng, J.S.; Lee, H.P.; Yen, C.Y.; Tsai, W.L.; Cheng, J.T.; Li, Y.J.; Huang, W.C.; Lee, C.H.; et al. Map1lc3b and Sqstm1 modulated autophagy for tumorigenesis and prognosis in certain subsites of oral squamous cell carcinoma. *J. Clin. Med.* **2018**, *7*, 478. [CrossRef] [PubMed]
63. Vignon, C.; Debeissat, C.; Georget, M.T.; Bouscary, D.; Gyan, E.; Rosset, P.; Herault, O. Flow cytometric quantification of all phases of the cell cycle and apoptosis in a two-color fluorescence plot. *PLoS ONE* **2013**, *8*, e68425. [CrossRef] [PubMed]
64. Huang, H.W.; Tang, J.Y.; Ou-Yang, F.; Wang, H.R.; Guan, P.Y.; Huang, C.Y.; Chen, C.Y.; Hou, M.F.; Sheu, J.H.; Chang, H.W. Sinularin selectively kills breast cancer cells showing G2/M arrest, apoptosis, and oxidative DNA damage. *Molecules* **2018**, *23*, 849. [CrossRef] [PubMed]
65. Bartkova, J.; Hamerlik, P.; Stockhausen, M.T.; Ehrmann, J.; Hlobilkova, A.; Laursen, H.; Kalita, O.; Kolar, Z.; Poulsen, H.S.; Broholm, H.; et al. Replication stress and oxidative damage contribute to aberrant constitutive activation of DNA damage signalling in human gliomas. *Oncogene* **2010**, *29*, 5095–5102. [CrossRef] [PubMed]
66. Nagy, S.; Kakasi, B.; Bercsenyi, M. Flow cytometric detection of oxidative DNA damage in fish spermatozoa exposed to cadmium—Short communication. *Acta Vet. Hung.* **2016**, *64*, 120–124. [CrossRef] [PubMed]
67. Mercer, J.R.; Gray, K.; Figg, N.; Kumar, S.; Bennett, M.R. The methyl xanthine caffeine inhibits DNA damage signaling and reactive species and reduces atherosclerosis in ApoE(-/-) mice. *Arter. Thromb. Vasc. Biol.* **2012**, *32*, 2461–2467. [CrossRef] [PubMed]

cancers

MDPI

Article

Ethanol-Mediated Stress Promotes Autophagic Survival and Aggressiveness of Colon Cancer Cells via Activation of Nrf2/HO-1 Pathway

Cesare Cernigliaro [1,†], Antonella D'Anneo [2,†], Daniela Carlisi [1], Michela Giuliano [2], Antonella Marino Gammazza [3,4], Rosario Barone [3,4], Lucia Longhitano [5], Francesco Cappello [3,4], Sonia Emanuele [1], Alfio Distefano [5], Claudia Campanella [3,4], Giuseppe Calvaruso [2] and Marianna Lauricella [1,*]

1 Department of Biomedicine, Neurosciences and Advanced Diagnostics (BIND), Institute of Biochemistry, University of Palermo, 90127 Palermo, Italy; cesare.cernigliaro@unipa.it (C.C.); daniela.carlisi@unipa.it (D.C.); sonia.emanuele@unipa.it (S.E.)

2 Department of Biological, Chemical and Pharmaceutical Sciences and Technologies (STEBICEF), Laboratory of Biochemistry, University of Palermo, 90127 Palermo, Italy; antonella.danneo@unipa.it (A.D.); michela.giuliano@unipa.it (M.G.); giuseppe.calvaruso@unipa.it (G.C.)

3 Department of Biomedicine, Neurosciences and Advanced Diagnostics (BIND), Institute of Human Anatomy, University of Palermo, 90127 Palermo, Italy; antonella.marino@hotmail.it (A.M.G.); rosario.barone@unipa.it (R.B.); francapp@hotmail.com (F.C.); claudiettacam@hotmail.com (C.C.)

4 Euro-Mediterranean Institute of Science and Technology, 90100 Palermo, Italy

5 Department of Biomedical and Biotechnological Sciences, University of Catania, I-95123 Catania, Italy; lucia.longhitano@unict.it (L.L.); distalfio@gmail.com (A.D.)

* Correspondence: marianna.lauricella@unipa.it; Tel.: +39-091-6552457

† These authors contributed equally to this work as first author.

Received: 11 February 2019; Accepted: 7 April 2019; Published: 10 April 2019

Abstract: Epidemiological studies suggest that chronic alcohol consumption is a lifestyle risk factor strongly associated with colorectal cancer development and progression. The aim of the present study was to examine the effect of ethanol (EtOH) on survival and progression of three different colon cancer cell lines (HCT116, HT29, and Caco-2). Our data showed that EtOH induces oxidative and endoplasmic reticulum (ER) stress, as demonstrated by reactive oxygen species (ROS) and ER stress markers Grp78, ATF6, PERK and, CHOP increase. Moreover, EtOH triggers an autophagic response which is accompanied by the upregulation of beclin, LC3-II, ATG7, and p62 proteins. The addition of the antioxidant N-acetylcysteine significantly prevents autophagy, suggesting that autophagy is triggered by oxidative stress as a prosurvival response. EtOH treatment also upregulates the antioxidant enzymes SOD, catalase, and heme oxygenase (HO-1) and promotes the nuclear translocation of both Nrf2 and HO-1. Interestingly, EtOH also upregulates the levels of matrix metalloproteases (MMP2 and MMP9) and VEGF. Nrf2 silencing or preventing HO-1 nuclear translocation by the protease inhibitor E64d abrogates the EtOH-induced increase in the antioxidant enzyme levels as well as the migration markers. Taken together, our results suggest that EtOH mediates both the activation of Nrf2 and HO-1 to sustain colon cancer cell survival, thus leading to the acquisition of a more aggressive phenotype.

Keywords: colon cancer cells; ethanol; Nrf2; HO-1; ER stress; autophagy; MMPs

1. Introduction

Colorectal cancer (CRC) is one of the most widespread cancers in the world [1]. Epidemiological data show the highest incidence of this tumor in countries characterized by high indices of development, such as

Australia, Europe, and North America [2]. This geographic variability has been linked to differences in environment and lifestyle [2]. Numerous risk factors have been correlated with the development of CRC, including genetic factors, inflammation, intestinal microflora composition, as well as harmful lifestyle habits, such as smoking, high consumption of red meats, and alcohol intake [3].

Chronic and heavy alcohol consumption, a common lifestyle habit of developed countries, has been associated with an increased risk of developing gastrointestinal cancers, including CRC [4,5]. The association between alcohol drinking and CRC is dose-dependent. A recent meta-analysis of literature from 1996 to 2013 shows that compared with non-occasional drinkers, the relative risk was 0.97 for subjects who drink low doses of ethanol ($\leq$12.5 g/day), 1.04 for moderate drinkers (12.6–49.9 g/day), and 1.21 for heavy drinkers ($\geq$50 g/day) [6]. Alcohol's influence can also vary based on individual differences in enzymes involved in alcohol metabolism. For example, the presence of ADH1B [7] or ADH1C*1—two polymorphisms of alcohol dehydrogenase (ADH)—has been associated with an increased risk of CRC [8]. In addition to act as a risk factor for carcinogenesis, many studies indicate that chronic alcohol consumption also promotes colon tumor progression [6–9].

The mechanisms proposed to explain the role of alcohol in CRC promotion include acetaldehyde mutagenic effects, oxidative stress increase, and folic acid deficiency [10,11]. Acetaldehyde, which is produced by ADH, cytochrome P450 2E1 (CYP2E1), or catalase, is considered the most potent carcinogenic metabolite of ethanol [12]. It is highly reactive and mutagenic by forming protein and DNA adducts, which result in DNA mutations and reactive oxygen species (ROS) production [10]. In addition to acetaldehyde, ROS have been also shown to contribute to colon carcinogenesis [13]. In alcoholics, ROS are predominantly generated through CYP2E1, which is induced by chronic alcohol consumption in the mucosa of the esophagus and colon [14]. CYP2E1-induced ROS production has been correlated with the generation of lipid peroxidation products, such as 4-hydroxynonenal and malondialdehyde. These compounds can bind to DNA forming etheno-DNA adducts which are highly carcinogenic [14,15].

It is well known that chronic inflammation also promotes colorectal cancer carcinogenesis. To this regard, it has been reported that the persistent exposure of enterocytes to an inflammatory environment induces molecular alterations favoring tumor development [16]. However, following chronic inflammation epithelial cells can activate adaptive mechanisms to reduce the deleterious action of oxidative stress and survival [17]. The major mechanism by which cancer cells increase their antioxidant proteins is through the activation of nuclear factor erythroid 2-related factor 2 (Nrf2), a transcription factor involved in the transcription of antioxidant and detoxifying factors in response to oxidative stress [18,19]. Nrf2 is normally sequestered in the cytoplasm in an inactive complex with kelch-like ECH-associated protein 1 (Keap1). ROS induce oxidation of critical cysteines of Keap1 and its dissociation from Nrf2 [20]. This event allows Nrf2 to migrate into the nucleus and form transcriptionally active complexes with other proteins, such as Mafs (musculoaponeurotic fibrosarcoma proteins). Such molecular cascades lead to the increase in the transcription of cytoprotective and antioxidant genes, such as superoxide dismutase (SOD), catalase, and heme oxygenase (HO-1), thus favoring cancer cell survival [21]. In addition, previous reports showed that Nrf2, beyond its antioxidant role, also exhibits a protumorigenic activity promoting proliferation and sustaining migration and invasiveness of cancer cells [22]. To this regard, overexpression of Nrf2 has been reported in colon cancer cells and has been related with tumor progression and poor prognosis [23]. Similarly, HO-1, the main target of Nrf2, can also exert a double role in cancer [24,25]. In fact, HO-1 may counteract ROS-mediated carcinogenesis by favoring heme breakdown into biliverdin; however, its overexpression has been shown to provide tumor cells with a more aggressive survival phenotype [26]. Finally, under oxidative stress conditions HO-1 translocates into the nucleus, where it interacts with Nrf2 preventing its GSK3β-mediated phosphorylation coupled with ubiquitin–proteasomal degradation, thereby prolonging Nrf2 nuclear accumulation [27].

The aim of the present research was to evaluate the role of Nrf2/HO-1 axis in promoting colon cancer survival and progression under ethanol stimulation. Our data suggest that ethanol induces both Nrf2 and HO-1 nuclear translocation in response to oxidative and endoplasmic reticulum (ER) stress activation and that this event confers resistance to oxidative stress and contributes to the acquisition of a more invasive phenotype.

2. Results

2.1. Effects of Ethanol in Colon Cancer Cell Viability

To evaluate the effect of ethanol (EtOH) on colon cancer cells, different colon cancer cell lines (HCT116, HT29, and Caco-2 cells) were exposed to a range of EtOH concentrations (30–300 mM) for different durations. Then, the viability was assessed by 3-(4,5-dimethyl- thiazol-2-yl)-2,5-diphenyltetrazolium bromide (MTT) assay which evaluates the mitochondrial dehydrogenase activity, as reported in methods. As shown in Figure 1A, compared with untreated cells, EtOH, also at high doses, displayed a scarce effect on cell viability at 72 h of treatment, but slightly reduced cell viability at 24 and 48 h, in particular in HT29 and Caco-2 cells. Our data also demonstrated that staining the cells with propidium iodide (PI), a membrane impermeant dye that is generally excluded from viable cells, evidenced that EtOH treatment did not induce cell death in this condition, in which we observed a reduction of cell viability. Thus, the reduced viability assessed by MTT assay in EtOH-treated cells for 24 or 48 h could be due to cell proliferation arrest or reduced metabolic activity of cells.

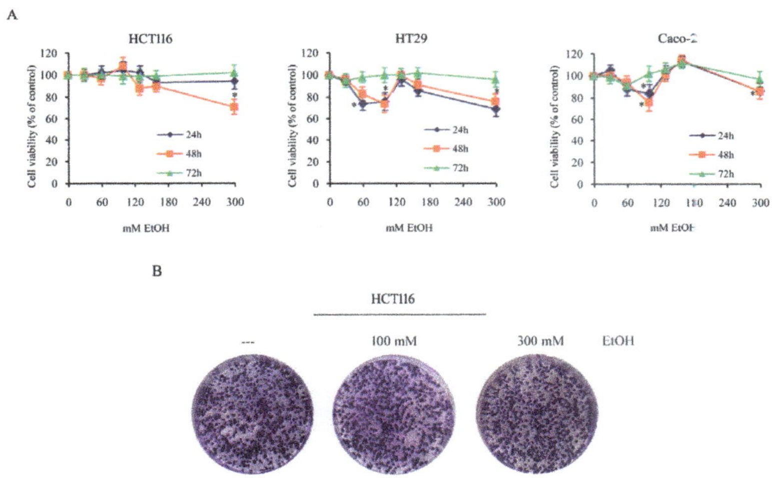

Figure 1. Ethanol effect on colon cancer cell viability and colony generation ability. (**A**) HCT116, HT29, and Caco-2 cells (7×10^3) were incubated in the presence of various doses of EtOH (30–300 mM) for the indicated times. Cell viability was assessed by 3-(4,5-dimethyl- thiazol-2-yl)-2,5-diphenyltetrazolium bromide (MTT) assay and expressed as the percentage of the viable control cells in untreated cultures. Values are the means of three independent experiments ± S.E. (*) $p < 0.05$ compared to the control. (**B**) The clonogenic assay was performed seeding a single cell suspension (200 cells/well) in 6-well plate and after 48 h treating it with different doses of EtOH. Clonogenic ability was evaluated after 14 days. Photographic images of cells after staining with crystal violet were reported. Results are representative of three independent experiments.

Morphological observations of cells confirmed the absence of cytotoxicity in all the conditions tested also prolonging EtOH treatment (1 week), leading to the conclusion that colon cancer cells survive even under treatment with high doses of ethanol.

To explore the ability of EtOH to counteract the production of colonies, HCT116 cells were plated with or without the addition of different doses of EtOH, and maintained in culture for additional 14 days to allow the formation of colonies. The results demonstrated no significant difference between control and EtOH-treated cells (Figure 1B), thus suggesting that the compound does not affect colony formation.

To evaluate the biochemical mechanisms linking high doses of alcohol consumption and colon cancer progression, the next experiments were performed using100 mM and 300 mM EtOH, two different high concentrations of alcohol.

2.2. Ethanol Treatment Induces Oxidative Stress in Colon Cancer Cells

EtOH metabolism produces acetaldehyde and reactive oxygen species (ROS), which have been correlated with the toxic effects of the compound [28]. To ascertain ROS generation in our model system, we performed fluorescence microscopy analysis by employing the fluorochrome H2DCFDA—a general indicator of intracellular ROS levels. As shown in Figure 2A, EtOH treatment rapidly increases the number of green fluorescent cells, which are indicative of intracellular ROS production. The effect appears at 10 min of exposure to EtOH, then rapidly decreases with the time of treatment.

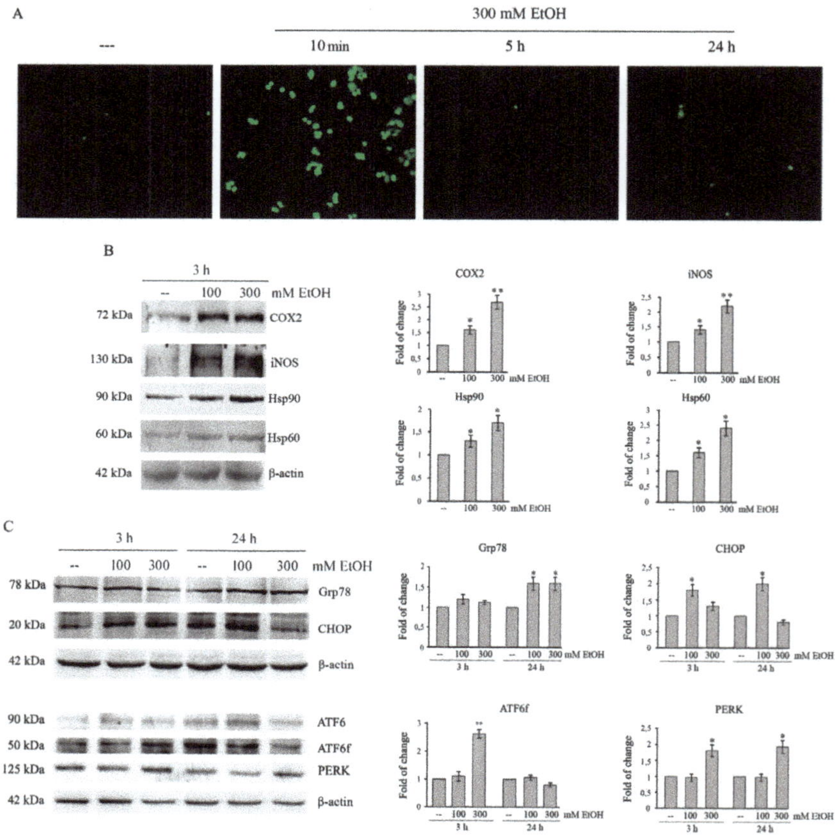

Figure 2. Ethanol induces oxidative and ER stress in colon cancer cells. (**A**) HCT116 cells (7×10^3) were treated with 300 mM EtOH for the indicated times. Reactive oxygen species (ROS) production was assayed by H2DCFDA staining. Images were taken by a Leica DC300F microscope (200× original magnification) using a FITC filter. Results are representative of three independent experiments. (**B**,**C**) Cells (1×10^5) were treated with 100 or 300 mMEtOH for the indicated times. Cell lysates (30 μg) were analyzed by Western blotting for COX2, iNOS, Hsp90, and Hsp60 (**B**) or Grp78, CHOP, ATF6 and its fragmented form (ATF6f) and PERK (**C**). Densitometric analysis of bands was carried out as reported in Material and Methods and data were normalized to β-actin expression. Results are from three independent experiments and statistical significance was determined using one-way ANOVA followed by Bonferroni's test. (*) $p < 0.05$ and (**) $p < 0.01$ compared to the untreated sample.

Cyclooxygenase 2 (COX2) and inducible nitric oxide synthase (iNOS) are two important enzymes expressed in response to a variety of stimuli that mediate inflammatory processes and tumor progression [29]. Western blotting analysis shows that COX2 and iNOS proteins are significantly up-regulated in EtOH-treated HCT116 cells. In particular, treatment for 3 h with 300 mM EtOH increases the levels of the enzymes respect to untreated control by 2.7 and 2.2 fold, respectively (Figure 2B, $p < 0.05$).

Moreover, we evaluated the effect of EtOH on the levels of Hsp90 and Hsp60, two heat shock proteins induced by cellular stress [30,31]. As shown in Figure 2B, the exposure to 100 and 300 mM EtOH for 3 h induces a significant increase of both Hsp90 and Hsp60 compared to the untreated condition ($p < 0.05$).

2.3. Ethanol Treatment Induces ER Stress in Colon Cancer Cells

Oxidative stress can be responsible for accumulation of unfolded-proteins and induction of ER stress [32]. To explore the ability of EtOH to activate ER stress in colon cancer cells, we analyzed the levels of glucose-regulated protein of 78 kDa (GRP78), which is a chaperone interacting with misfolded proteins and an indicator of ER stress, and CCAAT/enhancer-binding protein homologous protein (CHOP), a transcriptional factor which promotes apoptosis under prolonged ER stress [32]. Data reported in Figure 2C show that, in HCT116 cells, EtOH significantly upregulates the levels of Grp78 after 24 h ($p < 0.05$). In addition, EtOH upregulates CHOP level already after 3 h. After 24 h its level remains higher (2.2-fold, $p < 0.05$) in 100 mM EtOH-treated cells than untreated ones, while it lowers (0.8-fold) in cells exposed to 300 mM EtOH (Figure 2C). EtOH increases the levels of ER stress markers also in HT29 and Caco-2 cells, although the effects were observed after a prolonged time of treatment (24 h) (Figure S1).

We then evaluated the activation of ATF6 and PERK, two unfolding protein response (UPR) sensors which promotes adaptive signal transduction under ER stress [32]. When unfolded proteins increase, the full-length ATF6 is cleaved at Golgi into an active fragment (ATF6f), which migrates into the nucleus upregulating the expression of chaperones, including GRP78 and CHOP [32]. Treatment of HCT116 cells with EtOH decreases the ATF6 full-length protein at 3 h and concomitantly increases that of ATF6f, suggesting the activation of the factor. These effects are not observed at 24 h of EtOH treatment (Figure 2C). Moreover, EtOH treatment also promotes an early PERK increase that significantly ($p < 0.05$) occurs already at 3 h of treatment (Figure 2C).

2.4. Ethanol Stimulates A Prosurvival Effect Sustained by An Autophagic Flux In Colon Cancer Cells

Under ER stress condition, autophagy can be activated to degrade unfolded/aggregated proteins in order to maintain cell survival [33,34]. To assess whether this process occurs in our system model, the presence of autophagic vacuoles was evaluated using fluorescence microscopy by staining the cells with monodansylcadaverine (MDC). As Figure 3A shows, dot-likes structures appear in the cytoplasm of EtOH-treated HCT116 cells. MDC-positive fluorescent cells are already observed after 24 h of treatment with both 100 and 300 mM ethanol. The addition of the antioxidant N-acetylcysteine (NAC) to EtOH-treated cells markedly reduces the presence of autophagic vacuoles (Figure 3A), suggesting that autophagy is activated in response to oxidative stress as a prosurvival mechanism.

Then, we analyzed the levels of the autophagic markers by western blotting (Figure 3B). Microtubule-associated protein light chain 3 (LC3) can be present in two different forms: a cytosolic form (LC3-I) and an active lipidated form (LC3-II) [35], which is bound to the autophagosome membrane [36]. As shown in Figure 3B, in HCT116 cells EtOH treatment favors the conversion of LC3-I to LC3-II.

In addition to LC3, the level of beclin—a protein which plays an essential role in autophagosome formation—was evaluated. Data reported in Figure 3B demonstrate that beclin protein level is significantly higher after 3 h of treatment with 300 mMEtOH compared to untreated control ($p < 0.05$). The level remains elevated after 24 h.

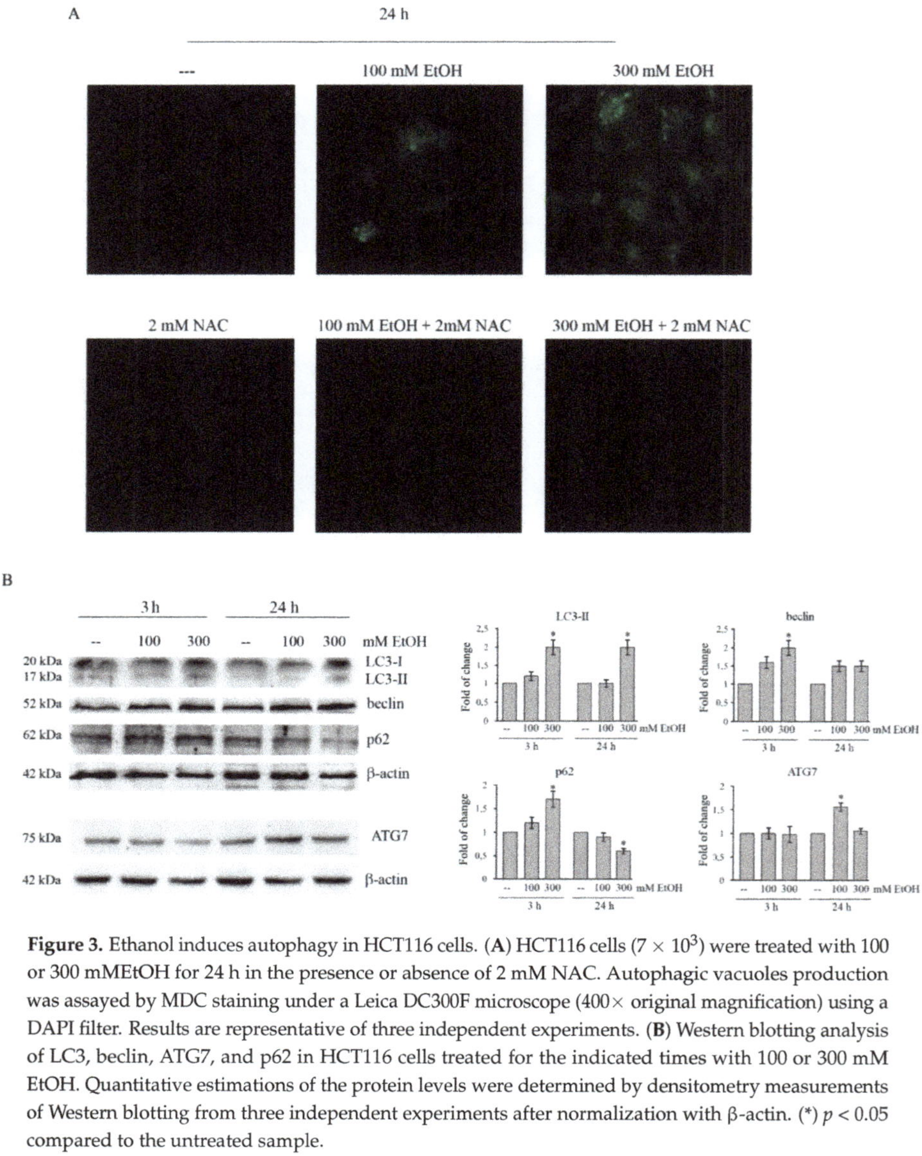

Figure 3. Ethanol induces autophagy in HCT116 cells. (**A**) HCT116 cells (7 × 10^3) were treated with 100 or 300 mMEtOH for 24 h in the presence or absence of 2 mM NAC. Autophagic vacuoles production was assayed by MDC staining under a Leica DC300F microscope (400× original magnification) using a DAPI filter. Results are representative of three independent experiments. (**B**) Western blotting analysis of LC3, beclin, ATG7, and p62 in HCT116 cells treated for the indicated times with 100 or 300 mM EtOH. Quantitative estimations of the protein levels were determined by densitometry measurements of Western blotting from three independent experiments after normalization with β-actin. (*) $p < 0.05$ compared to the untreated sample.

Moreover, EtOH treatment increases the level of ATG7, an essential regulator of autophagosome assembly [37]. As shown in the same Figure 3B, ATG7 level significantly ($p < 0.05$) increased after 24 h treatment with 100 mM EtOH with respect to the untreated sample (Figure 3B).

Finally, we also examined the level of p62, a multifunctional protein considered as a marker of autophagic flux. p62 is localized to the autophagosome via LC3 interaction and is constantly degraded by the autophagy–lysosome system [38]. The analysis of p62 shows that EtOH increases the content of this protein at 3 h of treatment, while its level lowers after 24 h (Figure 3B), thereby suggesting that HCT116 cells treated with ethanol undergo to a complete autophagic process.

2.5. Ethanol Activates Nrf2-Dependent Antioxidant Pathway

Despite the activation of oxidative and ER stress by EtOH, MTT cell viability and clonogenic assay, reported in Figure 1, show that colon cancer cells survive even after prolonged treatment with high doses of EtOH, thus suggesting the activation of an antioxidant response.

Nrf2 is one of the major transcription factors that promotes cellular defense against oxidative stress. Nrf2 is maintained in an inactive cytosolic complex with Keap1 [39]. Our data show that Nrf2 level is enhanced in HCT116 cells treated with EtOH. In fact, in comparison with control cells the level of Nrf2 protein increases with 300 mM EtOH approximately by 1.5-fold ($p < 0.05$) after 3 h of treatment (Figure 4). Data reported in the same Figure 4, also demonstrate that the increase of Nrf2 is accompanied by a significant decrease of Keap1 in cells treated for 3 h with 300 mM EtOH compared to untreated control ($p < 0.05$).

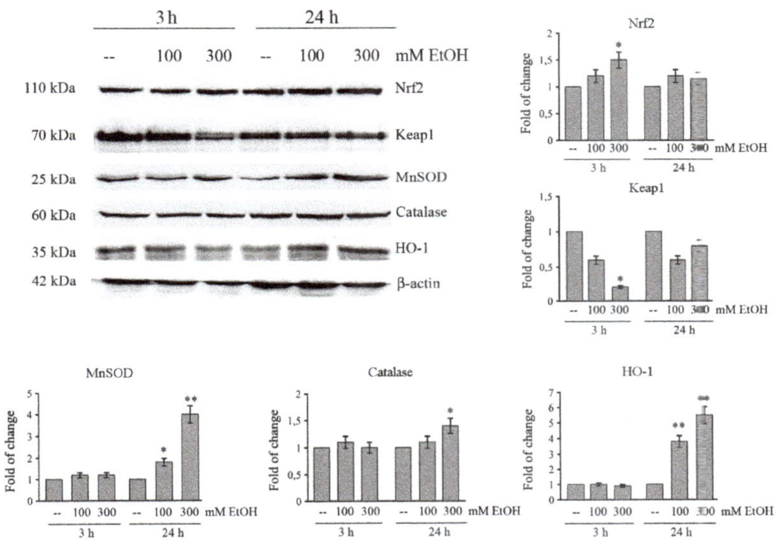

Figure 4. Effects of ethanol on Nrf2-dependent antioxidant pathway in HCT116 cells. Cells (1×10^5) were incubated in the presence of various doses of EtOH for different periods. The levels of Nrf2, Keap1, MnSOD, catalase, and HO-1 were assessed by Western blotting analysis. Immunoblot was quantified by densitometry and normalized against β-actin expression. Results are from three independent experiments and statistical significance was determined using one-way ANOVA followed by Bonferroni's test. (*) $p < 0.05$ and (**) $p < 0.01$ compared to the untreated sample.

It is known that Nrf2 controls the transcription of many antioxidant and detoxifying genes, such as superoxide dismutase (MnSOD), catalase, and heme oxygenase (HO-1) [21]. We demonstrated that the levels of these proteins transcriptionally regulated by Nrf-2 markedly increase after 24 h of treatment with 300 mMEtOH (Figure 4). In particular, the increase in the presence of 300 mM EtOH is estimated 1.4-fold for catalase, 4-fold for MnSOD ($p < 0.001$), and 5.5-fold for HO-1 ($p < 0.001$) after 24 h of treatment (Figure 4). An increase in the levels of Nrf-2 and HO-1 is also observed in both HT29 and Caco-2 cells after EtOH treatment (Figure S2).

2.6. Ethanol Promotes Nrf2 and HO-1 Translocation

Then we analyzed if the Nrf2 increase is accompanied by its nuclear translocation. Western blotting analysis showed that Nrf2 is present in both nuclear and cytosolic fraction in untreated HCT116 cells and that EtOH treatment increases the nuclear level of Nrf2, while, concomitantly, decreases its cytosolic fraction (Figure 5A). The analyses by confocal microscopy confirmed that in control cells Nrf2 is preferentially distributed in the cytoplasm, while in treated cells the immunoreactivity is mainly nuclear (Figure 5B).

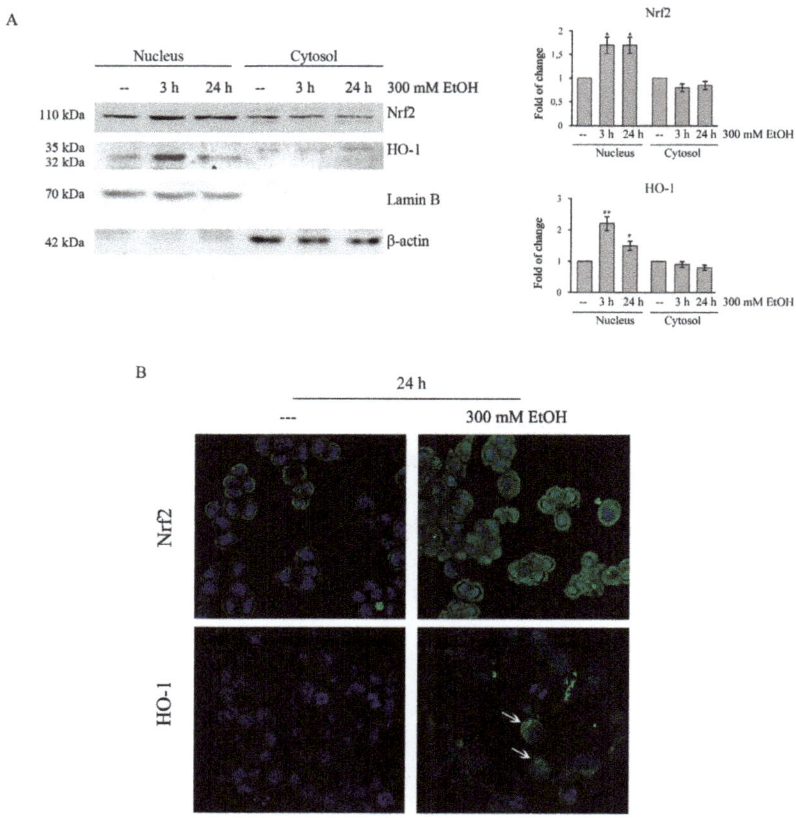

Figure 5. Ethanol induces the nuclear translocation of cytosolic Nrf2 and HO-1 in HCT116 cells. (**A**) Cells were treated for 3 h and 24 h with 300 mM EtOH. Equal amounts of nuclear or cytosolic proteins (30 µg) were analyzed by Western blotting and quantified by densitometry for Nrf2 and HO-1 expression normalized against Lamin B and β-actin. Representative blots of three independent experiments are shown. (*) $p < 0.05$ and (**) $p < 0.01$ compared to the untreated sample. (**B**) HCT116 cells (8×10^3) were grown on coverslips and treated for 24 h with 300 mM EtOH. EtOH-induced nuclear translocation was observed under confocal microscope TCS SP8 employing Nrf2 or HO-1 specific antibodies followed by incubation with a fluorescein isothiocyanate (FITC) conjugated IgG secondary antibody (green). The cells were also stained with Hoechst (blue fluorescence) to visualize nuclear morphology. Original magnification: 400×. The results are representative of three independent experiments.

It has been reported that HO-1 can undergo to intramembrane proteolysis and translocation into the nucleus to sustain tumor survival and invasiveness without depending on its enzymatic activity [40]. Interestingly, western blotting analyses showed that EtOH is able to increase nuclear HO-1 levels (Figure 5A). This effect was already observed after 3h of treatment with 300 mM EtOH ($p < 0.01$). Nuclear HO-1 translocation in EtOH-treated cells was also confirmed by confocal microscopy experiments (Figure 5B, white arrows).

2.7. Ethanol Increases MMPs and VEGF in Colon Cancer Cells

Given that metalloproteases (MMPs) are critical to cell invasion and metastasis [41], the expression and activity of MMP2 and MMP9 in EtOH-treated HCT116 cells were examined by both Western blotting analysis and gelatin zymography. Our results show that there is a significant increase in the protein level of active MMP2 (7-fold, $p < 0.001$) in HCT116 cells treated with 300 mM EtOH already after 24 h of treatment (Figure 6A). A minor increase (2.2-fold, $p < 0.01$) is observed for active MMP9 (Figure 6A). Moreover, gelatin zymography showed that 300 mM EtOH also increases the activity of both metalloproteinases (Figure 6B).

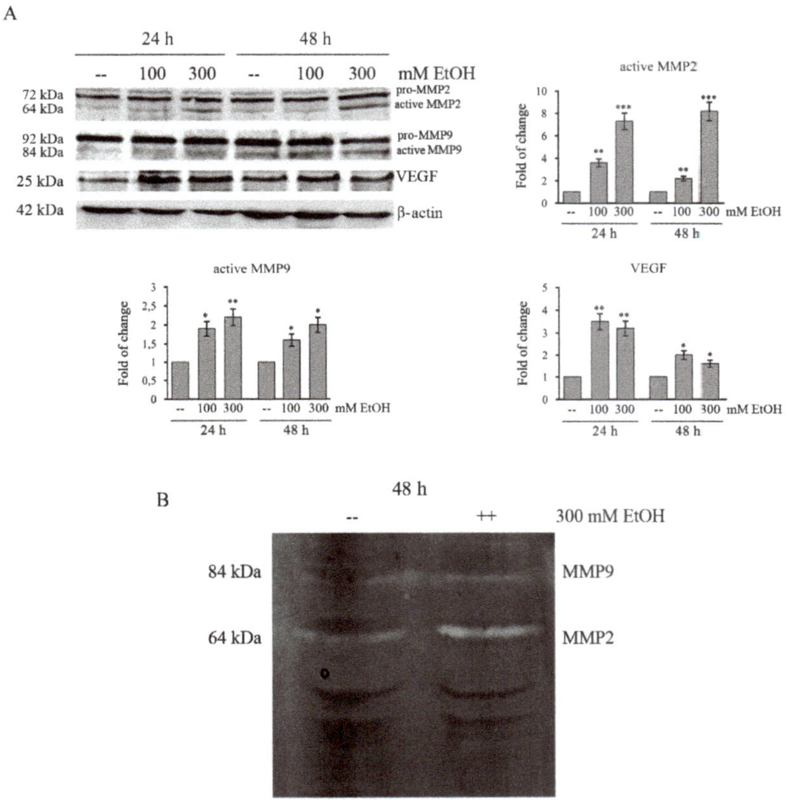

Figure 6. Ethanol increases the levels of VEGF and activates MMPs in HCT116 cells (**A**) Western blotting analysis of MMP2, MMP9, and VEGF in HCT116 cells treated for different times with 100 or 300 mMEtOH. Quantitative estimations of the protein levels were determined by densitometry measurements of western blotting from three independent experiments after normalization with β-actin. (*) $p < 0.05$, (**) $p < 0.01$, (***) $p < 0.001$ compared to the untreated sample. (**B**) The activity of MMP2 and MMP9 in HCT116 cells treated with 300 mM EtOH for 48 h was examined by gelatin zymography assay. The results are representative of three independent experiments.

Finally, we also demonstrated that EtOH increases the level of vascular endothelial growth factor (VEGF) in HCT116 cells. As shown in Figure 6A, in comparison with untreated cells, the level of VEGF protein significantly increases after 24 h in EtOH-treated HCT116 cells approximately by 3.5-fold ($p < 0.01$) with 100 mM and by 3.2 fold ($p < 0.01$) with 300 mM, respectively.

2.8. Role of Nrf2/HO-1 Axis in Colon Cancer Survival and Aggressiveness

To further demonstrate the contribution of Nrf2 in protecting colon cancer cells by oxidative damage induced by EtOH, Nrf2 was downregulated by transfecting HCT116 cells with a specific siRNA pool directed against Nrf2. Western blotting analysis showed that the levels of Nrf2 protein in siNrf2-transfected group are significantly decreased ($p < 0.05$) after 24 h compared with the siRNA control group (Figure 7). Interestingly, the increases in HO-1 and MnSOD protein level caused by 300 mM EtOH treatment is suppressed by Nrf2 siRNA (Figure 7A), thus confirming that the expression of such antioxidant proteins is under Nrf2 control. Moreover, cell viability assay showed that Nrf2 silencing reduces HCT116 cell viability (Figure 7B), thus suggesting a protective role of Nrf2 axis against EtOH-mediated toxic effect.

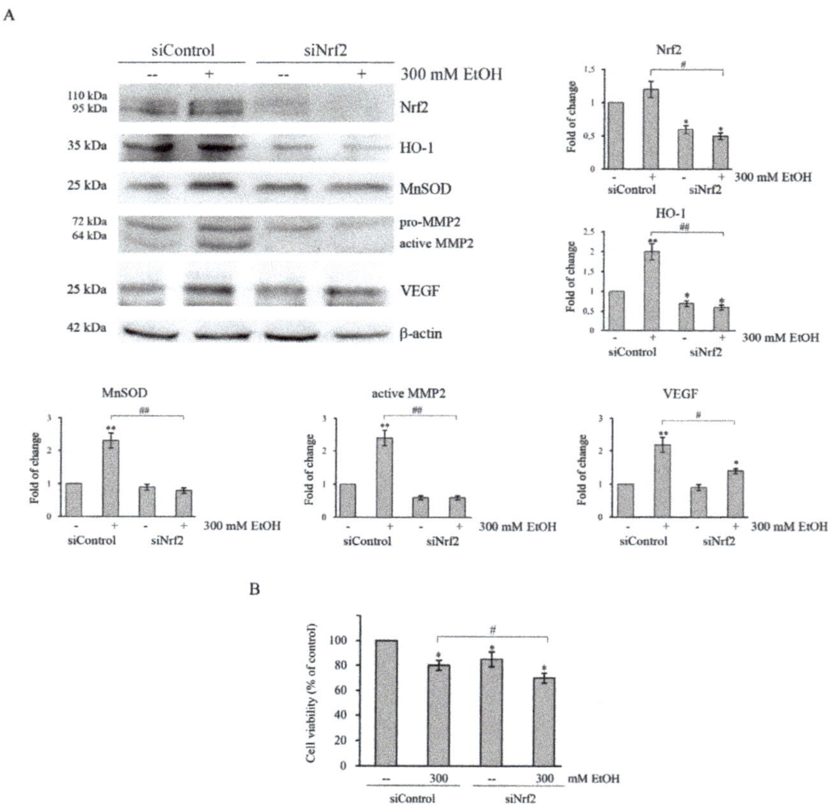

Figure 7. Nrf2-silencing represses the increase in the levels of antioxidant enzymes, MMP2 and VEGF induced by EtOH treatment. (**A**,**B**) HCT116 cells (1×10^5) were transiently transfected with a nonspecific siRNA (siControl) or with Nrf2 specific siRNA pool (siNrf2) and, 24 h after transfection, cells were treated with 300 mM EtOH for additional 24 h or 48 h. (**A**) The levels of HO-1, MnSOD, MMP2, and VEGF were analyzed by Western blotting after 24 h of transfection. The successful of Nrf2-silencing was verified by measuring the level of Nrf2 in transfected cells. Quantitative estimations of the protein levels were determined by densitometry measurements of western blotting from three independent experiments after normalization with β-actin (*) $p < 0.05$ and (**) $p < 0.01$ compared to the untreated sample. ($^{\#}$) $p < 0.05$, ($^{\#\#}$) $p < 0.01$. (**B**) Effects of Nrf2-silencing on the viability of HCT116 cells evaluated after 48 h of EtOH treatment by MTT assay and expressed as the percentage of the viable siControl cells in untreated cultures. Values are the means of three independent experiments ± S.E. ($^{\#}$) $p < 0.05$ between the two groups.

Furthermore, to evaluate whether also nuclear HO-1 translocation favors the survival of EtOH-treated colon cancer cells, HCT116 cells were incubated in the presence of E64d, an inhibitor of the protease responsible for the proteolytic cleavage of HO-1, an event necessary for its nuclear translocation [42]. Combination of E64d with 300 mM EtOHis accompanied by a significant reduction in nuclear localization of HO-1 (p <0.05) (Figure 8A). In addition, the data showed that inhibition of nuclear translocation of HO-1 by E64d significantly reduces the EtOH-induced increase of MnSOD (Figure 8B), thus supporting the conclusion that nuclear HO-1 could regulate MnSOD expression.

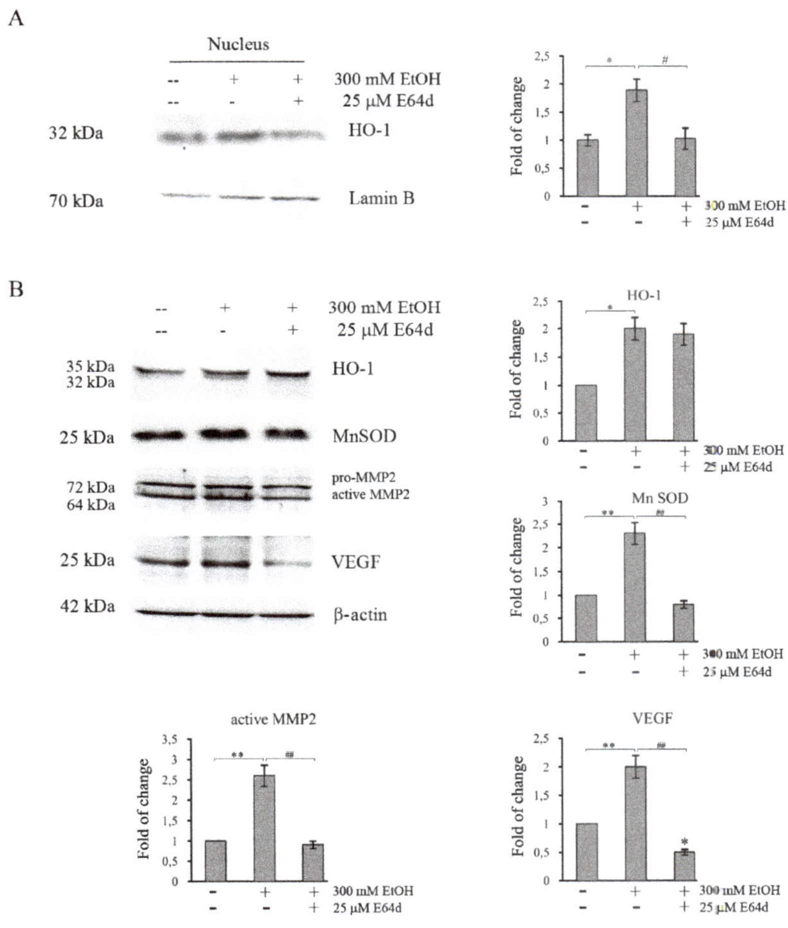

Figure 8. E64d counteracts the increase in antioxidant enzymes, MMP2 and VEGF induced by EtOH treatment in HCT116 cells. Cells were pretreated for 3 h with E64d, then 300 mM EtOH was added and the treatment was protracted for another 24 h. Nuclear and cytosolic fractions were prepared as reported in Materials and Methods section. Evaluation of the HO-1 level in nuclear fraction (**A**) and those of HO-1, MnSOD, MMP2, and VEGF in total fraction (**B**) by Western blotting analysis. Quantitative estimations of the protein levels were determined by densitometry measurements of western blotting from three independent experiments after normalization with β-actin. (*) $p < 0.05$, (**) $p < 0.01$ compared to the untreated sample; (#) $p < 0.05$; (##) $p < 0.01$ between the two groups.

It has been reported that both Nrf2 and HO-1 activation promote tumor progression by regulating proinvasive and angiogenetic factors [22,43]. Thus, further analyses were performed to examine whether the Nrf2/HO-1 axis also plays a role in the acquisition of a tumor aggressive phenotype in EtOH-treated colon cancer cells. As reported in Figures 7 and 8, in HCT116 cells the increase in MMP2 protein level induced by EtOH treatment at 300 mM is significantly ($p < 0.05$) suppressed by both Nrf2 siRNA (Figure 7A) and E64d (Figure 8B). Moreover, the increase in VEGF level in EtOH-treated cells is reduced by Nrf2 silencing (Figure 7A) and prevented by E64d (Figure 8B). These data indicate that activation of Nrf2/HO-1 axis could enhance colon cancer progression by inducing MMP2 and VEGF expression.

3. Discussion

Epidemiological studies highlighted that heavy alcohol drinking promotes colon cancer progression [44], although the underlying molecular mechanisms are still not clear. Data reported in this paper show that colon cancer cells survive even under treatment with high doses of ethanol (100–300 mM), which have been shown to exert toxic effects in other tumor cell lines [45,46]. It is interesting to note that concentrations ranging from 10 to 100 mM correspond to blood levels in humans that could result from moderate-to-heavy alcohol drinking [47].

Supportive signs of evidence suggest that EtOH is able to increase ROS level in different cell lines, including colon cancer cells, through both ADH and CYP2E1 activity [48]. Moreover, chronic EtOH exposure causes inflammation in different organs, like pancreas and liver, as indicated by the increase in proinflammatory cytokines and chemokines [49,50]. In line with these observations, we demonstrated that EtOH stimulates oxidative stress and an inflammatory response in colon cancer cells. This conclusion is supported by several pieces of evidence, such as rapid ROS production, the increase in the levels of two enzymatic markers of inflammation iNOS and COX2, and the upregulation of Hsp90 and Hsp60 induced by EtOH in colon cancer cells.

Our results also provided evidence that high doses of EtOH trigger ER stress as demonstrated by upregulation of the ER stress markers Grp78, CHOP, and PERK, as well as the increase in the active form of the transcription factor ATF6. These results are in line with the observation that ER stress contributes to alcoholic damage of major organs such as liver, pancreas, and brain [51,52].

Accumulating evidence suggest that activation of autophagy, which degrades proteins in organelles such as mitochondria and the ER, can play a protective role against the toxic effects of ER stress [53]. On the other hand, ER stress can activate autophagy and there is a considerable cross-talk between the ER and autophagy [54]. It has been reported that ER stress-activated PERK promotes the phosphorylation of eIF2α to induce the activation of LC3, an autophagosome marker [55]. Moreover, Shimodaira et al. [33] demonstrated that in colon cancer cells activation of ER marker CHOP promotes autophagy by activating inositol-requiring enzyme 1α (IRE1α). Our results provide evidence that EtOH induced autophagy in colon cancer cells as demonstrated by the augmented accumulation of acidic intracellular compartments. This effect is prevented by the addition of the antioxidant N-acetylcysteine, thus suggesting that autophagy is activated as a survival mechanism in response to oxidative stress. Our results are in line with the observation that EtOH activates autophagy in neuronal cells to prevent the oxidative damage of ROS [56]. The activation of autophagy in our system was also confirmed by the observation that EtOH promoted the cleavage of the cytosolic form of LC3-I to LC3-II, as well increasing the level of beclin and ATG7, two factors involved in the autophagosome formation [37]. Moreover, the observation that the level of p62 protein, a marker of autophagic degradation, decreases after 24 h of EtOH treatment, which suggested to us that the compound is able to trigger a complete autophagic flux.

In response to oxidative and ER stress production, we showed a significant activation of Nrf2, a transcription factor which acts as a key regulator of antioxidant-responsive genes [57]. Under our experimental conditions, EtOH promotes nuclear Nrf2 translocation, as suggested by the increase of nuclear Nrf2 content already at 3 h following EtOH treatment. This event is a consequence of the decrease of Keap1, a protein which in unstressed conditions sequesters Nrf2 in a cytoplasmic complex leading to its ubiquitination and consequent proteasomal degradation [58]. Nuclear translocation of

Nrf2 by EtOH could be promoted by oxidative events, in line with the observation that induction of oxidative stress triggers Nrf2 nuclear import through the oxidation of redox-sensitive cysteines within Keap1 [59]. Moreover, nuclear translocation could be also favored by ER stress activation. In fact, it has been shown that phosphorylation of Nrf2 by PERK—a kinase activated following the accumulation of unfolded proteins in ER—promotes dissolution of Nrf2/Keap1 complexes and Nrf2 nuclear import [60].

Activation of Nrf2 is responsible for transcription of a battery of genes encoding antioxidant enzymes such as MnSOD, catalase, and HO-1 [61]. Our data indicated that the level of both MnSOD and HO-1 is significantly upregulated after EtOH treatment in colon cancer cells, suggesting a protective role of these enzymes against EtOH-induced oxidative stress. The observation that Nrf2 silencing significantly reduces EtOH-induced HO-1 and MnSOD increase suggest to us that this effect is mediated by the activation of Nrf2 transcriptional activity.

Interestingly, our data also provide evidence that EtOH promotes HO-1 nuclear translocation already after 3 h of treatment. Nuclear expression of HO-1 has been detected in several tumors and it has been correlated with tumor growth and invasion [43]. Our data showed that inhibition of HO-1 nuclear translocation by the protease inhibitor E64d significantly reduces EtOH-induced MnSOD increase, suggesting that HO-1 nuclear translocation could cooperate with Nrf2 to stimulate antioxidant response and colon cancer cell survival. This suggestion is in accordance with some recent evidence that demonstrates that nuclear HO-1 modulates the activation of Nrf2, leading to activation of antioxidant genes [27].

Activation of the Nrf2/HO-1 axis represents a double-edged sword in cancer [61]. Increase in antioxidant enzymes by Nrf2 prevents the development of tumors by counteracting the genotoxic damage induced by ROS [62]. In line with this consideration, several dietary phytochemicals exert a cancer preventive effect by activating the Nrf2/HO-1 axis [63,64]. Moreover, activation of Nrf2 reduces chronic inflammation which has been correlated with CRC induction [65]. On the other hand, activation of the Nrf2/HO-1 antioxidant response in tumor cells can promote tumor survival by creating an optimal environment for cell growth [61]. Overexpression of Nrf2 has been detected in primary CRC and metastatic tissues relative to normal colon and contributes to chemoresistance in CRC cell lines [66]. In addition, it has been reported that Nrf2 increases CRC risk by promoting angiogenesis and uncontrolled proliferation [67]. Moreover, HO-1 overexpression has been associated with a more aggressive behavior of tumors and poor prognosis in various cancers [68,69]. Western blotting and zymography analyses demonstrated that EtOH increases both the levels and the activity of MMP-2 and 9—two enzymes involved in tumor progression—and upregulates VEGF, the main angiogenetic factor. Interestingly, Nrf2 silencing prevented EtOH-induced MMP-2 increase and reduced that of VEGF. Moreover, inhibition of HO-1 nuclear translocation by E64d counteracted the effect of EtOH on both MMP-2 and VEGF, thus markedly suggesting a role of Nrf2/HO-1 axis in colon cancer progression.

Collectively, our findings demonstrate that high doses of EtOH enhance autophagy and activation Nrf2/HO-1 axis in colon cancer cells. These events sustain the survival of cancer cells protecting them from oxidative and ER stress induced by EtOH. Moreover, we show, for the first time, that the activation of Nrf2/HO-1 axis could be also responsible for colon cancer progression through the acquisition of a metastatic behavior, as demonstrated by the increase in the levels of MMP-2 and VEGF. A schematic model of EtOH effects on colon cancer cells is shown in Figure 9. This study provides a novel mechanistic link between ethanol-induced activation of Nrf2/HO-1 pathway and increased survival and aggressiveness of colon cancer cells in in vitro models. However, to open to this new scenario, further in vivo investigations are strongly required to sustain the role of Nrf2/HO-1 activation in CRC progression. Detection of Nrf2 and HO-1 overexpression in CRC biopsies of alcohol drinkers could be, indeed, used as potential novel biomarkers to monitor CRC progression.

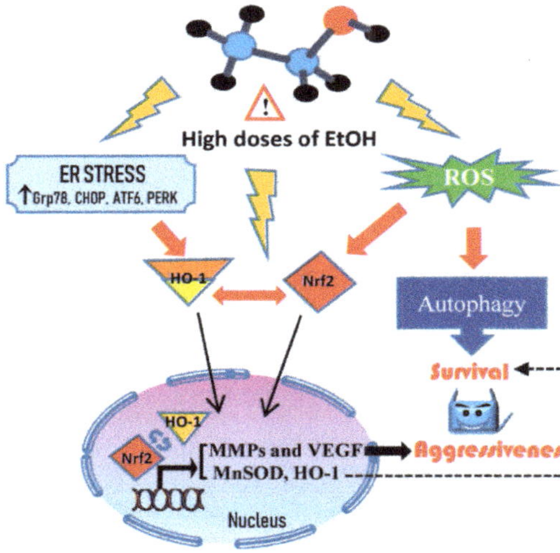

Figure 9. Scheme of the mechanism activated by high doses of ethanol. High doses of ethanol induce ROS generation and ER stress responsible for the induction of a prosurvival autophagic process and the aggressive tumor phenotype. Such events are sustained by the nuclear translocation of Nrf2 and HO-1, which activate antioxidant response systems and promote the upregulation of MMP2 and VEGF.

4. Materials and Methods

4.1. Cell Cultures and Chemicals

The human colon cancer lines HCT116, HT29, and Caco-2 (Interlab Cell Line Collection, ICLC, Genova, Italy) were grown in monolayer in flasks of 75 cm^2 in RPMI 1640 medium, supplemented with 10% (v/v) heat-inactivated fetal bovine serum (FBS), 2 mM L-glutamine, 100 U/mL penicillin, and 50 µg/mL streptomycin in a humidified atmosphere of 5% CO_2 in air at 37 °C. To study the effects of ethanol (EtOH), cells were detached using trypsin-EDTA (0.5 mg/mL trypsin and 0.2 mg/mL EDTA) and plated in accordance to the experimental conditions, as described in the paragraphs below. Cells were allowed to adhere for 24 h and then treated with different concentrations of EtOH at different times.

All the reagents used for cell cultures were purchased from Euroclone (Pero, Italy). EtOH, E64d and all chemicals, except when stated otherwise, were supplied by Sigma-Aldrich (Milan, Italy).

4.2. Cell Viability

To evaluate the effect of EtOH on cell viability the MTT (3-(4,5-dimethylthiazol-2-yl)-2,5-diphenyl tetrazolium bromide) colorimetric assay was used as previously reported [70]. In brief, HCT116, HT29, and Caco-2 cells (7×10^3/200 µL/well) were plated in 96-wells and treated with various concentrations of EtOH (30–300 mM) for different times. Fresh ethanol-containing medium was added to cells daily. Then, 20 µL MTT (11 mg/mL) was added and cells were incubated at 37 °C for 4 h. Finally, the medium was removed and 100 µL of lysis buffer (20% sodium dodecyl sulfate in 50% N,N-dimethylformamide) was added. At the end, the absorbance of the formazan was measured directly at 490 nm with 630 nm as a reference wavelength using an automatic ELISA plate reader (OPSYS MR, Dynex Technologies, Chantilly, VA, USA). Cell viability was expressed as the percentage of the OD value of EtOH-treated cells compared with untreated samples used as control. Each experiment was performed in triplicate. The viability of cells was also assessed through the use of propidium iodide (PI) dye exclusion assay. The intact membrane of live cells excludes a variety of dyes that easily penetrate the damaged, permeable membranes of dead cells. After incubation with ethanol, 2 µg/mL

PI was added and the incubation was protracted for 15 min before the visualization of cell morphology by a Leica DC 300F microscope (Leica microsystems, Wetzlar, Germany) equipped with a rhodamine filter (excitation wavelength of 596 nm and emission wavelength of 620 nm). Also, cell morphology was visualized using a Leica DC 300F microscope inverted.

4.3. Clonogenic Assay

For the colony formation assay, a single cell suspension (200 cells/well) was plated into each well of a 6-well plate and incubated at 37 °C for 2 weeks. The colonies were fixed and stained with a dye solution containing crystal violet as reported by Raffa et al. [71]. All assays were replicated three times.

4.4. Western Blotting Analysis

Cell lysates were prepared as reported [72] and protein concentration was determined by Bradford Protein Assay (Bio-Rad Laboratories S.r.l., Segrate, Milan, Italy). Protein extracts (30 μg/sample) were subjected to SDS-polyacrylamide gel electrophoresis. Then, proteins were blotted on nitrocellulose membranes. Primary antibodies used for the identification of catalase, Lamin B, MnSOD, Nrf2, iNOS, COX2, Grp78, Chop, MMP2, MMP9, and PERK (1:200) were purchased from Santa Cruz Biotechnology (St. Cruz, CA, USA); β-actin (1:1000) from Sigma Aldrich; Hsp60, Hsp90, and HO-1 (1:1000) from Enzo Life Sciences, (Milan, Italy), ATG7 (1:1000) from Cell Signaling Tecnology (Beverly, MA, USA); and ATF6 (1:300) from Novus Biologicals (Milano, Italy). Membranes were then incubated with HRP-conjugated secondary antibody (1:5000) (Pierce, Thermo Fisher Scientific, Loughborough, UK) and the signals were detected using enhanced chemiluminescence (ECL) reagents (Cyanagen, Bologna, Italy). The signal obtained was visualized and photographed with ChemiDoc XRS (Bio-Rad, Hercules, CA, USA). After analysis, membranes were stripped (200 mM glycine, 0,1% SDS, 1% Tween 20, pH 2.2) and reincubated. The intensity of the bands was performed using Quantity One software (Bio-Rad). After checking that β-actin content was not modified by ethanol treatment vs untreated sample, in accordance with Lamichhane et al. [73], β-actin was used as a loading control and for band normalization. All the blots shown are representative of at least three separate experiments.

4.5. Extraction of Cytosolic and Nuclear Fraction

HCT116 cells were seeded in 100-mm tissue culture dishes (1×10^6 cells/5 mL culture medium) and, after treatment with EtOH, were lysed as reported [74]. In particular, cells were washed in PBS and scraped with lysis buffer (250 mM Sucrose, 20 mM HEPES, 10 mM KCl, 1.5 mM $MgCl_2$, 1 mM EDTA, 1 mM EGTA, 1 mM DTT, and protease inhibitors, pH 7.4). Next, cells were passed 10 times through a needle of 25 g on ice for 20 min. The homogenates were centrifuged at 1000× g for 10 min at 4 °C. The pellets were resuspended in lysis buffer and passed 10 times through a needle of 25 g and centrifuged at 1000× g for 10 min at 4 °C. The pellets of the second centrifugation (nuclear fraction) were lysed with RIPA buffer (1% NP-40, 0.5% sodium deoxycholate, 0.1% SDS, inhibitors of proteases: 25 μg/mL aprotinin, 1 mM PMSF, 25 μg/mL leupeptin, and 0.2 mM sodium pyrophosphate) and sonicated. The supernatants obtained from the first centrifugation were centrifuged at 10,000× g for 30 min at 4 °C. The supernatants obtained were considered as cytosolic fraction. Nuclear and cytosolic fractions were used to evaluate Nrf2 and HO-1. β-actin and LaminB were used as cytoplasmic and nuclear markers, respectively.

4.6. Monodansylcadaverine Test

To evaluate the formation of autophagic vacuoles monodansylcadaverine (MDC) test was employed as reported [75]. HCT116 cells (7×10^3/200 μL culture medium) were plated in 96-wells plates and treated with EtOH. After treatment, cells were incubated with 50 μM MDC for 10 min at 37 °C in the darkness. Then, cells were washed with PBS and analysed by fluorescence microscopy using a Leica DMR (Leica Microsystems, Milan, Italy) microscope equipped with a DAPI filter system (excitation wavelength of 372 nm and emission wavelength of 456 nm). Images were acquired by computer imaging system (Leica DC300F camera, Milan, Italy). Three different visual fields were examined for each condition.

4.7. *Nrf2 siRNA Transfection*

RNA interference of Nrf2 was performed using FlexiTube siRNA (SI03246950, SI03246614 Qiagen, Hilden, Germany) targeting Nrf2. A nonsilencing siRNA (SI03650318, Qiagen) was used as a negative control. As previously reported [76] for transfection, HCT116 cells were seeded (2×10^5 cells/well) in 6-well plates and cultured in antibiotic-free RPMI 1640 medium supplemented with 10% FBS for 24 h to reach approximately 60–80% confluence before transfection. Specific siRNAs (50 nM final concentration) and negative siRNA control (50 nM) were transfected for 5 h into the cells in the presence of 5 μL Lipofectamine 2000 (Invitrogen, Carlsbad, CA, USA) in a final volume of 1 mL serum/antibiotic-free RPMI 1640 medium. The reaction was stopped by replacing the culture medium with complete RPMI 1640 medium. After 24 h of transfection, cells were treated with EtOH for other 24 h or 48 h. Then, cells were employed for immunoblotting analysis or MTT.

4.8. *Measurement of Intracellular ROS Content*

Intracellular ROS production was detected using the cell-permeant 2′,7′-dichlorodihydrofluorescein diacetate (H2DCFDCA Molecular Probes; Eugene, OR, USA) as described [77]. HCT116 cells (7×10^3/200 μL) were seeded in 96-well plates and incubated with 300 mM EtOH for different times. After treatment, cells were washed with PBS and incubated with a 10 μM H2DCFDA for 15 min at 37 °C in the dark. Finally, cells were resuspended in PBS and analyzed by fluorescence microscopy using a Leica DMR (Leica Microsystems S.r.l., Wetzlar, Germany) inverted microscope equipped with a FITC filter system (excitation wavelength of 485 nm and emission wavelength of 530 nm). Images were acquired by computer imaging system (Leica DC300F camera). Three different visual fields were examined for each condition.

4.9. *Gelatin Zymography*

HCT116 cells were seeded in 100-mm tissue culture dishes (5×10^5 cells/ 5 mL culture medium). After 48 h of EtOH treatment, cells were washed in PBS and scraped with lysis buffer. The lysates were centrifuged at 800× *g* for 10 min. The samples (50 μg of proteins prepared in sample buffer: 50 mM Tris-HCl, 2% SDS, 0.1% Bromophenol Blue, 40% Glycerol, pH 6.8) were loaded on polyacrylamide gels (10%) with 10× gelatin and subjected to electrophoresis. After, the gel was washed for 1 h with enzyme renaturing buffer (200 mM NaCl, 5 mM $CaCl_2$, 5 μM $ZnCl_2$, 2,5% (*v*/*v*) Triton X-100 and 50 mM Tris-HCl, pH 7.5) and incubated overnight at 37 °C with developing buffer (50 mM Tris base, 200 mM NaCl, 5 mM $CaCl_2$, pH 7.5). Then, the gel was incubated for 30 min at room temperature with staining solution (0.125% Coomassie brilliant blue R-250, 50% methanol, 20% acetic acid) and washed with destaining solution (30% methanol, 0,01% formic acid) until clear bands of MMP activity are visible in the blue background.

4.10. *Immunofluorescence*

HCT116 cells (8×10^3) were plated on coverslips and treated with 300 mM EtOH for different times. Immunofluorescence was performed as previously described [78,79]. Briefly, after washing twice in PBS, cells were fixed in methanol for 30 min at room temperature. After fixation, cells were washed three times in PBS for 5 min and treated with a blocking solution (3% BSA in PBS) for 30 min. Subsequently, the cells were washed twice in PBS and incubated with the primary antibody directed against HO-1 (anti-rabbit, Enzo Life Sciences) or against Nrf2 (anti-rabbit, Santa Cruz Biotecnology, St Cruz, CA, USA) at a dilution 1:100, overnight at 4 °C. Then, cells were washed three times in PBS for 5 min and incubated for 1h with a conjugated secondary antibody: anti-rabbit IgG–FITC produced in goat (Sigma-Aldrich) at dilution 1:200. Nuclei were stained with Hoechst Stain Solution (1:1000,

Hoechst 33258, Sigma-Aldrich). The images were captured using a Leica Confocal Microscope TCS SP8 (Leica Microsystems). Ten random visual fields were examined for each condition.

4.11. Statistical Analysis

Data were represented as mean ± S.E. and analysis was performed using the Student's *t*-test and one-way analysis of variance. Comparisons between the control (untreated) vs. all treated samples were made. If a significant difference was detected by ANOVA analyses, this was re-evaluated by post-hoc Bonferroni's test. GraphPadPrismTM 4.0 software (Graph PadPrismTM Software Inc., San Diego, CA, USA) was used for statistical calculations. The statistical significance threshold was fixed at $p < 0.05$.

5. Conclusions

In summary, cell culture experiments performed in three different colon cancer cell lines demonstrate that nuclear translocation and consequent activation of both Nrf2 and HO-1 in response to high doses of ethanol exert a protective effect against the toxic effects of alcohol-induced oxidative and ER stress. Moreover, Nrf2/HO-1 activation also favors the acquisition of a more aggressive phenotype through the upregulation of proinvasive and angiogenetic factors like MMPs and VEGF. This might represent a causative link between alcohol consumption and an increased risk of CRC progression.

Supplementary Materials: The following are available online at http://www.mdpi.com/2072-6694/11/4/505/s1, Figure S1: Ethanol induces ER stress in HT29 and Caco-2 colon cancer cells, Figure S2: Effects of ethanol on Nrf2 and HO-1 in HT29 and Caco-2 colon cancer cells.

Author Contributions: Conceptualization, M.L., A.D. (Antonella D'Anneo), and C.C. (Cesare Cernigliaro); Methodology, C.C. (Cesare Cernigliaro), D.C.; Investigation, C.C. (Cesare Cernigliaro), D.C., A.M.G., R.B., A.D. (Alfio Distefano), and L.L.; Data Curation, C.C. (Cesare Cernigliaro), D.C.; Software, S.E., C.C. (Claudia Campanella); Writing—Original Draft Preparation, M.L., D.C, and A.D. (Antonella D'Anneo). Writing—Review and Editing, All Authors; Supervision, M.L. and M.G; Funding Acquisition, G.C. and F.C.

Funding: This work has been carried out with the financial support from Gruppo Azione Locale (GAL) of Golfo di Castellammare, Italy (Progetto Operativo n.17/2015, misura 313B). Part of this work was funded by the Italian National Operational Programme (PON) «Imprese e Competitività» 2014–2020 FESR, grant awarded by the Italian Ministry of Economic Development to the project titled «Gestione di un servizio integrato multi centrico di diagnostica e terapia personalizzata in oncologia» (Project code: F/090012/01-02/X36).

Acknowledgments: We are grateful to Patrizia Guarneri for providing ATF6 antibody and Marta Di Carlo for providing PERK antibody.

Conflicts of Interest: The authors declare no conflicts of interest.

References

1. Haggar, F.A.; Boushey, R.P. Colorectal cancer epidemiology: Incidence, mortality, survival, and risk factors. *Clin. Colon Rectal Surg.* **2009**, *22*, 191–197. [CrossRef] [PubMed]
2. Seitz, H.K.; Maurer, B.; Stickel, F. Alcohol consumption and cancer of the gastrointestinal tract. *Dig. Dis.* **2005**, *23*, 297–303. [CrossRef]
3. Tuan, J.; Chen, Y.-X. Dietary and Lifestyle Factors Associated with Colorectal Cancer Risk and Interactions with Microbiota: Fiber, Red or Processed Meat and Alcoholic Drinks. *Gastrointest. Tumors* **2016**, *3*, 17–24. [CrossRef]
4. Thygesen, L.C.; Wu, K.; Grønbæk, M.; Fuchs, C.S.; Willett, W.C.; Giovannucci, E. Alcohol Intake and Colorectal Cancer: A Comparison of Approaches for Including Repeated Measures of Alcohol Consumption. *Epidemiology* **2008**, *19*, 258–264. [CrossRef] [PubMed]
5. Rossi, M.; Jahanzaib Anwar, M.; Usman, A.; Keshavarzian, A.; Bishehsari, F. Colorectal Cancer and Alcohol Consumption-Populations to Molecules. *Cancers* **2018**, *10*, 38. [CrossRef] [PubMed]
6. Cai, S.; Li, Y.; Ding, Y.; Chen, K.; Jin, M. Alcohol drinking and the risk of colorectal cancer death: A meta-analysis. *Eur. J. Cancer Prev.* **2014**, *23*, 532–539. [CrossRef] [PubMed]

7. Chen, C.; Wang, L.; Liao, Q.; Xu, L.; Huang, Y.; Zhang, C.; Ye, H.; Xu, X.; Ye, M.; Duan, S. Association Between Six Genetic Polymorphisms and Colorectal Cancer: A Meta-Analysis. *Gen. Test. Mol. Biomark.* **2014**, *18*, 187–195. [CrossRef] [PubMed]
8. Homann, N.; König, I.R.; Marks, M.; Benesova, M.; Stickel, F.; Millonig, G.; Mueller, S.; Seitz, H.K. Alcohol and Colorectal Cancer: The Role of Alcohol Dehydrogenase 1C Polymorphism. *Alcohol. Clin. Exp. Res.* **2009**, *33*, 551–556. [CrossRef]
9. Zhao, H.; Chen, D.; Cao, R.; Wang, S.; Yu, D.; Liu, Y.; Jiang, Y.; Xu, M.; Luo, J.; Wang, S. Alcohol consumption promotes colorectal carcinoma metastasis via a CCL5-induced and AMPK-pathway-mediated activation of autophagy. *Sci. Rep.* **2018**, *8*, 8640. [CrossRef]
10. Tsuruya, A.; Kuwahara, A.; Saito, Y.; Yamaguchi, H.; Tenma, N.; Inai, M.; Takahashi, S.; Tsutsumi, E.; Suwa, Y.; Totsuka, Y.; et al. Major Anaerobic Bacteria Responsible for the Production of Carcinogenic Acetaldehyde from Ethanol in the Colon and Rectum. *Alcohol Alcohol.* **2016**, *51*, 395–401. [CrossRef]
11. Van Engeland, M.; Weijenberg, M.P.; Roemen, G.M.J.M.; Brink, M.; de Bruïne, A.P.; Goldbohm, R.A.; van den Brandt, P.A.; Baylin, S.B.; de Goeij, A.F.P.M.; Herman, J.G. Effects of dietary folate and alcohol intake on promoter methylation in sporadic colorectal cancer: The Netherlands cohort study on diet and cancer. *Cancer Res.* **2003**, *63*, 3133. [PubMed]
12. Nieminen, M.; Salaspuro, M. Local Acetaldehyde—An Essential Role in Alcohol-Related Upper Gastrointestinal Tract Carcinogenesis. *Cancers* **2018**, *10*, 11. [CrossRef]
13. Bishehsari, F. Epidemiological transition of colorectal cancer in developing countries: Environmental factors, molecular pathways, and opportunities for prevention. *World J. Gastroenterol.* **2014**, *20*, 6055–6072. [CrossRef]
14. Linhart, K.; Bartsch, H.; Seitz, H.K. The role of reactive oxygen species (ROS) and cytochrome P-450 2E1 in the generation of carcinogenic etheno-DNA adducts. *Redox Biol.* **2014**, *3*, 56–62. [CrossRef]
15. Koehler, B.C.; Arslic-Schmitt, T.; Peccerella, T.; Scherr, A.-L.; Schulze-Bergkamen, H.; Bruckner, T.; Gdynia, G.; Jäger, D.; Mueller, S.; Bartsch, H.; et al. Possible Mechanisms of Ethanol-Mediated Colorectal Carcinogenesis: The Role of Cytochrome P 4502E1, Etheno-DNA Adducts, and the Anti-Apoptotic Protein Mcl-1. *Alcohol. Clin. Exp. Res.* **2016**, *40*, 2094–2101. [CrossRef]
16. Lucas, C.; Barnich, N.; Nguyen, H.T.T. Microbiota, Inflammation and Colorectal Cancer. *Int. J. Mol. Sci.* **2017**, *18*, 1310. [CrossRef]
17. Emanuele, S.; D'Anneo, A.; Calvaruso, G.; Cernigliaro, C.; Giuliano, M.; Lauricella, M. The Double-Edged Sword Profile of Redox Signaling: Oxidative Events as Molecular Switches in the Balance between Cell Physiology and Cancer. *Chem. Res. Toxicol.* **2018**, *31*, 201–210. [CrossRef]
18. Liu, S.; Tian, L.; Chai, G.; Wen, B.; Wang, B. Targeting heme oxygenase-1 by quercetin ameliorates alcohol-induced acute liver injury via inhibiting NLRP3 inflammasome activation. *Food Funct.* **2018**, *9*, 4184–4193. [CrossRef]
19. Carlisi, D.; Lauricella, M.; D'Anneo, A.; Buttitta, G.; Emanuele, S.; di Fiore, R.; Martinez, R.; Rolfo, C.; Vento, R.; Tesoriere, G. The Synergistic Effect of SAHA and Parthenolide in MDA-MB231 Breast Cancer Cells: The effect of PN/SAHA combination in MDA-MB231 cells. *J. Cell. Physiol.* **2015**, *230*, 1276–1289. [CrossRef]
20. Kim, S.; Lee, H.-G.; Park, S.-A.; Kundu, J.K.; Keum, Y.-S.; Cha, Y.-N.; Na, H.-K.; Surh, Y.-J. Keap1 Cysteine 288 as a Potential Target for Diallyl Trisulfide-Induced Nrf2 Activation. *PLoS ONE* **2014**, *9*, e85984. [CrossRef]
21. Bajpai, V.K.; Alam, M.B.; Quan, K.T.; Kwon, K.-R.; Ju, M.-K.; Choi, H.-J.; Lee, J.S.; Yoon, J.-I.; Majumder, R.; Rather, I.A.; et al. Antioxidant efficacy and the upregulation of Nrf2-mediated HO-1 expression by (+)-lariciresinol, a lignan isolated from Rubia philippinensis, through the activation of p38. *Sci. Rep.* **2017**, *7*, 46035. [CrossRef]
22. Sadeghi, M.R.; Jeddi, F.; Soozangar, N.; Somi, M.H.; Samadi, N. The role of Nrf2-Keap1 axis in colorectal cancer, progression, and chemoresistance. *Tumor Biol.* **2017**, *39*. [CrossRef]
23. Kitamura, H.; Motohashi, H. NRF2 addiction in cancer cells. *Cancer Sci.* **2018**, *109*, 900–911. [CrossRef]
24. Tibullo, D.; Barbagallo, I.; Giallongo, C.; Vanella, L.; Conticello, C.; Romano, A.; Saccone, S.; Godos, J.; Di Raimondo, F.; Volti, G.L. Heme oxygenase-1 nuclear translocation regulates bortezomib-induced cytotoxicity and mediates genomic instability in myeloma cells. *Oncotarget* **2016**, *7*, 28868–28880. [CrossRef]
25. Chiang, S.-K.; Chen, S.-E.; Chang, L.-C. A Dual Role of Heme Oxygenase-1 in Cancer Cells. *Int. J. Mol. Sci.* **2018**, *20*, 39. [CrossRef]

26. Miyata, Y.; Kanda, S.; Mitsunari, K.; Asai, A.; Sakai, H. Heme oxygenase-1 expression is associated with tumor aggressiveness and outcomes in patients with bladder cancer: A correlation with smoking intensity. *Transl. Res.* **2014**, *164*, 468–476. [CrossRef]
27. Biswas, C.; Shah, N.; Muthu, M.; La, P.; Fernando, A.P.; Sengupta, S.; Yang, G.; Dennery, P.A. Nuclear Heme Oxygenase-1 (HO-1) Modulates Subcellular Distribution and Activation of Nrf2, Impacting Metabolic and Antioxidant Defenses. *J. Biol. Chem.* **2014**, *289*, 26882–26894. [CrossRef]
28. Yang, C.-X. Phase I/II enzyme gene polymorphisms and esophageal cancer risk: A meta-analysis of the literature. *World J. Gastroenterol.* **2005**, *11*, 2531–2538. [CrossRef]
29. Blanco, A.M.; Pascual, M.; Valles, S.L.; Guerri, C. Ethanol-induced iNOS and COX-2 expression in cultured astrocytes via NF-kappa B. *Neuroreport* **2004**, *15*, 681–685. [CrossRef]
30. Cappello, F.; Marino Gammazza, A.; Palumbo Piccionello, A.; Campanella, C.; Pace, A.; Conway de Macario, E.; Macario, A.J.L. Hsp60 chaperonopathies and chaperonotherapy: Targets and agents. *Expert Opin. Ther. Targets* **2014**, *18*, 185–208. [CrossRef]
31. Tomasello, G.; Sciumé, C.; Rappa, F.; Rodolico, V.; Zerilli, M.; Martorana, A.; Cicero, G.; De Luca, R.; Damiani, P.; Accardo, F.M.; et al. Hsp10, Hsp70, and Hsp90 immunohistochemical levels change in ulcerative colitis after therapy. *Eur. J. Histochem.* **2011**, *55*, e38. [CrossRef]
32. Cao, S.S.; Kaufman, R.J. Endoplasmic Reticulum Stress and Oxidative Stress in Cell Fate Decision and Human Disease. *Antioxid. Redox Signal.* **2014**, *21*, 396–413. [CrossRef]
33. Shimodaira, Y.; Takahashi, S.; Kinouchi, Y.; Endo, K.; Shiga, H.; Kakuta, Y.; Kuroha, M.; Shimosegawa, T. Modulation of endoplasmic reticulum (ER) stress-induced autophagy by C/EBP homologous protein (CHOP) and inositol-requiring enzyme 1α (IRE1α) in human colon cancer cells. *Biochem. Biophys. Res. Commun.* **2014**, *445*, 524–533. [CrossRef]
34. Emanuele, S.; Oddo, E.; D'Anneo, A.; Notaro, A.; Calvaruso, G.; Lauricella, M.; Giuliano, M. Routes to cell death in animal and plant kingdoms: From classic apoptosis to alternative ways to die—A review. *Rend. Lincei Sci. Fis. Nat.* **2018**, *29*, 397–409. [CrossRef]
35. Tanida, I.; Ueno, T.; Kominami, E. LC3 and Autophagy. *Methods Mol. Biol.* **2008**, *445*, 77–88.
36. Kabeya, Y. LC3, a mammalian homologue of yeast Apg8p, is localized in autophagosome membranes after processing. *EMBO J.* **2000**, *19*, 5720–5728. [CrossRef]
37. Xiong, J. Atg7 in development and disease: Panacea or Pandora's Box? *Protein Cell* **2015**, *6*, 722–734. [CrossRef]
38. Pugsley, H.R. Assessing Autophagic Flux by Measuring LC3, p62, and LAMP1 Co-localization Using Multispectral Imaging Flow Cytometry. *J. Vis. Exp.* **2017**, *125*. [CrossRef]
39. Motohashi, H.; Yamamoto, M. Nrf2–Keap1 defines a physiologically important stress response mechanism. *Trends Mol. Med.* **2004**, *10*, 549–557. [CrossRef]
40. Vanella, L.; Barbagallo, I.; Tibullo, D.; Forte, S.; Zappalà, A.; Volti, G.L. The non-canonical functions of the heme oxygenases. *Oncotarget* **2016**, *7*, 69075–69086. [CrossRef]
41. Kleiner, D.E.; Stetler-Stevenson, W.G. Matrix metalloproteinases and metastasis. *Cancer Chemother. Pharmacol.* **1999**, *43*, S42–S51. [CrossRef]
42. Tibullo, D.; Barbagallo, I.; Giallongo, C.; La Cava, P.; Parrinello, N.; Vanella, L.; Stagno, F.; Palumbo, G.A.; Li Volti, G.; Di Raimondo, F. Nuclear Translocation of Heme Oxygenase-1 Confers Resistance to Imatinib in Chronic Myeloid Leukemia Cells. *Curr. Pharm. Des.* **2013**, *19*, 2765–2770. [CrossRef]
43. Nitti, M.; Piras, S.; Marinari, U.; Moretta, L.; Pronzato, M.; Furfaro, A. HO-1 Induction in Cancer Progression: A Matter of Cell Adaptation. *Antioxidants* **2017**, *6*, 29. [CrossRef]
44. Baan, R.; Straif, K.; Grosse, Y.; Secretan, B.; El Ghissassi, F.; Bouvard, V.; Altieri, A.; Cogliano, V.; WHO International Agency for Research on Cancer Monograph Working Group. Carcinogenicity of alcoholic beverages. *Lancet Oncol.* **2007**, *8*, 292–293. [CrossRef]
45. Vignesh, R.C.; Sitta Djody, S.; Jayasudha, E.; Gopalakrishnan, V.; Ilangovan, R.; Balaganesh, M.; Veni, S.; Sridhar, M.; Srinivasan, N. Effect of ethanol on human osteosarcoma cell proliferatation, differentiation and mineralization. *Toxicology* **2006**, *220*, 63–70. [CrossRef]
46. Yeo, E.-J.; Lim, H.K.; Park, S.C. Effect of short-term ethanol on the proliferative response of Swiss 3T3 cells to mitogenic growth factors. *Exp. Mol. Med.* **2000**, *32*, 161–169. [CrossRef]
47. Singletary, K.W.; Frey, R.S.; Yan, W. Effect of ethanol on proliferation and estrogen receptor-alpha expression in human breast cancer cells. *Cancer Lett.* **2001**, *165*, 131–137. [CrossRef]

48. Haorah, J.; Ramirez, S.H.; Floreani, N.; Gorantla, S.; Morsey, B.; Persidsky, Y. Mechanism of alcohol-induced oxidative stress and neuronal injury. *Free Radic. Biol. Med.* **2008**, *45*, 1542–1550. [CrossRef]
49. Kawaratani, H.; Tsujimoto, T.; Douhara, A.; Takaya, H.; Moriya, K.; Namisaki, T.; Noguchi, R.; Yoshiji, H.; Fujimoto, M.; Fukui, H. The Effect of Inflammatory Cytokines in Alcoholic Liver Disease. *Mediat. Inflamm.* **2013**, *2013*, 495156. [CrossRef]
50. Wang, Z.; Dou, X.; Li, S.; Zhang, X.; Sun, X.; Zhou, Z.; Song, Z. Nuclear factor (erythroid-derived 2)-like 2 activation-induced hepatic very-low-density lipoprotein receptor overexpression in response to oxidative stress contributes to alcoholic liver disease in mice. *Hepatology* **2014**, *59*, 1381–1392. [CrossRef]
51. Zhang, Z.; Zhang, L.; Zhou, L.; Lei, Y.; Zhang, Y.; Huang, C. Redox signaling and unfolded protein response coordinate cell fate decisions under ER stress. *Redox Biol.* **2018**. [CrossRef]
52. Ji, C. Mechanisms of Alcohol-Induced Endoplasmic Reticulum Stress and Organ Injuries. *Biochem. Res. Int.* **2012**, *2012*, 216450. [CrossRef]
53. Bartolome, A.; Guillen, C.; Benito, M. Autophagy plays a protective role in endoplasmic reticulum stress-mediated pancreatic β cell death. *Autophagy* **2012**, *8*, 1757–1768. [CrossRef]
54. Song, S.; Tan, J.; Miao, Y.; Zhang, Q. Crosstalk of ER stress-mediated autophagy and ER-phagy: Involvement of UPR and the core autophagy machinery. *J. Cell. Physiol.* **2018**, *233*, 3867–3874. [CrossRef]
55. Sakaki, K.; Kaufman, R.J. Regulation of ER stress-induced macroautophagy by protein kinase C. *Autophagy* **2008**, *4*, 841–843. [CrossRef]
56. Luo, J. Autophagy and ethanol neurotoxicity. *Autophagy* **2014**, *10*, 2099–2108. [CrossRef]
57. Kim, J.; Keum, Y.-S. NRF2, a Key Regulator of Antioxidants with Two Faces towards Cancer. *Oxid. Med. Cell. Longev.* **2016**, *2016*, 2746457. [CrossRef]
58. Dodson, M.; Redmann, M.; Rajasekaran, N.S.; Darley-Usmar, V.; Zhang, J. KEAP1–NRF2 signalling and autophagy in protection against oxidative and reductive proteotoxicity. *Biochem. J.* **2015**, *469*, 347–355. [CrossRef]
59. Kaspar, J.W.; Niture, S.K.; Jaiswal, A.K. Nrf2:INrf2 (Keap1) signaling in oxidative stress. *Free Radic. Biol. Med.* **2009**, *47*, 1304–1309. [CrossRef]
60. Cullinan, S.B.; Zhang, D.; Hannink, M.; Arvisais, E.; Kaufman, R.J.; Diehl, J.A. Nrf2 is a direct PERK substrate and effector of PERK-dependent cell survival. *Mol. Cell. Biol.* **2003**, *23*, 7198–7209. [CrossRef]
61. Furfaro, A.L.; Traverso, N.; Domenicotti, C.; Piras, S.; Moretta, L.; Marinari, U.M.; Pronzato, M.A.; Nitti, M. The Nrf2/HO-1 Axis in Cancer Cell Growth and Chemoresistance. *Oxid. Med. Cell. Longev.* **2016**, *2016*, 1958174. [CrossRef]
62. Ma, Q. Role of Nrf2 in Oxidative Stress and Toxicity. *Ann. Rev. Pharm. Toxicol.* **2013**, *53*, 401–426. [CrossRef]
63. Kou, X.; Kirberger, M.; Yang, Y.; Chen, N. Natural products for cancer prevention associated with Nrf2–ARE pathway. *Food Sci. Hum. Wellness* **2013**, *2*, 22–28. [CrossRef]
64. Lauricella, M.; Emanuele, S.; Calvaruso, G.; Giuliano, M.; D'Anneo, A. Multifaceted Health Benefits of Mangifera indica L. (Mango): The Inestimable Value of Orchards Recently Planted in Sicilian Rural Areas. *Nutrients* **2017**, *9*, 525. [CrossRef]
65. Ahmed, S.M.U.; Luo, L.; Namani, A.; Wang, X.J.; Tang, X. Nrf2 signaling pathway: Pivotal roles in inflammation. *Biochim. Biophys. Acta Mol. Basis Dis.* **2017**, *1863*, 585–597. [CrossRef]
66. Gonzalez-Donquiles, C.; Alonso-Molero, J.; Fernandez-Villa, T.; Vilorio-Marqués, L.; Molina, A.J.; Martín, V. The NRF2 transcription factor plays a dual role in colorectal cancer: A systematic review. *PLoS ONE* **2017**, *12*, e0177549. [CrossRef]
67. Zhou, S.; Ye, W.; Zhang, M.; Liang, J. The effects of nrf2 on tumor angiogenesis: A review of the possible mechanisms of action. *Crit. Rev. Eukaryot. Gene Expr.* **2012**, *22*, 149–160. [CrossRef]
68. Berberat, P.O.; Dambrauskas, Z.; Gulbinas, A.; Giese, T.; Giese, N.; Künzli, B.; Autschbach, F.; Meuer, S.; Büchler, M.W.; Friess, H. Inhibition of heme oxygenase-1 increases responsiveness of pancreatic cancer cells to anticancer treatment. *Clin. Cancer Res.* **2005**, *11*, 3790–3798. [CrossRef]
69. Hu, J.L.; Li, Z.Y.; Liu, W.; Zhang, R.G.; Li, G.L.; Wang, T.; Ren, J.H.; Wu, G. Polymorphism in heme oxygenase-1 (HO-1) promoter and alcohol are related to the risk of esophageal squamous cell carcinoma on Chinese males. *Neoplasma* **2010**, *57*, 86–92. [CrossRef]
70. Lauricella, M.; Carlisi, D.; Giuliano, M.; Calvaruso, G.; Cernigliaro, C.; Vento, R.; D'Anneo, A. The analysis of estrogen receptor-α positive breast cancer stem-like cells unveils a high expression of the serpin proteinase inhibitor PI-9: Possible regulatory mechanisms. *Int. J. Oncol.* **2016**, *49*, 352–360. [CrossRef]

71. Raffa, D.; D'Anneo, A.; Plescia, F.; Daidone, G.; Lauricella, M.; Maggio, B. Novel 4-(3-phenylpropionamido), 4-(2-phenoxyacetamido) and 4-(cinnamamido) substituted benzamides bearing the pyrazole or indazole nucleus: Synthesis, biological evaluation and mechanism of action. *Bioorg. Chem.* **2019**, *83*, 367–379. [CrossRef]
72. D'Anneo, A.; Carlisi, D.; Emanuele, S.; Buttitta, G.; Di Fiore, R.; Vento, R.; Tesoriere, G.; Lauricella, M. Parthenolide induces superoxide anion production by stimulating EGF receptor in MDA-MB-231 breast cancer cells. *Int. J. Oncol.* **2013**, *43*, 1895–1900. [CrossRef]
73. Lamichhane, T.N.; Leung, C.A.; Douti, L.Y.; Jay, S.M. Ethanol Induces Enhanced Vascularization Bioactivity of Endothelial Cell-Derived Extracellular Vesicles via Regulation of MicroRNAs and Long Non-Coding RNAs. *Sci. Rep.* **2017**, *7*, 13794. [CrossRef]
74. D'Anneo, A.; Carlisi, D.; Lauricella, M.; Emanuele, S.; Di Fiore, R.; Vento, R.; Tesoriere, G. Parthenolide induces caspase-independent and AIF-mediated cell death in human osteosarcoma and melanoma cells. *J. Cell. Physiol.* **2013**, *228*, 952–967. [CrossRef]
75. Emanuele, S.; Notaro, A.; Palumbo Piccionello, A.; Maggio, A.; Lauricella, M.; D'Anneo, A.; Cernigliaro, C.; Calvaruso, G.; Giuliano, M. Sicilian Litchi Fruit Extracts Induce Autophagy versus Apoptosis Switch in Human Colon Cancer Cells. *Nutrients* **2018**, *10*, 1490. [CrossRef]
76. Carlisi, D.; De Blasio, A.; Drago-Ferrante, R.; Di Fiore, R.; Buttitta, G.; Morreale, M.; Scerri, C.; Vento, R.; Tesoriere, G. Parthenolide prevents resistance of MDA-MB231 cells to doxorubicin and mitoxantrone: The role of Nrf2. *Cell Death Discov.* **2017**, *3*, 17078. [CrossRef]
77. Campanella, C.; D'Anneo, A.; Gammazza, A.M.; Bavisotto, C.C.; Barone, R.; Emanuele, S.; Cascio, F.L.; Mocciaro, E.; Fais, S.; De Macario, E.C.; et al. The histone deacetylase inhibitor SAHA induces HSP60 nitration and its extracellular release by exosomal vesicles in human lung-derived carcinoma cells. *Oncotarget* **2016**, *7*, 28849–28867. [CrossRef]
78. Barone, R.; Macaluso, F.; Sangiorgi, C.; Campanella, C.; Marino Gammazza, A.; Moresi, V.; Coletti, D.; Conway de Macario, E.; Macario, A.J.; Cappello, F.; et al. Skeletal muscle Heat shock protein 60 increases after endurance training and induces peroxisome proliferator-activated receptor gamma coactivator 1 α1 expression. *Sci. Rep.* **2016**, *6*, 19781. [CrossRef]
79. Marino Gammazza, A.; Campanella, C.; Barone, R.; Caruso Bavisotto, C.; Gorska, M.; Wozniak, M.; Carini, F.; Cappello, F.; D'Anneo, A.; Lauricella, M.; et al. Doxorubicin anti-tumor mechanisms include Hsp60 post-translational modifications leading to the Hsp60/p53 complex dissociation and instauration of replicative senescence. *Cancer Lett.* **2017**, *385*, 75–86. [CrossRef]

cancers

MDPI

Article

Targeted Nano-Drug Delivery of Colchicine against Colon Cancer Cells by Means of Mesoporous Silica Nanoparticles

Khaled AbouAitah [1,2,*], Heba A. Hassan [3], Anna Swiderska-Sroda [1], Lamiaa Gohar [4], Olfat G. Shaker [5], Jacek Wojnarowicz [1], Agnieszka Opalinska [1], Julita Smalc-Koziorowska [6], Stanislaw Gierlotka [1] and Witold Lojkowski [1]

1 Laboratory of Nanostructures, Institute of High Pressure Physics, Polish Academy of Sciences, Sokolowska 29/37, 01-142 Warsaw, Poland; a.swiderska-sroda@labnano.pl (A.S.-S.); j.wojnarowicz@labnano.pl (J.W.); a.opalinska@labnano.pl (A.O.); xray@unipress.waw.pl (S.G.); w.lojkowski@labnano.pl (W.L.)

2 Medicinal and Aromatic Plants Research Department, Pharmaceutical and Drug Industries Research Division, National Research Centre (NRC), Dokki, Giza 12622, Egypt

3 Therapeutic Chemistry Department, Pharmaceutical and Drug Industries Research Division, National Research Centre (NRC), Dokki, Giza 12622, Egypt; hassan_heba59@yahoo.com

4 Pharmacognosy Department, Pharmaceutical and Drug Industries Research Division, National Research Centre (NRC), Dokki, Giza 12622, Egypt; la_gohar@hotmail.com

5 Medical Biochemistry and Molecular Biology Department, Faculty of Medicine, Cairo University, Cairo 11511, Egypt; olfatshaker@yahoo.com

6 Laboratory of Semiconductor Characterization, Institute of High Pressure Physics, Polish Academy of Sciences, Sokolowska 29/37, 01-142 Warsaw, Poland; julita@unipress.waw.pl

* Correspondence: k.abouaitah@labnano.pl; Tel.: +48-22-888-0429

Received: 3 December 2019; Accepted: 3 January 2020; Published: 7 January 2020

Abstract: Antimitotics are important anticancer agents and include the natural alkaloid prodrug colchicine (COL). However, a major challenge of using COL as an anticancer drug is its cytotoxicity. We developed a novel drug delivery system (DDS) for COL using mesoporous silica nanoparticles (MSNs). The MSNs were functionalized with phosphonate groups, loaded with COL, and coated with folic acid chitosan-glycine complex. The resulting nanoformulation, called MSNsPCOL/CG-FA, was tested for action against cancer and normal cell lines. The anticancer effect was highly enhanced for MSNsPCOL/CG-FA compared to COL. In the case of HCT116 cells, 100% inhibition was achieved. The efficiency of MSNsPCOL/CG-FA ranked in this order: HCT116 (colon cancer) > HepG2 (liver cancer) > PC3 (prostate cancer). MSNsPCOL/CG-FA exhibited low cytotoxicity (4%) compared to COL (~60%) in BJ1 normal cells. The mechanism of action was studied in detail for HCT116 cells and found to be primarily intrinsic apoptosis caused by an enhanced antimitotic effect. Furthermore, a contribution of genetic regulation (metastasis-associated lung adenocarcinoma transcript 1 (MALAT 1), and microRNA (mir-205)) and immunotherapy effects (angiopoietin-2 (Ang-2 protein) and programmed cell death protein 1 (PD-1) was found. Therefore, this study shows enhanced anticancer effects and reduced cytotoxicity of COL with targeted delivery compared to free COL and is a novel method of developing cancer immunotherapy using a low-cost small-molecule natural prodrug.

Keywords: colchicine alkaloid; colon cancer cells; mesoporous silica nanoparticles; targeted delivery system; apoptosis; PD-1 immune checkpoint inhibitor and cancer immunotherapy

1. Introduction

Colchicine (COL) is a natural alkaloid compound derived mainly from the medicinal plant Colchicum automnale and has been used in the clinic for treating gout and familial Mediterranean fever.

COL has also shown some benefit in primary biliary cirrhosis [1], amyloidosis [2], and condyloma acuminate treatments [3]. COL is an antimitotic drug, which interferes strongly with cell division by affecting microtubule assembly and disassembly during mitosis. Most of the antimitotic drugs are toxic anticancer agents, which preferentially kill cancer cells, as they divide much faster than normal cells. However, the major challenge for COL is its toxicity, which causes severe side effects to patients [4].

Despite COL not yet being used clinically for cancer therapy because of its toxicity to normal cells, it is used as a lead compound in generating potent anticancer agents [5–9]. To reduce the side effects of COL on normal cells, many types of research have included analogs of COL [10,11] and combination treatments with other drugs [12]. Regarding the mechanism underlying the antimitotic effects of COL, several actions are associated with antimitotic and post-antimitotic responses. Gupta and Dudani [13] proposed that the mechanism of action for antimitotic drugs, such as COL, includes blockade of cell growth at metaphase upon the binding of antimitotic drugs to tubulin (for COL through the colchicine binding site) due to cell-cycle arrest (e.g., at G2/M). Following this action, the microtubules cannot exert any cellular functions [13]. In the presence of antimitotic drugs, cells either die during mitosis or exit mitosis. When they exit mitosis, several post-mitotic responses occur that can lead to cell death, including cell-cycle arrest and apoptosis [14]. In a report on the mitotic cell death that occurs during mitosis, Castedo et al. [15] proposed that the mitotic cell death catastrophe results from a combination of deficient cell-cycle checkpoints and cellular damage. Cell death occurring during mitosis is characterized by activation of caspase-2 in response to DNA damage, or caspase-9, caspase-3, and cytochrome c in response to mitochondrial membrane permeabilization. These effectors make up the molecular hallmarks of apoptosis. Thus, mitotic catastrophe is controlled by molecular players, including cell cycle-specific kinases, cell-cycle checkpoint proteins, caspases, and some proteins of the Bcl-2 family (e.g., Bax, Bcl-2), among others. Qi et al. provided evidence that, by arresting cell cycle progression in the presence of antimitotic drugs, mitotic spindles are disrupted, and cancer cells directly undergo apoptosis via the mitotic catastrophe [16].

Recently, antimitotic agents such as COL were reported to have a regulatory effect on most immune cell types, leading to the development of effective cancer immunotherapies [17]. With respect to cancer immunotherapy, targeting programmed cell death protein 1 (PD-1) and programmed cell death-ligand 1 (PD-L1), among other immune checkpoints, is important. PD-1 is an inducible immune modulatory receptor expressed on surface-activated T cells, and its ligand PD-L1 is expressed on cancer cells [18]. Binding PD-1 to PD-L1 leads to prevent from the immune antitumor effects by T cells against cancer cells [19]. However, recent studies have revealed the intrinsic expression of PD-1 in many cancers, along with immune T cells [20,21]. Therefore, it is considered a new potential target for cancer immunotherapy [19,21]. Thus, checkpoint blockade immunotherapy has revolutionized treatment for many tumors. In the next few years, scientists will be able to focus on immunotherapy research and broaden target cancers with different strategies [22]. Recently, the regulatory action by which antimitotic drugs inhibit PD-1 over-expression on the surface of T-cells was highlighted, and that PD-1 is strongly inhibited in the presence of the COL-binding site (CBS) of tubulin [23].

Traditional treatments for cancer, including surgery, radiation, cryosurgery, and chemotherapy, can be used alone or in combination. These methods have several limitations, such as toxicity, side effects, and expense. When chemotherapy is considered, normal treatment protocols include anticancer drugs alone or in combination to inhibit/kill cancer cells by affecting cell division and proliferation through different mechanisms. Drugs can produce various side effects for patients (e.g., neutropenia, liver and gastrointestinal toxicity, anemia, mucositis, and others) [24]. Therefore, cancer-targeting delivery systems with different nano-platforms have gained attention [25–27].

Despite the promising anticancer effects of COL, a few COL drug delivery systems (DDSs) have been tested to improve its therapeutic efficiency through cancer-targeting [28,29]. We focused on developing a novel DDS for COL by facilitating the active targeting of cancer cells. In the current study, we constructed a DDS for COL using three-dimensional fibrous dendritic mesoporous silica nanoparticles with a spherical shape (MSNs) known as KCC-1 type [26,30]. MSNs have been investigated as a drug delivery carrier for several drugs and biomolecules [27,31,32]. The tailored DDS comprises MSNs functionalized with phosphonate functional groups (MSNsP), and then loaded with COL MSNsPCOL, with the latter product subsequently coated with chitosan-glycine conjugated to folic acid (FA) to obtain a nanoformulation called MSNsPCOL/CG-FA. Glycine was employed as a source of amino groups that require for cancer growth [33]. FA was employed as the main targeting ligand, which is known for its binding potential to folate receptors, which are over-expressed on many cancers and facilitate the endocytosis pathway [26,34–36].

2. Results and Discussion

2.1. Synthesis and Characterization of the Targeted Delivery System

The targeted drug delivery system for COL, MSNsPCOL/CG-FA, comprised COL-loaded MSNsP subsequently coated with chitosan-glycine conjugated to folic acid. The schematic representation of the process is shown in Figure 1.

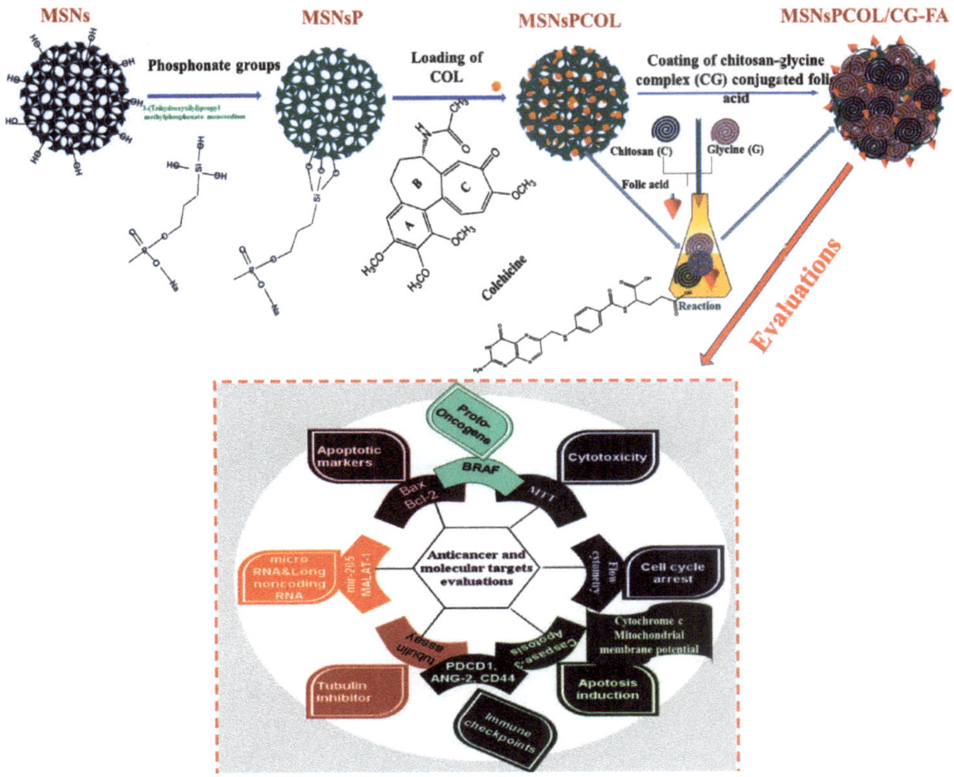

Figure 1. Schematic representation of the steps to obtain the proposed drug delivery system with the final product (MSNsPCOL/CG-FA) and various biological evaluations in vitro.

2.2. Electron Microscopy of Materials

To observe the structural changes between the prepared materials, we used TEM STEM and FE-SEM techniques. The FE-SEM images (Figure 2A) show the 3D dendritic mesoporous structure of MSNs with a spherical shape that is uniform in size. No aggregation was seen. In TEM images (Figure 2B), there were no detectable differences between MSNs, MSNsP, and MSNsPCOL. However, in MSNsPCOL/CG-FA, the central part appeared gray-white in color due to the coating complex (chitosan-glycine). This observation is similar to that seen in STEM images, where MSNsPCOL/CG-FA was white in color (Figure 2C). Thus, the coating was confirmed by STEM and TEM observations.

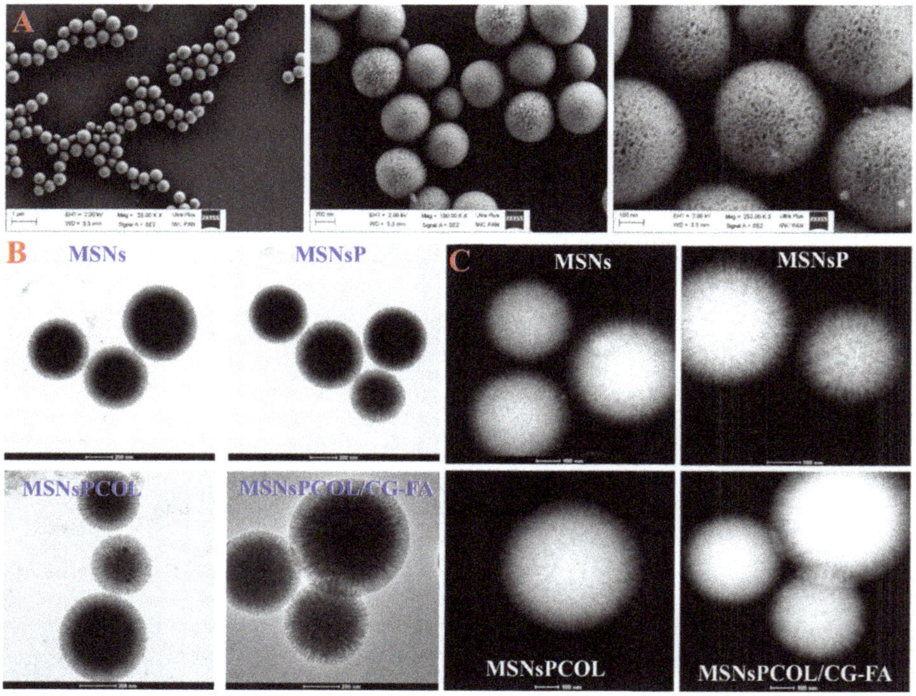

Figure 2. Morphological structures of the materials. (**A**) Field emission scanning electron microscopy (FE-SEM) of prepared mesoporous silica nanoparticles (MSNs) at different magnifications. (**B**) High-resolution transmission electron microscopy (HR-TEM) of prepared materials at different stages: Before and after of modification, colchicine (COL) loading, and coating. (**C**) Scanning transmission electron microscopy (STEM) of prepared materials at different stages: before and after of modification, COL loading, and coating.

2.3. Surface Area Characteristics

The specific surface area and pore volume characteristics were measured (Table 1). Via modification with phosphonates, COL loading, and coating with CG-FA, both surface area and total pore volume were decreased compared to MSNs (380.1 m^2/g and 0.772 cm^3/g of specific surface area and total pore volume, respectively). This observation indicates successful preparation, in agreement with previous results [26,27,37,38].

Table 1. Physicochemical properties of MSNs before and after colchicine loading and polymer coating.

Sample Code	S_{BET} (m^2/g)	Total Pore Volume [a] (cm^3/g)	P, COL, CG-FA Content Calculation from Weight Loss (wt%) [b]
MSNs	380.1	0.772	3.3
MSNsP	202.1	0.489	6.98 as P
MSNsPCOL	181.8	0.467	3.60 as COL
MSNsPCOL/CG-FA	89.5	0.352	33.48 as CG-FA

[a] Pore volume from nitrogen adsorption-desorption measurements at 0.999 P/P°. [b] Calculated from the thermogravimetric analysis. MSNs: mesoporous silica nanoparticles; P: phosphate groups; COL: colchicine; CG-FA: chitosan-glycine-folic acid.

2.4. Elemental Content Analysis

Energy-dispersive X-ray spectroscopy allowed us to determine the changes in the elemental content of prepared materials (Figure S1A, Supporting Information). Prior to modification, MSNs were composed of 53.18 wt% Si and 46.82 wt% O. Additional modification with organic phosphonate groups changed the elemental content, resulting in the presence of new elements, including 0.24 wt% P, 7.40 wt% C, and 0.95 wt% Na, as seen for MSNsP, confirming the modification. With further loading of COL on MSNsP, the C content increased and P content decreased in MSNsPCOL as a result of COL loading because it is an organic compound. Further polymer coating of MSNsPCOL relatively increased the C amount to 9.92 wt% and N amount to 2.77 wt% in MSNsPCOL/CG-FA. This observation is expected because both chitosan and glycine are composed of amino groups, which is important for confirmation of the coating process.

2.5. Particle Size Measurement

The particle size distribution was measured by nanoparticle tracking analysis (NTA) in aqueous solution (Figure S1B, Supporting Information). The mean size distribution for MSNs was 324 ± 33.2 nm. For MSNsP, the size increased to 407 ± 13.9 nm. For MSNsPCOL, the size slightly decreased to 391 ± 3.9 nm but was an insignificant difference. Unexpectedly, MSNsPCOL/CG-FA size decreased to 330 ± 22.2 nm compared to MSNsPCOL. An explanation may be the coating used for most of the particles with a smaller size compared to particles with a larger size, leading to an increase in their average size in the sample. This observation is in agreement with the two peaks appearing at 10–200 nm, which has not been observed previously.

2.6. Functional Group Determination

FTIR was used to identify surface functional groups (Figure 3). MSNs had several peaks at 450, 800, and 1056 cm^{-1} because of the siliceous mesostructured framework. As their surface was functionalized with phosphonate groups in MSNsP, we observed a new band at 953 cm^{-1} and a broad peak from 3000 to 3600 cm^{-1}, reflecting phosphonate groups [39]. Two new peaks at 1639 cm^{-1}, and an intensive peak centered at 3355 cm^{-1} were presented in MSNsPCOL, corresponding to shifted peaks at 1542 and 3629 cm^{-1} for COL. Additional coating resulted in several new peaks, with bands at 698 and 946 cm^{-1} attributed to chitosan, glycine, or folic acid. A band at 1315 cm^{-1} could correspond to chitosan, a band at 1436 cm^{-1} could correspond to either chitosan or folic acid, but a band at 2900 cm^{-1} could correspond to folic acid. A peak centered at 1034 cm^{-1} became broader, which ascribes all components for coating.

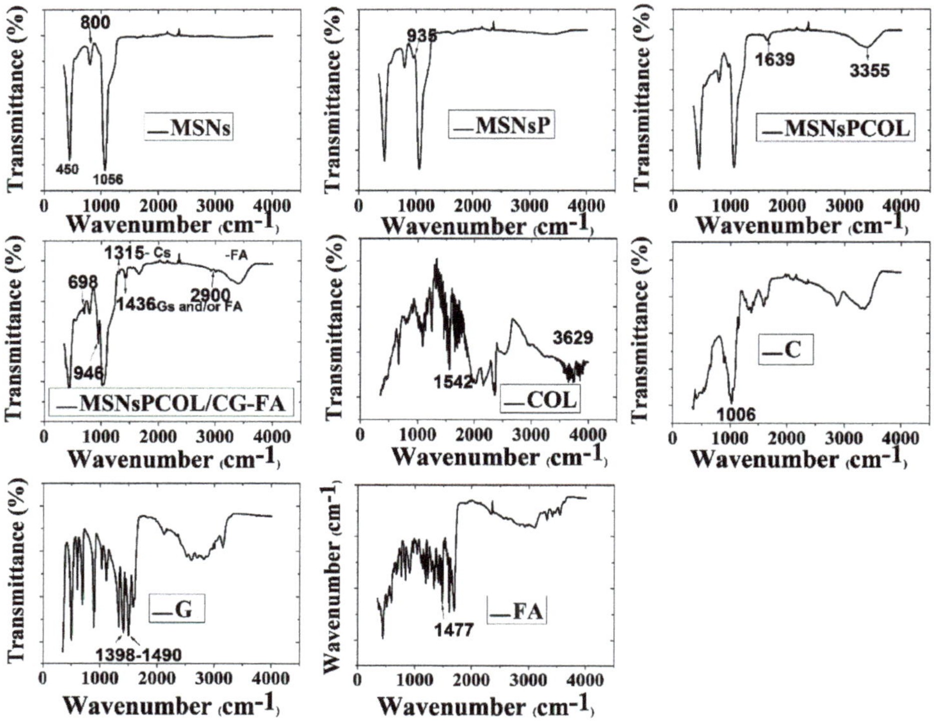

Figure 3. Fourier transform infrared spectroscopy (FTIR) spectra of MSNs before and after modification and COL loading, as well as chitosan (C), and glycine (G), folic acid (FA), and COL.

2.7. Simultaneous Thermal Analysis (STA-DSC) and XRD Characterization

To identify the mass fraction of phosphonate groups, COL, and polymer, thermal analysis was performed by calculating the weight loss values over their thermal decomposition (Figure 4A and Table 1). MSNs lost ~3.32 wt% because of moisture content. The weight loss further increased with MSNsP resulting from the decomposition of phosphonate groups. The heating of MSNsPCOL resulted in increased loss compared to MSNsP, which was calculated as COL loaded on MSNsP (3.60 wt%). The highest weight loss was recorded for MSNsPCOL/CG-FA, which was the result of decomposition of the coating complex (calculated to be 33.4 wt%), which confirms successful fabrication of the coating.

The results of the DTG thermograms in Figure 4B are consistent with STA, providing clear data that confirm the degradation. In MSNsP, a peak at 63 °C showed moisture, and another at 500 °C was attributed to organic decomposition. The degradation of COL was confirmed with the peak in the range of 340 to 525 °C in MSNsPCOL based on the two maximum peaks at 342 °C and 550 °C for free COL. A peak recorded at 106 °C in MSNsPCOL/CG-FA corresponded to decomposition of the coating material.

To further understand whether COL loaded on the surface of or inside particles, differential scanning calorimetry (DSC) analysis was conducted. Figure 4C shows no peaks in the MSNsPCOL and MSNsPCOL/CG-FA spectrum corresponding to free COL melting (409 to 537 °C). This observation shows that COL was enclosed in the particles.

To further verify the DSC results, we used XRD analysis. Figure 4D illustrates that no peaks appeared in the MSNsPCOL pattern corresponding to COL. We propose two reasons for this finding: Low loading and entrapment in pores. To confirm that we created a physical mixture of MSNsP and COL, no peaks were observed corresponding to COL. The MSNsPCOL/CG-FA pattern did not represent any peaks corresponding to chitosan, glycine, or folic acid, confirming the DSC data These two complementary techniques show the characteristics of the meso-porosity of MSNs allowing the accommodation of various molecules, including functional groups, polymers, and drugs.

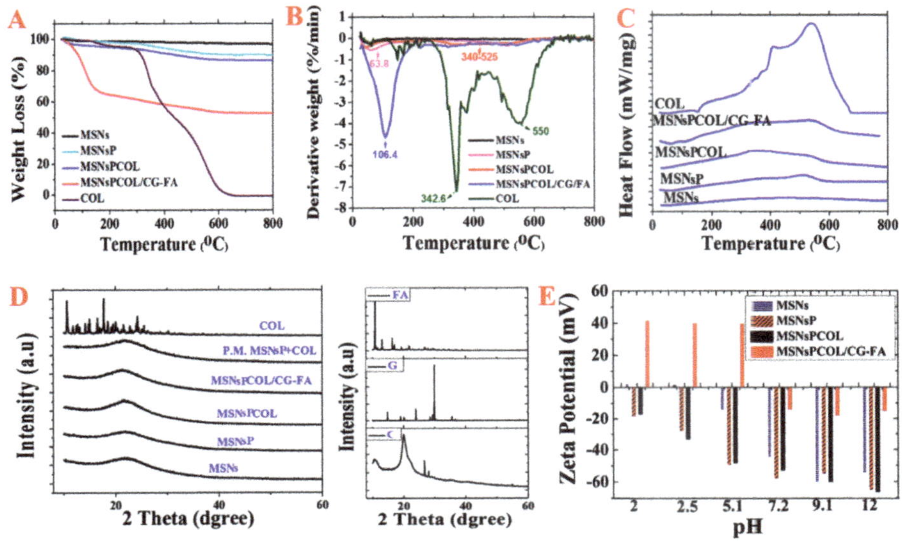

Figure 4. Thermal analysis of materials. (**A**) Simultaneous thermal analysis (STA) before and after surface modification, COL loading, and coating. (**B**) DTG analysis before and after surface modification, COL loading, and coating. (**C**) Differential scanning calorimetry (DSC) analysis before and after surface modification, COL loading, and coating. (**D**) X-ray diffraction (XRD) analysis before and after surface modification, COL loading, and coating. (**E**) Zeta potential measurements in aqueous solution before and after surface modification, COL loading, and coating.

2.8. Zeta Potential Measurements

Figure 4E shows the differences in zeta potential for the materials and pH conditions. MSNs had a very low positive value (<+2 mV) in acidic media and very negative in alkaline medium (−55 mV). As expected, MSNsP had negative zeta values compared to MSNs associated with phosphonate groups. MSNsPCOL exhibited negative values at various pH values, probably due to the low COL content in particles. MSNsPCOL/CG-FA exhibited high positive values (>+40 mV) when medium was acidic (pH 2, 2.5 and 5.1). This observation is important for the cellular uptake of these particles by cancer cells to reach tumor sites. Negative values were found for MSNsPCOL/CG-FA particles in neutral or alkaline medium (pH 7.2, 9.9 and 12). The high positive zeta values for MSNsPCOL/CG-FA particles suggest that they can enter cells via endocytosis. The surface charge of particles (neutral, anionic, and cationic charges) determines their internalization into cells. Anionic nanoparticles are less efficiently internalized than cationic and neutral particles [40–43]. These results suggest that the developed DDS leads to higher cellular uptake in cancer cells.

2.9. MSN Cytotoxicity Evaluation

Figure 5 presents the dependence of cell inhibition on concentration, time, cell line, and MSNs, with significant differences at $p < 0.5$. A gradual cell inhibition effect was found only when cells were treated with either MSNs or MSNsP at an increased concentration of 1000 µg/mL and incubation for 72 h. Higher cytotoxicity was recorded for HCT116 cells than PC3 and HepG2 cells, with 1000 µg/mL MSN and MSNsP treatment of HCT116 cells resulting in 85.9 ± 6.0% and 77.4 ± 4.7% inhibition, respectively. In contrast, normal BJ1 cells were less inhibited than cancer cells under the same treatment conditions.

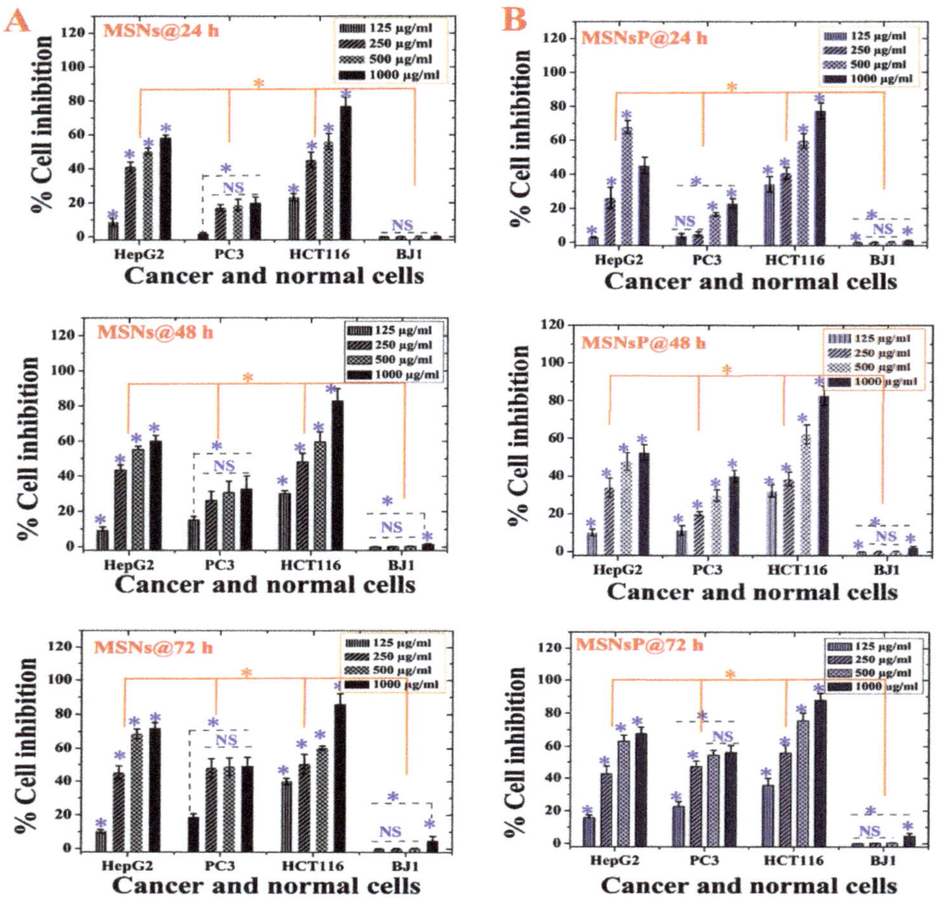

Figure 5. In vitro cytotoxicity (as percent inhibition) of MSNs and MSNs functionalized with phosphonate functional groups (MSNsP) for biocompatibility evaluations in cancer and normal cell lines after 24, 48, and 72 h of incubation with cancer cells (liver, HepG2; prostate, PC3; and colon, HCT116) and normal fibroblasts (BJ1). (**A**) Cytotoxicity of MSNs towards cell lines. (**B**) Cytotoxicity of MSNsP towards cell lines. Note: A blue asterisk (*) indicates significant ($p < 0.05$) differences between tested concentrations, whereas an orange asterisk (*) indicates significant differences between cell lines. NS, not significant. All data are expressed as mean ± SD.

The toxicity differences between MSNs and MSNsP varied according to cell line in response to concentration and time (Table S1 in Supplementary Information). With the IC50 value, it is possible to identify the differences in cytotoxicity; MSNs had a more toxic effect on HepG2 and HCT116 cells after 48 h compared to other incubations. In contrast, MSNsP had a more toxic effect on HCT116 cells after 24 and 72 h compared to 48 h. In addition, HCT116 cells were more sensitive than other cancer cell lines. Both types of nanoparticles had nearly equal IC50 values in PC3 cells after 24 and 48 h. Negligible cytotoxicity (IC50 > 1000 µg/mL) was observed for normal BJ1 cells in response to both types of nanoparticles. The negligible cytotoxicity on BJ1 normal cells can be related to the low internalization of nanoparticles in BJ1 normal cells. There is evidence in literature that cancer cells allow higher nanoparticles internalization compared normal cells due to the enhanced permeation and retention effect [44]. This, because of the vasculature of tumors, is often leaky, leading to accumulating nanoparticles in the bloodstream compared to normal tissue [45]. This finding agrees with previously published data for MCF-7 cells and BJ cells treated with MSNs and phosphonate-functionalized MSNs [39]. They mentioned that cancer cells uptake more MSNs than normal cells, and MSNs are more cytotoxic for cancer cells compared normal cells. Therefore, either MSNs or MSNsP is a promising nanocarrier for COL delivery.

2.10. In Vitro Anticancer Effects against Cancer Cells

We studied the anticancer activity in terms of cell inhibition and found that it was significantly dependent on the cell line, concentration, incubation time, and delivery method. For HepG2 cells (Figure 6A), high inhibition was observed after 72 h and 200 μg/mL of all treatments. Regarding the role of the delivery route, MSNsPCOL/CG-FA exhibited high inhibition (80–82%), especially at 100 and 200 μg/mL, compared to MSNsPCOL and COL. This finding was also confirmed by IC50 values, with lower values detected for three incubation times with MSNsPCOL/CG-FA (Table S1 in Supplementary Information). Obviously, these results indicate that the anticancer activity against HepG2 cells was ranked in the following order: MSNsPCOL/CG-FA > COL > MSNsPCOL/CG-FA.

As shown in Figure 6B, PC3 cells were significantly inhibited by increasing dose and incubation time. Notably, MSNsPCOL/CG-FA had a gradual inhibitory effect compared to carriers; it had lower inhibition at 24 h and increased after 72 h, with a maximum inhibition of 80% when treated at 200 μg/mL. The anticancer effect in PC3 cells was ranked in this order: MSNsPCOL/CG-FA > COL > MSNsPCOL. This effect was also confirmed by IC50 values (Table S1 in Supplementary Information), with lower values obtained for MSNsPCOL/CG-FA compared to other treatments.

Interestingly, HCT116 cancer cells were highly inhibited compared to HepG2 and PC3 (Figure 6C). We observed that at a high concentration of 200 μg/mL MSNsPCOL, MSNsPCOL/CG-FA, and COL inhibited 100% at 24, 48, and 72 h. However, differences were seen among the three treatments when the concentration was decreased. MSNsPCOL/CG-FA potentially inhibited HCT116 growth >90% after 72 h compared to other carriers. Clear differences were confirmed by IC50 calculations, as MSNsPCOL/CG-FA had lower values than MSNsPCOL and COL and reached 19.7 μg/mL after 24 h, 17.4 μg/mL after 48 h, and 17.0 μg/mL after 72 h. The anticancer activity pattern against HCT116 was ranked in this order: MSNsPCOL/CG-FA > COL > MSNsPCOL. Thus, the findings concerning anticancer effects in three cancer cell lines show that HCT116 cells were more sensitive than HepG2 and PC3 cancer cells. Therefore, we selected HCT116 cancer cells for further investigation in our study.

For normal BJ1 cells, we observed that the inhibitory effect depended on the concentration and delivery route (Figure 6D). MSNsPCOL/CG-FA inhibited 4% after 72 h, compared to 60% for COL, which indicates negligible effects. As seen in the IC50 values (Table S1 in Supplementary Information), the delivery COL in the nano or free form had IC50 values > 100 μg/mL. MSNsPCOL/CG-FA significantly inhibited all tested cancer cells lines than COL ranking in this order: HCT116 > HepG2 > PC3. Also, MSNsPCOL/CG-FA had less or negligible toxic effects on normal BJ1 cells versus cancer cells than COL. Thus, this effect is the most likely to be developed as a DDS for cancer therapy.

From the observations, the proposed delivery route has an enhanced anticancer effect with a longer duration to 72 h. The reason for this is that the anticancer effect is in response to COL release from nanoformulations because it is controlled by a coating that needs some time for degradation. The polymeric coating controls the release of guest molecules from MSNs depending on time, pH, and other factors [27,31,46]. The enhanced anticancer effects with MSNsPCOL/CG-FA compared to free COL supports a cancer-targeting effect. This is possible through the interaction of folate receptors in cells and folic acid in nanoparticles. The sensitivity of HCT116 cancer cells to MSNsPCOL/CG-FA compared to HepG2 and PC3 is because folate receptors are overexpressed among cancer cell types [36]. It was shown that expression levels of folate receptors vary depending on type of cancer and the colorectal cancers show higher level than liver and prostate cancers [47]. However, the anticancer effect of free COL or enhanced delivery route is in response to inherited antimitotic effects from COL.

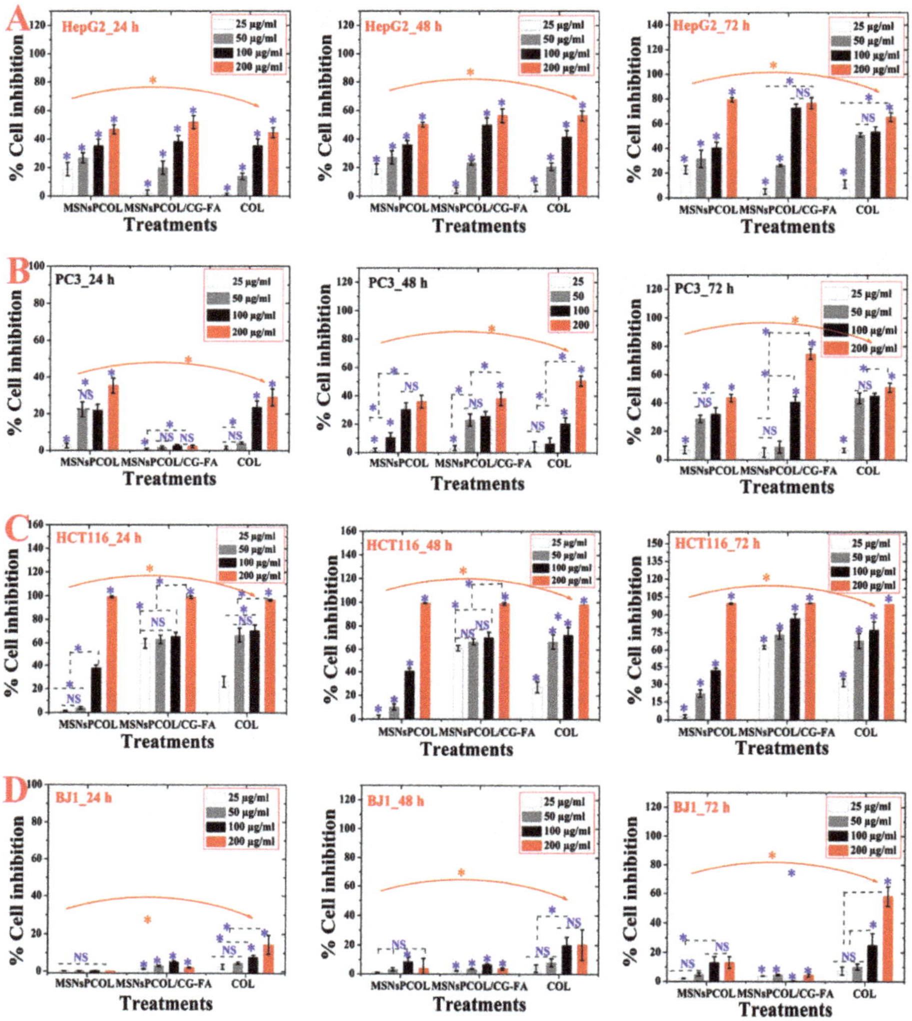

Figure 6. In vitro cytotoxicity (as percent inhibition) of the proposed delivery system in cancer and normal cells after 24, 48, and 72 h of incubation with cells. (**A**) Anticancer effects on HepG2 cancer cells. (**B**) Anticancer effects on PC3 cancer cells. (**C**) Anticancer effects on HCT116 cancer cells. (**D**) Anticancer effects on BJ1 normal cells. Note: A blue asterisk (*) indicates significant ($p < 0.05$) differences between tested concentrations, whereas an orange asterisk (*) indicates significant differences between tested samples (nanoformulations and COL). NS, not significant. All data are expressed as mean ± SD.

2.11. Inhibition of Tubulin Activity in HCT116 Cancer Cells

As shown in Figure 7A, MSNs and MSNsP significantly inhibited tubulin in response to increasing the concentration used to treat cells. In this context, MSNsP inhibited ~12% compared to MSNs ~7% at 1000 μg/mL. Figure 7B shows that MSNsPCOL/CG-FA inhibited 90% more than COL at 50 μg/mL. Thus, the developed DDS is the most likely reason for enhanced tubulin inhibition because of the COL tubulin inhibitor agent [5,48,49]. We propose that the action for MSNsPCOL/CG-FA is the release of COL molecules into cells after folate receptor–folic acid interaction; the COL binds to tubulin in cells through colchicine binding sites, destabilizing tubulin [50], and further interferes with microtubule dynamics.

The later response causes effective mitotic action against HCT116 cells. Further post-antimitotic response can arrest the cell cycle at the G2/M phase [51] and cause apoptotic cell death [52].

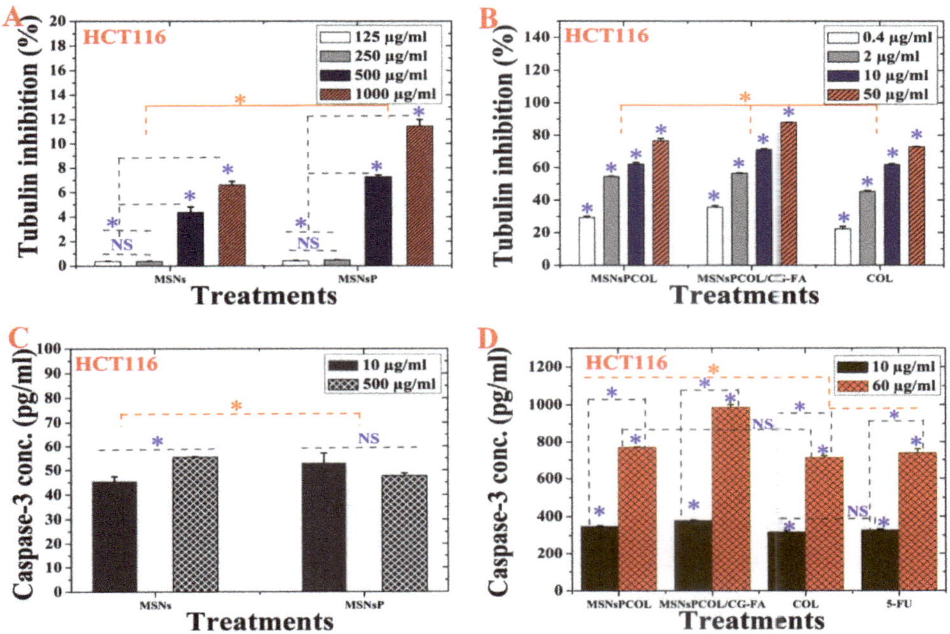

Figure 7. (**A**) Tubulin inhibition activity of HCT116 cells treated with MSNs and MSNsP as a function of concentration. (**B**) Tubulin inhibition activity HCT116 cells treated with the proposed delivery system as a function of concentration. (**C**) Caspase-3 activity of HCT116 cells treated with MSNs and MSNsP as a function of concentration. (**D**) Caspase-3 activity of HCT116 cells treated with the proposed delivery system and compared to the model anticancer drug (5-FU) as a function of concentration. Note: A blue asterisk (*) indicates significant ($p < 0.05$) differences between tested concentrations, whereas an orange asterisk (*) indicates significant differences between tested samples. NS, not significant. All data are expressed as mean ± SD.

2.12. Cell Cycle Arrest at the G2/M Phase in HCT116 Cancer Cells

The G2/M checkpoint inhibitor class of drugs leads to DNA damage, preventing cells from passing mitosis and stopping their proliferation. As a result, cells cannot enter mitosis prior to repairing their DNA damage, resulting in apoptosis or death after their division [53] through various molecular signaling pathways [54,55]. Several reports have demonstrated that COL can arrest at G2/M in different cancer cells [7,56].

The cell cycle analysis was dependent on incubation time (Figure S2A,B in Supplementary Information). After 24 h, MSNsPCOL/CG-FA resulted in the maximum accumulation of cells at G2/M (14.8%), followed by 5-FU (12.4%), MSNsPCOL (11.2%), and COL (8.7%), with the lowest accumulation for MSNs (4.9%), MSNsP (4.8%), and untreated control (4.76%). In contrast, accumulation of cells at S phase was inhibited by MSNsPCOL/CG-FA (18.8%), MSNsPCOL (21.1%), 5-FU (21.2%), COL (23.3%), MSNsP (27.4%), and MSNs (27.6%) compared to control (27.7%). After 72 h, the number of cells increased at G2/M and deceased at S phase compared to 24 h. This cycle analysis pattern shows the endorsement of HCT116 cancer cell arrest at G2/M phase together with evidence that cells could not enter the S phase.

2.13. Enhancement of Apoptosis Induction in HCT116 Cancer Cells

Apoptosis is a distinct or intrinsic occurrence relating to different physiological and pathological responses [55] and usually contributes to efficient antitumor action for most anticancer drugs [54,57]. To investigate whether the treatments induce apoptosis, HCT116 cells were stained with Annexin V-FITC and propidium iodide (PI), and then analyzed by flow cytometry. The parameters of late apoptosis, early apoptosis, and necrosis were quantified. Total apoptosis induction (early plus late) increased with treatment compared to untreated control cells after 72 h compared to 24 h (Figure S3A,B in Supplementary Information). The effects on total apoptosis were in this order: MSNsPCOL/CG-FA (15.0%) > MSNsPCOL (14.2%) > 5-FU (13.2%) > COL (10.6%) > MSNsP (0.5%) > MSNs (0.43). Compared to untreated control (0.5%), MSNs and MSNsP had no effects on apoptosis, reflecting the importance of DDS other than COL and 5-FU drugs. In addition to the induction of apoptosis, MSNsPCOL/CG-FA had a 1.4% necrosis effect. As MSNsPCOL/CG-FA promoted apoptosis (early and late apoptosis) (Figure 8B), the G2/M arrest triggers apoptosis in HCT116 cancer cells [58,59].

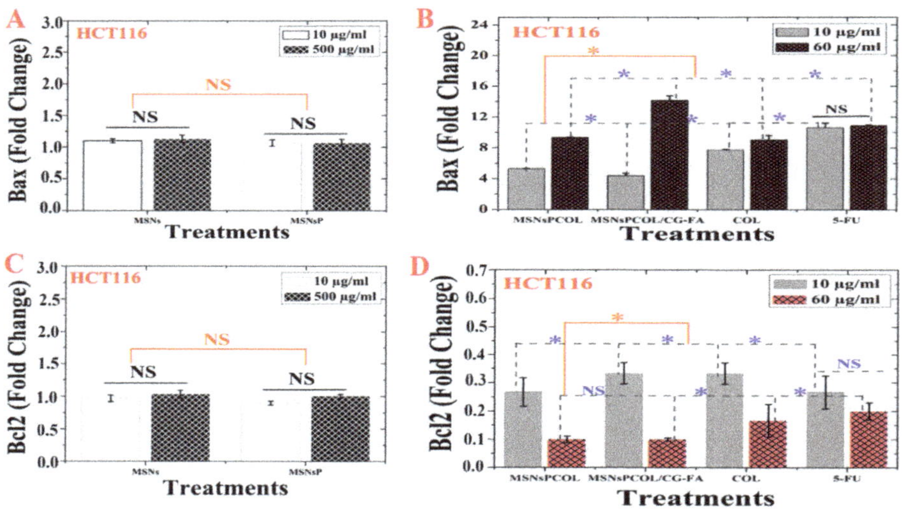

Figure 8. (**A**) Bax activity of HCT116 cells treated with MSNs and MSNsP as a function of concentration. (**B**) Bax activity of HCT116 cells treated with the proposed delivery system and compared to the model anticancer drug (5-FU) as a function of concentration. (**C**) Bcl-2 inhibition activity of HCT116 cells treated with MSNs and MSNsP as a function of concentration. (**D**) Bcl-2 inhibition activity of HCT116 cells treated with the proposed delivery system and compared to the model anticancer drug (5-FU) as a function of concentration. Note: A blue asterisk (*) indicates significant ($p < 0.05$) differences between tested concentrations, whereas an orange asterisk (*) indicates significant differences between tested samples. NS, not significant. All data are expressed as mean ± SD.

2.14. Activation of Caspase-3 in HCT116 Cancer Cells

The activation of intracellular caspase is one of the main characteristics of the apoptosis cell death pathway leading to the cleavage and inactivation of many cellular proteins, leading to the occurrence of apoptotic cell death in most cancer cell types [60]. As shown in Figure 7C, a significant effect was detected among MSNs and MSNsP. Treatment of cells with 500 μg/mL MSNs significantly increased activity compared to 10 μg/mL. In contrast, no significant differences were detected in either concentrations of MSNsP. Maximal activation of both treatments did not reach >55 pg/mL, which is a low enhancement effect. The caspase-3 activity was dependent on concentration and delivery method (Figure 7D), with high activity resulting from increased concentration to 60 μg/mL compared to 10 μg/mL. Maximal caspase-3 activity (>1000 pg/mL) was recorded for MSNsPCOL/CG-FA. Caspase-3 activation was enhanced in this order: MSNsPCOL/CG-FA > MSNsPCOL > 5-FU > COL. These results agree with our previous results using MSNs [26,27].

2.15. Modulation of Proapoptotic Bax/Bcl-2 in HCT116 Cancer Cells

Figure 8A shows no significant effect of MSNs and MSNsP on Bax, but these nanoparticles enhanced it to a little over 1 FLD. Figure 8B shows the significant difference for Bax at 60 μg/mL compared to 10 μg/mL. In addition, MSNsPCOL/CG-FA enhanced Bax to ~14 FLD, and the enhancement ranked in the following order: MSNsPCOL/CG-FA > 5-FU > COL > MSNsPCOL. In particular, MSNsPCOL/CG-FA strongly enhanced Bax compared to the clinically used 5-FU drug.

Though MSNs and MSNsP had not significant effect, they slightly inhibited BCL-2 ~1.2 FLD (Figure 8C). As shown in Figure 8D, the inhibition of BCL-2 was affected by concentration. Treatment of cells at 10 μg/mL resulted in low inhibition and ranked in effect as follows: MSNsPCOL and 5-FU (>0.2 FLD) > MSNsPCOL/CG-FA and COL (~0.35 FLD). Upon further increasing the concentration to 60 μg/mL, BCL-2 highly inhibited to ~0.1 FLD (for MSNsPCOL/CG-FA and MSNsPCOL), <0.2 FLD (for COL), and ~0.2 FLD. These findings show significantly better molecular targeting of nanoformulations in HCT116 cells via Bax enhancement and BCL-2 inhibition than free COL and 5-FU. Furthermore, this effect strongly requires efficient upregulation of Bax and down-regulation of BCL-2, allowing progression of the apoptotic pathway in HCT116 cancer cells. Inhibition of BCL-2 is important because it is considered a potent death suppressor protein; it promotes cell survival by blocking apoptosis and is upregulated in several cancers. BCL-2 is classified as an anti-apoptotic protein in many cancers, including colorectal cancer [61,62], inducing cancer resistance to drugs. Bax is a pro-apoptotic protein that promotes apoptosis in cells but is always present in inactivated states in many cancers, including colon cancer [57,63]. Therefore, these results confirm that the enhancement of apoptosis by MSNsPCOL/CG-FA interconnects with reduced Bcl-2 and increased Bax.

2.16. Inhibition of BRAF Expression in HCT116 Cancer Cells

Colon cancer is associated with multiple processes through various genetic alterations [64], including RAF genes, which mediate several cellular responses [65]. They contribute to carcinogenesis by upregulating the anti-apoptotic RAS/RAF/MEK/ERK pathway [66]. High expression of BRAF (anti-apoptotic protein) inhibits apoptosis by activating the BRAF/MEK/ERK route, which interferes with apoptosis through reduction of caspase and cytochrome c [67].

BRAF inhibition was concentration- and time-dependent. MSNs and MSNsP (at 1000 μg/mL) reduced BRAF ~3% to 3.5% for cells incubated for 24 and 48 h, respectively (Figure 9A,B). Significant inhibition of BRAF was dependent on concentration, time, and delivery route (Figure 9C,D). It was decreased in cells incubated for a longer time. For example, at 50 μg/mL treatment, the inhibition was ~90%, ~87%, ~82%, and ~80% with MSNsPCOL/CG-FA, MSNsPCOL, COL, and 5-FU, respectively. This pattern shows that MSNsPCOL/CG-FA is efficient in inhibiting BRAF in HCT116 cells. Our findings on MSNsPCOL/CG-FA confirm the killing of HCT116 cancer cells was enhanced through apoptosis mechanisms.

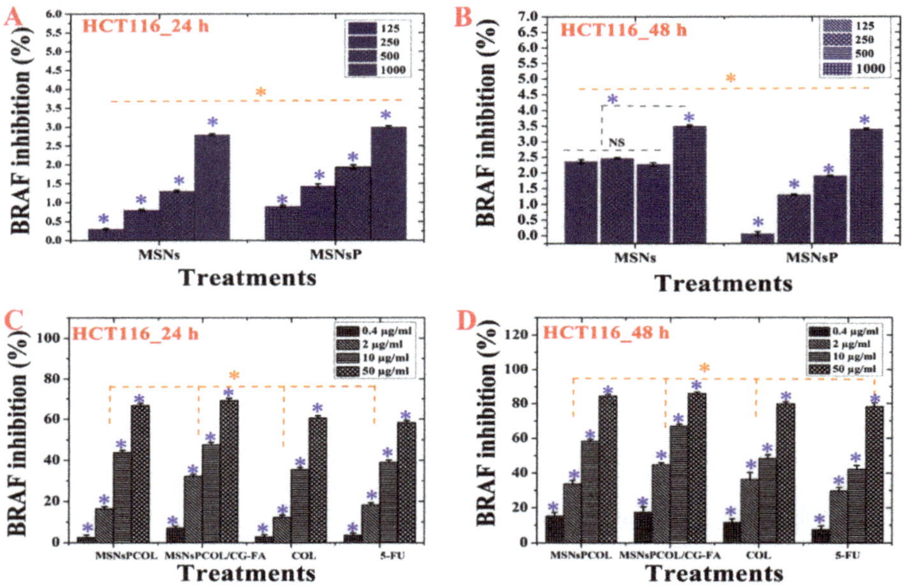

Figure 9. (**A**) BRAF activity of HCT116 cells treated with MSNs and MSNsP as a function of concentration. (**B**) BRAF activity of HCT116 cells treated with the proposed delivery system and compared to the model anticancer drug (5-FU) as a function of concentration. (**C**) BRAF inhibition activity of HCT116 cells treated with MSNs and MSNsP as a function of concentration. (**D**) BRAF inhibition activity of HCT116 cells treated with the proposed delivery system and compared to the model anticancer drug (5-FU) as a function of concentration. Note: A blue asterisk (*) indicates significant ($p < 0.05$) differences between tested concentrations, whereas an orange asterisk (*) indicates significant differences between tested samples. All data are expressed as mean ± SD.

2.17. Enhancement of Cytochrome c Triggers in HCT116 Cancer Cells

As shown in Figure 10A, no significant effect was detected for MSNs and MSNsP treated at 500 μg/mL for 72 h. Significant effect was detected in HCT116 cells treated with the nanoformulations compared to COL (Figure 10B). Triggering of cytochrome c was found in this order: MSNsPCOL/CG-FA > MSNsPCOL > COL > 5-FU. Our results agree with the results by Zhang et al. that COL triggers cytochrome c from the mitochondria to the cytoplasm in human gastric cancer cells [68]. These results confirm that the delivery system for COL activates apoptosis via the intrinsic apoptotic signaling pathway.

2.18. Reduction of Mitochondrial Membrane Potential in HCT116 Cancer Cells

No significant effect was observed among MSNs and MSNsP treated at 500 μg/mL for 72 h (Figure 10C). Importantly, treatment of HCT116 cells with MSNsPCOL/CG-FA significantly decreased mitochondrial membrane potential compared to MSNsPCOL, COL, 5-FU, and control (Figure 10D). Minimal mitochondrial membrane potential was found in this order: MSNsPCOL/CG-FA < 5-FU < COL < MSNsPCOL. It was reported that free COL decreased mitochondrial membrane potential in HT-29 colon cells, which led to intrinsic apoptotic cell death [6].

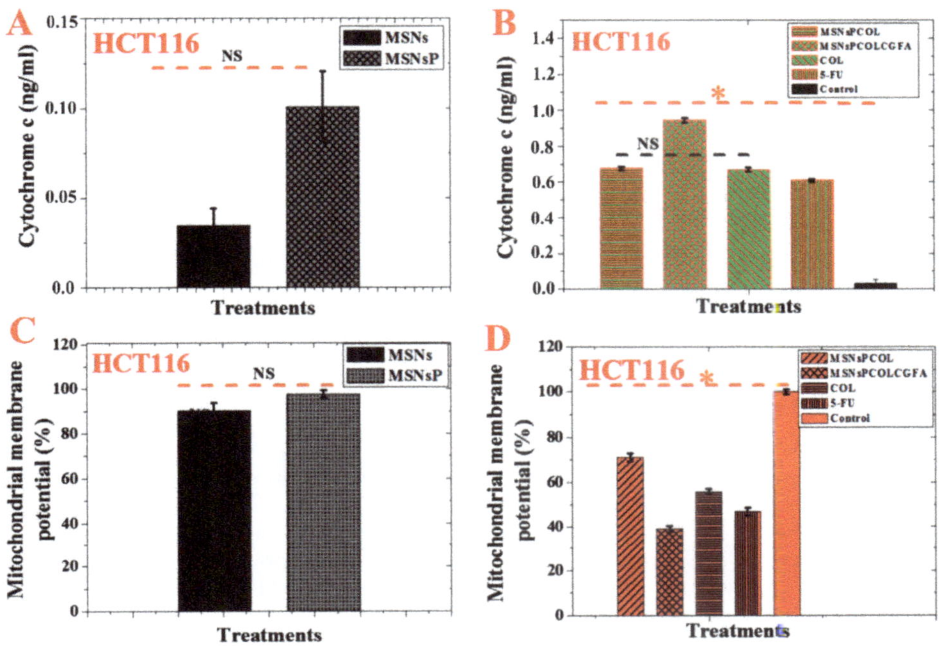

Figure 10. (**A**) Cytochrome c triggering of HCT116 cells treated with MSNs and MSNsP at 500 µg/mL. (**B**) Cytochrome c triggering of HCT116 cells treated at 60 µg/mL with the proposed delivery system for 72 h and compared to the model anticancer drug (5-FU) and control (without any treatment). (**C**) Mitochondrial membrane potential (Δψm) of HCT116 cells treated with MSNs and MSNsP at 500 µg/mL. (**D**) Mitochondrial membrane potential of HCT116 cells treated at 60 µg/mL with the proposed delivery system for 72 h and compared to the model anticancer drug (5-FU) and control. Note: An orange asterisk (*) indicates significant ($p < 0.05$) differences between tested samples. NS, not significant. All data are expressed as mean ± SD.

2.19. Inhibition of CD44 Expression in HCT116 Cancer Cells

As CD44 is a transmembrane glycoprotein, it can take part in different cellular processes, including growth, survival, cell differentiation, resistance to apoptosis [69–71], and tumorigenesis of colon cancer, such as HCT-116 cells [72]. CD44 is well-recognized as a robust marker (via overexpression) of colon cancer initiation because it promotes cell adhesion, maintains cell–matrix interactions [73], and induces anti-apoptotic properties [71,74]. Therefore, CD44 is a potential therapeutic target in colon cancer.

MSNsPCOL/CG-FA, MSNsPCOL, and COL highly attenuated the CD44 concentration in HCT116 compared to positive control cells (185.5 ± 15.4 ng/mL; Figure 11A). We found no significant differences between MSNsPCOL/CG-FA (33.3 ± 3.5 ng/mL), MSNsPCOL (34.1 ± 3.3 ng/mL), and COL (37.3 ± 3.2 ng/mL). However, they significantly inhibited CD44 compared to 5-FU (63.7 ± 8.5 ng/mL). Concerning the anti-apoptotic effect of CD44 expression in colon cancer, Lakshman et al. previously reported that it prevents the apoptosis killing pathway because it promotes cell transformation into a malignant phenotype with the help of other anti-apoptotic factors in the tumor microenvironment [75]. We recorded the maximum inhibition for MSNsPCOL/CG-FA, supporting the results obtained for apoptosis induction. In a previous study of colon cancer cells, Park et al. reported that knockdown of CD44 leads to inhibition of cell proliferation and induction of apoptosis [71].

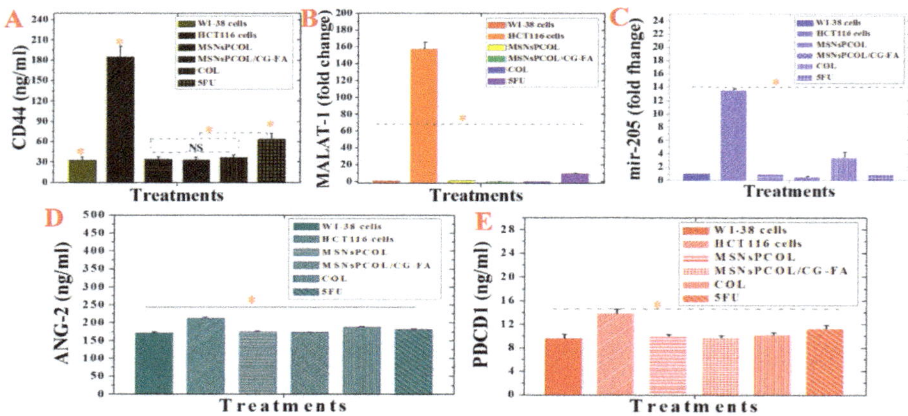

Figure 11. (**A**) MALAT-1 protein expression in normal WI-38 cells, HCT116 cells, and HCT116 cells treated with proposed delivery and 5-FU. (**B**) mir-205 expression in normal WI-38 cells, HCT116 cells, and HCT116 cells treated with proposed delivery and 5-FU. (**C**) Ang-2 protein expression in normal WI-38 cells, HCT116 cells, and HCT116 cells treated with proposed delivery and 5-FU. (**D**) CD44 protein expression in normal WI-38 cells, HCT116 cells, and HCT116 cells treated with proposed delivery and 5-FU. (**E**) PD-1 protein expression in normal WI-38 cells, HCT116 cells, and HCT116 cells treated with proposed delivery and 5-FU. Note: An orange asterisk (*) indicates significant ($p < 0.05$) differences between tested samples, whereas a blue asterisk (*) indicates significant differences between specific samples. NS, not significant. All data are expressed as mean ± SD.

2.20. Inhibition of MALAT-1 Expression in HCT116 Cancer Cells

In recent years, emerging indications are that lncRNAs, non-protein coding transcripts longer than 200 nucleotides, are responsible for a broad spectrum of biological impacts with gene regulation and other functions in many diseases [76,77]. Among others, MALAT1 increases tumor formation in many cancers, including gastric, gallbladder, and lung cancer in vivo [77–80]. Importantly, MALAT1 has been shown in several studies to promote colorectal cancer cell development via proliferation, migration, and invasion [81,82], and is considered a potential therapeutic target for colon cancer. MALAT-1 was significantly inhibited by all delivery methods and free drugs (Figure 11B) compared to positive control cells. MSNsPCOL/CG-FA had a maximum inhibition effect with 0.3 ± 0.05-fold change, followed by COL (0.46 ± 0.05-fold change), MSNsPCOL (1.5 ± 0.15-fold change), and 5-FU (10.06 ± 0.57-fold change) compared to positive control cells (~160-fold change). Importantly, when HCT116 cells were treated with MSNsPCOL/CG-FA, the inhibition level was close to the level of normal cells. These results point out that COL and its nanoformulations are more efficient than 5-FU in hindering the expression of MALAT-1 in HCT116 cells, which is necessary for therapeutic targeting of colon cancer. To the best of our knowledge, no data are available yet on the effects of COL on MALAT1 inhibition or promotion.

2.21. Attenuation of mir-205 Expression in HCT116 Cancer Cells

Micro-RNAs are small non-coding RNAs discovered in 1993 that play crucial roles in cancer, including in cell viability, proliferation, invasion, metastasis, tumor suppressors, and oncogenes [83,84], allowing them to be used in cancer diagnosis and treatment [85,86]. Jing et al. showed that the expression of mir-205 is upregulated in plasma from colon cancer patients, permitting the occurrence and development of the cancer. Thus, mir-205 may be a potential tumor marker and therapeutic target [87]. However, mir-205 has also been reported to be downregulated in colon cancer [88].

Treating HCT116 cells with MSNsPCOL/CG-FA, MSNsPCOL, 5-FU, and COL decreased mir-205 expression. The inhibition values were 0.53 ± 0.05, 0.97 ± 0.05, 10 ± 0.3, and 3.4 ± 0.7-fold change, respectively, compared to the 14-fold change in positive control cells. This provides of evidence of possible mir-205 targeting in colon cancer (Figure 11C). To the best of our knowledge, there are no data available yet on the effects of COL in either downregulating or upregulating mir-205 in cancer. Our results may pave the way for further deep investigations.

2.22. Inhibition of Ang-2 Expression in HCT116 Cancer Cells

One characteristic of cancer is pathological angiogenesis, which plays a crucial role in selecting a therapy. Ang-2 is a pro-angiogenic cytokine that maintains angiogenesis and restricts the antitumor immune response from attacking cancer cells. In a study by Schmittnaegel et al., inhibition of Ang-2 together with vascular endothelial growth factor A (VEGFA) enhanced the antitumor immunity, allowing PD-1 checkpoint inhibition [89]. In addition, Kim et al. indicated that Ang-2 maybe play a crucial role as an oncogene in colorectal carcinogenesis, supporting tumor progression as a prognostic marker [90]. Therefore, we sought to explore the proposed DDS for the possibility of inhibiting Ang-2 and further explain the PD-1 checkpoint inhibitor.

A significant difference was detected between all treatments and positive control cells (HCT116 cells with no treatment; Figure 11D). Notably, treated HCT116 cells had strongly inhibited expression of Ang-2, which was slightly like its expression level in normal cells (172.1 ± 2.5). A maximum inhibition of Ang-2 recorded for MSNsPCOL/CG-FA (174.0 ± 0.75 ng/mL) > MSNsPCOL (175.7 ± 1.0 ng/mL) > 5-FU (182.3 ± 1.15 ng/mL) > COL (188.6 ± 1 ng/mL) compared to positive control cells (213.3 ± 3.0 ng/mL). These results suggest that the anticancer activity of the proposed DDS could promise to target Ang-2 supporting the immune checkpoints inhibition [89,91].

2.23. Inhibition of PD-1 Expression in HCT116 Cancer Cells

Figure 11E shows the suppression of PD-1 checkpoint with significant changes compared control cells; the maximum inhibition level was reached at 9.8 ± 0.3 ng/mL and 10 ± 0.3 ng/mL for MSNsPCOL/CG-FA and MSNsPCOL, respectively, compared to 14 ± 0.5 ng/mL with positive control cells. This was similar to the PD-1 concentration in normal BJ1 cells (9.6 ± 0.5), which is significant. PD-1 reached 10.3 ± 0.2 and 11.2 ± 0.5 with free COL and 5-FU, respectively.

Recent studies have revealed the intrinsic expression of PD-1 in many cancers, along with T immune cells [20,21]. Therefore, it is considered as a new potential for cancer therapy [19] and efficient treatments [92]. Checkpoint blockade immunotherapy has revolutionized treatment for tumors and, in the next years, scientists are expected to focus on immunotherapy research and broaden its scope to target cancers by different strategies [22]. PD-1 immune checkpoint inhibitors act by blocking the PD-1 protein and activate the immune system to treat many tumors [22,93]. MSNsPCOL/CG-FA inhibited PD-1 in HCT116 cells to the range close to normal BJ1 cells, which is important. These are the first results for COL or a DDS as a small drug molecule. PD-1 inhibitors are usually large molecules, such as antibodies, antigens, peptides, and therapeutic proteins. Although many approved therapies are efficient in cancer (e.g., nivolumab and pembrolizumab), they have many limitations, including side effects, toxicity, internal response changes in the immune system, and expense [94,95]. The results create an opportunity to develop a new approach of PD-1 inhibitors through further ongoing research.

In summary, the main findings in the current study were enhancement of the antimitotic effects of COL by means of a targeted delivery system (MSNsPCOL/CG-FA). The enhanced antimitotic effects block cancer cell growth due to the binding of COL to tubulin in cells, leading to cell death. The main mechanism of action is an apoptosis cell death mechanism through different molecular pathways known for COL action on cancer cells. In addition, genetic regulation and immunotherapeutic effects have a role. A schematic representation of the nanoformulation mechanism of action is shown in Figure 12, and Table 2 outlines the results of the current study. The following factors explain the feasibility of cancer-targeting via folate receptors using folic acid ligands as developed here:

(1) The developed nano-delivery system efficiently targets the interaction of folic acid (on the surface of MSNsPCOL/CG-FA) with folate receptors over-expressed on cancer cells. In a next step, it

releases COL into cancer cells, which can bind to the tubulin forming the microtubules in the cellular skeleton, resulting in tubulin inhibition. This effect is important because microtubules enable cells to undergo mitosis or subsequent intracellular post-antimitotic responses. Maximum tubulin inhibition (~90%) was achieved when HCT116 cancer cells were treated with MSNsPCOL/CG-FA.

(2) By producing the antimitotic effect via tubulin inhibition, the cells died. The cell-killing was cell line-, concentration-, time-, and delivery method-dependent. We observed that HCT116 cells were more sensitive to treatments than HepG2 and PC3 cells. In the case of HepG2 and PC3 cells, MSNsPCOL/CG-FA efficiently inhibited the cells compared to COL at all concentrations used. In the case of HCT116 colon cells, a strong effect was observed when incubating cells for 72 h. MSNsPCOL/CG-FA and COL had equal inhibition (100%) at 200 μg/mL. At lower concentrations, MSNsPCOL/CG-FA significantly inhibited HCT116 cells compared to COL. Thus, at all concentrations, MSNsPCOL/CG-FA was more efficient than COL in HCT116 cells. Concerning IC50, MSNsPCOL/CG-FA had a lower value (17.0 μg/mL) than free COL after 72 h. In addition, MSNsPCOL/CG-FA had less toxicity towards normal cells (10%) than COL (60%).

(3) Because COL inhibits tubulin, the post-antimitotic response resulted in various molecular pathways, including cell-cycle arrest, apoptosis, and genetic regulation [14]. We investigated several molecular, genetic, and immunology pathways to explore the mechanism of action. The results confirm that tubulin was markedly inhibited, especially with MSNsPCOL/CG-FA compared to COL. Because of the inhibition of tubulin and the resulting cell cycle arrest, we investigated cell cycle analysis by flow cytometry. MSNsPCOL/CG-FA enhanced the cell-cycle arrest at G2/M phase compared to COL. This observation confirms the tubulin inhibition effect. Arresting cells at the G2/M phase results in apoptosis. The maximum induction of apoptosis was detected for MSNsPCOL/CG-FA. Therefore, this observation confirms that the apoptosis cell death mechanism occurs in HCT116 cancer cells. To further confirm whether apoptosis is the main mechanism of action, we tested many molecular pathways. We observed that caspase-3 was enhanced when cells were treated with MSNsPCOL/CG-FA compared to other formulation, and this is one of the main routes for apoptosis. The modulation of pro-apoptotic proteins is shared with apoptosis; treating cells with MSNsPCOL/CG-FA promoted Bax and inhibited Bcl-2 protein. The anti-apoptotic BRAF protein and CD44 cellar protein levels were highly inhibited after cells were treated with MSNsPCOL/CG-FA. The observations provide strong evidence of the intrinsic apoptosis mechanism due to enhanced cytochrome c triggering and reduce the mitochondrial membrane potential, in agreement with the literature.

(4) New effects were obtained in our study. The antimitotic drugs act by inducing various genetic regulations that may or may not be related to the apoptosis mechanism. Our results show that MSNsPCOL/CG-FA inhibited MALAT 1 and mir-205 in treated cells. Regarding the importance of cancer immunotherapy because of the post-antimitotic effects, MSNsPCOL/CG-FA markedly inhibited Ang-2 protein and PD-1 in HCT116 cancer cells compared to normal WI-38 cells. To the best of our knowledge, the results for MALAT1, mir-205, and PD-1 with COL and its delivery in HCT116 cancer cells were obtained for the first time. The findings suggest that the killing of cancer cells is in response to the effects of the post-antimitotic response of the COL delivery route. We propose that the anticancer mechanism is mainly apoptosis cell death, with the contribution of genetic regulation and immunotherapy effects.

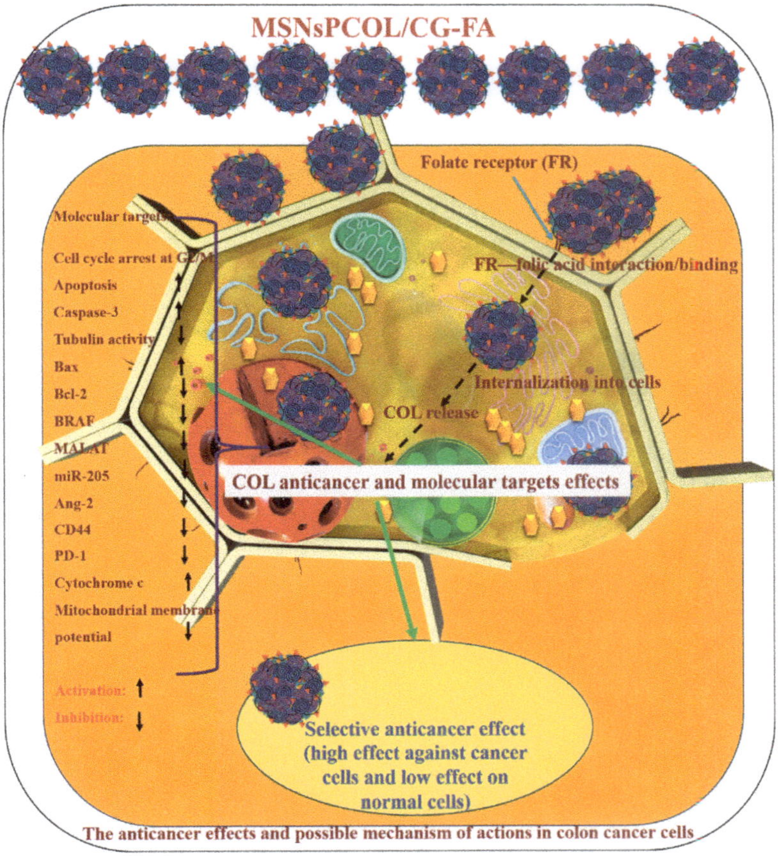

Figure 12. Schematic representation of cancer cell killing and possible anticancer mechanisms by which the proposed drug delivery system for colchicine (MSNsPCOL/CG-FA nanoformulation) acts in vitro against HCT116 colon cancer cells.

Table 2. Summary of the results obtained for colchicine (COL) delivery methods.

Colchicine Delivery Method	Structure	Cancer and Normal Cell Lines	Targeting Specifications
COL (free)	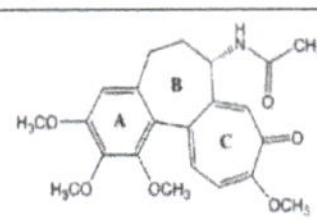 Colchicine	Normal cell line Cancer cell lines	High cytotoxicity (~60%) High anticancer effect at high concentration Non-targeting
MSNsPCOL COL-loaded silica spheres	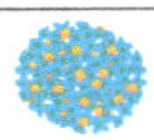	Normal cell line Cancer cell lines	Less cytotoxicity (~15%) High anticancer effect Not enough cancer targeting
MSNsPCOL/CG-FA COL-loaded silica spheres coated with chitosan and glycine and conjugated to folic acid		Normal cell line Cancer cell lines	Negligible cytotoxicity (~5%) High anticancer effect High cancer-targeting Specific killing of HCT116 colon cancer (achieved 100% inhibition), PC3 prostate, and HepG2 liver cells (~80% inhibition) Cancer targeting through folate receptors Enhanced apoptosis mechanism, genetic regulation, and cancer immunotherapy effects

3. Materials and Methods

3.1. Materials

COL, 5-fluorouracil (5-FU), tetraethyl orthosilicate (TEOS), cetylpyridinium bromide (CPB), cyclohexane, isopropanol, urea, glycine, folic acid, insulin, penicillin G, streptomycin, and MTT assay kits were purchased from Sigma-Aldrich (St. Louis, MO, USA), and the Caspase-3 (active) Human ELISA kit from Invitrogen (Camarillo, CA, USA). 3-(Trihydroxysilyl)propyl methylphosphonate monosodium salt solution was purchased from Santa Cruz Biotechnology, Texas, USA. Methanol, ethanol, and acetic acid were obtained from Fisher Scientific UK (Loughborough, UK). Dimethyl sulfoxide (DMSO) was obtained from Tedia (Fairfield, OH, USA). 1-(3-dimethylaminopropyl)-3-ethylcarbodiimide hydrochloride (EDC), *N*-hydroxysuccinimide (NHS), and chitosan (100,000-300,000 Da) were obtained from Acros Organics (Geel, Belgium). Phosphate-buffered saline (PBS), Dulbecco's modified Eagle medium (DMEM), Roswell Park Memorial Institute medium (RPMI 1640), and fetal bovine serum (FBS) were obtained from Gibco/Life Technologies (Thermo Fisher Scientific, Langenselbold, Germany). Insulin (Novo Nordisk, Bagsvaerd, Denmark) and trypsin versene (Vacsera, Giza, Egypt) were also used. The well plates were obtained from Greiner Bio-One GmbH (Frickenhausen, Germany). The Annexin V-FITC Apoptosis Detection Kit was from BioVision (Mountain View, CA, USA), the flow cytometry kit for cell cycle analysis (ab139418) from Abcam® (Cambridge, UK), Script One-Step RT-PCR Kit with SYBR® Green from BIO-RAD (Hercules, CA, USA), Human Angiopoietin 2 ELISA Kit and Human Programmed Cell Death Protein 1 ELISA Kit from Bioassay Technology Laboratory Systems (Shanghai, China), Human CD44 ELISA kit from Gen-Probe Diaclone SAS (Besançon, France), and the miRNeasy extraction kit from Qiagen (Valencia, CA, USA). Enzyme-linked immunosorbent assays (ELISAs) used a kit for tubulin β (TUBb; SEB870Hu, Cloud-Clone Corp., Houston, TX, USA) and human B-RAF/B-Raf sandwich ELISA (LifeSpan BioSciences, Seattle, WA, USA). TMRE Mitochondrial Membrane Potential Assay Kit (Cymans Chemical, Ann Arbor, MI, USA). Cytochrome c Human ELISA Kit (abcam, Austria). Ultrapure water (18.2 MΩ; Milli-Q® system, Millipore, Darmstadt, Germany) was used in all prepared solutions. All other reagents were of analytical grade.

3.2. Methods

3.2.1. Synthesis and Modification of Mesoporous Silica Nanoparticles

We prepared MSNs of KCC-1 type nanospheres with a 3D fibrous dendritic structure using the detailed synthesis methods according to Polshettiwar et al. [30] and AbouAitah et al. [26]. For further surface modification through the post-synthesis route, to graft phosphonate groups, the MSNs were dried at 50 °C for 5 h to remove the physically adsorbed water. Next, 1 g of dried MSNs was suspended in 100 mL of ultra-pure water with the aid of sonication (Elma GmbH, Singen, Germany) for 30 min, and then 1.5 mL of 3-(trihydroxysilyl)propyl methylphosphonate monosodium salt was added drop-wise over with stirring. The mixture solution was left at room temperature under reflux conditions for 24 h. The material was collected by centrifugation (Cooling Sigma 16K, Laborzentrifugen GmbH, Osterode am Harz, Germany) at 10,000 rpm for 10 min, and then washed three times with methanol. Finally, the material was dried in an oven at 50 °C for 6 h to obtain MSNsP.

3.2.2. Colchicine Loading

To load COL into MSNsP, we followed our previous method with some modification [38]: 100 mg COL was dissolved in 10 mL deionized water (pH 7), and then 300 mg of MSNsP was added and stirred (DAIHAN Scientific, Seoul, Korea) for 24 h at room temperature. The solution was centrifuged (Cooling Sigma 16K, Laborzentrifugen GmbH, Osterode am Harz, Germany) for 10 min and washed with deionized water once. Finally, the collected material was heated in an oven at 50 °C for 12 h and labeled as MSNsPCOL.

3.2.3. Coating and Conjugation with Chitosan-Glycine and Folic Acid

Coating with chitosan-glycine and conjugating with folic acid was achieved via the following steps. First, the chitosan-glycine was prepared based on a coacervate and EDC/NHS coupling reaction [27] by dissolving 250 mg of chitosan in 20 mL (2%) acetic acid and stirred for 2 h at 60 °C (solution A). In another beaker, 100 mg of glycine was dissolved in 10 deionized water, followed by the addition of 60 mg EDC and 50 mg NHS and stirring for 2 h at room temperature (solution B). Solution B was introduced to solution A drop-wise and left to stir for 4 h at 50 °C. Hereafter, the mixture solution is referred to as the CG complex solution. Second, the activation of folic acid was carried out by dissolving 85 mg of folic acid, 70 mg of EDC, and 50 mg of NHS in 20 mL DMSO, stirring the mixture for 20 h at room temperature. Third, the activated folic acid solution was added drop-wise into the CG complex solution and stirred for 4 h at 50 °C to obtain CG-FA complex solution. The solution was kept at −20 °C until further use. Fourth, COL-loaded nanoparticles were coated with chitosan-glycine and conjugated to folic acid by dispersing 600 mg of MSNsPCOL into 15 mL of CG-FA complex solution and stirring at room temperature for 24 h. The MSNsPCOL-CG-FA product was collected by centrifugation and washed with ethanol and ultra-pure water, then dried in an oven at 50 °C for 12 h.

3.2.4. Characterization Techniques

The following techniques were used to characterize the obtained materials: Field emission scanning electron microscopy (FE-SEM; Ultra Plus, Zeiss, Germany) equipped with QUANTAX EDS (Bruker); scanning transmission electron microscopy (STEM; FEI TECNAI G2 F20 S-TWIN, Thermo Fisher Scientific, Waltham, MA, USA); powder X-ray diffraction (XRD; X'PertPRO System, PANalytical) using CuKα radiation in the 2θ range of 10–100; and Brunauer, Emmett, and Teller (BET) specific surface area analysis using a Gemini 2360 instrument (Micromeritics) according to ISO 9277:2010. Before the density and SSA measurements were carried out, the powders were dried at 150 °C (without drug) or 50 °C (with drug and polymer) for 24 h under a constant flow of helium (FlowPrep 060 desorption station by Micromeritics). Fourier transformed infrared (FTIR) spectroscopy (Bruker Optics Tensor 27, Bruker Corporation, Billerica, MA, USA) was performed with attenuated total reflectance (ATR, model Platinum ATR-Einheit A 255). Simultaneous thermal analysis (STA)-coupled differential scanning calorimetry (DSC) (STA-DSC) analysis was performed using the STA 499 F1Jupiter (NETZSCH-Feinmahltechnik GmbH, Germany). Samples weighing 10–20 mg were inserted into the alumina pan of the STA unit, and before measurements, helium was flowed through the STA furnace chamber for 30 min. The experimental parameters were programmed to reach 850 °C with a heating rate of 10 °C/min under a helium/air mixture. Zeta potential measurements using a Malvern ZetaSizer (NanoZS, UK) were performed based on the water suspension of nanoparticles at 24 °C. Water suspensions were used for nanoparticle tracking analysis (NTA) using NanoSight NS500 instrument (Malvern Instruments Ltd., Malvern, UK) and data analyzed by NanoSight software (Malvern Instruments Ltd., Malvern, UK).

3.2.5. In Vitro Cytotoxicity Assessment

For the estimation of in vitro cytotoxic potency, the 3-4,5-dimethylthiazol-Z-yl-2,5-diphenyltetrazolium bromide (MTT) assay was conducted according to Mosmann [96]. Human prostate adenocarcinoma PC3 cells (ATCC®CRL-1435 ™), human colon colorectal carcinoma HCT116 cells (ATCC®CCL-247™), human hepatic carcinoma HepG2 cells (ATCC® HB-8065™), and human fibroblast BJ1 cells (ATCC®CRL-2522™) were grown in RPMI supplemented with 10% fetal bovine serum (Cologne, Germany) and 1% penicillin-streptomycin solution (penicillin 10,000 IU/mL; streptomycin 10,000 µg/mL. The cell lines were cultured at 37 °C in 95% humidity and 5% CO_2, and subcultured twice weekly using trypsin versene 0.15%. The cells were seeded in flat-bottom 96-well plates at a density of 10,000 cells per well for 24 h. The serum enriched medium was used during the bioassay to keep the cells alive through the long duration of the bioassay and to better simulate in vivo conditions. Cells were treated with the different samples diluted in medium at 125, 250, 500, and 1000 μg/mL (for MSNs and MSNsP) to test the biocompatibility of nanoparticles. For anticancer activity, cells were treated (as equivalent amount to COL in nanoformulations, the equivalent amount used throughout all studies for investigated nanoformulations) at 25, 50, 100, 200 μg/mL (for MSNsPCOL, MSNsPCOL/CG-FA, COL, 5-FU). After treatment for 24, 48, or 72 h, 10 μL of MTT was added, and after 5 h of incubation the color was measured at 495 nm against the 690 nm reference. Finally, the percent cytotoxicity was calculated according to [1-(av(S)/(av(NC))] × 100, where av(NC) is the average absorbance of the three negative control wells measured at 495 nm (reference 690 nm) and av(S) is the average absorbance of the three sample wells measured at 495 nm (reference 690 nm). IC50 and IC90 values were calculated using Probit analysis and SPSS for Windows statistical analysis software package, version 9 (1989, SPSS Inc., Chicago, IL, USA).

3.2.6. Apoptosis Detection and Cell Cycle Analysis with Flow Cytometry

To analyze apoptosis and the cell cycle, we followed manufacture protocols and our pervious study [27]; HCT116 cells were seeded on a six-well plate at a density of 2 × 10^5 cells/per well in RPMI 1640 supplemented with 10% FBS, 1% penicillin/streptomycin and incubated at 37 °C in a 5% CO_2 atmosphere. After 24 h, the medium was replaced with fresh, treated with different samples at 40 μg/mL (100 μL), and incubated for 24 h or 72 h. Control cells received no treatments. For cell cycle analysis, cells were trypsinized, washed with cold PBS, and fixed with 70% ethanol. The fixed cells were rinsed with PBS, and then labeled with propidium iodide (PI) following the manufacturer's instructions (Abcam®, Cambridge, UK). Finally, cells were analyzed by flow cytometry (FACSCalibur, Becton Dickinson, NJ, USA). Cell cycle analysis was performed with an FL2-A histogram of single cells. The Annexin V-FITC Apoptosis Detection Kit (BioVision, CA, USA) was used to detect apoptosis according to the manufacturer's instructions. After treating the cells as mentioned, the cells were trypsinized, fixed in 70% ethanol, washed with cold PBS, and suspended in binding buffer (500 μL). Next, 5 μL of Annexin V-FITC and 5 μL of PI were added and incubated for 10 min in the dark, and then immediately analyzed by flow cytometry (FACSCalibur, Becton Dickinson, NJ, USA).

3.2.7. Caspase-3 Activity Assay

The caspase-3 activity was determined according to the manufacturer's instructions using the human active caspase-3 content assay kit. HCT116 cells were cultured on 96-well plates (1.2–1.8 × 10,000 cells/well) in 100 μL of RPMI 1640 containing 10% FBS, 1% penicillin/streptomycin at 37 °C. The cells were treated at 10 and 500 μg/mL (for MSNs and MSNsP) or 10 and 60 μg/mL (for MSNsPCOL, MSNsPCOL/CG-FA, COL, 5-FU) with 100 μL sample volume per well and incubated for 72 h prior to the assay. The procedures are reported in detail in our previous study [27]. The absorbance was measured at 450 nm using the Robonik P2000 ELISA reader. The assay was performed in triplicate, and data are expressed as mean ± SD.

3.2.8. Tubulin Assay

To assess tubulin polymerization, the ELISA kit for TUBb (SEB870Hu, Cloud-Clone Corp., Houston, TX, USA) was used following the manufacturer's instructions. HCT116 cells were seeded in 96-well plates at a cell density of 1.2–1.8 × 10,000 cells/well in 100 μL growth medium (DMEM); 100 μL of each sample was added per well. MSNs and MSNsP were tested at 125, 250, 500, and 1000 μg/mL, whereas MSNsPCOL, MSNsPCOL/CG-FA, and COL were used at 0.4, 2, 10, and 50 μg/mL. After 48 h of incubation, the solution was removed, cells were detached by trypsinization, washed with cold PBS buffer, suspended in PBS, and followed by three freeze/thaw cycles to lyse the cells. Cell lysates were centrifuged in a cooling centrifuge for 10 min, which permits the removal of cellular debris to detect β-tubulin in the supernatant. The assay was performed as instructed in the kit. Finally, ELISA reader ROBONIK P2000 (Robonik India PVT LTD, Thane, India) was used to measure the color at 450 nm in triplicate. Data were calculated as percent inhibition.

3.2.9. Expression of Bax and Bcl-2

Cell Culture Treatment and RNA Extraction

HCT116 cells were cultured at a density at 1 x 10^6 and incubated for 48 h at 10 and 500 µL (for MSNs and MSNsP) or 10 and 60 µg/mL (for MSNsPCOL, MSNsPCOL/CG-FA, COL, and 5-FU). Cells were then collected for RNA extraction using the RNeasy extraction kit according to the manufacturer's protocol (Qiagen, Hilden, Germany). Cells were disrupted in buffer RLT, homogenized, and disrupted before adding ethanol to the lysates to create conditions that subsequently promoted the selective binding of RNA to the RNeasy membrane. A total of 100 µl of sample lysate was added to a RNeasy Mini spin column, with total RNA binding to the membrane. High-quality RNA was eluted using RNase-free water. Centrifugation in a micro-centrifuge was used during all steps (binding, washing, elution).

Quantitative Determination by RT-PCR

Bax and Bcl-2 expression was investigated by real-time polymerase chain reaction (RT-PCR) using the BIORAD iScript™ One-Step RT-PCR Kit with SYBR® Green (Bio-Rad, Hercules, CA) according to the manufacturer's instructions and as described by Labib et al. [97]. The RT-PCR reactions were performed using the following primers for the BAX, BCL-2, and β-actin genes: Bax F, 5′-GTTTCA TCC AGG ATC GAG CAG-3′; Bax R, 5′-CATCTT CTT CCA GAT GGT GA-3′; Bcl-2 F, 5′-CCTGTG GAT GAC TGA GTA CC-3′; Bcl-2 R, 5′-GAGACA GCC AGG AGA AAT CA-3′; β-actin F, 5′-GTGACATCCACACCCAGAGG-3′; and β-actin R, 5′-ACAGGATGTCAAAACTGCCC-3′. The reaction and amplifications protocol done according to Labib et al. [97]. A reaction mix (50 µL) was used, prepared as following: 2X Sybr Green RT-PCR Master (25 µl), forward primer—10 µM (1.5 10 µL), Reverse primer—10 µM (1.5 µL), nuclease-free H2O (11 µl), RNA template (1 pg to 100 ng total RNA) (10 µL), and iScript Reverse Transcriptase for One-Step RT-PCR (1 µL). The amplification protocol was performed as follows: cDNA synthesis: 50 °C (10 min), iScript Reverse transcriptase inactivation (95 °C, 5 min), PCR cycling and detection (40 cycle) (95 °C, 10 s), data collection step (60 °C, 30 s), melt curve analysis (95 °C, 1 min, 55 °C (1 min) and 55 °C (10 s) (80 cycles, increased 0.5 °C for each cycle). The reactions were performed in triplicate on a Rotor-Gene 3000 RT-PCR system. The data were analyzed by Rotor-Gene Series Software 1.7 (Build 87).

3.2.10. BRAF Assay

The BRAF ELISA kit was used with cell lysates according to the manufacturer's protocol (LifeSpan BioSciences "LSBio", Seattle, WA, USA). HCT116 cells were cultured at density of 1 × 10^6 and incubated for 48 h at 125, 250, 500, and 1000 µl (for MSNs and MSNsP) or 0.4, 2, 10, and 50 µg/mL (for MSNsPCOL, MSNsPCOL/CG-FA, COL, and 5-FU). The cells were then collected and pelleted by centrifugation to remove the supernatant before washing three times with PBS. Next, the cells were resuspended in PBS, lysed, and centrifuged at 1500× *g* for 10 min with cooling centrifugation to remove cellular debris. The supernatant was collected for assay, 100 µL added to the plate reader and incubated for 90 min at 37 °C, and the liquid removed. Subsequently, 100 µL of 1× Biotinylated Detection Antibody was added to each well and incubated for 1 h at 37 °C, followed by removal of the liquid and washing three times with wash buffer. Next, 100 µL of 1× HRP conjugate working solution was added to each well and incubated for 30 min at 37 °C, then replaced with 90 µL of TMB substrate solution and incubated for 15 min at 37 °C. Finally, 50 µL of stop solution was added and the absorbance measured at 450 nm using the microplate reader ROBONIK P2000 (Robonik India PVT LTD, Thane, India). Measurements were made in triplicate.

3.2.11. Cytochrome c Assay

The cytochrome c was measured using the human cytochrome c ELISA kit in cell lysates according to the manufacturer's protocol (Abcam, Austria). HCT116 cells were cultured at density of 1.2–1.8 × 10,000 cells/well and incubated for 72 h at 500 µL (for MSNs and MSNsP) or 60 µg/mL (for MSNsPCOL,

MSNsPCOL/CG-FA, COL, and 5-FU). The cells were then collected and pelleted by centrifugation to remove the supernatant before washing with PBS. Next, several steps were procced as the kit protocol. Finally, absorbance was measured at 450 nm using the microplate reader (Robonik India PVT LTD, Thane, India). Measurements were made in triplicate.

3.2.12. Measurement of Mitochondrial Membrane Potential

The mitochondrial membrane potential was done by means of flow cytometry with cell lysates according to the manufacturer's protocol TMRE mitochondrial potential assay (Cyman chemical, Ann Arbor, MI, USA). HCT116 cells were cultured at density of 1.2–1.8 × 10,000 cells/well and incubated for 72 h at 500 μL (for MSNs and MSNsP) or 60 μg/mL (for MSNsPCOL, MSNsPCOL/CG-FA, COL, and 5-FU) with volume of 100 μg/mL. The cells were then collected and pelleted by centrifugation to remove the supernatant before washing with PBS. Next, we resuspended it in 100 μL of assay buffer assay, followed by adding 100 μL of TMRE buffer, and incubated for 30 min. Then, it was centrifuged, and resuspended in assay buffer, and the data collected by flow cytometry (FACSCalibur, Becton Dickinson, Franklin Lakes, NJ, USA). Finally, measurements were made in triplicate.

3.2.13. Assays for MALAT-1, mir-205, Ang-2-CD44, and PD-1

Cell Culture and Treatment

HCT116 colon cancer cells were obtained from American Type Culture Collection (ATCC® CCL-247), cultured in 96-well plates (cells density 1.2–1.8 × 10,000 cells/well) in RPMI 1640 medium supplemented with 10% FBS, 10 μg/mL of insulin, and 1% penicillin-streptomycin, and allowed to attach and grow for 24 h. The cell culture was treated with different samples, and control cells were left untreated. To prepare cell culture supernatants, after incubation, cells were harvested after detaching with trypsin and lysates collected by centrifugation. The cells were lysed with cell lysis buffer and centrifuged at 1500g for 10 min at 2–8 °C to exclude cell debris. The expression levels of angiopoietin-2 (Ang-2), PD-1, CD44, metastasis-associated lung adenocarcinoma transcript 1 (MALAT1, and miR-205 were measured in the prepared cell culture supernatants by RT-PCR or ELISA.

The supernatant was used to measure Ang-2 and PD-1 by ELISA using an ELISA plate reader (Model stat fax 2100, Awareness, Ramsey, MN, USA) according to the manufacturer's instructions. For quantitative detection of total soluble human CD44, normal and variant isoforms were measured by ELISA. For measurement of long non-coding RNAs (lncRNAs) for MALAT1 and mir-205, cells were collected, and RNA extracted using the miRNeasy extraction kit. Total RNA including non-coding RNAs was extracted from supernatants using the miRNeasy extraction kit (Qiagen, Valencia CA, USA) and QIAzollysis reagent according to the manufacturer's instructions. The concentration of RNA was determined using NanoDrop2000, which is very accurate for measuring even the smallest quantities of RNA (NanoDrop2000, Thermo Scientific, Wilmington, NC, USA). Reverse transcription was carried out on extracted RNA in a final volume of 20 μL using the RT2 First Strand kit (Qiagen) according to the manufacturer's instructions. The expression levels of the studied lncRNAs were evaluated using GAPDH, which is widely used as an internal control for serum lncRNAs in numerous studies [98,99] according to the manufacturer's protocol. The MALAT1 Ref Seq no. was NR 002819.2. The primer sequences for GAPDH were 5′-CCCTTCATTGACCTCAACTA-3′ (forward) and 5′-TGGAAGATGGTGAT GGGATT-3′ (reverse). RT-PCR was done in a 20 Ml reaction mixture using the Rotor gene Q System (ROTOR-Gene Q, SN R1211164, Qiagen, Hilden, Germany) with the following conditions: 95 °C for 10 min, followed by 45 cycles at 9 °C for 15 s and 60 °C for 60 s. The cycle threshold (Ct) method was used to quantify target genes relative to their endogenous control. The ΔCt of microRNAs was calculated by subtracting the Ct values of SNORD 68 from miR-205. The ΔCt of lncRNAs was calculated by subtracting the Ct value of GAPDH from that of MALAT1. The fold change in miR-205 and MALAT1 expression levels were calculated using the equation $2^{-\Delta\Delta Ct}$. Gene expression was calculated relative to the internal control (2−Ct). The fold change was calculated using 2−Ct for relative quantitation [100].

3.3. Statistical Analysis

Data for biological evaluations are expressed as mean ± SD. Significance differences were calculated using the Student t-test, Mann Whitney U test, and analysis of variance (ANOVA) analysis at $p < 0.05$. All statistical calculations were performed in triplicate using computer program IBM SPSS (Statistical Package for the Social Science; IBM Corp, Armonk, NY, USA) release 22 for Microsoft Windows.

4. Conclusions

We successfully designed a novel DDS for COL prodrug that efficiently targets cancer cells. The DDS was fabricated by loading COL into spherical mesoporous silica nanoparticles and their subsequent modification with phosphonate groups. They were subsequently coated with chitosan-glycine complex conjugated to folic acid, which acted as a targeting ligand for cancers. Full inhibition of HCT116 colon cancer cells was observed. A weaker effect was observed in HepG2 liver and PC3 prostate cancer cells. The most important characteristic of the DDS was its negligible cytotoxicity in normal cells. We observed, after 72 h of incubation with MSNsPCOL/CG-FA, low inhibition in normal BJ1 cells (4%) compared to free COL (~60%). Apoptosis (intrinsic) was found to be the main mechanism of action occurring as a consequence of the strong antimitotic effects. MSNsPCOL/CG-FA more strongly inhibited tubulin than free COL. It also increased the cell cycle at G2/M, caspase-3 activation, and Bax expression compared to COL. On the other hand, MSNsPCOL/CG-FA inhibited anti-apoptotic proteins (Bcl-2, BRAF, and CD44) more strongly than clinically used COL and 5-FU anticancer drugs. New effects of the DDS on genetic regulation and cancer related immuno-effects were found. MSNsPCOL/CG-FA significantly inhibited MALAT1, mir-205 expression, Ang-2 protein, and PD-1 compared to COL and 5-FU. We expect that the tailored DDS for COL has the potential to become a nanomedical platform for cancer treatment.

Supplementary Materials: The following are available online at http://www.mdpi.com/2072-6694/12/1/144/s1, Figure S1: Energy-dispersive X-ray spectroscopy analysis for elemental content (**A**), and particle size analysis by NTA in aqueous solution (**B**), Figure S2: Cell cycle analysis in HCT116 colon cancer cells: after 25 h (**A**), and after 72 h (**B**), Figure S3: Apoptosis analysis by flow cytometry measurements without and with treatments as a function of incubation time in HCT116 cells, Table S1: IC50 of MSNs before and after modification, COL loading, Coating, and folic acid conjugation after 24 h, 48 h, and 72 h incubation with cell lines.

Author Contributions: Conceptualization, K.A.; funding acquisition, K.A. and W.L.; investigation, K.A., H.A.H., A.S.-S., L.G., O.G.S., J.W., A.O., J.S.-K., and S.G.; methodology, K.A., H.A.H., and O.G.S.; resources, K.A., H.A.H., O.G.S., and W.L.; supervision, K.A. and W.L.; writing-original draft, K.A.; writing-review and editing, K.A. and W.L. All authors have read and agreed to the published version of the manuscript.

Funding: The authors would like to thank the National Research Centre (NRC, Egypt) for supporting this work. Many thanks to the National Center for Research and Development, Poland (STRATEGMED3/306888/3/NCBR/2017, project iTE, Poland). This research was carried out using equipment funded by the CePT project, reference: POIG.02.02.00-14-024/08, financed by the European Regional Development Fund within the Operational Programme "Innovative Economy" for 2007–2013.

Acknowledgments: K.A. would like to thank Adam Presz from the Laboratory of Nanostructures, Institute of High Pressure Physics, Polish Academy of Sciences, for the FE-SEM images. The authors would like to thank Essam R. Ahmed, head of the confirmatory diagnostic unit, VACSERA, Dokki, Egypt for his efforts in several molecular assays performed there.

Conflicts of Interest: The authors declare no conflict of interest.

References

1. Kaplan, M.M.; Alling, D.W.; Zimmerman, H.J.; Wolfe, H.J.; Sepersky, R.A.; Hirsch, G.S.; Elta, G.H.; Glick, K.A.; Eagen, K.A. A prospective trial of colchicine for primary biliary cirrhosis. *N. Engl. J. Med.* **1986**, *315*, 1448–1454. [CrossRef]

2. Cohen, A.S.; Rubinow, A.; Anderson, J.J.; Skinner, M.; Mason, J.H.; Libbey, C.; Kayne, H. Survival of patients with primary (AL) amyloidosis: Colchicine-treated cases from 1976 to 1983 compared with cases seen in previous years (1961 to 1973). *Am. J. Med.* **1987**, *82*, 1182–1190. [CrossRef]
3. Gigax, J.H.; Robison, J.R. The Successful Treatment of Intraurethral Condyloma Acuminata with Colchicine. *J. Urol.* **1971**, *105*, 809–811. [CrossRef]
4. Sampedro-Núñez, M.; Serrano-Somavilla, A.; Adrados, M.; Cameselle-Teijeiro, J.M.; Blanco-Carrera, C.; Cabezas-Agricola, J.M.; Martínez-Hernández, R.; Martín-Pérez, E.; Muñoz de Nova, J.L.; Díaz, J.Á.; et al. Analysis of expression of the PD-1/PD-L1 immune checkpoint system and its prognostic impact in gastroenteropancreatic neuroendocrine tumors. *Sci. Rep.* **2018**, *8*, 17812. [CrossRef] [PubMed]
5. Bhattacharyya, B.; Panda, D.; Gupta, S.; Banerjee, M. Anti-mitotic activity of colchicine and the structural basis for its interaction with tubulin. *Med. Res. Rev.* **2008**, *28*, 155–183. [CrossRef] [PubMed]
6. Huang, Z.; Xu, Y.; Peng, W. Colchicine induces apoptosis in HT29 human colon cancer cells via the AKT and c-Jun N-terminal kinase signaling pathways. *Mol. Med. Rep.* **2015**, *12*, 5939–5944. [CrossRef]
7. Zhang, L.; Yang, Z.; Granieri, L.; Pasculescu, A.; Datti, A.; Asa, S.L.; Xu, Z.; Ezzat, S. High-throughput drug library screening identifies colchicine as a thyroid cancer inhibitor. *Oncotarget* **2016**, *7*, 19948–19959. [CrossRef]
8. Cho, J.H.; Joo, Y.H.; Shin, E.Y.; Park, E.J.; Kim, M.S. Anticancer Effects of Colchicine on Hypopharyngeal Cancer. *Anticancer Res.* **2017**, *37*, 6269–6280. [CrossRef]
9. Lin, Z.Y.; Kuo, C.H.; Wu, D.C.; Chuang, W.L. Anticancer effects of clinically acceptable colchicine concentrations on human gastric cancer cell lines. *Kaohsiung J. Med. Sci.* **2016**, *32*, 68–73. [CrossRef]
10. Kumar, A.; Sharma, P.R.; Mondhe, D.M. Potential anticancer role of colchicine-based derivatives: An overview. *Anti Cancer Drugs* **2017**, *28*, 250–262. [CrossRef]
11. Kumar, A.; Singh, B.; Sharma, P.R.; Bharate, S.B.; Saxena, A.K.; Mondhe, D.M. A novel microtubule depolymerizing colchicine analogue triggers apoptosis and autophagy in HCT-116 colon cancer cells. *Cell Biochem. Funct.* **2016**, *34*, 69–81. [CrossRef] [PubMed]
12. Choi, A.R.; Kim, J.H.; Cheon, J.H.; Kim, H.S.; Yoon, S. Attenuation of Colchicine Toxicity in Drug-resistant Cancer Cells by Co-treatment with Anti-malarial Drugs. *Anticancer Res.* **2016**, *36*, 5859–5866. [CrossRef] [PubMed]
13. Gupta, R.S.; Dudani, A.K. Mechanism of action of antimitotic drugs: A new hypothesis based on the role of cellular calcium. *Med. Hypotheses* **1989**, *28*, 57–69. [CrossRef]
14. Gascoigne, K.E.; Taylor, S.S. How do anti-mitotic drugs kill cancer cells? *J. Cell Sci.* **2009**, *122*, 2579–2585. [CrossRef] [PubMed]
15. Castedo, M.; Perfettini, J.L.; Roumier, T.; Andreau, K.; Medema, R.; Kroemer, G. Cell death by mitotic catastrophe: A molecular definition. *Oncogene* **2004**, *23*, 2825–2837. [CrossRef]
16. Qi, C.; Wang, X.; Shen, Z.; Chen, S.; Yu, H.; Williams, N.; Wang, G. Anti-mitotic chemotherapeutics promote apoptosis through TL1A-activated death receptor 3 in cancer cells. *Cell Res.* **2018**, *28*, 544–555. [CrossRef]
17. Fong, A.; Durkin, A.; Lee, H. The Potential of Combining Tubulin-Targeting Anticancer Therapeutics and Immune Therapy. *Int. J. Mol. Sci.* **2019**, *20*, 586. [CrossRef]
18. Fourcade, J.; Sun, Z.; Benallaoua, M.; Guillaume, P.; Luescher, I.F.; Sander, C.; Kirkwood, J.M.; Kuchroo, V.; Zarour, H.M. Upregulation of Tim-3 and PD-1 expression is associated with tumor antigen-specific CD8+ T cell dysfunction in melanoma patients. *J. Exp. Med.* **2010**, *207*, 2175–2186. [CrossRef]
19. Silva, R.; Gullo, I.; Carneiro, F. The PD-1:PD-L1 immune inhibitory checkpoint in Helicobacter pylori infection and gastric cancer: A comprehensive review and future perspectives. *Porto Biomed. J.* **2016**, *1*, 4–11. [CrossRef]
20. Yao, H.; Wang, H.; Li, C.; Fang, J.-Y.; Xu, J. Cancer Cell-Intrinsic PD-1 and Implications in Combinatorial Immunotherapy. *Front. Immunol.* **2018**, *9*, 1774. [CrossRef]
21. Berntsson, J.; Eberhard, J.; Nodin, B.; Leandersson, K.; Larsson, A.H.; Jirström, K. Expression of programmed cell death protein 1 (PD-1) and its ligand PD-L1 in colorectal cancer: Relationship with sidedness and prognosis. *Oncoimmunology* **2018**, *7*, e1465165. [CrossRef] [PubMed]
22. Syn, N.L.; Teng, M.W.L.; Mok, T.S.K.; Soo, R.A. De-novo and acquired resistance to immune checkpoint targeting. *Lancet Oncol.* **2017**, *18*, e731–e741. [CrossRef]

23. Franchini, D.-M.; Lanvin, O.; Tosolini, M.; Patras de Campaigno, E.; Cammas, A.; Péricart, S.; Scarlata, C.-M.; Lebras, M.; Rossi, C.; Ligat, L.; et al. Microtubule-Driven Stress Granule Dynamics Regulate Inhibitory Immune Checkpoint Expression in T Cells. *Cell Rep.* **2019**, *26*, 94–107. [CrossRef] [PubMed]
24. Ades, S. Adjuvant chemotherapy for colon cancer in the elderly: Moving from evidence to practice. *Oncology* **2009**, *23*, 162–167. [PubMed]
25. Banerjee, A.; Pathak, S.; Subramanium, V.D.; Dharanivasan, G.; Murugesan, R.; Verma, R.S. Strategies for targeted drug delivery in treatment of colon cancer: Current trends and future perspectives *Drug Discov. Today* **2017**, *22*, 1224–1232. [CrossRef] [PubMed]
26. AbouAitah, K.; Swiderska-Sroda, A.; Farghali, A.A.; Wojnarowicz, J.; Stefanek, A.; Gierlotka, S.; Opalinska, A.; Allayeh, A.K.; Ciach, T.; Lojkowski, W. Folic acid-conjugated mesoporous silica particles as nanocarriers of natural prodrugs for cancer targeting and antioxidant action. *Oncotarget* **2018**, *9*, 26466–26490. [CrossRef] [PubMed]
27. Shahein, S.A.; Aboul-Enein, A.M.; Higazy, I.M.; Abou-Elella, F.; Lojkowski, W.; Ahmed, E.R.; Mousa, S.A.; AbouAitah, K. Targeted anticancer potential against glioma cells of thymoquinone delivered by mesoporous silica core-shell nanoformulations with pH-dependent release. *Int. J. Nanomed.* **2019**, *14*, 5503–5526. [CrossRef]
28. Cauda, V.; Engelke, H.; Sauer, A.; Arcizet, D.; Brauchle, C.; Radler, J.; Bein, T. Colchicine-loaded lipid bilayer-coated 50 nm mesoporous nanoparticles efficiently induce microtubule depolymerization upon cell uptake. *Nano Lett.* **2010**, *10*, 2484–2492. [CrossRef]
29. Muvaffak, A.; Gurhan, I.; Hasirci, N. Prolonged cytotoxic effect of colchicine released from biodegradable microspheres. *J. Biomed. Mater. Res. Part B Appl. Biomater.* **2004**, *71*, 295–304. [CrossRef]
30. Polshettiwar, V.; Cha, D.; Zhang, X.; Basset, J.M. High-surface-area silica nanospheres (KCC-1) with a fibrous morphology. *Angew. Chem. Int. Ed. Engl.* **2010**, *49*, 9652–9656. [CrossRef]
31. Chang, B.; Chen, D.; Wang, Y.; Chen, Y.; Jiao, Y.; Sha, X.; Yang, W. Bioresponsive Controlled Drug Release Based on Mesoporous Silica Nanoparticles Coated with Reductively Sheddable Polymer Shell. *Chem. Mater.* **2013**, *25*, 574–585. [CrossRef]
32. Castillo, R.R.; Lozano, D.; González, B.; Manzano, M.; Izquierdo-Barba, I.; Vallet-Regí, M. Advances in mesoporous silica nanoparticles for targeted stimuli-responsive drug delivery: An update. *Expert Opin. Drug Deliv.* **2019**, *16*, 415–439. [CrossRef] [PubMed]
33. Lukey, M.J.; Katt, W.P.; Cerione, R.A. Targeting amino acid metabolism for cancer therapy. *Drug Discov. Today* **2017**, *22*, 796–804. [CrossRef] [PubMed]
34. Fernández, M.; Javaid, F.; Chudasama, V. Advances in targeting the folate receptor in the treatment/imaging of cancers. *Chem. Sci.* **2018**, *9*, 790–810. [CrossRef] [PubMed]
35. Sabharanjak, S.; Mayor, S. Folate receptor endocytosis and trafficking. *Adv. Drug Deliv. Rev.* **2004**, *56*, 1099–1109. [CrossRef]
36. Zwicke, G.L.; Ali Mansoori, G.; Jeffery, C.J. Utilizing the folate receptor for active targeting of cancer nanotherapeutics. *Nano Rev.* **2012**, *3*, 18496. [CrossRef]
37. AbouAitah, K.; Farghali, A.; Swiderska-Sroda, A.; Lojkowski, W.; Razin, A.; Khedr, M.K. Mesoporous silica materials in drug delivery system: pH/glutathione-responsive release of poorly water-soluble pro-drug quercetin from two and three-dimensional pore-structure nanoparticles. *J. Nanomed. Nanotechnol.* **2016**, *7*. [CrossRef]
38. AbouAitah, K.E.; Farghali, A.; Swiderska-Sroda, A.; Lojkowski, W.; Razin, A.; Khedr, M.K. pH-controlled release system for curcumin based on functionalized dendritic mesoporous silica nanoparticles. *J. Nanomed. Nanotechnol.* **2016**, *7*. [CrossRef]
39. Menon, N.; Leong, D.T. Cytotoxic Effects of Phosphonate-Functionalized Mesoporous Silica Nanoparticles. *ACS Appl. Mater. Interfaces* **2016**, *8*, 2416–2422. [CrossRef]
40. Villanueva, A.; Canete, M.; Roca, A.G.; Calero, M.; Veintemillas-Verdaguer, S.; Serna, C.J.; Morales Mdel, P.; Miranda, R. The influence of surface functionalization on the enhanced internalization of magnetic nanoparticles in cancer cells. *Nanotechnology* **2009**, *20*, 115103. [CrossRef]
41. Yue, Z.-G.; Wei, W.; Lv, P.-P.; Yue, H.; Wang, L.-Y.; Su, Z.-G.; Ma, G.-H. Surface Charge Affects Cellular Uptake and Intracellular Trafficking of Chitosan-Based Nanoparticles. *Biomacromolecules* **2011**, *12*, 2440–2446. [CrossRef] [PubMed]

42. Fröhlich, E. The role of surface charge in cellular uptake and cytotoxicity of medical nanoparticles. *Int. J. Nanomed.* **2012**, *7*, 5577–5591. [CrossRef] [PubMed]
43. Zhang, S.; Gao, H.; Bao, G. Physical Principles of Nanoparticle Cellular Endocytosis. *ACS Nano* **2015**, *9*, 8655–8671. [CrossRef] [PubMed]
44. Xin, Y.; Yin, M.; Zhao, L.; Meng, F.; Luo, L. Recent progress on nanoparticle-based drug delivery systems for cancer therapy. *Cancer Biol. Med.* **2017**, *14*, 228–241. [CrossRef] [PubMed]
45. Tran, S.; DeGiovanni, P.-J.; Piel, B.; Rai, P. Cancer nanomedicine: A review of recent success in drug delivery. *Clin. Transl. Med.* **2017**, *6*, 44. [CrossRef]
46. Chen, T.; Wu, W.; Xiao, H.; Chen, Y.; Chen, M.; Li, J. Intelligent Drug Delivery System Based on Mesoporous Silica Nanoparticles Coated with an Ultra-pH-Sensitive Gatekeeper and Poly(ethylene glycol). *ACS Macro Lett.* **2016**, *5*, 55–58. [CrossRef]
47. Cheung, A.; Bax, H.J.; Josephs, D.H.; Ilieva, K.M.; Pellizzari, G.; Opzoomer, J.; Bloomfield, J.; Fittall, M.; Grigoriadis, A.; Figini, M.; et al. Targeting folate receptor alpha for cancer treatment. *Oncotarget* **2016**, *7*, 52553–52574. [CrossRef]
48. Andreu, J.M.; Perez-Ramirez, B.; Gorbunoff, M.J.; Ayala, D.; Timasheff, S.N. Role of the colchicine ring A and its methoxy groups in the binding to tubulin and microtubule inhibition. *Biochemistry* **1998**, *37*, 8356–8368. [CrossRef]
49. Jordan, M.A.; Wilson, L. Microtubules as a target for anticancer drugs. *Nat. Rev. Cancer* **2004**, *4*, 253–265. [CrossRef]
50. Lu, Y.; Chen, J.; Xiao, M.; Li, W.; Miller, D.D. An overview of tubulin inhibitors that interact with the colchicine binding site. *Pharm. Res.* **2012**, *29*, 2943–2971. [CrossRef]
51. Blajeski, A.L.; Phan, V.A.; Kottke, T.J.; Kaufmann, S.H. G(1) and G(2) cell-cycle arrest following microtubule depolymerization in human breast cancer cells. *J. Clin. Investig.* **2002**, *110*, 91–99. [CrossRef] [PubMed]
52. Kulshrestha, A.; Katara, G.K.; Ibrahim, S.A.; Patil, R.; Patil, S.A.; Beaman, K.D. Microtubule inhibitor, SP-6-27 inhibits angiogenesis and induces apoptosis in ovarian cancer cells. *Oncotarget* **2017**, *8*, 67017–67028. [CrossRef] [PubMed]
53. Cuddihy, A.R.; O'Connell, M.J. Cell-cycle responses to DNA damage in G2. *Int. Rev. Cytol.* **2003**, *222*, 99–140. [PubMed]
54. Johnstone, R.W.; Ruefli, A.A.; Lowe, S.W. Apoptosis: A Link between Cancer Genetics and Chemotherapy. *Cell* **2002**, *108*, 153–164. [CrossRef]
55. Hengartner, M.O. The biochemistry of apoptosis. *Nature* **2000**, *407*, 770–776. [CrossRef]
56. Kim, S.-K.; Cho, S.-M.; Kim, H.; Seok, H.; Kim, S.-O.; Kwon, T.K.; Chang, J.-S. The colchicine derivative CT20126 shows a novel microtubule-modulating activity with apoptosis. *Exp. Mol. Med.* **2013**, *45*, e19. [CrossRef]
57. Lowe, S.W.; Lin, A.W. Apoptosis in cancer. *Carcinogenesis* **2000**, *21*, 485–495. [CrossRef]
58. Boonstra, J.; Post, J.A. Molecular events associated with reactive oxygen species and cell cycle progression in mammalian cells. *Gene* **2004**, *337*, 1–13. [CrossRef]
59. Nagappan, A.; Lee, W.S.; Yun, J.W.; Lu, J.N.; Chang, S.-H.; Jeong, J.-H.; Kim, G.S.; Jung, J.-M.; Hong, S.C. Tetraarsenic hexoxide induces G2/M arrest, apoptosis, and autophagy via PI3K/Akt suppression and p38 MAPK activation in SW620 human colon cancer cells. *PLoS ONE* **2017**, *12*, e0174591. [CrossRef]
60. McIlwain, D.R.; Berger, T.; Mak, T.W. Caspase functions in cell death and disease. *Cold Spring Harb. Perspect. Biol.* **2013**, *5*, a008656. [CrossRef]
61. Vaux, D.L.; Cory, S.; Adams, J.M. Bcl-2 gene promotes haemopoietic cell survival and cooperates with c-myc to immortalize pre-B cells. *Nature* **1988**, *335*, 440–442. [CrossRef] [PubMed]
62. Scherr, A.-L.; Gdynia, G.; Salou, M.; Radhakrishnan, P.; Duglova, K.; Heller, A.; Keim, S.; Kautz, N.; Jassowicz, A.; Elssner, C.; et al. Bcl-x_L is an oncogenic driver in colorectal cancer. *Cell Death Dis.* **2016**, *7*, e2342. [CrossRef] [PubMed]
63. Rampino, N.; Yamamoto, H.; Ionov, Y.; Li, Y.; Sawai, H.; Reed, J.C.; Perucho, M. Somatic Frameshift Mutations in the BAX Gene in Colon Cancers of the Microsatellite Mutator Phenotype. *Science* **1997**, *275*, 967–969. [CrossRef] [PubMed]
64. Vogelstein, B.; Fearon, E.R.; Hamilton, S.R.; Kern, S.E.; Preisinger, A.C.; Leppert, M.; Nakamura, Y.; White, R.; Smits, A.M.; Bos, J.L. Genetic alterations during colorectal-tumor development. *N. Engl. J. Med.* **1988**, *319*, 525–532. [CrossRef]

65. Peyssonnaux, C.; Eychène, A. The Raf/MEK/ERK pathway: New concepts of activation. *Biol. Cell* **2001**, *93*, 53–62. [CrossRef]
66. Ikehara, N.; Semba, S.; Sakashita, M.; Aoyama, N.; Kasuga, M.; Yokozaki, H. BRAF mutation associated with dysregulation of apoptosis in human colorectal neoplasms. *Int. J. Cancer* **2005**, *115*, 943–950. [CrossRef]
67. Erhardt, P.; Schremser, E.J.; Cooper, G.M. B-Raf inhibits programmed cell death downstream of cytochrome c release from mitochondria by activating the MEK/Erk pathway. *Mol. Cell. Biol.* **1999**, *19*, 5308–5315. [CrossRef]
68. Zhang, T.; Chen, W.; Jiang, X.; Liu, L.; Wei, K.; Du, H.; Wang, H.; Li, J. Anticancer effects and underlying mechanism of Colchicine on human gastric cancer cell lines in vitro and in vivo. *Biosci. Rep.* **2019**, 39. [CrossRef]
69. Nagano, O.; Saya, H. Mechanism and biological significance of CD44 cleavage. *Cancer Sci.* **2004**, *95*, 930–935. [CrossRef]
70. Chen, C.; Zhao, S.; Karnad, A.; Freeman, J.W. The biology and role of CD44 in cancer progression: Therapeutic implications. *J. Hematol. Oncol.* **2018**, *11*, 64. [CrossRef]
71. Park, Y.S.; Huh, J.W.; Lee, J.H.; Kim, H.R. shRNA against CD44 inhibits cell proliferation, invasion and migration, and promotes apoptosis of colon carcinoma cells. *Oncol. Rep.* **2012**, *27*, 339–346. [CrossRef] [PubMed]
72. Zhou, J.-Y.; Chen, M.; Ma, L.; Wang, X.; Chen, Y.-G.; Liu, S.-L. Role of CD44(high)/CD133(high) HCT-116 cells in the tumorigenesis of colon cancer. *Oncotarget* **2016**, *7*, 7657–7666. [CrossRef] [PubMed]
73. Sahlberg, S.H.; Spiegelberg, D.; Glimelius, B.; Stenerlow, B.; Nestor, M. Evaluation of cancer stem cell markers CD133, CD44, CD24: Association with AKT isoforms and radiation resistance in colon cancer cells. *PLoS ONE* **2014**, *9*, e94621. [CrossRef] [PubMed]
74. Schneider, M.; Huber, J.; Hadaschik, B.; Siegers, G.M.; Fiebig, H.H.; Schuler, J. Characterization of colon cancer cells: A functional approach characterizing CD133 as a potential stem cell marker. *BMC Cancer* **2012**, *12*, 96. [CrossRef] [PubMed]
75. Lakshman, M.; Subramaniam, V.; Rubenthiran, U.; Jothy, S. CD44 promotes resistance to apoptosis in human colon cancer cells. *Exp. Mol. Pathol.* **2004**, *77*, 18–25. [CrossRef]
76. Ponting, C.P.; Oliver, P.L.; Reik, W. Evolution and Functions of Long Noncoding RNAs. *Cell* **2009**, *136*, 629–641. [CrossRef]
77. Zhao, M.; Wang, S.; Li, Q.; Ji, Q.; Guo, P.; Liu, X. MALAT1: A long non-coding RNA highly associated with human cancers. *Oncol. Lett.* **2018**, *16*, 19–26. [CrossRef]
78. Okugawa, Y.; Toiyama, Y.; Hur, K.; Toden, S.; Saigusa, S.; Tanaka, K.; Inoue, Y.; Mohri, Y.; Kusunoki, M.; Boland, C.R.; et al. Metastasis-associated long non-coding RNA drives gastric cancer development and promotes peritoneal metastasis. *Carcinogenesis* **2014**, *35*, 2731–2739. [CrossRef]
79. Wu, X.S.; Wang, X.A.; Wu, W.G.; Hu, Y.P.; Li, M.L.; Ding, Q.; Weng, H.; Shu, Y.J.; Liu, T.Y.; Jiang, L.; et al. MALAT1 promotes the proliferation and metastasis of gallbladder cancer cells by activating the ERK/MAPK pathway. *Cancer Biol. Ther.* **2014**, *15*, 806–814. [CrossRef]
80. Gutschner, T.; Hammerle, M.; Eissmann, M.; Hsu, J.; Kim, Y.; Hung, G.; Revenko, A.; Arun, G.; Stentrup, M.; Gross, M.; et al. The noncoding RNA MALAT1 is a critical regulator of the metastasis phenotype of lung cancer cells. *Cancer Res.* **2013**, *73*, 1180–1189. [CrossRef]
81. Yang, M.-H.; Hu, Z.-Y.; Xu, C.; Xie, L.-Y.; Wang, X.-Y.; Chen, S.-Y.; Li, Z.-G. MALAT1 promotes colorectal cancer cell proliferation/migration/invasion via PRKA kinase anchor protein 9. *Biochim Biophys Acta* **2015**, *1852*, 166–174. [CrossRef]
82. Wu, C.; Zhu, X.; Tao, K.; Liu, W.; Ruan, T.; Wan, W.; Zhang, C.; Zhang, W. MALAT1 promotes the colorectal cancer malignancy by increasing DCP1A expression and miR203 downregulation. *Mol. Carcinog.* **2018**, *57*, 1421–1431. [CrossRef]
83. Esquela-Kerscher, A.; Slack, F.J. Oncomirs—MicroRNAs with a role in cancer. *Nat. Rev. Cancer* **2006**, *6*, 259–269. [CrossRef]
84. Lee, R.C.; Feinbaum, R.L.; Ambros, V. The C. elegans heterochronic gene lin-4 encodes small RNAs with antisense complementarity to lin-14. *Cell* **1993**, *75*, 843–854. [CrossRef]
85. Schetter, A.J.; Okayama, H.; Harris, C.C. The role of microRNAs in colorectal cancer. *Cancer J.* **2012**, *18*, 244–252. [CrossRef]

86. Cekaite, L.; Eide, P.W.; Lind, G.E.; Skotheim, R.I.; Lothe, R.A. MicroRNAs as growth regulators, their function and biomarker status in colorectal cancer. *Oncotarget* **2016**, *7*, 6476–6505. [CrossRef]
87. Jing, N.; Yin, L.; Sun, J.; Cao, Z.; Mao, W. Expression levels of miR-205 and miR-506 in colon cancer tissues and their relationships with clinicopathological features. *Oncol. Lett.* **2018**, *16*, 4331–4336. [CrossRef]
88. Orang, A.V.; Safaralizadeh, R.; Hosseinpour Feizi, M.A.; Somi, M.H. Diagnostic and prognostic value of miR-205 in colorectal cancer. *Asian Pac. J. Cancer Prev.* **2014**, *15*, 4033–4037. [CrossRef]
89. Schmittnaegel, M.; Rigamonti, N.; Kadioglu, E.; Cassara, A.; Wyser Rmili, C.; Kiialainen, A.; Kienast, Y.; Mueller, H.J.; Ooi, C.H.; Laoui, D.; et al. Dual angiopoietin-2 and VEGFA inhibition elicits antitumor immunity that is enhanced by PD-1 checkpoint blockade. *Sci. Transl. Med.* **2017**, 9. [CrossRef]
90. Kim, H.; Ahn, T.S.; Kim, C.-J.; Bae, S.B.; Kim, H.J.; Lee, C.-S.; Kim, T.H.; Im, J.; Lee, S.H.; Son, M.W.; et al. Oncogenic function of angiopoietin-2 in vitro and its modulation of tumor progression in colorectal carcinoma. *Oncol. Lett.* **2017**, *14*, 553–560. [CrossRef]
91. Fukumura, D.; Kloepper, J.; Amoozgar, Z.; Duda, D.G.; Jain, R.K. Enhancing cancer immunotherapy using antiangiogenics: Opportunities and challenges. *Nat. Rev. Clin. Oncol.* **2018**, *15*, 325–340. [CrossRef]
92. Khalil, D.N.; Smith, E.L.; Brentjens, R.J.; Wolchok, J.D. The future of cancer treatment: Immunomodulation, CARs and combination immunotherapy. Nature reviews. *Clin. Oncol.* **2016**, *13*, 273–290. [CrossRef]
93. Webb, E.S.; Liu, P.; Baleeiro, R.; Lemoine, N.R.; Yuan, M.; Wang, Y.-H. Immune checkpoint inhibitors in cancer therapy. *J. Biomed. Res.* **2018**, *32*, 317–326. [CrossRef]
94. Ottaviano, M.; De Placido, S.; Ascierto, P.A. Recent success and limitations of immune checkpoint inhibitors for cancer: A lesson from melanoma. *Virchows Arch. Int. J. Pathol.* **2019**, *474*, 421–432. [CrossRef]
95. Seidel, J.A.; Otsuka, A.; Kabashima, K. Anti-PD-1 and Anti-CTLA-4 Therapies in Cancer: Mechanisms of Action, Efficacy, and Limitations. *Front. Oncol.* **2018**, *8*, 86. [CrossRef]
96. Mosmann, T. Rapid colorimetric assay for cellular growth and survival: Application to proliferation and cytotoxicity assays. *J. Immunol. Methods* **1983**, *65*, 55–63. [CrossRef]
97. Labib, M.B.; Philoppes, J.N.; Lamie, P.F.; Ahmed, E.R. Azole-hydrazone derivatives: Design, synthesis, in vitro biological evaluation, dual EGFR/HER2 inhibitory activity, cell cycle analysis and molecular docking study as anticancer agents. *Bioorg. Chem.* **2018**, *76*, 67–80. [CrossRef]
98. Shaker, O.G.; Senousy, M.A.; Elbaz, E.M. Association of rs6983267 at 8q24, HULC rs7763881 polymorphisms and serum lncRNAs CCAT2 and HULC with colorectal cancer in Egyptian patients. *Sci. Rep.* **2017**, *7*, 16246. [CrossRef]
99. Duan, W.; Du, L.; Jiang, X.; Wang, R.; Yan, S.; Xie, Y.; Yan, K.; Wang, Q.; Wang, L.; Zhang, X.; et al. Identification of a serum circulating lncRNA panel for the diagnosis and recurrence prediction of bladder cancer. *Oncotarget* **2016**, *7*, 78850–78858. [CrossRef]
100. Livak, K.J.; Schmittgen, T.D. Analysis of relative gene expression data using real-time quantitative PCR and the $2^{-\Delta\Delta CT}$ Method. *Methods* **2001**, *25*, 402–408. [CrossRef]

cancers

MDPI

Review

Shaping the Innate Immune Response by Dietary Glucans: Any Role in the Control of Cancer?

Manuela Del Cornò, Sandra Gessani * and Lucia Conti

Center for Gender-specific Medicine, Istituto Superiore di Sanità, 00161 Rome, Italy; manuela.delcorno@iss.it (M.D.C.); lucia.conti@iss.it (L.C.)

* Correspondence: sandra.gessani@iss.it

Received: 5 December 2019; Accepted: 20 December 2019; Published: 8 January 2020

Abstract: β-glucans represent a heterogeneous group of naturally occurring and biologically active polysaccharides found in many kinds of edible mushrooms, baker's yeast, cereals and seaweeds, whose health-promoting effects have been known since ancient times. These compounds can be taken orally as food supplements or as part of daily diets, and are safe to use, nonimmunogenic and well tolerated. A main feature of β-glucans is their capacity to function as biological response modifiers, exerting regulatory effects on inflammation and shaping the effector functions of different innate and adaptive immunity cell populations. The potential to interfere with processes involved in the development or control of cancer makes β-glucans interesting candidates as adjuvants in antitumor therapies as well as in cancer prevention strategies. Here, the regulatory effects of dietary β-glucans on human innate immunity cells are reviewed and their potential role in cancer control is discussed.

Keywords: innate immunity; β-glucans; nutrition; immunotherapy; cancer

1. Introduction

The immune system has an active role in all phases of carcinogenesis, exerting multifaceted functions that range from antitumoral to protumoral activities. Tumors are populated by a vast and diverse array of immune components: innate sentinels including phagocytes [macrophages, neutrophils and dendritic cells (DC)], natural killer (NK), natural killer T (NKT) and $\gamma\delta$ T cells, as well as adaptive leukocytes, including naive, memory, and effector B- and T-lymphocytes [1]. The degree of immune infiltration and the composition of the infiltrate, as well as of the circulating immune cell pool can vary markedly across tumor types and stages [1,2], and can represent important correlates of cancer prognosis and treatment responsiveness [1–3]. Although the immune system can elicit an antitumor response that leads to tumor destruction in some cases, the successful development of such a response is often hampered by a plethora of factors. Indeed, immune cell populations are conditioned by soluble factors, enzymes and metabolites produced by nearby tumor and stromal cells within the tumor microenvironment (TME) that contribute to dampen the antitumor immune response [4]. Moreover, the immunosuppressive TME can even shape immune cell functions toward a tumor-promoting response [5].

Cancer is frequently associated with chronic inflammation, which is considered a requirement for maintaining an immunosuppressive network. Innate immune cells are the main actors in the inflammatory response by virtue of their expression of pathogen recognition receptors (PRR). The engagement of these receptors not only induces the secretion of proinflammatory cytokines, thereby further supporting inflammation, but also activates antigen presenting cell functions, thus favoring the recognition and killing of tumor cells and the orchestration of a specific adaptive response.

Shifting the balance towards a protective response has been the goal of many anticancer strategies. An adequate response has been achieved in several human cancers with the use of nonspecific immunotherapies, vaccines, adoptive cell therapy and, more recently, through the blockade of immune checkpoints. Nevertheless, the high antigen-specific T-cell-mediated response measured in some tumor cases does not correlate with clinical benefit [6], suggesting that resistance mechanisms within the TME might orchestrate tumor escape, and that an appropriate activation of innate immunity cells may be the most critical determinant of therapeutic success.

In recent years, a clear-cut association between nutrition and cancer has been highlighted by several epidemiological and research studies [7]. Dietary patterns and bioactive food-derived compounds have been associated with either increased or decreased risk of several human noncommunicable diseases, including cancer, and their capacity to influence the immune response under physiological and pathological conditions has been widely described (World Cancer Research Fund. Continuous Update Project Expert Report 2018. Diet, Nutrition, Physical Activity and Colorectal Cancer. Available at: dietandcancerreport.org. 2018). There is currently a growing interest in studying the mechanisms underlying the capacity of bioactive food compounds to modulate processes implicated in either the promotion of or in the protection against carcinogenesis. In this regard, the ability of natural food compounds to regulate innate cell functions and to promote or attenuate inflammation is receiving great attention. Among a vast array of compounds naturally occurring polysaccharides, including β-glucans, have been widely studied for their effects on human health and disease in both western and eastern medicine [8,9]. Indeed, despite their structural function, some β-glucans exert important biological activities.

In this review, we will overview the immunomodulatory effects of β-glucans derived from edible mushrooms, baker's yeast and cereals, assumed as dietary supplements in the form of purified compounds or as extracts. We will focus on human studies and describe the in vitro and in vivo effects of these compounds on innate immunity cells involved in cancer surveillance. The mechanisms underlying the capacity of β-glucans to modulate immune response and how this may impact on tumor development/progression will be discussed.

2. Dietary β-glucans: Main Features and Sources

Dietary fibers have been consumed for a long time, due to their beneficial effects on health, as part of the carbohydrate fraction within food [10]. All existing definitions recognize fibers as a group of carbohydrate polymers and oligomers (and lignin) that are resistant to digestion and absorption in the human small intestine where they are, partially or completely, fermented by the gut microbiota. They are present in two main forms depending on their solubility in water (i.e., soluble and insoluble fibers) [8]. Among soluble fibers, β-glucans are homopolysaccharides composed of D-glucopyranosyl residues linked through β-glycosidic bonds comprising a heterogeneous group of naturally occurring biologically active compounds found in the bran of some cereal grains (oat, barley), in the cell wall of baker's yeast, in many kinds of edible mushrooms (fruit bodies and cultured mycelium) and seaweeds [11,12]. β-glucans of different origins differ in their macromolecular structure and β-linkage, with mushroom- and yeast-derived β-glucans exhibiting primarily β-(1,3/1,6) linkages, while those from cereals have β-(1,3/1,4) linkages [9]. They have the peculiarity of being encountered by humans either as part of their diet or as pathogen associated molecular patterns (PAMP), since they also constitute the cell wall of some pathogenic yeasts and bacteria, and strongly contribute to microorganism recognition and clearance [13]. β-glucans greatly vary in their macromolecular structure, solubility, molecular weight, degree of branching and charge of polymers, as well as in receptor recognition/binding affinity, and their biological activity strongly depends on these features [14,15]. β-glucans derived from different sources appear to show diverse activities. While β-glucans originated from mushrooms are effective as antitumor defense and immunity boost, those found in cereals are more active in lowering cholesterol and blood sugars [15,16].

Upon oral ingestion as dietary components, they reach, in an undigested form, the small intestine where intestinal epithelial and/or M cells internalize and present them to the immune cell populations within the Peyer's patches [17]. β-glucan particles can also reach distant lymphoid organs via blood or lymph [18–20]. The recognition by innate immunity cells occurs via ligation of specific PRR, such as Toll-like (TLR) and C-type lectin-like receptors [15,21]. Among the latter, Dectin-1 is the best characterized receptor, reported to bind β-glucan from different sources, and is expressed on the surface of monocytes, macrophages, neutrophils, DC and T lymphocytes [22]. Other PRR, including lactosylceramide receptor, mannose receptor, complement and scavenger receptors were reported to directly bind β-glucan or to cooperate with Dectin-1 for its recognition [15,21,23]. Recently, β-glucan was shown to stimulate NK cell cytotoxic activity through direct binding to the NKp30 activating receptor [24]. Receptor binding on innate immunity cells elicits a number of cellular responses through the modulation of inflammasome and transcription factor activation, and the production of immune response modifiers [cytokines, chemokines, reactive oxygen species (ROS)], finally shaping adaptive immune response.

3. Immunomodulatory Effects of β-glucans

β-glucans from different sources have been extensively studied for their anti-inflammatory, anti-allergic, anti-obesity and anti-osteoporotic activities [9]. The regulatory effects of β-glucans on immune response have been identified as one of the main biological functions that make these compounds attractive targets for therapeutic interventions. Strong evidence on the immunomodulatory activities of β-glucans has been proved by in vitro, as well as by animal- and human-based clinical studies, highlighting their ability to protect against infections and to improve the immunogenicity of vaccines, their antitumor activity and, more recently, their therapeutic potential when combined with other cancer therapies [25–28]. Indeed some β-glucan preparations have been approved as adjunctive therapeutic drugs for cancer treatment. The safety and lack of toxicity of their oral or intravenous administration have been assayed in phase I–II clinical trials carried out in healthy volunteers, as well as virus-infected or cancer patients [29–34]. The size and biochemical composition of β-glucans isolated from different sources have been reported to affect their immunomodulatory properties with the molecular pathways activated and the features of the immune response generated depending on the cell type and receptor triggered [15,35]. Furthermore, β-glucans derived from different sources differ in their capacity to function as PAMP and, as a consequence, in their immunoregulatory efficacy. In this regard, β-glucans derived from baker's yeast and mushrooms are by far the best-documented immunostimulators, while those from cereals fall short as immune regulators, as they are structurally different and not recognized as PAMP [36,37]. The ensemble of these variables can result in different pharmacologic activities and bioavailability, thus explaining the discrepancies observed when comparing different studies.

The main evidence on the in vitro effects of β-glucans on human innate immunity cells, as well as the clinical studies assessing their in vivo effects on the same cell populations in either healthy subjects or cancer patients, are reviewed in the next sections and summarized in Tables 1–3.

3.1. In Vitro Effects on Innate Immunity Cells

In vitro studies have shown that β-glucans from yeasts, mushrooms or cereals are able to enhance the responsiveness or function of human primary immune cells, eliciting potent immune responses through their recognition by a variety of PRR, particularly Dectin-1 and complement receptor 3 (CR3). Within the innate immune system, the targeted cells of β-glucans include macrophages, neutrophils, monocytes, NK cells and DC. Specifically, β-glucans can enhance the functional activity of monocytes/macrophages and DC and activate antimicrobial activity of mononuclear cells and neutrophils in vitro. This enhanced immune response is accomplished by an increased proinflammatory cytokine and chemokine production and an enhanced oxidative burst, as mainly demonstrated in mouse models [38].

Table 1. In vitro studies evaluating the immunomodulatory effects of β-glucans on human innate immunity cells.

Compound (Concentration Range [1])	Cell Type	Effects	Molecular Mechanisms	Refs
Agaricus brasiliensis acid-treated polysaccharide-rich fraction (50 μg/mL)	Monocytes	↑ Adherence ↑ Phagocytosis ↑ TNFα , IL-1β, IL-10	↑ TLR2 and TLR4 = βGR or MR	[39]
Flammulina velutipes extract	Monocytes Macrophages	↑ Cytokine production ↑ Phagocytosis ↑ ROS		[40]
Agaricus blazei Murill extract (1–15%)	Monocytes	↑ IL-8, TNFα, IL-1β, IL-6		[41]
Agaricus blazei Murill extract (0.5–15%)	Monocytes	↑ Phagocytosis	↑ CD11b ↓ CD62L	[42]
Pleurotus citrinopileatus polysaccharide (PCPS, 0.5 μg/mL)	Monocytes Macrophages	Differentiation of monocytes toward macrophages (IFNγ + LPS) with reduced proinflammatory capacity: ↓ TNFα, IL-6 and CCR2 mRNA ↑ IL-10, CCL2 and CCL8 mRNA	Dectin-1 and TLR2 signaling pathways	[43]
Piptoporus betulinus extract	Monocytes	↓ Apoptosis ↑ IL-8		[44]
	MDDC	↑ Maturation ↑ IL-8		
Pleurotus citrinopileatus polysaccharide (PCPS, 0.01–5 μg/mL)	MDDC	↑ CD80, CD86, HLA-DR ↑ Pro- and anti-inflammatory cytokines (TNFα, IL-1β, IL-6, IL-12, IL-10) ↑ mRNA: CCL2, CCL3, CCL8, CXCL9, CXCL10, and LTA	Dectin-1, TLR2 and TLR4 signaling pathways	[45]
Armillariella mellea water-soluble components (2–20 μg/mL)	MDDC	↑ CD80, CD83, CD86, MHC class I and II, CD205 ↓ CD206 ↓ Endocytic capacity = TNFα, IL-12, IL-10		[46]
Hericium erinaceum water-soluble components (2–20 μg/mL)	MDDC	↑ CD80, CD83, CD86, MHC class I and II, CD205 ↓ CD206 ↓ Endocytic capacity = TNFα, IL-12		[47]
Agaricus blazei Murill Extract (10% ABM = 2.8 g of β-glucan/100 g)	MDDC	↑ IL-8, G-CSF, TNFα, IL-1β, IL-6, CCL4		[48]
Various higher Basidiomycetes exctracts (0.0005–5 mg/mL)	Neutrophils	↑ ROS		[49]
Pleurotus ostreatus β-glucans extracted from fruiting bodies (5 mg/mL)	NK	↑ Cytotoxic effects against lung and breast cancer cell lines	↑ KIR2DL genes ↑ NKG2D, IFNγ, NO	[50]
Grifola frondosa polysaccharide (10 mg/L), Lentinan (500 mg/L), Yeast glucan (100 mg/L)	NK	↑ Cytotoxicity, IFNγ, perforin secretion	↑ NKp30 expression	[51]
Saccaromyces cerevisiae glucan from baker's yeast and zymosan (10 or 100 μg/mL)	Macrophages	↑ IL-1β transcription and secretion	Dectin 1/Syk signaling pathway NLPR3 activation	[52]

Table 1. *Cont.*

Compound (Concentration Range [1])	Cell Type	Effects	Molecular Mechanisms	Refs
Saccaromyces cerevisiae whole β-glucan particles (100 μg/mL)	MDDC	↑ CD40, CD86, HLA-DR ↑ IL12, IL-2, TNF, IFNγ ↑ CD8 T cell priming ↑ Tumor-specific CTL activity	PI3K/Akt signalling	[53]
Saccaromyces cerevisiae baker's yeast	MDDC	↑ Th17 cells ↑ Adhesion and migration	IL-1α, IFNγ	[54]
Saccharomyces cerevisiae zymosan (1 mg/mL)	MDDC	↑ IL-23	LTB4, PAF	[55]
Saccharomyces cerevisiae zymosan (1 mg/mL)	MDDC Macrophages	↑ p-STAT3 ↑ mRNA: IL-10, IL-23, INF1B, CSF1, CSF2 and CSF3	PGE2	[56]
Saccaromyces cerevisiae Imprime PGG (10 μg/mL)	Monocytes	↑ ADCP ↑ C5a ↑ IL-8, CCL2 ↑ CD11b ↓ CD62L, CD88, CXCR2 ↑ Phenotypic and functional activation	Formation of an immune complex with naturally occurring ABA	[57]
	Neutrophils	↑ ROS ↑ IL-8, CCL2 ↑ CD11b ↓ CD62L, CD88, CXCR2		
Saccaromyces cerevisiae Betafectin PGG (0–300 μg/mL)	Neutrophils	↑ Chemotaxis toward C5a ↓ Chemotaxis toward IL-8	CR3-dependent	[58]
Barley polysaccharides (100 μg/mL)	MDDC	↑ Phenotypic and functional maturation of DC ↑ IL-12, IL-10		[59]
Barley β-glucan	Umbilical cord blood-generated DC	↓ CCL2 ↑ CD83 cells		[60]

Abbreviations: anti-β glucan antibodies, ABA; antibody-dependent cellular phagocytosis, ADCP; β-glucan receptor, βGR; cytotoxic T lymphocyte, CTL; dendritic cells, DC; Killer immunoglobulin receptor, KIR; Lymphotoxin a, LTA; leukotriene B4, LTB4; monocyte-derived dendritic cells, MDDC; mannose receptor, MR; natural killer, NK; nitric oxide, NO; platelet-activating factor, PAF; prostaglandin E2, PGE2; reactive oxygen species, ROS. [1] β-glucan concentrations are shown when available.

As shown in Table 1, biologically active fungal β-glucans found in edible mushrooms (e.g., *Agaricus brasiliensis*, *Pleurotus citrinopileatus*, *Agaricus blazei*, *Flammulina velutipes*) can enhance inflammatory cytokine production in monocytes [39–41,43,44] as well as in monocyte-derived macrophages (MDM) [40]. Furthermore, Minato and colleagues have recently explored the functional effect of *Pleurotus citrinopileatus* polysaccharide (PCPS) on monocyte-to-macrophage differentiation, finding that PCPS can direct Dectin-1 and TLR2-mediated differentiation of monocytes toward a macrophage cell population with reduced proinflammatory capacity [43]. Similarly, the function of DC can be influenced by β-glucans from mushroom extracts in terms of secretion of proinflammatory cytokines/chemokines [45,48], through Dectin-1 signaling and Syk/Raf-1-dependent pathways [45]. In addition, it has been reported that β-glucans derived from *Armillariella mellea*, *Hericium erinaceum* and *Piptoporus betulinus*, enhance phenotypic and functional maturation of monocyte derived dendritic cells (MDDC), with significant interleukin (IL)-12 and IL-10 production, and upregulation of surface maturation markers [44–48]. In general, β-glucans of a larger size and more branching complexity have higher immunomodulating potency [59], while the effect of fruit body extracts of various higher Basidiomycetes are slightly higher than those observed for mycelium extracts of the same species [49].

Additional studies indicated that also β-glucans isolated from yeasts overall enhance the immune response by modulating proinflammatory cytokine production in macrophages via Dectin-1/Syk signaling pathway and NLPR3 inflammasome activation, as well as in MDDC via PI3K/AKT signaling pathway [52–54]. Moreover, yeast-derived β-glucans induce DC maturation, significantly increase tumor-specific cytotoxic T lymphocyte (CTL) responses [53] and program DC to express cell adhesion and migration mediators, antimicrobial molecules, and Th17-polarizing factors [54]. Furthermore, the role of secondary mediators, which include lipids, in the induction of the cytokine signature was demonstrated in DC stimulated with zymosan, a cell-wall extract from *Saccharomyces cerevisiae* that contains a β-glucan component, providing mechanistic clues that β-glucans can modulate the immune response by acting on the lipid mediator cascade and by activating STAT3 signaling [55,56]. The regulation of additional cellular responses by mushroom-derived β-glucans in monocytes and DC has been also described. Specifically, β-glucans from *Agaricus brasiliensis*, *Agaricus blazei* or *Flammulina velutipes* extracts with immunomodulatory and antitumor activities, significantly increase the adherence and phagocytosis by monocytes [39–41] and enhance ROS production in monocytes and macrophages [40], while water-soluble components from *Armillariella mellea* or *Hericium erinaceum* decrease the MDDC endocytic capacity [46,47]. Besides β-glucans from yeasts and mushrooms, also those derived from cereals influence MDDC functions. The interaction of barley β-glucan with DC enhances maturation and cytokine production, leading to changes in the balance of Th1/Th2 cytokines in autologous T cells [60]. However, the effects of this compound on DC maturation/activation are rather weak as compared to those of β-glucans derived from mushrooms [59].

Large β-glucans can be degraded by macrophages into smaller fractions that, when released, prime CR3 on neutrophils and NK cells, triggering antimicrobial activity. Indeed, poly-[1–6]-D-glucopyranosyl-[1–3]-D-glucopyranose (PGG) glucan from yeast *Saccharomyces cerevisiae* (Imprime PGG) as well as aqueous extracts from fruit bodies and mycelia of various higher Basidiomycetes, enhance ROS production and antibody-dependent cellular phagocytosis by neutrophils [49,57,61] and monocytes [57], respectively. Mechanistically, Imprime binding and functional activation of these innate effector cells require first the formation of an immune complex with naturally occurring anti-β-glucan antibodies and subsequent opsonization by complement proteins [57,62], the activation of NF-kB [61], which results in enhanced cell migration toward the chemoattractant C5a [58]. Conversely, studies investigating the biological effects of β-glucans on NK cells have been rarely performed. Nevertheless, the effects of fungal- or yeast-derived β-glucans on primary NK cells demonstrate that these polysaccharides markedly enhance NK cell cytotoxicity by stimulating interferon (IFN)γ and perforin secretion and increasing the expression of the activating receptor NKp30, with CR3 as a critical receptor [51]. Moreover, β-glucans present in *Pleurotus ostreatus* extracts have the ability to induce NK-cell mediated cytotoxicity against cancer cells and to enhance cytokine production. The cytotoxic effects are mediated by NKG2D and IFNγ upregulation and are enhanced in the presence of IL-2, while the activation for cytokine secretion is associated with upregulation of KIR2DL genes [50].

In summary, these studies suggest that β-glucans, derived from edible mushrooms, yeasts or cereals, have the potential to regulate a plethora of immune functions. β-glucans binding to specific receptors on DC and macrophages trigger their activation and maturation, enhance the production of proinflammatory cytokines that in turn stimulate the polarization toward Th1 or Th17 responses, and induce the activation of antigen-specific $CD8^+$ CTL. Furthermore, β-glucans can enhance the antimicrobial activity of neutrophils and promote NK cell cytotoxicity.

3.2. *In Vivo Effects in Healthy Subjects*

A number of human studies involving healthy volunteers have examined the effects of dietary supplementation with β-glucans of different origins on innate immune responses (Table 2).

Table 2. Main clinical studies evaluating the immunomodulatory effects of β-glucans in healthy subjects.

Compound (Concentration Range [1])	Subjects/Study Type	Control Group	β-glucan Group	Major Findings	Refs.
Pleurotus ostreatus β-glucan (Pleuran, 100 mg/day)	Regularly training athletes/Randomized	Vitamin C (n = 25)	β-glucan + vitamin C (n = 25)	↑ NK cell frequency ↑ PMN-mediated phagocytosis ↓ URTI symptom incidence	[63]
Pleurotus ostreatus β-glucan (Pleuran, 100 mg/day)	Elite athletes/Randomized	Fructose + vitamin C (n = 11)	β-glucan + vitamin C (n = 9)	Restrained high intensity PA-induced reduction of NK cell number and activity = Monocyte and granulocyte counts	[64]
Pleurotus cornucopiae water extract (β-glucan 24 mg/meal)	Healthy volunteers/ Randomized	Water, tea, oyster souce, caffeine-free coffee (n = 21)	β-glucan + water, tea, oyster souce, caffeine-free coffee (n = 20)	↑ NK cell activity ↑ Th1-type response	[65]
Agaricus blazei Murill extract (AndoSan, 60 mL/day = 5.7 g β-glucan)	Healthy volunteers/ Intervention study	None	Mushroom extract (n = 10)	↓ Intracellular ROS in monocytes and granulocytes vs baseline	[66]
Grifola frondosa extract (6 mg/kg/day)	Myelodysplastic syndrome patients/ Non randomized phase II trial	None	Mushroom extract (n = 21)	↑ Neutrophil and monocyte functions (ROS production)	[67]
Oat soluble β-glucan (5.6 g)	Trained male cyclists (on intense exercise)/ Randomized	Cornstarch + Gatorade (n = 20)	β-glucan (Oatvantge) + Gatorade (n = 20)	No rescue of NK cell activity No rescue of PNM-RBA No effect on URTI symptom incidence	[68]
Saccharomyces cerevisiae β-glucan (Purified, Imuneks, 20 mg/day)	Subjects with seasonal allergic rhinitis (allergen sensitized)/ Randomized	Nihil (n = 12)	β-glucan (n = 12)	↓ Eosinophil frequency in the nasal fluid lavage	[69]
Saccharomyces cerevisiae β-glucan (insoluble, 1 g/day)	Healthy volunteers/ Intervention study	Nihil (n = 5)	β-glucan (n = 10)	No effect on phagocyte functions (cytokine production + microbicide activity)	[70]

Abbreviations: Natural Killer, NK; Physical activity, PA; Polymorphonuclear leukocyte, PMN; Reactive oxygen species, ROS; Respiratory burst activity, RBA; Subjects' number, N; T helper, Th; Upper respiratory tract infection, URTI. [1] β-glucan concentrations are shown when available.

In keeping with the in vitro studies, these compounds were reported to affect either the abundance/frequency or the effector functions of different innate cell populations. In a study performed on regularly training athletes (from different sport disciplines) receiving a three months supplementation with insoluble, particulate β-glucan (Pleuran, Imunoglukan) extracted from the fruit bodies of *Pleurotus ostreatus*, the number of circulating NK cells significantly increased in the treated as compared to the placebo group [63]. In agreement with data from in vitro studies [49], a concomitant enhancement of polymorphonuclear leukocyte (PMN)-mediated phagocytosis was observed in treated subjects, together with a decreased incidence of upper respiratory tract infection (URTI) symptoms [63]. Moreover, a two-month oral supplementation with the same compound (Pleuran) was found to counteract the decrease of blood NK cell number and activity (NKCA), but not that of granulocytes and monocytes, induced by short-term high-intensity exercise in elite athletes [64]. This evidence points to a role of *Pleurotus ostreatus* insoluble β-glucan not only as immune modifier and adjuvant of regular physical activity but also in the prevention of immunosuppression induced by acute exhausting physical load. Moreover, these studies, together with in vitro evidence on the enhancement of NK cell cytotoxicity against tumor cells [50], highlight the peculiar feature of this compound to boost innate immunity mainly by acting on NK cells. β-glucan-mediated increase of NKCA and Th1-type immune responses, although not statistically significant, was also observed in a Japanese clinical trial in which the subjects ingested water extracts (β-glucan content 24 mg/meal) from Oyster mushroom Pleurotus *Cornucopiae* (Tomogitake) for 8 weeks [65]. In contrast, a beverage supplement containing soluble β-glucan extracted from oat did rescue neither NKCA and PMN respiratory burst activity, nor the incidence of URTI symptoms, with respect to a placebo over an 18-day intake in trained male cyclists [68,71].

Regulatory effects on blood granulocytes and monocytes (i.e., increased frequency or reduction of proinflammatory cytokine and ROS production) have been described after oral administration of extracts from the higher Basidiomycetes *Grifola frondosa* and *Agaricus blazei* Murill (AbM, AndoSan) [66,67]. However, although these extracts are characterized by a high content of β-glucan, the specific contribution of the glucan fraction to the immunomodulatory activity has not been demonstrated in these studies [66,67]. Finally, β-glucan purified from the cell wall of *Saccharomyces cerevisiae* (Imuneks) was reported to reduce some immunopathogenic processes characterizing subjects with allergic rhinitis [69]. In particular, a 12-week oral supplementation with β-glucan in allergen-sensitized subjects resulted in a significant decrease in the frequency of eosinophils, crucial effectors in allergic reactions, in the nasal lavage fluid [69]. Conversely, in a pilot intervention study on healthy volunteers, dietary supplementation with *Saccharomyces cerevisiae* insoluble β-glucan for 7 days did not affect cytokine production and microbicide activity of peripheral blood phagocytes [70].

With the exception of some studies on animal models, the impact of dietary whole mushroom consumption on immunity has been only poorly investigated in humans. In a study by Dai and colleagues, a regular consumption of dried *Lentinula edodes* mushrooms (5 or 10 g/day) for four weeks was found to significantly increase the activation and proliferative potential of innate lymphocytes, particularly γδ T and NKT cells, concomitantly with a decrease of systemic inflammatory markers [72]. Regardless of the mechanism(s) and bioactive fraction(s) responsible for this activity, these results suggest that regular dietary intake of mushrooms could improve innate immunity and dampen inflammation.

3.3. *In Vivo Effects in Cancer Patients*

The effects of dietary supplementation with β-glucans of different origins have been investigated in a number of clinical trials carried out in cancer patients either undergoing conventional anticancer therapy, or in the absence of any other concomitant therapeutic option (Table 3).

Table 3. Main clinical studies evaluating the immunomodulatory effects of β-glucans in cancer patients.

Compound (Concentration Range [1])	Cancer Type	Conventional Therapy	Treated Patients (*N*)	Major Findings	Refs.
Agaricus blazei Murill extract	Gynecological	Yes	39	↑ NK cell activity = LAK and monocyte activity ↓ Chemotherapy-induced side effects	[73]
Agaricus blazei Murill extract (AndoSan, 60 mL/day = 5.7 g β-glucan)	Multiple myeloma	Yes	19	↑ T_{reg} and pDC numbers ↑ IL-1Ra, IL-5, IL-7 ↑ Ig, KIR, HLA gene expression	[74]
Lentinula edodes mycelia extract (1.8 g/day)	Advanced breast	Yes	10	Restrained chemotherapy-induced reduction of NK and LAK cell activity and of white blood cell/neutrophil counts ↑ QOL	[75]
Lentinula edodes mycelia extract (1.8 g/day)	Breast, gastric, colorectal, esophageal	Yes	7	↑ NK cell and LAK activity ↑ QOL ↓ IAP	[76]
Lentinula edodes β-glucan Lentinan (1 mg/every other day)	Esophageal	Yes	25	↓ Chemotherapy side effects ↑ QOL ↑ IL-12, IL-2, IL-6 ↓IL-4, IL-5, IL-10	[77]
Lentinula edodes β-glucan Lentinan	Gastric	Yes	20	↑ QOL	[78]
Lentinula edodes β-glucan Lentinan (2 mg/Kg/week)	Unresectable or recurrent gastric	Yes	147	= Leukocyte and neutrophil counts = Side-effects = QOL	[79]
Yeast β-glucan (Purified, Imuneks, 20 mg/day)	Advanced breast	Yes	15	Restrained chemotherapy-induced reduction of white blood cells = Neutrophil and monocyte counts ↑ IL-12 ↓IL-4 ↑ QOL	[80,81]
Yeast β-glucan (Purified, Imuneks, 20 mg/day)	Advanced breast	Yes	8	↑ $CD14^+$ monocyte number ↑ CD95 and CD45RA expression in monocytes	[82]
Agaricus bisporus powder (4–14 g/day)	Recurrent prostate	No	36	↓ MDSC numbers ↑ IL-15	[83]
Grifola frondosa D-Fraction (40–150 mg/day)	Advanced lung and breast	No	10	↑ NK cell activity	[84]
Grifola frondosa D-Fraction (0.1–5 mg/twice/day	Breast	No	34	↑ NKT and T_{reg} cell numbers ↑ Response of ex-vivo immune cells	[85]
Yeast β-glucan (500 mg/day)	Newly diagnosed NSCLC	No	23	↓ MDSC numbers	[86]

Abbreviations: Patients' number, N; Immunoglobulin, Ig; Immunosuppressive acidic protein, IAP; Killer immunoglobulin receptor, KIR; Lymphokine-activated killer, LAK; Myeloid-derived suppressor cell, MDSC; Natural killer, NK; Non-small cell lung carcinoma, NSCLC; Plasmacytoid DC, pDC; Quality of life, QOL; Regulatory T, T_{reg}. [1] β-glucan concentrations are shown when available.

The majority of these studies investigated changes in patients' immune system by monitoring surrogate markers such as leukocyte counts or cytokine profiles as well as by measuring chemotherapy-associated side effects and indexes of quality of life (QOL). One of the most prominent effect described for the oral administration of β-glucans is the attenuation of chemotherapy-induced fall of leukocyte counts paralleling numerical or functional changes of some specific immune cell populations. As shown in Table 3, *Lentinula edodes* extracts and yeast-derived β-glucan (Imuneks) counteract the chemotherapy induced decrease in NK and lymphokine-activated killer (LAK) cell activity when administered to advanced breast cancer patients [75,80]. Likewise, NK, NKT and LAK cell cytotoxic activity was reported to be higher in cancer patients administered with *Agaricus blazei* or *Lentinula edodes* extracts [73,75,76]. Conversely, the administration of *Agaricus blazei* Murill extracts (AndoSan) to multiple myeloma patients undergoing high dose chemotherapy resulted in the expansion of different immune cell populations such as regulatory T (T_{reg}) cells and plasmacytoid DC, and in enhanced serum levels of IL-Ra, IL-5 and IL-7. Furthermore, the expression of immunoglobulin and NK cell-related genes was also observed [74]. The capacity of β-glucans to numerically/functionally influence specific subsets of immune cells has been also documented by studies assessing the effects of yeast-derived β-glucan (Imuneks) in patients affected by advanced breast cancer. These studies reported that β-glucan stimulates the proliferation and activation of $CD14^+$ monocytes expressing high levels of CD95 and CD45RA surface antigens [82].

The ability of β-glucan to counteract, at least in part, the negative effect of chemotherapy on blood cell counts mainly relies on its hematopoiesis-enhancing capacity. In this regard, administration of carboxymethyl-glucan (CM-G), a water-soluble derivative of *Saccharomyces cerevisiae*, to advanced prostate cancer patients, has been reported to enhance hematopoietic regeneration. After CM-G administration, the total leukocyte, red blood cell and platelet counts, as well as hematocrit and hemoglobin increased significantly independently of changes in the lifestyle habits of patients [87].

Numerical and functional changes of immune cell populations have been also observed following β-glucan supplementation to cancer patients in the absence of any other therapeutic option, either in newly diagnosed patients or in those who have already completed their therapeutic pathway. In this regard, in non small cell lung carcinoma patients, the administration of yeast-derived particulate β-glucan soon after diagnosis decreases the frequency of circulating $CD33^+HLA\text{-}DR^-$ myeloid-derived suppressor cells (MDSC) with improved effector function [86]. A similar decrease in the percentage of $CD33^+HLA\text{-}DR^-$ MDSC was also observed in a phase I trial involving patients with biochemically recurrent prostate cancer following supplementation with *Agaricus bisporus* powder [83]. Other clinical studies highlighted the effects of the administration of β-glucan enriched D-Fraction, extracted from the fruit body of the Maitake mushroom *Grifola frondosa*, to either advanced lung or breast cancer patients [84] as well as to breast cancer patients free of disease after initial treatments [85]. While in patients at advanced stages of disease the administration of D-fraction promoted NK cell activation [84], in those free of disease the main changes were observed at the level of $CD3^+$ $CD56^+$ NKT and $CD4^+$ $CD25^+$ T_{reg} cells, whose frequency was increased in the β-glucan treated group [85].

Some studies also reported modulations of cytokine serum levels as well as functional changes of ex-vivo immune cell populations upon in vitro stimulation. In particular, increased secretion of IL-12 paralleling a decrease in IL-4 content was observed in advanced breast cancer patients administered with Imuneks [80] as well as in esophageal cancer patients supplemented with Lentinan, a purified β-glucan from *Lentinula edodes* [77]. Interestingly, a significant increase in the serum content of IL-15, a cytokine promoting NK cell differentiation and optimizing NK cell function, was detected upon supplementation with *Agaricus bisporus* [83]. Likewise, functional changes were observed in ex-vivo immune cells from breast cancer patients free of disease including granulocyte response to phorbol myristate acetate (PMA) stimulation, IL-10 production by PMA stimulated $CD14^+$ cells, and constitutive IL-2 secretion by $CD3^+$ $CD56^+$ NKT cells [85].

The most promising evidence to date in clinical trials has come from studies investigating the benefits of β-glucan, in combination with conventional cancer treatment, on therapy side effects, QOL and survival of cancer patients. Indeed, a significant reduction of chemotherapy-associated side effects such as loss of appetite, alopecia, emotional instability, and general weakness was reported in β-glucan supplemented patients, leading to a general improvement of QOL [73,75–78,81]. In this regard, a recent systematic review of clinical applications of Lentinan in treating lung cancer during the past 12 years in China reported a solid effect of Lentinan on improving QOL and on promoting the efficacy of chemotherapy and radiation therapy [88]. In contrast with these results, Yoshino and colleagues failed to detect any change in leukocyte/neutrophil counts or any improvement in chemotherapy-induced side effects and QOL in unresectable or recurrent gastric cancer patients upon administration of Lentinan. These discrepancies likely rely not only on differences in the compound administered but also in the type of cancer and related chemotherapeutic options. However, it is of interest that a subpopulation of patients with Lentinan-binding monocytes ≥ 2% in the peripheral blood showed a longer survival time when this compound was administered together with chemotherapy, suggesting that Lentinan-binding monocytes could represent a potential biomarker predicting chemotherapy response [79].

Taken together, these observations suggest that β-glucan provides an additional alternative therapeutic modality to maintain innate immunity cell activity and, in particular, to reduce many severe side effects caused by chemotherapy in cancer patients by stimulating hematopoiesis and assisting with an overall improvement in quality of life.

4. Conclusions

Nowadays, dietary β-glucans consumed in a daily diet or as a food supplement are well studied for their multifaceted activities with important implications for human health. Multiple effects have been reported for these compounds, including metabolic and anti-obesity activities, anti-osteoporotic, anti-allergic and anti-inflammatory effects, promotion of hematopoiesis, and regulation of gut commensal flora (Figure 1). The anticarcinogenic effects of β-glucans are clearly evidenced in the scientific literature and include: (i) the direct control of cancer cell growth, (ii) the modulation of TME, both by bridging the innate and adaptive arms of the immune system and by orienting immunosuppressive cells toward an immunostimulatory phenotype, (iii) their synergism with conventional anticancer therapy. In this regard, preclinical studies have demonstrated that the administration of certain β-glucans can effectively manipulate TME, for example by inducing the phenotypic conversion of tumor associated macrophages, leading to a significant reduction of primary tumor and distant metastases [89,90].

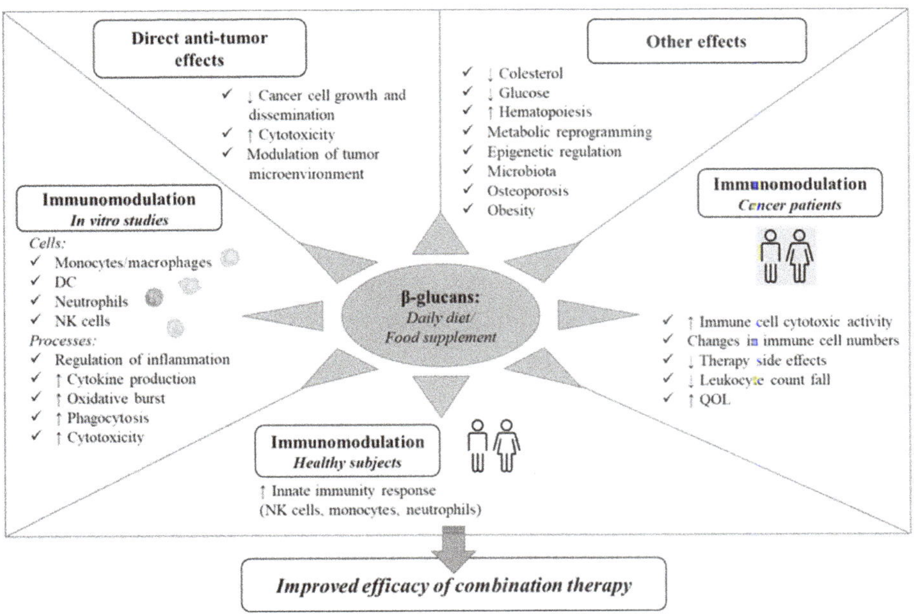

Figure 1. A schematic representation of the potential role of β-glucan in the control of cancer.

In this review, we focused on the immunomodulatory effects of β-glucans indirectly affecting tumor development/progression that may be relevant for their potential application in the fight against cancer in combination therapies. As schematically represented in Figure 1, multiple innate immunity cell subsets and biological processes are modulated by β-glucans, ranging from secretion of immune mediators to induction of phagocytosis and cytotoxic functions. Furthermore, the capacity of these compounds to attenuate chemotherapy-induced side effects, to promote hematopoiesis and to improve QOL has been reported in cancer patients. The ensemble of effects on innate immunity cells, key players in cancer surveillance, observed in vitro as well as in preclinical and clinical studies, clearly highlights the potential of β-glucans to modulate key processes involved in carcinogenesis and immune-mediated control of cancer, and make these compounds attractive as adjuvants in cancer therapies.

Nevertheless, the emerging success of combinatorial approaches to cancer treatment involving Imprime, a yeast-derived soluble β-glucan in clinical development for cancer, as an intravenously administered immunotherapy, in combination with immune checkpoint inhibitors (CPI), expanded the clinical benefit of CPI therapy by enhancing the anticancer immune response. The combinatorial use of β-glucan with immunotherapy or conventional cancer therapy is beginning to show great promise in improving patient morbidity and mortality, strongly suggesting that β-glucans may play an essential role in future strategies to inhibit tumor growth and to improve the clinical responses to current therapies. To this aim, studies that deeply characterize how β-glucans with different structures and size interact with specific receptors and regulated molecular pathways—as well as the use of β-glucan molecules with precisely defined biochemical properties—are necessary for the full exploitation of these food-derived biological response modifiers in the fight against cancer.

Author Contributions: M.D.C., S.G. and L.C. contributed to the conception, writing, and editing of this manuscript. All authors have read and agreed to the published version of the manuscript.

Funding: This research was funded the Italian Association for Cancer Research (AIRC), IG2013 grant number 14185 to SG.

Conflicts of Interest: The authors declare no conflict of interest.

References

1. Bindea, G.; Mlecnik, B.; Tosolini, M.; Kirilovsky, A.; Waldner, M.; Obenauf, A.C.; Angell, H.; Fredriksen, T.; Lafontaine, L.; Berger, A.; et al. Spatiotemporal dynamics of intratumoral immune cells reveal the immune landscape in human cancer. *Immunity* **2013**, *39*, 782–795. [CrossRef] [PubMed]
2. Gentles, A.J.; Newman, A.M.; Liu, C.L.; Bratman, S.V.; Feng, W.; Kim, D.; Nair, V.S.; Xu, Y.; Khuong, A.; Hoang, C.D.; et al. The prognostic landscape of genes and infiltrating immune cells across human cancers. *Nat. Med.* **2015**, *21*, 938–945. [CrossRef] [PubMed]
3. Wculek, S.K.; Cueto, F.J.; Mujal, A.M.; Melero, I.; Krummel, M.F.; Sancho, D. Dendritic cells in cancer immunology and immunotherapy. *Nat. Rev. Immunol.* **2019**, 10.1038/s41577-019-0210-z. [CrossRef] [PubMed]
4. Ribeiro Franco, P.I.; Rodrigues, A.P.; de Menezes, L.B.; Pacheco Miguel, M. Tumor microenvironment components: Allies of cancer progression. *Pathol. Res. Pract.* **2019**, 152729. [CrossRef]
5. Lei, X.; Lei, Y.; Li, J.K.; Du, W.X.; Li, R.G.; Yang, J.; Li, J.; Li, F.; Tan, H.B. Immune cells within the tumor microenvironment: Biological functions and roles in cancer immunotherapy. *Cancer Lett.* **2019**. [CrossRef]
6. Woo, S.R.; Corrales, L.; Gajewski, T.F. The STING pathway and the T cell-inflamed tumor microenvironment. *Trends Immunol.* **2015**, *36*, 250–256. [CrossRef]
7. Mayne, S.T.; Playdon, M.C.; Rock, C.L. Diet, nutrition, and cancer: Past, present and future. *Nat. Rev. Clin. Oncol.* **2016**, *13*, 504–515. [CrossRef]
8. Fuller, S.; Beck, E.; Salman, H.; Tapsell, L. New Horizons for the Study of Dietary Fiber and Health: A Review. *Plant Foods Hum. Nutr.* **2016**, *71*, 1–12. [CrossRef]
9. Bashir, K.M.I.; Choi, J.S. Clinical and Physiological Perspectives of beta-Glucans: The Past, Present, and Future. *Int. J. Mol. Sci.* **2017**, *18*, 1906. [CrossRef]

10. Veronese, N.; Solmi, M.; Caruso, M.G.; Giannelli, G.; Osella, A.R.; Evangelou, E.; Maggi, S.; Fontana, L.; Stubbs, B.; Tzoulaki, I. Dietary fiber and health outcomes: An umbrella review of systematic reviews and meta-analyses. *Am. J. Clin. Nutr.* **2018**, *107*, 436–444. [CrossRef]
11. Nakashima, A.; Yamada, K.; Iwata, O.; Sugimoto, R.; Atsuji, K.; Ogawa, T.; Ishibashi-Ohgo, N.; Suzuki, K. beta-Glucan in Foods and Its Physiological Functions. *J. Nutr. Sci. Vitaminol.* **2018**, *64*, 8–17. [CrossRef] [PubMed]
12. Ahmad, A.; Anjum, F.M.; Zahoor, T.; Nawaz, H.; Dilshad, S.M. Beta glucan: A valuable functional ingredient in foods. *Crit. Rev. Food Sci. Nutr.* **2012**, *52*, 201–212. [CrossRef] [PubMed]
13. Sukhithasri, V.; Nisha, N.; Biswas, L.; Anil Kumar, V.; Biswas, R. Innate immune recognition of microbial cell wall components and microbial strategies to evade such recognitions. *Microbiol. Res.* **2013**, *168*, 396–406. [CrossRef] [PubMed]
14. Zekovic, D.B.; Kwiatkowski, S.; Vrvic, M.M.; Jakovljevic, D.; Moran, C.A. Natural and modified (1–>3)-beta-D-glucans in health promotion and disease alleviation. *Crit. Rev. Biotechnol.* **2005**, *25*, 205–230. [CrossRef]
15. Jin, Y.; Li, P.; Wang, F. beta-glucans as potential immunoadjuvants: A review on the adjuvanticity, structure-activity relationship and receptor recognition properties. *Vaccine* **2018**, *36*, 5235–5244. [CrossRef]
16. Zhu, F.; Du, B.; Bian, Z.; Xu, B. Beta-glucans from edible and medicinal mushrooms: Characteristics, physicochemical and biological activities. *J. Food Compos. Anal.* **2015**, *41*, 165–173. [CrossRef]
17. Batbayar, S.; Lee, D.H.; Kim, H.W. Immunomodulation of Fungal beta-Glucan in Host Defense Signaling by Dectin-1. *Biomol. Ther.* **2012**, *20*, 433–445. [CrossRef]
18. Rice, P.J.; Adams, E.L.; Ozment-Skelton, T.; Gonzalez, A.J.; Goldman, M.P.; Lockhart, B.E.; Barker, L.A.; Breuel, K.F.; Deponti, W.K.; Kalbfleisch, J.H.; et al. Oral delivery and gastrointestinal absorption of soluble glucans stimulate increased resistance to infectious challenge. *J. Pharmacol. Exp. Ther.* **2005**, *314*, 1079–1086. [CrossRef]
19. Hong, F.; Yan, J.; Baran, J.T.; Allendorf, D.J.; Hansen, R.D.; Ostroff, G.R.; Xing, P.X.; Cheung, N.K.; Ross, G.D. Mechanism by which orally administered beta-1,3-glucans enhance the tumoricidal activity of antitumor monoclonal antibodies in murine tumor models. *J. Immunol.* **2004**, *173*, 797–806. [CrossRef]
20. Sandvik, A.; Wang, Y.Y.; Morton, H.C.; Aasen, A.O.; Wang, J.E.; Johansen, F.E. Oral and systemic administration of beta-glucan protects against lipopolysaccharide-induced shock and organ injury in rats. *Clin. Exp. Immunol.* **2007**, *148*, 168–177. [CrossRef]
21. Legentil, L.; Paris, F.; Ballet, C.; Trouvelot, S.; Daire, X.; Vetvicka, V.; Ferrieres, V. Molecular Interactions of beta-(1–>3)-Glucans with Their Receptors. *Molecules* **2015**, *20*, 9745–9766. [CrossRef] [PubMed]
22. Taylor, P.R.; Brown, G.D.; Reid, D.M.; Willment, J.A.; Martinez-Pomares, L.; Gordon, S.; Wong, S.Y. The beta-glucan receptor, dectin-1, is predominantly expressed on the surface of cells of the monocyte/macrophage and neutrophil lineages. *J. Immunol.* **2002**, *169*, 3876–3882. [CrossRef] [PubMed]
23. Li, X.; Zhang, X.; Pang, L.; Yao, L.; ShangGuan, Z.; Pan, Y. Agaricus bisporus-derived beta-glucan enter macrophages and adipocytes by CD36 receptor. *Nat. Prod. Res.* **2019**, 1–4. [CrossRef]
24. Li, S.S.; Ogbomo, H.; Mansour, M.K.; Xiang, R.F.; Szabo, L.; Munro, F.; Mukherjee, P.; Mariuzza, R.A.; Amrein, M.; Vyas, J.M.; et al. Identification of the fungal ligand triggering cytotoxic PRR-mediated NK cell killing of Cryptococcus and Candida. *Nat. Commun.* **2018**, *9*, 751. [CrossRef]
25. de Graaff, P.; Govers, C.; Wichers, H.J.; Debets, R. Consumption of beta-glucans to spice up T cell treatment of tumors: A review. *Expert Opin. Biol. Ther.* **2018**, *18*, 1023–1040. [CrossRef]
26. Geller, A.; Shrestha, R.; Yan, J. Yeast-Derived beta-Glucan in Cancer: Novel Uses of a Traditional Therapeutic. *Int. J. Mol. Sci.* **2019**, *20*, 3618. [CrossRef]
27. Pan, P.; Huang, Y.W.; Oshima, K.; Yearsley, M.; Zhang, J.; Arnold, M.; Yu, J.; Wang, L.S. The immunomodulatory potential of natural compounds in tumor-bearing mice and humans. *Crit. Rev. Food Sci. Nutr.* **2019**, *59*, 992–1007. [CrossRef]
28. Vetvicka, V.; Vannucci, L.; Sima, P. beta-glucan as a new tool in vaccine development. *Scand. J. Immunol.* **2019**, e12833. [CrossRef]
29. Halstenson, C.E.; Shamp, T.; Gargano, M.A.; Walsh, R.M.; Patchen, M.L. Two randomized, double-blind, placebo-controlled, dose-escalation phase 1 studies evaluating BTH1677, a 1, 3-1,6 beta glucan pathogen associated molecular pattern, in healthy volunteer subjects. *Investig. New Drug* **2016**, *34*, 202–215. [CrossRef]

30. Zent, C.S.; Call, T.G.; Bowen, D.A.; Conte, M.J.; LaPlant, B.R.; Witzig, T.E.; Ansell, S.M.; Weiner, G.J. Early treatment of high risk chronic lymphocytic leukemia with alemtuzumab, rituximab and poly-(1-6)-beta-glucotriosyl-(1-3)-beta- glucopyranose beta- glucan is well tolerated and achieves high complete remission rates. *Leuk Lymphoma* **2015**, *56*, 2373–2378. [CrossRef]
31. Weitberg, A.B. A phase I/II trial of beta-(1,3)/(1,6) D-glucan in the treatment of patients with advanced malignancies receiving chemotherapy. *J. Exp. Clin. Cancer Res.* **2008**, 27. [CrossRef] [PubMed]
32. Lehne, G.; Haneberg, B.; Gaustad, P.; Johansen, P.W.; Preus, H.; Abrahamsen, T.G. Oral administration of a new soluble branched beta-1,3-D-glucan is well tolerated and can lead to increased salivary concentrations of immunoglobulin A in healthy volunteers. *Clin. Exp. Immunol.* **2006**, *143*, 65–69. [CrossRef] [PubMed]
33. Gordon, M.; Bihari, B.; Goosby, E.; Gorter, R.; Greco, M.; Guralnik, M.; Mimura, T.; Rudinicki, V.; Wong, R.; Kaneko, Y. A placebo-controlled trial of the immune modulator, lentinan, in HIV-positive patients: A phase I/II trial. *J. Med.* **1998**, *29*, 305–330. [PubMed]
34. Ohno, S.; Sumiyoshi, Y.; Hashine, K.; Shirato, A.; Kyo, S.; Inoue, M. Phase I Clinical Study of the Dietary Supplement, Agaricus blazei Murill, in Cancer Patients in Remission. *Evid. Based Compl. Alt.* **2011**, 1–9. [CrossRef]
35. Synytsya, A.; Novak, M. Structural diversity of fungal glucans. *Carbohydr. Polym.* **2013**, *92*, 792–809. [CrossRef] [PubMed]
36. Mueller, A.; Raptis, J.; Rice, P.J.; Kalbfleisch, J.H.; Stout, R.D.; Ensley, H.E.; Browder, W.; Williams, D.L. The influence of glucan polymer structure and solution conformation on binding to (1 -> 3)-beta-D-glucan receptors in a human monocyte-like cell line. *Glycobiology* **2000**, *10*, 339–346. [CrossRef]
37. Adams, E.L.; Rice, P.J.; Graves, B.; Ensley, H.E.; Yu, H.; Brown, G.D.; Gordon, S.; Monteiro, M.A.; Papp-Szabo, E.; Lowman, D.W.; et al. Differential high-affinity interaction of dectin-1 with natural or synthetic glucans is dependent upon primary structure and is influenced by polymer chain length and side-chain branching. *J. Pharmacol. Exp. Ther.* **2008**, *325*, 115–123. [CrossRef]
38. Williams, D.L. Overview of (1–>3)-beta-D-glucan immunobiology. *Mediat. Inflamm.* **1997**, *6*, 247–250. [CrossRef]
39. Martins, P.R.; de Campos Soares, A.M.V.; da Silva Pinto Domeneghini, A.V.; Golim, M.A.; Kaneno, R. Agaricus brasiliensis polysaccharides stimulate human monocytes to capture Candida albicans, express toll-like receptors 2 and 4, and produce pro-inflammatory cytokines. *J. Venom. Anim. Toxins Incl. Trop. Dis.* **2017**, *23*, 17. [CrossRef]
40. Kashina, S.; Villavicencio, L.L.; Zaina, S.; Ordaz, M.B.; Sabanero, G.B.; Fujiyoshi, V.T.; Lopez, M.S. Activity of Extracts from Submerged Cultured Mycelium of Winter Mushroom, Flammulina velutipes (Agaricomycetes), on the Immune System In Vitro. *Int. J. Med. Mushrooms* **2016**, *18*, 49–57. [CrossRef]
41. Bernardshaw, S.; Hetland, G.; Ellertsen, L.K.; Tryggestad, A.M.; Johnson, E. An extract of the medicinal mushroom Agaricus blazei Murill differentially stimulates production of pro-inflammatory cytokines in human monocytes and human vein endothelial cells in vitro. *Inflammation* **2005**, *29*, 147–153. [CrossRef] [PubMed]
42. Bernardshaw, S.; Lyberg, T.; Hetland, G.; Johnson, E. Effect of an extract of the mushroom Agaricus blazei Murill on expression of adhesion molecules and production of reactive oxygen species in monocytes and granulocytes in human whole blood ex vivo. *Apmis* **2007**, *115*, 719–725. [CrossRef]
43. Minato, K.I.; Laan, L.C.; van Die, I.; Mizuno, M. Pleurotus citrinopileatus polysaccharide stimulates anti-inflammatory properties during monocyte-to-macrophage differentiation. *Int. J. Biol. Macromol.* **2019**, *122*, 705–712. [CrossRef] [PubMed]
44. Grunewald, F.; Steinborn, C.; Huber, R.; Wille, R.; Meier, S.; Alresly, Z.; Lindequist, U.; Grundemann, C. Effects of Birch Polypore Mushroom, Piptoporus betulinus (Agaricomycetes), the "Iceman's Fungus", on Human Immune Cells. *Int. J. Med. Mushrooms* **2018**, *20*, 1135–1147. [CrossRef] [PubMed]
45. Minato, K.I.; Laan, L.C.; Ohara, A.; van Die, I. Pleurotus citrinopileatus polysaccharide induces activation of human dendritic cells through multiple pathways. *Int. Immunopharmacol.* **2016**, *40*, 156–163. [CrossRef] [PubMed]
46. Kim, S.K.; Im, J.; Yun, C.H.; Son, J.Y.; Son, C.G.; Park, D.K.; Han, S.H. Armillariella mellea induces maturation of human dendritic cells without induction of cytokine expression. *J. Ethnopharmacol.* **2008**, *119*, 153–159. [CrossRef]

47. Kim, S.K.; Son, C.G.; Yun, C.H.; Han, S.H. Hericium erinaceum induces maturation of dendritic cells derived from human peripheral blood monocytes. *Phytother. Res.* **2010**, *24*, 14–19. [CrossRef]
48. Forland, D.T.; Johnson, E.; Tryggestad, A.M.; Lyberg, T.; Hetland, G. An extract based on the medicinal mushroom Agaricus blazei Murill stimulates monocyte-derived dendritic cells to cytokine and chemokine production in vitro. *Cytokine* **2010**, *49*, 245–250. [CrossRef]
49. Shamtsyan, M.; Konusova, V.; Maksimova, Y.; Goloshchev, A.; Panchenko, A.; Simbirtsev, A.; Petrishchev, N.; Denisova, N. Immunomodulating and anti-tumor action of extracts of several mushrooms. *J. Biotechnol.* **2004**, *113*, 77–83. [CrossRef]
50. El-Deeb, N.M.; El-Adawi, H.I.; El-Wahab, A.E.A.; Haddad, A.M.; El Enshasy, H.A.; He, Y.W.; Davis, K.R. Modulation of NKG2D, KIR2DL and Cytokine Production by Pleurotus ostreatus Glucan Enhances Natural Killer Cell Cytotoxicity Toward Cancer Cells. *Front. Cell Dev. Biol.* **2019**, *7*, 165. [CrossRef]
51. Huyan, T.; Li, Q.; Yang, H.; Jin, M.L.; Zhang, M.J.; Ye, L.J.; Li, J.; Huang, Q.S.; Yin, D.C. Protective effect of polysaccharides on simulated microgravity-induced functional inhibition of human NK cells. *Carbohydr. Polym.* **2014**, *101*, 819–827. [CrossRef] [PubMed]
52. Kankkunen, P.; Teirila, L.; Rintahaka, J.; Alenius, H.; Wolff, H.; Matikainen, S. (1,3)-beta-glucans activate both dectin-1 and NLRP3 inflammasome in human macrophages. *J. Immunol.* **2010**, *184*, 6335–6342. [CrossRef] [PubMed]
53. Ding, J.; Feng, T.; Ning, Y.; Li, W.; Wu, Q.; Qian, K.; Wang, Y.; Qi, C. beta-Glucan enhances cytotoxic T lymphocyte responses by activation of human monocyte-derived dendritic cells via the PI3K/AKT pathway. *Hum. Immunol.* **2015**, *76*, 146–154. [CrossRef] [PubMed]
54. Cardone, M.; Dzutsev, A.K.; Li, H.; Riteau, N.; Gerosa, F.; Shenderov, K.; Winkler-Pickett, R.; Provezza, L.; Riboldi, E.; Leighty, R.M.; et al. Interleukin-1 and interferon-gamma orchestrate beta-glucan-activated human dendritic cell programming via IkappaB-zeta modulation. *PLoS ONE* **2014**, *9*, e114516. [CrossRef]
55. Rodriguez, M.; Marquez, S.; Montero, O.; Alonso, S.; Frade, J.G.; Crespo, M.S.; Fernandez, N. Pharmacological inhibition of eicosanoids and platelet-activating factor signaling impairs zymosan-induced release of IL-23 by dendritic cells. *Biochem. Pharmacol.* **2016**, *102*, 78–96. [CrossRef]
56. Rodriguez, M.; Marquez, S.; de la Rosa, J.V.; Alonso, S.; Castrillo, A.; Sanchez Crespo, M.; Fernandez, N. Fungal pattern receptors down-regulate the inflammatory response by a cross-inhibitory mechanism independent of interleukin-10 production. *Immunology* **2017**, *150*, 184–198. [CrossRef]
57. Chan, A.S.; Jonas, A.B.; Qiu, X.; Ottoson, N.R.; Walsh, R.M.; Gorden, K.B.; Harrison, B.; Maimonis, P.J.; Leonardo, S.M.; Ertelt, K.E.; et al. Imprime PGG-Mediated Anti-Cancer Immune Activation Requires Immune Complex Formation. *PLoS ONE* **2016**, *11*, e0165909. [CrossRef]
58. Tsikitis, V.L.; Albina, J.E.; Reichner, J.S. Beta-glucan affects leukocyte navigation in a complex chemotactic gradient. *Surgery* **2004**, *136*, 384–389. [CrossRef]
59. Chan, W.K.; Law, H.K.; Lin, Z.B.; Lau, Y.L.; Chan, G.C. Response of human dendritic cells to different immunomodulatory polysaccharides derived from mushroom and barley. *Int. Immunol.* **2007**, *19*, 891–899. [CrossRef]
60. Bermudez-Brito, M.; Sahasrabudhe, N.M.; Rosch, C.; Schols, H.A.; Faas, M.M.; de Vos, P. The impact of dietary fibers on dendritic cell responses in vitro is dependent on the differential effects of the fibers on intestinal epithelial cells. *Mol. Nutr. Food Res.* **2015**, *59*, 698–710. [CrossRef]
61. Wakshull, E.; Brunke-Reese, D.; Lindermuth, J.; Fisette, L.; Nathans, R.S.; Crowley, J.J.; Tufts, J.C.; Zimmerman, J.; Mackin, W.; Adams, D.S. PGG-glucan, a soluble beta-(1,3)-glucan, enhances the oxidative burst response, microbicidal activity, and activates an NF-kappa B-like factor in human PMN: Evidence for a glycosphingolipid beta-(1,3)-glucan receptor. *Immunopharmacology* **1999**, *41*, 89–107. [CrossRef]
62. Bose, N.; Chan, A.S.; Guerrero, F.; Maristany, C.M.; Qiu, X.; Walsh, R.M.; Ertelt, K.E.; Jonas, A.B.; Gorden, K.B.; Dudney, C.M.; et al. Binding of Soluble Yeast beta-Glucan to Human Neutrophils and Monocytes is Complement-Dependent. *Front. Immunol.* **2013**, *4*, 230. [CrossRef] [PubMed]
63. Bergendiova, K.; Tibenska, E.; Majtan, J. Pleuran (beta-glucan from Pleurotus ostreatus) supplementation, cellular immune response and respiratory tract infections in athletes. *Eur. J. Appl. Physiol.* **2011**, *111*, 2033–2040. [CrossRef] [PubMed]
64. Bobovcak, M.; Kuniakova, R.; Gabriz, J.; Majtan, J. Effect of Pleuran (beta-glucan from Pleurotus ostreatus) supplementation on cellular immune response after intensive exercise in elite athletes. *Appl. Physiol. Nutr. Metab.* **2010**, *35*, 755–762. [CrossRef] [PubMed]

65. Tanaka, A.; Nishimura, M.; Sato, Y.; Sato, H.; Nishihira, J. Enhancement of the Th1-phenotype immune system by the intake of Oyster mushroom (Tamogitake) extract in a double-blind, placebo-controlled study. *J. Tradit. Complementary Med.* **2016**, *6*, 424–430. [CrossRef]
66. Johnson, E.; Forland, D.T.; Hetland, G.; Saetre, L.; Olstad, O.K.; Lyberg, T. Effect of AndoSan on expression of adhesion molecules and production of reactive oxygen species in human monocytes and granulocytes in vivo. *Scand. J. Gastroenterol.* **2012**, *47*, 984–992. [CrossRef]
67. Wesa, K.M.; Cunningham-Rundles, S.; Klimek, V.M.; Vertosick, E.; Coleton, M.I.; Yeung, K.S.; Lin, H.; Nimer, S.; Cassileth, B.R. Maitake mushroom extract in myelodysplastic syndromes (MDS): A phase II study. *Cancer Immunol. Immunother.* **2015**, *64*, 237–247. [CrossRef]
68. Nieman, D.C.; Henson, D.A.; McMahon, M.; Wrieden, J.L.; Davis, J.M.; Murphy, E.A.; Gross, S.J.; McAnulty, L.S.; Dumke, C.L. Beta-glucan, immune function, and upper respiratory tract infections in athletes. *Med. Sci. Sports Exerc.* **2008**, *40*, 1463–1471. [CrossRef]
69. Kirmaz, C.; Bayrak, P.; Yilmaz, O.; Yuksel, H. Effects of glucan treatment on the Th1/Th2 balance in patients with allergic rhinitis: A double-blind placebo-controlled study. *Eur. Cytokine Netw.* **2005**, *16*, 128–134.
70. Leentjens, J.; Quintin, J.; Gerretsen, J.; Kox, M.; Pickkers, P.; Netea, M.G. The effects of orally administered Beta-glucan on innate immune responses in humans, a randomized open-label intervention pilot-study. *PLoS ONE* **2014**, *9*, e108794. [CrossRef]
71. Nieman, D.C. Immunonutrition support for athletes. *Nutr. Rev.* **2008**, *66*, 310–320. [CrossRef] [PubMed]
72. Dai, X.; Stanilka, J.M.; Rowe, C.A.; Esteves, E.A.; Nieves, C., Jr.; Spaiser, S.J.; Christman, M.C.; Langkamp-Henken, B.; Percival, S.S. Consuming Lentinula edodes (Shiitake) Mushrooms Daily Improves Human Immunity: A Randomized Dietary Intervention in Healthy Young Adults. *J. Am. Coll. Nutr.* **2015**, *34*, 478–487. [CrossRef] [PubMed]
73. Ahn, W.S.; Kim, D.J.; Chae, G.T.; Lee, J.M.; Bae, S.M.; Sin, J.I.; Kim, Y.W.; Namkoong, S.E.; Lee, I.P. Natural killer cell activity and quality of life were improved by consumption of a mushroom extract, Agaricus blazei Murill Kyowa, in gynecological cancer patients undergoing chemotherapy. *Int. J. Gynecol. Cancer: Off. J. Int. Gynecol. Cancer Soc.* **2004**, *14*, 589–594. [CrossRef]
74. Tangen, J.M.; Tierens, A.; Caers, J.; Binsfeld, M.; Olstad, O.K.; Troseid, A.M.; Wang, J.; Tjonnfjord, G.E.; Hetland, G. Immunomodulatory effects of the Agaricus blazei Murrill-based mushroom extract AndoSan in patients with multiple myeloma undergoing high dose chemotherapy and autologous stem cell transplantation: A randomized, double blinded clinical study. *BioMed Res. Int.* **2015**, *2015*, 718539. [CrossRef]
75. Nagashima, Y.; Maeda, N.; Yamamoto, S.; Yoshino, S.; Oka, M. Evaluation of host quality of life and immune function in breast cancer patients treated with combination of adjuvant chemotherapy and oral administration of Lentinula edodes mycelia extract. *Oncotargets Ther.* **2013**, *6*, 853–859. [CrossRef]
76. Yamaguchi, Y.; Miyahara, E.; Hihara, J. Efficacy and safety of orally administered Lentinula edodes mycelia extract for patients undergoing cancer chemotherapy: A pilot study. *Am. J. Chin. Med.* **2011**, *39*, 451–459. [CrossRef]
77. Wang, J.L.; Bi, Z.; Zou, J.W.; Gu, X.M. Combination therapy with lentinan improves outcomes in patients with esophageal carcinoma. *Mol. Med. Rep.* **2012**, *5*, 745–748. [CrossRef]
78. Kataoka, H.; Shimura, T.; Mizoshita, T.; Kubota, E.; Mori, Y.; Mizushima, T.; Wada, T.; Ogasawara, N.; Tanida, S.; Sasaki, M.; et al. Lentinan with S-1 and paclitaxel for gastric cancer chemotherapy improve patient quality of life. *Hepato Gastroenterol.* **2009**, *56*, 547–550.
79. Yoshino, S.; Nishikawa, K.; Morita, S.; Takahashi, T.; Sakata, K.; Nagao, J.; Nemoto, H.; Murakami, N.; Matsuda, T.; Hasegawa, H.; et al. Randomised phase III study of S-1 alone versus S-1 plus lentinan for unresectable or recurrent gastric cancer (JFMC36-0701). *Eur. J. Cancer* **2016**, *65*, 164–171. [CrossRef]
80. Ostadrahimi, A.; Ziaei, J.E.; Esfahani, A.; Jafarabadi, M.A.; Movassaghpourakbari, A.; Farrin, N. Effect of beta glucan on white blood cell counts and serum levels of IL-4 and IL-12 in women with breast cancer undergoing chemotherapy: A randomized double-blind placebo-controlled clinical trial. *Asian Pac. J. Cancer Prev.* **2014**, *15*, 5733–5739. [CrossRef]
81. Ostadrahimi, A.; Esfahani, A.; Asghari Jafarabadi, M.; Eivazi Ziaei, J.; Movassaghpourakbari, A.; Farrin, N. Effect of Beta glucan on quality of life in women with breast cancer undergoing chemotherapy: A randomized double-blind placebo-controlled clinical trial. *Adv. Pharm. Bull.* **2014**, *4*, 471–477. [CrossRef] [PubMed]

82. Demir, G.; Klein, H.O.; Mandel-Molinas, N.; Tuzuner, N. Beta glucan induces proliferation and activation of monocytes in peripheral blood of patients with advanced breast cancer. *Int. Immunopharmacol.* **2007**, *7*, 113–116. [CrossRef] [PubMed]
83. Twardowski, P.; Kanaya, N.; Frankel, P.; Synold, T.; Ruel, C.; Pal, S.K.; Junqueira, M.; Prajapati, M.; Moore, T.; Tryon, P.; et al. A phase I trial of mushroom powder in patients with biochemically recurrent prostate cancer: Roles of cytokines and myeloid-derived suppressor cells for Agaricus bisporus-induced prostate-specific antigen responses. *Cancer* **2015**, *121*, 2942–2950. [CrossRef] [PubMed]
84. Kodama, N.; Komuta, K.; Nanba, H. Effect of Maitake (Grifola frondosa) D-Fraction on the activation of NK cells in cancer patients. *J. Med. Food* **2003**, *6*, 371–377. [CrossRef] [PubMed]
85. Deng, G.; Lin, H.; Seidman, A.; Fornier, M.; D'Andrea, G.; Wesa, K.; Yeung, S.; Cunningham-Rundles, S.; Vickers, A.J.; Cassileth, B. A phase I/II trial of a polysaccharide extract from Grifola frondosa (Maitake mushroom) in breast cancer patients: Immunological effects. *J. Cancer Res. Clin. Oncol.* **2009**, *135*, 1215–1221. [CrossRef]
86. Albeituni, S.H.; Ding, C.; Liu, M.; Hu, X.; Luo, F.; Kloecker, G.; Bousamra, M., 2nd; Zhang, H.G.; Yan, J. Yeast-Derived Particulate beta-Glucan Treatment Subverts the Suppression of Myeloid-Derived Suppressor Cells (MDSC) by Inducing Polymorphonuclear MDSC Apoptosis and Monocytic MDSC Differentiation to APC in Cancer. *J. Immunol.* **2016**, *196*, 2167–2180. [CrossRef]
87. Magnani, M.; Castro-Gomez, R.H.; Aoki, M.N.; Gregorio, E.P.; Libos, F.; Watanabe, M.A.E. Effects of carboxymethyl-glucan from Saccharomyces cerevisiae on the peripheral blood cells of patients with advanced prostate cancer. *Exp. Ther. Med.* **2010**, *1*, 859–862. [CrossRef]
88. Zhang, M.; Zhang, Y.; Zhang, L.; Tian, Q. Mushroom polysaccharide lentinan for treating different types of cancers: A review of 12 years clinical studies in China. *Prog. Mol. Biol. Transl. Sci.* **2019**, *163*, 297–328. [CrossRef]
89. Zhang, M.; Yan, L.; Kim, J.A. Modulating mammary tumor growth, metastasis and immunosuppression by siRNA-induced MIF reduction in tumor microenvironment. *Cancer Gene Ther.* **2015**, *22*, 463–474. [CrossRef]
90. Liu, M.; Luo, F.; Ding, C.; Albeituni, S.; Hu, X.; Ma, Y.; Cai, Y.; McNally, L.; Sanders, M.A.; Jain, D.; et al. Dectin-1 activation by a natural product β-glucan converts immunosuppressive macrophages into an M1-like phenotype. *J. Immunol.* **2015**, *195*, 5055–5065. [CrossRef]

cancers

Review

Reversal of Epithelial–Mesenchymal Transition by Natural Anti-Inflammatory and Pro-Resolving Lipids

Chang Hoon Lee

College of Pharmacy, Dongguk University, Seoul 100-715, Korea; uatheone@dongguk.edu; Tel.: +82-109-755-1746

Received: 23 October 2019; Accepted: 19 November 2019; Published: 21 November 2019

Abstract: Epithelial mesenchymal transition (EMT) is a key process in the progression of malignant cancer. Therefore, blocking the EMT can be a critical fast track for the development of anticancer drugs. In this paper, we update recent research output of EMT and we explore suppression of EMT by natural anti-inflammatory compounds and pro-resolving lipids.

Keywords: epithelial mesenchymal transition; inflammation; malignant cancer; natural anti-inflammatory compounds; pro-resolving lipids

1. Introduction

The epithelial–mesenchymal transition (EMT) is defined as a phenomenon that epithelial cells transform into mesenchymal cells [1]. EMT plays a key role in cancer progression and fibrosis. Many researchers and pharmaceutical companies have tried to develop novel EMT blockers due to its importance in such diseases [2–4].

Inflammation is one of ten in cancer hallmark [5]. It is a critical factor of tumor microenvironment affecting EMT. Several reviews have emphasized role of inflammation in EMT [6]. However, there are a few reviews that deal with the blocking of EMT by natural anti-inflammatory compounds and pro-resolving lipids.

In this review, I have dealt with ingredients derived from natural products that were not covered in the 2018 review on EMT [6]. I have also added the story of inflammasomes, which play an essential role in the early steps of inflammation, and how they are involved in EMT. Naturally derived compounds that control these inflammasome-related molecules in the EMT have been discussed. About the resolution of inflammation, the newly discovered pro-resolving lipids including RvTs are added and the receptors specifically acting on pro-resolving lipids have also been discussed. Understanding the action of natural anti-inflammatory compounds and pro-resolving lipids with anti-EMT activities might provide a new armory to suppress the progression of cancer.

2. EMT in Cancers

EMT shows the reduced expression of epithelial markers including E-cadherin and keratins and the increased expression of mesenchymal marker proteins such as vimentin and N-cadherin via actions of transcription factors including as SNAIL1 and ZEB1 (Figure 1) [6]. The molecular mechanism of EMT process has been well explained in the report (references in it) [1]. In this part, we will briefly update the concept that reflects recent achievements for EMT.

EMT occurs by a various mediators from tumor microenvironments via receptor through signal transduction. EMT-related transcription factors blocks the expression of epithelial cell-marker genes and evoke mesenchymal-marker genes. E-cadherin, keratin, ZO-1, miR-34, and miR-200 belong to the epithelial markers, and N-cadherin, vimentin, fibronectin, SNAIL, ZEB1, TWIST, Brachyury, Foxq1, Runx2, GATA, and SOX belong to the mesenchymal markers. P-cadherin is the marker of partial EMT. The dot triangle indicates the relative ratio of p-cadherin expression. Modified from Lee's report [6].

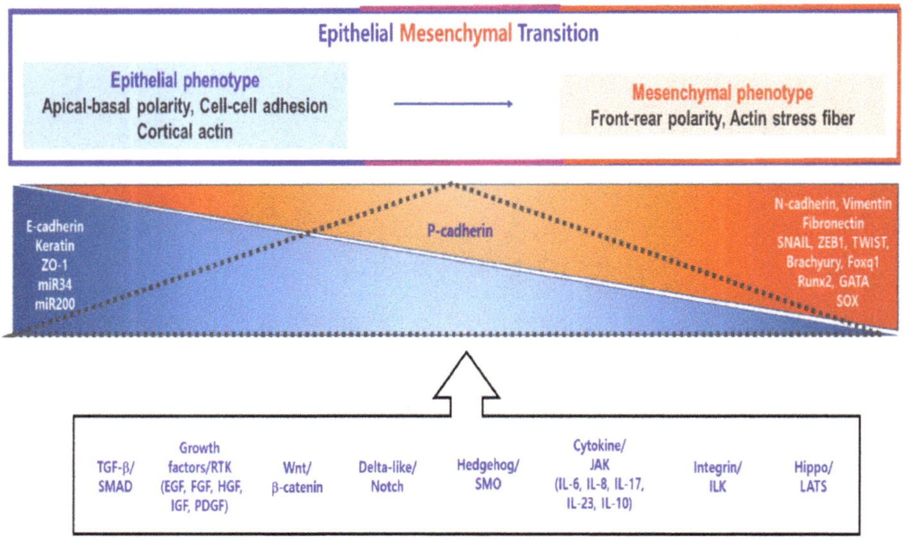

Figure 1. Epithelial–mesenchymal transition (EMT).E-cadherin: Epithelial cadherin; EGF: Epidermal Growth Factor; FGF: Fibroblast Growth Factor; GATA: GATA Binding Protein; HGF: Hepatocyte Growth Factor; IGF: Insulin Like Growth Factor; ILK: Integrin Linked Kinase; JAK: Janus Kinase; LATS: Large Tumor Suppressor Kinase; miR34: microRNA 34; miR200: microRNA 200; N-cadherin: Neural cadherin; P-cadherin: Placental cadherin; PDGF: Platelet Derived Growth Factor; Runx2: RUNX Family Transcription Factor 2; SMAD: Sma- And Mad-Related Protein; SMO: Smoothened, Frizzled Class Receptor; SOX: SRY-Box Transcription Factor; TGF-β: Transforming Growth Factor Beta; Wnt: Wingless-Type MMTV Integration Site Family; ZEB1: Zinc Finger E-Box Binding Homeobox 1; ZO-1: Zona Occludens 1.

2.1. Adaptation to New Concepts of EMT

2.1.1. Signaling Pathways in the EMT Process

A typical signaling pathway of EMT is the transforming growth factor-β1 (TGF-β1) pathway. TGF-β1 induces EMT via SMAD-dependent or non-SMAD signaling pathway [7]. Growth factors including FGF, HGF, IGF1, EGF, and PDGF via receptor tyrosine kinase can induce EMT via signaling pathway of PI3K-AKT and ERK MAPK [8–11]. Wnt signaling, hedgehog signaling, Notch signaling, hypoxia, and inflammatory tumor microenvironment also involves in EMT [5]. Recently, it has been shown that hippo signaling is also involved in EMT [12]. YAP and TAZ can enhance EMT through upregulation of EMT transcription factors such as forkhead box C2 (FOXC2), snail family zinc finger 1/2 (SNAIL1, SLUG), twist-related protein 1 (TWIST1), and ZEB1 [12–15].

2.1.2. Transcription Factors Involved in EMT

Novel players are newly recognized as regulatory transcription factor in the EMT. Brachyury, the T-box transcription factor, is a novel transcription factor implicated in the EMT of cancer cells [16]. Brachyury is known as the target gene of WNT, one of the major signaling pathways of EMT [17]. Foxq1, one of forkhead transcription factor, has also regarded as a novel transcription factor mediating the EMT of gastric cancer [6,18]. Runt-related transcription factor 2 (Runx2) belongs to the runt-related transcription factor family [19]. Runx2 plays a key role in EMT of hepatocellular carcinoma (HCC) [20]. GATA transcription factors are also implicated in the EMT of cancer cells [21]. Serine 161 and serine 187 phosphorylated GATA1 by PAK5 can promote EMT of breast cancer cells by recruiting histone deacetylase 3/4 to E-cadherin promoter [22]. Other players of EMT include SRY-box (SOX) transcription factors [23]. Sox4 acts as a master regulator in EMT of cancerous breast epithelial cells [24].

2.1.3. Partial EMT

EMT could be not defined as a dichotomous transition from epithelial status to mesenchymal one of cells [6,25].

Partial EMT with both epithelial and mesenchymal cell markers was proposed (Figure 1) [25]. Cells that have undergone with partial EMT have the capability showing collective sheet or cluster migration [26]. Partial EMTed cells have the competitiveness in that not all cells need to respond to EMT signals. Thus, they can far more efficiently execute plasticity in converting to a colonization state of metastasis via MET [25].

2.1.4. Parallelism between Cancer Stem Cell and EMT

A cancer stem cell (CSC) is a cancer cell having the ability of self-renewal and differentiation. It divides to progenitor cancer cells. It is a culprit of cancer recurrence and metastasis [27]. Several EMT transcription factors and inducers can evoke the expression of cancer stem cells markers, thereby enhancing the capability to initiate cancer, a typical characteristic of cancer stem cells [28]. TGF-β promotes the de-differentiation of human basal breast non-CSCs into CSCs via ZEB1, suggesting that the activation of EMT in cancer cells by TGF-β/ZEB1 is closely linked to the de-differentiation of cancer cells into the CSC state [29].

2.2. *Focus on Chemoresistance and Immune Evasion of EMT in Cancers*

In the course of EMT, epithelial cancer cells can lose contacts between cells and apicobasal polarity but gain enhanced migration and invasion [1,6]. EMT has also strong influences on several hallmarks of cancer including cancer initiation, immune evasion, proliferation, survival, and resistance to therapeutics [6,30]. We will briefly update EMT's roles in cancer such as chemoresistance and immune evasion.

2.2.1. Chemoresistance

Chemoresistance to anticancer chemotherapeutics implies that cancer cells can survive despite the administration of an anticancer drug in a dose that can usually kill cancer cells. A significant correlation has been found between EMT-related gene expression and chemo-resistance to anticancer therapy [6,30,31]. Although the role of EMT in metastasis is disputable by some group, EMT is crucial for anticancer drug resistance [32,33]. For example, TWIST-mediated EMT is related to sorafenib resistance to advanced HCC [34]. Therefore, EMT should be understood as a predictor of chemoresistance for anticancer drugs.

2.2.2. Immune Evasion

EMT has been understood as a key mechanism of immune escape of cancer cells. When snail1, an EMT transcription factor, is ectopically expressed in MCF7 breast cancer cells, cancer cell lysis executed by TNF-α-induced CTL is curtailed [35]. When snail1 is overexpressed in B16 melanoma cells, CTL-induced lysis is reduced and maturation of dendritic cells is inhibited while inhibitory Treg-like $CD4^+$ $Foxp3^+$ cells are expanded [36].

Enhanced EMT properties in cancer cells (A549, MCF7, and HepG2) by the TGF-β, IFN-γ, and TNF-α may affect differentiation and death of natural killer (NK), T, and B cells [37]. Reduced miR-200 and enhanced ZEB1 expression in lung cancer cells not only can evoke EMT, but also can lead to enhanced expression of PD-L1, which is related to the exhaustion of $CD8^+$ T lymphocytes in lung cancer tissues [38]. In contrast, activated $CD8^+$ T cells provoke mammary epithelial tumor cells to experience EMT, thus obtaining cancer-initiating power of breast cancer stem cells [6,39].

3. Induction of EMT by Mediators from the Chronic Inflammatory Tumor Microenvironment

3.1. *Upgrade of Inflammation Concept: From Initiation (alpha) to Resolution (omega)*

Inflammation is intrinsically a protective process via microcirculation. Local or systemic inflammatory reactions delete the causing stimuli and reboot repair and healing processes of tissue [40].

Acute inflammation has two phases: initiation (alpha) and resolution (omega). Inflammation starts by the soluble inflammatory mediators such as complement, cytokines including chemokines, free radicals, vasoactive amines, and eicosanoids (including prostaglandins) by adjacent cells of the infected or injured part in the body [6,40,41].

Inflammasome is a multiprotein oligomer responsible for the activation of inflammatory responses and consists of NLRP protein such as NLRP3, ASC, and procaspase-1 [42,43]. The inflammasome can promote the maturation and secretion of interleukin 1β (IL-1β) and IL-18 [44].

Inflammasomes are involved in the EMT of cancer or other epithelial cells. For example, knockdown of NLRP3 alleviates high glucose or TGF-β1-induced EMT in human renal tubular cells [45]. NLRP3 regulates cellular proliferation and metastasis via EMT and the PTEN/AKT signaling pathway [46]. NLRP3 inhibition can attenuate silica-induced EMT in human bronchial epithelial cells [47]. NLRP3 also participates in the regulation of EMT in bleomycin-induced pulmonary fibrosis [48]. Uric acid can activate NLRP3 inflammasome in the EMT in the kidney of rats [49]. NLRP3 appears to be important for EMT since inflammasome-independent NLRP3 is enough to EMT in colon cancer cells [50].

The resolution (omega) phase of inflammation releases specialized lipid mediators that can actively prevent further progress of inflammation and enhance resolution of inflammation [51]. A new specialized group of lipids that can actively terminate inflammation has been found by Serhan et al. [52]. These kinds of lipids include lipoxins (Lx), resolvins (Rvs), protectins (PDs), and maresins (MaRs; Figure 2). They exhibit inflammation-suppressing action with pro-resolving effect that can promote efferocytosis [41,52].

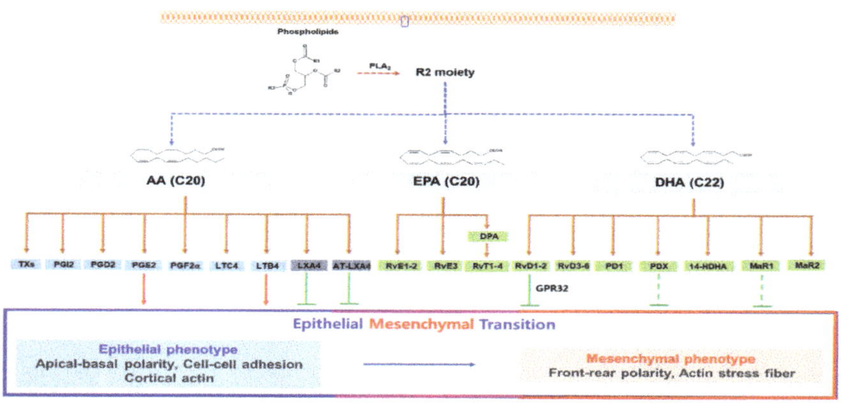

Figure 2. Inflammatory contribution to EMT. PGE2 and LTB4 produced from arachidonic acid can induce EMT in epithelial cancer cells. In contrast, LxA4 and AT-LxA4 from arachidonic acid can repress EMT in cancer cells. RvD1, RvD2, PDX, and MaR1 from DHA can suppress EMT in cancer cells or immortalized cells. AT-LxA4 (15(R)-lipoxin A4) is produced from 15(R)-HETE from arachidonic acid by aspirin trigger. Modified from Lee's report [6]. 14-HDHA: 14-hydroxy Docosahexaenoic Acid; AA: arachidonic acid; AT-LxA4: Aspirin-triggered lipoxin A4; EPA: Eicosapentaenoic acid; DHA: Docosahexaenoic acid; DPA: Docosapentaenoic acid; MaR1: Maresin 1; MaR2: Maresin 2; PD1: Protectin 1; PDX: Protectin X; PGD2: Prostaglandin D2; PGF2a: ProstaglandinF2a; PGI2: Prostaglandin I2; PLA2:Phospholipase A2; LTB4: Leukotriene B4; LTC4: Leukotriene C4; LxA4: LIpoxin A4; RvD1-2: Resolvin D1-2; RvD3-6: Resolvin D3-6; RvE1-2: Resolvin E1-E2; RvE3: Resolvin E3; RvT1-4: 13-series resolvins; TXs: Thromboxane.

13-series resolvins (RvT1-4) derived from docosapentaenoic acid (DPA) have been newly discovered [53]. Briefly describing their production, human platelets pre-treated with aspirin or atorvastatin convert omega-3 DPA (DPAn-3) to a 13S-hydroperoxy intermediate via aspirin-treated or atorvastatin-treated COX-2. This intermediate is converted into four RvTs via ALOX5 enzyme activity exerted on the nearby neutrophils [53]: RvT1 (7,13R,20-trihydroxy-DPAn-3); RvT2 (7,8,13R-trihydroxy-DPAn-3); RvT3 (7,12,13R-trihydroxy-8Z,10E,14E,16Z,19Z-DPAn-3); and RvT4 (7,13R-dihydroxy-DPAn-3). Four RvTs are also formed by a mixture of human neutrophils and vascular endothelium cells, which are found in infected rodent and human tissues. [52].

3.2. EMT Inducers from Chronic Inflammatory Tumor Microenvironments

The tumor microenvironment mainly influences the progression of cancers via secretion of various factors that cause EMT [54,55]. Cancer-related chronic inflammation is described as a chaotic state where both pro-inflammatory and anti-inflammatory signals are present to permit tumor growth and immune evasion [6]. Besides, the tumor microenvironment contributes to the cancer heterogeneity. Therefore, the tumor microenvironment has been regarded as a promising target for the cure of cancer. Thus, Vanneman and Dranoff have demonstrated a novel way of curing cancer by re-educating the tumor microenvironment [56].

Here, we will briefly introduce EMT inducers from chronic inflammatory tumor microenvironment (Figure 3).

TNF-α is a critical determiner of inflammatory responses [57]. Serum concentration of TNF-α was determined as 1.47 pg/mL in invasive breast cancer patients and 0.98 $\pm$ 0.37 pg/mL in the control cohort [6,58]. TNF-α produced by macrophages can accelerate TGF- 1-induced EMT [59].

IL-6 serves as either a pro-inflammatory or anti-inflammatory cytokine [6]. The mean serum concentration of IL-6 was observed as 31.7 pg/mL in patients with breast cancer and 3.3 pg/mL in the normal cohort [6,60]. IL-6 induces EMT of human breast cancer cells [61]. IL-6/STAT3-induced expression of lncTCF7 can promote EMT of liver cancer cells [62]. The IL-6 pathway induces EMT in biliary tract cancer via cross-talking to the SMAD4 in the TGF- 1 pathway [63].

IL-8 is a chemokine mainly secreted by macrophages [64]. The mean serum concentration of IL-8 was found as 40.1 pg/mL in patients with breast cancer and 5.3 pg/mL in the normal group [60]. IL-8 expression is highly increased in TGF- 1-induced EMT in colon carcinoma and nasopharyngeal carcinoma [65]. IL-8 is also involved in mast cell-induced EMT of human lung and thyroid cancer cells [66,67]. JAK2/STAT3/Snail pathway is involved in the IL-8-induced EMT of HCC cells [68]. Brachyury-induced EMT of the tumor is mediated by IL-8/IL-8R signaling pathway [69].

IL-17 is a pro-inflammatory cytokine and mainly released from Th17 cells and macrophages [70]. IL-17 induces EMT of prostate cancers via MMP7 [71]. IL-17 induces EMT through STAT3 in the lung adenocarcinoma [72]. IL-17 can evoke self-renewal of CD133^{+} cancer cells in ovarian cancer [73].

High-mobility group box 1 (HMGB1) is a nuclear DNA-binding protein and released to the outside from macrophages, NK cells, dendritic cells, necrotic cells, and apoptotic cells according to infection, injury, and inflammation [74]. The mean serum HMGB1 level was 4.64 ng/mL in patients with malignant breast cancer, which was remarkably higher than in patients with benign breast cancer (1.32 ng/mL) or in healthy subjects (1.36 ng/mL) [75]. HMGB1 (2 μg/mL) induces EMT of colorectal and prostate cancer cells via the RAGE/NF-κB pathway [76,77].

IL-10 is a potent anti-inflammatory cytokine that suppresses T cell/macrophage cytokine synthesis and blocks their antigen-presenting capacity [78]. In vitro generated M1- and M2-macrophages both can induce EMT of pancreatic cancer cells via the IL-10 signaling pathway [79].

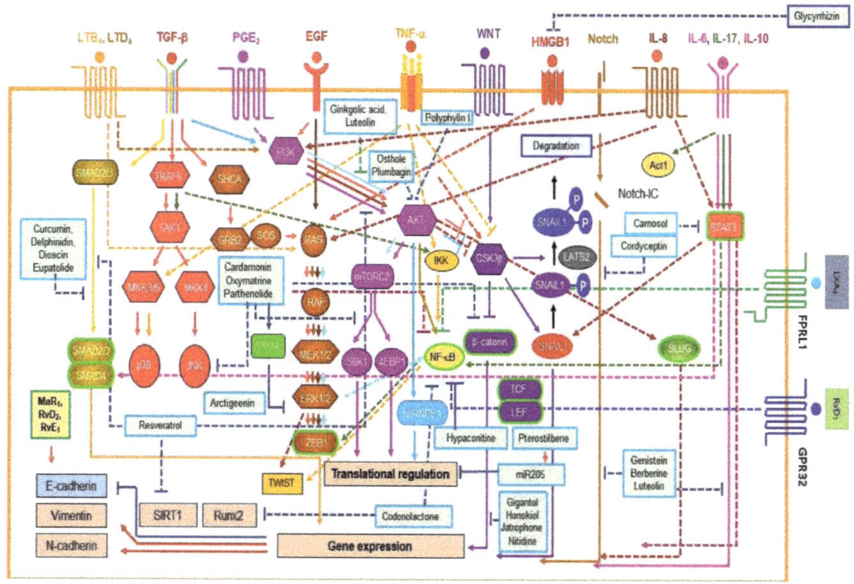

Figure 3. EMT inducers from the chronic tumor microenvironment and EMT repressors from natural anti-inflammatory compounds and pro-resolving lipids. EMT signaling process is simplified due to space limitation. Dot line (–) means indirect effects. Arrow (→) means promotion or induction. Bar line (–I) indicates suppression. Inbox demonstrates the Lats2-mediated snail1 degradation. Receptors here have several subtypes coupled to different signaling pathway. For instance, EP4 use the PI3K/AKT pathway. This figure only covers limited parts of EMT receptor signaling. Modified from Lee's report [6]. 4EBP1: Eukaryotic Translation Initiation Factor 4E Binding Protein 1; Act1: NF-κB activator 1; Akt: AKR mouse thymoma; ERK: Extracellular Signal-Regulated Kinase; FPRL1: Formyl Peptide Receptor-Like 1; GSK3: Glycogen Synthase Kinase 3 Beta; GRB2: Growth Factor Receptor Bound Protein 2; hnRNPE2: Heterogeneous Nuclear Ribonucleoprotein E2; IKK: Inhibitor Of Nuclear Factor Kappa B Kinase; JNK: JUN N-Terminal Kinase; LATS2: Large Tumor Suppressor Kinase; LEF: Lymphoid Enhancer Binding Factor; MEK: MAPK/ERK Kinase; MKK: Mitogen-Activated Protein Kinase Kinase; mTORC: mammalian Target of Rapamycin Complex; NF-κB: Nuclear Factor Kappa B; PI3K: Phosphatidylinositol-4,5-Bisphosphate 3-Kinase; PP2A: Protein Phosphatase 2A; RAF: Rapidly Accelerated Fibrosarcoma; RAS: Rat Sarcoma Viral Oncogene homolog; Runx2: RUNX Family Transcription Factor 2; S6K1: Ribosomal protein S6 kinase beta-1; SHCA: Src Homology 2 Domain Containing Adaptor protein; SIRT: Sirtuin; SMAD: Sma- and Mad-Related Protein; SOS: Son of Sevenless; STAT3: Signal Transducer And Activator of Transcription 3; TAK1: TGF- Activated Kinase 1; TCF: Transcription Factor; TRAF6: TNF Receptor Associated Factor 6; TWIST: Twist Family BHLH Transcription Factor; ZEB1: Zinc Finger E-Box Binding Homeobox 1.

TGF- 1 is a potent anti-inflammatory cytokines [80]. Plasma TGF- 1 levels were significantly higher in stage IIIB/IV breast cancer patients (2.40 ng/mL) than those in healthy controls (1.30 ng/mL) [81]. It is a typical EMT inducer involved in cancer progression [6]. Please read other review for a detailed information about TGF- 1-induced EMT [6,82]. TGF- 1 promotes the production of IL-10 in macrophages from mouse cancer [83]. HMGB1 promotes expression of TGF- 1 via RAGE pathway to mediate TGF-1-induced EMT [84].

PGE2 is biosynthesized from arachidonic acids (Figure 2). The mean serum level of PGE2 was 6.324 pg/mL in patients with brain cancer and 1.677 pg/mL in the compared normal cohort [85]. PGE2 acts through G protein-coupled receptors such as EP1-EP4 [86]. PGE2 (2–10 μg/mL) works in an autocrine or paracrine manner, leading to stimulation of EMT through the expression of SNAIL and ZEB1 [87]. In turn, SNAIL induces blocking of degradation of PGE2 by repressing prostaglandin dehydrogenase, generating a positive loop that promotes cancer progression [88]. PGE2 (5 μM) enhances invasion of HCC cells via EP1-mediated expression of YB-1, which induces TGF- 1-induced EMT by AKT activation [89]. PGE2 (0.01–1 μM) inhibits fibroblast chemotaxis but stimulates chemotaxis of epithelial cells in the airway [90]. PGE2's inhibitory action against TGF-1-induced EMT seems to be via the EP2 pathway since EP2 agonist can block TGF- 1-induced EMT [91]. As such, the effect of PGE2 on EMT varies depending on concentration and in cell types.

Leukotriene B4 (LTB4) can be produced via the 5-lipoxygenase pathway. Significantly higher levels of LTB4 have been found in the whole blood of lung cancer patients than those in the control group (44.1 vs. 17.9 pg/mL) [92]. LTB4 acts via two distinct GPCR called BLT1 and BLT2. BLT2 is involved in keratin phosphorylation and perinuclear reorganization, which is a prelude of EMT [93]. Accordingly, BLT2 is involved in the ras-promoted TGF-1-induced EMT [94]. As expected, LTB4 (100 nM) can induce EMT leading to vimentin expression through the BLT2/ERK2 activation [95].

Cysteinyl leukotrienes (CysLTs) include LTC4, LTD4, and LTE4 [96]. Mean serum levels of LTD4 found in HCC patients and healthy groups were 174.95 and 10.75 pg/mL, respectively [97]. Actions of CysLTs are mediated via GPCRs such as CysLT1 and CysLT2 [98]. LTD4 (100 nM) can suppress E-cadherin expression in cancer cells through enhanced translocation of -catenin to the nucleus while montelukast (0.1 mM) suppresses eosinophil-induced EMT in bronchial epithelial cells [99]. Recently, exosomes and cells from ascites in lung cancer patients can convert LTC4 to LTD4 to promote cancer cell migration and invasion via CysLT1 [100].

4. Reversal of EMT by Anti-inflammatory and Pro-Resolving Natural Compounds

The 2018 review by us did not cover the regulation of EMT by natural products [6]. In this part, we describe anti-inflammatory natural compounds and pro-resolving lipids that can prohibit EMT stimulated by mediators released in the tumor microenvironment.

4.1. Reversal of EMT by Anti-Inflammatory Natural Compounds

Diverse natural products can inhibit EMT of cancer cells or epithelial cells. We focused on EMT inhibitors found from natural compounds that can inhibit the EMT of cancer cells caused by EMT inducers from the chronic inflammatory tumor microenvironment due to limited space (Figure 4, Table 1).

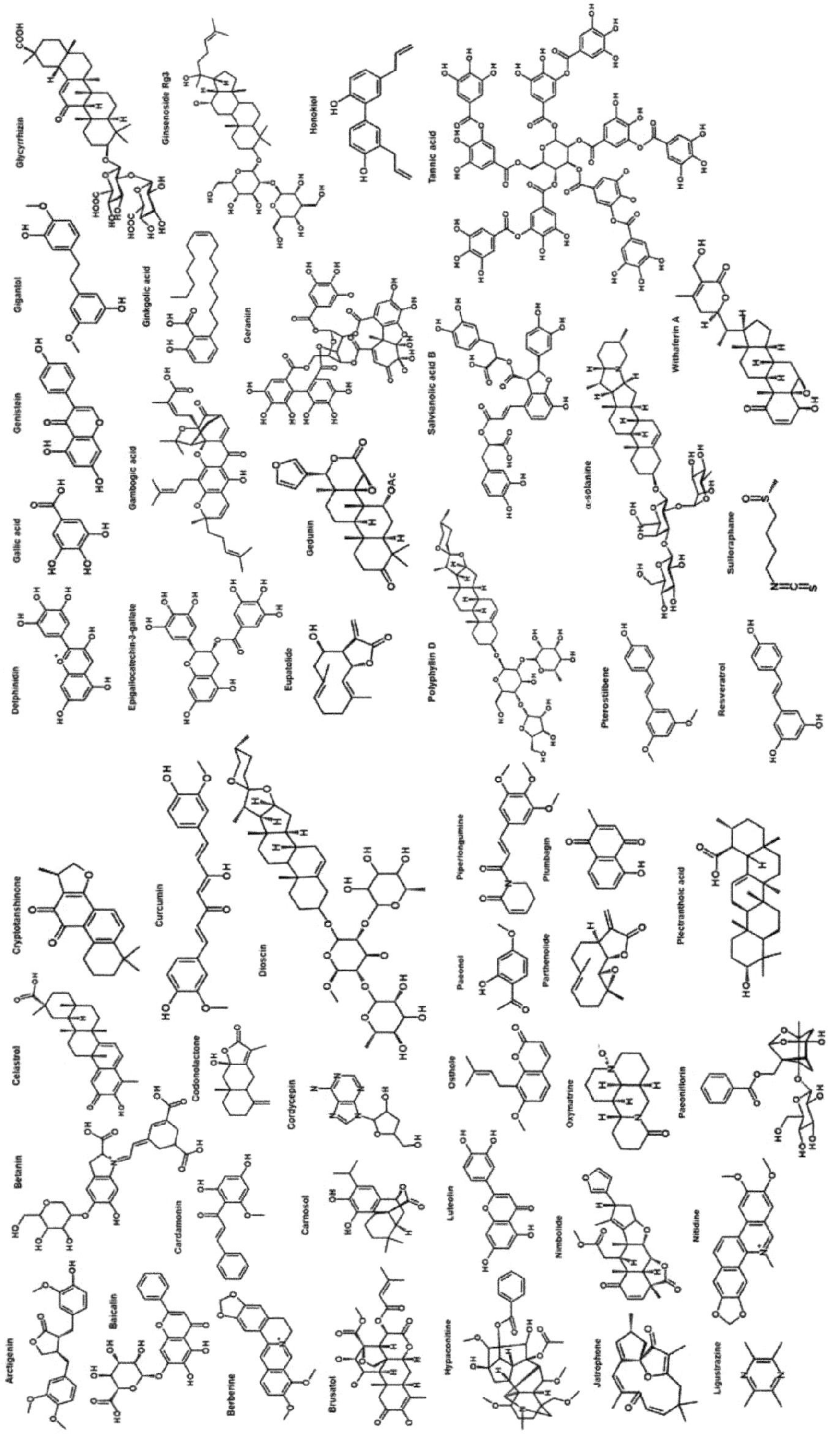

Figure 4. Structure of EMT inhibitors from natural sources.

Table 1. Lists of EMT inhibitors from natural sources.

Name	Source	Name	Source
Arctigenin	*Astraceae*	Glycyrrhizin	*Glycyrrhiza glabra*
Baicalin	*Scutellaria baicalensis, Scutellaria lateriflora*	Honokiol	*Magnolia*
Berberine	*Berberis*	Hypaconitine	*Aconitum*
Betanin	beets	Jatrophone	*Jatropha isabellei*
Brusatol	*Brucea sumatrana*	Ligustrazine	Nattō, fermented cocoa beans
Cardamonin	*Alpinia katsumadai*	Luteolin	*Reseda luteola*
Carnosol	*Rosmarinus officinalis, Salvia pachyphylla*	Nimbolide	*Azadirachta indica*
Celastrol	*Tripterygium wilfordii, Celastrus regelii*	Nitidine chloride	*Zanthoxylum nitidium*
Codonolactone	*Atractylodes lancea*	Osthole	*Cnidium monnieri*
Cordycepin	*Cordyceps militaris*	Oxymatrine	*Sophorae flavescentis*
Cryptotanshinone	*Salvia miltiorrhiza*	Paeoniflorin	*Paeonia lactiflora*
Curcumin	*Zingiberaceae*	Paeonol	*Paeonia suffruticosa*
Dioscin	*Dioscorea villosa*	Parthenolide	*Tanacetum parthenium*
Delphinidin	*Viola, Delphinium*	Plectranthoic acid	*Ficus microcarpa*
Epigallocatechin-3-gallate	Green tea	Piperlongumine	*Piper longum*
Eupatolide	*Inula britannica*	Plumbargin	*Plumbago*
Galic acid	gallnuts, sumac, witch hazel, tea leaves	Polyphyllin I	Rhizoma of Paris
Gambogic acid	*Garcinia hanburyi*	Pterostilbene	blueberries
Gedunin	*Azadirachta indica*	Resveratrol	grapes
Genistein	*Genista tinctoria*	Salvianolic acid	*Salvia miltiorrhiza*
Geraniin	Geraniums	α–Solanine	*Solanum*
Gigantol	*Cymbidium goeringii*	Sulforaphane	cruciferous vegetables
Ginkgolic acid	*Ginkgo biloba*	Tannic acid	*Caesalpinia spinosa*
Ginsenosides	Ginseng	Withaferin A	Solanaceae

Arctigenin from Asteraceae has anti-inflammatory effects [101]. Arctigenin (12–50 μM) can suppress TGF–induced EMT of human lung cancer cells, thus blocking invasion [102]. Arctigenin (0.25–1 μM) can inhibit the expression of MCP-1 and subsequent EMT induced by ROS-dependent ERK/NF-κB pathway of renal tubular epithelial cells [103].

Baicalin found in *Scutellaria baicalensis* and *Scutellaria lateriflora* has well-known anti-inflammatory effects [104]. Baicalin (2 μM) can suppress TGF-1-mediated EMT in MCF10A cells by reducing the expression of slug [105]. Baicalin (12.5–25 μM) can inhibit the expression of TGF-1-induced EMT-related transcription factors in osteosarcoma cells and inhibit the aggressive metastasis of breast cancer by blocking EMT via inhibiting the activation of -catenin [106,107]

Berberine found in *Berberis* can reduce the secretion of IL-1 and TNF-α [108,109]. Berberine (5–20 μM) can reverse EMT in uterine cancer, leading to suppression of cancer metastasis [110]. Berberine can inhibit the metastatic ability of prostate cancer cells by suppressing EMT-associated genes [111]. Berberine (50 μM) can make nasopharyngeal carcinoma cells sensitive to radiation through EMT inhibition [112].

Betanin from beets can decrease the production of superoxide anion and cytokines TNF-α and IL-1 [113,114].

Betanin (25–50 μM) can inhibit high glucose-induced EMT of renal proximal tubular cells [115]. However, activity of betanin for EMT of cancer cells has not been reported yet.

Brusatol from the seeds of *Brucea sumatrana* can inhibit the response of cultured beta-cells to pro-inflammatory cytokines in vitro [116]. Brusatol (2 μM) can inhibit the EMT of pancreatic cancer cells [117].

Cardamonin, one of major component of *Alpinia katsumadai* has anti-tumor, anti-inflammatory, anti-nociceptive, and anti-itching activities [118–120]. Cardamonin (1–10 μM) can suppress TGF-1-stimulated EMT of A549 cells by restoring protein phosphatase 2A expression [120]. Cardamonin (5–20 μM) can block the invasiveness of human triple negative breast cancer cell by downregulation of Wnt/-catenin signaling pathway and induce the reversal of EMT [121]. Cardamonin (5–25 μM) also inhibits transglutaminase-2, one players in EMT, leading to JNK activation and NF-κB pathway [122].

Carnosol, found in *Rosmarinus officinalis* and *Salvia pachyphylla*, can block UV-induced inflammation through inhibition of STAT3 [123–125]. For a more in-depth look at various anti-inflammatory effects of carnosol, please refer to the 2017 review [126]. Carnosol (0.1–10 μM) controls the human glioblastoma stemness features by modulating EMT and inducing cancer stem cell apoptosis [127]. Carnosol (5–10 μM)-mediated SIRT1 activation inhibits the enhancer of zeste homolog 2 to attenuate liver fibrosis [128].

Celastrol (0.1–1 μM), from the root extracts of *Tripterygium wilfordii* and *Celastrus regelii*, can suppress experimental autoimmune encephalomyelitis [129]. Celastrol can inhibit the expression of snail and increased the expression of E-cadherin in the lung cancer cells [130]. Many studies have reported the effect of celastrol on EMT of diseases other than cancer (please refer the ref 128 by Kashyap et al.) [131].

Codonolactone, a major component of *Atractylodes lancea*, exhibits anti-allergic activity, anti-inflammatory, anticancer, gastroprotective, and neuroprotective activities [132,133]. Codonolactone (10–40 μM) can inhibit EMT in breast cancer cells by downregulating the transcriptional activity of Runx2 [134].

Cordycepin from the fungus *Cordyceps militaris* can suppress LPS-induced cytokine production by increasing heme oxygenase-1 expression [135,136]. Cordycepin (25–100 μM) can inhibit cancer stemness of TGF- induced chemo-resistant ovarian cancer cells [137]. Metronomic cordycepin therapy (25 mg/kg and 50 mg/kg) can prolong the survival of oral cancer-bearing mice and it (50 μM) inhibit EMT [138]. Cordycepin (100–200 μM) suppresses integrin/FAK signaling and EMT in HCC [139].

Cryptotanshinone, obtained from the root of *Salvia miltiorrhiza*, can protect against IL-1-induced inflammation in human osteoarthritis chondrocytes [140]. Cryptotanshinone (5–10 μM) targets tumor-initiating cells through down-regulation of stemness genes expression [141].

Curcumin, a phenolic compound found in *Zingiberaceae* turmeric, has strong anti-inflammatory, antioxidant, and antitumor properties [142]. A more extensive and detailed review of curcumin's EMT has been reported recently [143]. Curcumin (25–50 μM) can inhibit metastasis in human papillary thyroid carcinoma cells by negatively regulating TGF-1-mediated Smad2/3 signaling pathway [142]. Curcumin (15 μM) can inhibit TNF-α-induced EMT in melanoma [144]. It can decrease EMT in cervical cancer cells by a pirin-dependent mechanism [145]. Pirin is a coregulatory of NF-κB involved in EMT [146]. Curcumin (30 μM) can suppress paraquat-induced EMT by blocking TGF- in A549 cells [147]. It (8 μM) can reverse oxaliplatin resistance caused by EMT in colorectal cancer through inhibition of the TGF-/Smad2/3 pathway [148].

Dioscin from roots of wild yam (*Dioscorea villosa*) shows potent anti-inflammatory effects via suppression of TNF-α-induced NF-κB-mediated VCAM-1, and ICAM-1 expression [149,150]. Dioscin (3 μM) can suppress TGF-1-induced EMT in A549 and HepG2 cells [149,151]. Dioscin (1–10 μM) also reverses HMT-induced EMT by down-regulating mdm2 and vimentin [152]. Diosgenin (400 μM), an aglycone of dioscin inhibits breast cancer cells with stem cell like properties by attenuation of the Wnt-catenin signaling [153].

Delphinidin, an anthocyanin, can reduce levels of inflammatory mediators including IL-6 and TNF-α induced by LPS [154]. Delphinidin (10–50 μM) inhibits TGF-1-induced EMT through a TGF-1/Smad2 signaling pathway in glioblastoma cells and EGF-induced EMT in HCC cells [155,156].

Epigallocatechin-3-gallate found in green tea can regulates anti-inflammatory action through laminin receptor-mediated tollip signaling induction in LPS-stimulated human intestinal epithelial cells [157]. Epigallocatechin-3-gallate (25–50 μM) can also suppress nicotine-induced migration and invasion by blocking angiogenesis and EMT of non-small cell lung carcinoma (NSCLC) cells [158]. It (10–60 μM) also suppresses EMT and invasion in anaplastic thyroid carcinoma cells by blocking TGF-1/Smad pathways [159].

Eupatolide from the *Inula britannica* is used to treat bronchitis, disorders of the digestive system and inflammation [160]. Eupatolide can prohibit LPS-stimulated COX-2 and iNOS expression of RAW264.7 cells by evoking proteasomal degradation of TRAF6 [161]. Eupatolide (3–10 μM) can inhibit TGF-1-induced migration of breast cancer cells via down-regulation of SMAD3 phosphorylation and transcriptional repression of ALK5 [160].

Gallic acid is a trihydroxybenzoic acid found in gallnuts, sumac, witch hazel, and tea leaves [162]. For various anti-inflammatory effects of gallic acid, please refer to the 2017 review [163]. Gallic acid-coated sliver nanoparticle (50 μg/mL) can alter the expression of radiation-induced EMT in NSCLC [164]. Black tea polyphenols (10–40 μM) can also reverse EMT and inhibit invasion of human oral cancer cells [165].

Gambogic acid from the brownish or orange resin of *Garcinia hanburyi* can enhance the expression of heme oxygenase-1 through Nrf2 pathway and inhibit NF-κB and MAPK activation to mitigate inflammation in LPS-activated RAW264.7 cells [166]. Gambogic acid (0.5–1 μM) can suppress cancer invasion and migration by inhibiting TGF-1-induced EMT [167]. It (2 μM) can induce cleavage of vimentin in HeLa cells [168].

Gedunin, one of the main chemical compounds in the neem tree, can protect TLR-mediated inflammation by suppression of inflammasome activation and cytokine production [169,170]. Gedunin (15 μM) suppresses EMT of pancreatic cancer by inhibiting sonic hedgehog signaling pathway [171].

Genistein first isolated from *Genista tinctoria* can suppress psoriasis-related inflammation through a STAT3/NF-κB-dependent mechanism in keratinocytes [172,173]. For more detailed information about genistein's anti-inflammatory action or anti-EMT, please refer to a previous review by Spagnulo et al. and Lee et al. [174,175]. Genistein (200 μM) can induce apoptosis of colon cancer cells by reversal of EMT via a notch1/NF-κB/slug/E-cadherin pathway [176]. miR-223 inhibitor and genistein (20 μM) can synergistically reverse in EMT of gemcitabine-resistant pancreatic cancer cells [177].

Geraniin found in geraniums can ameliorate experimental acute reflux esophagitis via NF-κB-regulated anti-inflammatory activities in rats [178,179]. It (15–20 μM) can also suppress TGF-1-induced EMT, migration, invasion and anoikis resistance in A549 lung cancer cells [180].

Gigantol from the *Cymbidium goeringii* can suppress LPS-stimulated iNOS and COX-2 expression through NF-κB inactivation in RAW 264.7 cells [181]. It (5–20 μM) can attenuate cancer stem cell-like phenotypes and induce anoikis in human lung cancer H460 cells [182,183].

Ginkgolic acid from *Ginkgo biloba* can significantly inhibit the production of NO, PGE_2, and pro-inflammatory cytokines in ox-LDL-stimulated HUVECs cells [184]

Ginkgolic acid (100 μM) can inhibit TGF-1-induced EMT of lung cancer cells through PI3K/AKT/mTOR inactivation [185]. PPAR-γ might be involved in the suppression of EMT since ginkgolic acid is a PPAR-γ modulator.

Ginsenosides Rh1, Rg3, Rb1, Rg5, and Rg1 from ginseng can block inflammatory responses by inhibiting the activation of NLRP3, NLRP1, and AIM [186]. Rg3 (25–100 μM) can suppress EMT and invasion in lung cancer cells by reducing expression of FUT4 [187]. It (75 μg/mL) can also sensitize hypoxic lung cancer cells to cisplatin via blocking of NF-κB mediated EMT [188]. Rb1 (160 μg/mL) can inhibit hypoxia-induced EMT in ovarian cancer cells by regulating miR-25 [189]. Downregulation of HDAC3 by Rg3 (25 and 50 μg/mL) can inhibit EMT of cutaneous squamous cell carcinoma through c-Jun acetylation [190]. Rg3 (75 μg/mL) can inhibit growth and EMT of human oral squamous carcinoma cells by down-regulating miR-221 [191].

Glycyrrhizin (50–200 μM) from the roots of *Glycyrrhiza glabra* (Licorice) has anti-inflammatory and antiviral activities. It is also a novel pharmacological inhibitor of HMGB1 [192]. Glycyrrhizin can attenuate the EMT of prostate cancer cells by suppressing HMGB1-involved signaling pathway [192].

Honokiol from the *Magnolia* possesses anti-inflammatory activity by blocking downstream signaling of MEKK-1 in NF-κB activation pathway [193,194]. It (30 μM) can also inhibit EMT-mediated migration of human NSCLC cells in vitro by targeting c-FLIP and EMT of breast cancer cells by targeting STAT3/Zeb1/E-cadherin axis [195,196]. Honokiol (20 μM) can inhibit the metastasis of renal cancer cells by blocking EMT through regulating miR-141/ZEB2 pathway [197]. It (5–20 μM) can also inhibit the invasion of U87MG human glioblastoma cell via regulation of EMT [198].

Hypaconitine from the root of *Aconitum* species can suppress 0.1% histamine-induced acute inflammation without showing an ulcerogenic effect [199]. Hypaconitine (8 μM) can inhibit TGF-1-evoked EMT of A549 lung cancer cells possibly by blocking NF-κB activation [200].

Jatrophone from *Jatropha isabellei* has anti-nociceptive and anti-inflammatory activities [201]. Jatrophone (2 μM) can interfere with Wnt/-catenin signaling and reverses EMT of human triple-negative breast cancer [202].

Ligustrazine found in nattō and in fermented cocoa beans can significantly decrease CCL3, CCL19, CCl21, IL-4, IL-5, and IL-17A in bronchoalveolar lavage fluid of ovalalbumin-induced mice [203,204]. There are many reports of anti-inflammatory effects of ligustrazine [205]. Ligustrazine (100 μM) can suppress EMT of renal cell carcinoma cells by blocking MMP9 and TGF-1 [206].

Luteolin is a natural flavonoid that possesses anti-inflammatory and anti-cancer activities [207]. Luteolin (40 μM) inhibits TGF-1-induced EMT of A549 lung cancer cells through interfering with PI3K/AKT/NF-κB/Snail pathway [207]. Luteolin (10 μM) can suppress EMT and negatively regulating -catenin expression in breast cancer cells [208,209]. It (5–20 μM) can inhibit metastasis of melanoma cells by decreasing HIF-1α/VEGF signaling-mediated EMT [210]. It (5 μM) can also inhibit EMT of colorectal cancer cell by suppressing CREB1 expression [211]. Luteolin (20 μM) can inhibit the invasion of cervical cancer by blocking EMT signaling [212]. It (30 μM) can block gastric cancer progression by reversing EMT through inhibition of the notch signaling [213]. It (15.6–31.3 μM) can also inhibit EMT in paclitaxel-resistant ovarian cancer cells [214]. Luteolin (20–160 μM) can also block IL-6-induced EMT in pancreatic cancer cells by inhibiting STAT3 signaling [215].

Nimbolide, from the neem tree (*Azadirachta indica*), can inhibit NF-κB pathway in intestinal epithelial cells and macrophages, resulting in alleviation of experimental colitis in mice [216]. Nimbolide (5 μM) can suppress pancreatic cancer growth and metastasis through inhibition of EMT [217]. It (1–6 μM) can also suppress NSCLC cell invasion and migration via manipulation of DUSP4 expression and ERK1/2 signaling [218].

Nitidine chloride, a pentacyclic alkaloid isolated from the root of *Zanthoxylum nitidium*, can suppress LPS-induced interleukin production via MAPK and NF-κB in RAW 264.7 cells [219]. Hedgehog pathway is implicated in nitidine chloride (2.5 μM)-induced blocking of EMT of breast cancer cells [220]. Nitidine chloride (5 μM) can also inhibit EMT of osteosarcoma cell via Akt/GSK-3/snail signaling pathway [221].

Osthole, the major natural coumarin from *Cnidium monnieri* (L.) Cuss, exerts anti-inflammatory effects by blocking of the activation of the NF-κB and MAPK/p38 pathways [222]. Osthole (20 μM) can suppress HGF-induced EMT via repression of the c-Met/Akt/mTOR pathway in human breast cancer cells [223]. Osthole (20–40 μM) can also inhibit IGF-1-induced EMT by inhibiting PI3K/Akt signaling pathway in human brain cancer cells [224]. By inhibiting snail signaling and miR-23a-3p, osthole (20–80 μM) can suppress EMT-mediated metastatic ability in prostate cancer [225]. Osthole (5–20 μM) an also inhibit TGF -induced EMT by suppressing NF-κB mediated snail activation in A549 cells [226].

Oxymatrine, the active component from Radix *Sophorae flavescentis*, is well known for its anti-inflammatory activity [227]. Oxymatrine (1.5–6 μM) can reverse EMT of breast cancer cells by depressing αV3 integrin/FAK/PI3K/Akt signaling activation [228]. It (0.25–0.75 μM) can inhibit EMT of colorectal cancer cells by suppressing NF-κB signaling [229]. Chronic oxymatrine treatment can induce resistance and EMT of colorectal cancer cells [230].

Paeoniflorin from *Paeonia lactiflora* has anti-inflammatory effects [231,232]. Paeoniflorin (5–10 μM) can suppress EMT of human colorectal cancer cells and glioblastoma cells and prevent hypoxia-induced EMT of human breast cancer cells [233,234].

Paeonol found in *Paeonia suffruticosa* (moutan cortex) can suppress LPS-induced HMGB1 translocation from the nucleus to the cytoplasm in RAW264.7 cells [235,236]. It (60–120 μM) can attenuate aging of MRC-5 cells and inhibit EMT of HaCaT cells induced by aging MRC-5 cell-conditioned medium [237].

Parthenolide from *Tanacetum parthenium* has well-known anti-inflammatory activities [238]. Parthenolide (5 μM) can inhibit TGF-1-induced EMT of colorectal cancer cells [239]. Parthenolide (10–20 μM) can suppress HIF-1α signaling and hypoxia-induced EMT in colorectal cancer [240]. Parthenolide binds Gly-Leu-Ser/Lys- "co-adaptation pocket" to inhibit EMT of lung cancer cell [241].

Plectranthoic acid isolated from *Ficus microcarpa*, can alleviate the symptoms of type 2 diabetes mellitus by inhibiting dipeptidyl peptidase 4 [242]. Plectranthoic acid is a novel activator of AMPK can induce apoptotic death in prostate cancer cells [243]. Plectranthoic acid (20–40 μM) can suppress EMT of prostate cancer [244].

Piperlongumine, a constituent of the fruit of the long pepper (*Piper longum*) can inhibit neuroinflammation via regulating NF-κB signaling pathways in LPS-stimulated BV2 microglia cells [245,246]. Piperlongumine (1–5 μM) inhibits TGF–induced EMT by modulating the expression of E-cadherin, Snail1, and Twist1 [247].

Plumbargin from *Plumbago* genus can attenuate the expression of inflammatory cytokine in LPS-activated BV-2 cells [248]. Plumbagin (0.5–1 μM) can inhibit EMT of human cervical carcinoma cells and inhibit tumor invasion of endocrine-resistant breast cancer through EMT [249]. It (0.1–0.5 μM) suppresses EMT via inhibiting Nrf2-mediated signaling pathway in human tongue squamous cell carcinoma cells [250]. Plumbagin (1–5 μM) can inhibit PI3K/Akt/mTOR-mediated EMT in human pancreatic cancer cells [251]. Plumbagin (1–5 μM) shows differential proteomic responses to EMT of PC-3 and DU145 prostate cancer cells [252].

Polyphyllin I, a component in the Rhizoma of Paris, can improve collagen-induced arthritis by blocking the inflammation response in macrophages through the NF-κB Pathway [253]. Polyphyllin I (0.3 μM) can overcome EMT-associated resistance to erlotinib in lung cancer cells via IL-6/STAT3 pathway inhibition [254].

Pterostilbene (5–10 μM) from blueberries can effectively suppress the generation of cancer stem cells and metastatic potential under the influence of M2 TAMs by modulating EMT associated signaling pathways, specifically the NF-κB/miR488 circuit [255]. Pterostilbene (10 μM) can also inhibit triple-negative breast cancer metastasis by inducing miR205 expression and negatively modulates EMT [256]. Long non-coding RNAs such as MEG3, TUG1, H19, and DICER1-AS1 contribute to the inhibitory effect of pterostilbene (1–50 μM) on proliferation and EMT of human breast cancer cells [257].

Resveratrol, a constituent of grapes and various other plants, is an activator of PPAR and SIRT1 [258]. Resveratrol (12 μM) can suppress TGF-1-induced EMT in colorectal cancer through the TGF-1/SMADs signaling pathway [259]. Resveratrol (20 μM) can inhibit TGF-1-induced EMT and suppress lung cancer invasion and metastasis [260]. Resveratrol (50 μM) can inhibit the EMT of pancreatic cancer cells through suppression of the PI3K/Akt/NF-κB pathway [261]. Resveratrol might inhibit EMT via downregulation of COX and SIRT1 activation, although roles of SIRT1 in EMT of cancer cells show conflicting results [262,263].

Salvianolic acid, an active compound present in *Salvia miltiorrhiza*, can suppress CCL-20 expression in TNF-α-treated macrophages [264]. Salvianolic acid B (1–100 μM)-induced expression of miR-106b-25 can suppress EMT of HK-2 cells [265]. Salvianolic acid B (1–10 μM) can prevent EMT through the TGF-1 signal transduction pathway [266].

α–Solanine is a glycoalkaloid poison found in species of the nightshade family within the genus *Solanum*, such as the potato and the eggplant [267]. A chloroform fraction of *Solanum nigrum* can suppresses nitric oxide and TNF-α in LPS-stimulated mouse peritoneal macrophages through inhibition of p38, JNK and ERK1/2 [268]. α-Solanine (4–12 μM) can suppress the invasion of human prostate cancer cell by inhibiting EMT and MMPs expression [269].

Sulforaphane from cruciferous vegetables may directly impair the formation of NLRP3 inflammasome by inhibiting ASC or caspase-1 [270]. Sulforaphane (1–5 μM) can also inhibit the EMT and metastasis of human lung cancer through miR616-5p-involved GSK3/-catenin pathways [271]. Sulforaphane (20–40 μM) can suppress TGF-1-induced EMT of HCC cells via the ROS-dependent pathway [272]. It (5–20 μM) can block the EMT of human bladder cancer cells via COX-2/MMP2,9/SNAIL, ZEB1, and miR200c/ZEB1 pathways [273].

Tannic acid is a type of polyphenol inhibits NLRP3 inflammasome-mediated IL-1 production via blocking NF-κB signaling in macrophages [274]. Tannic acid (25 μM) attenuates TGF-1-induced EMT in lung epithelial cells [275].

Withaferin A from the Solanaceae family attenuates bleomycin-induced scleroderma by targeting FoxO3a and NF-κB signaling [276]. Withaferin A (0.5 μM) can inhibit the EMT of NSCLC cells [277]. It (2 μM) can also inhibit the EMT of MCF10A cells and suppress vimentin expression in breast tumors [278]. Extracts of root in *Withania somnifera* can suppress mammary EMT and cancer metastasis [279].

4.2. Reversal of EMT by Natural Pro-resolving Lipids

Pro-resolving lipids can be useful agent against cancer and EMT of cancer has been studied since cancer is understood as a non-resolving disease [6,280]

Here, we briefly introduced the effect of pro-resolving lipids on EMT of cancer cells and emphasized natural sources of pro-resolving lipids (Table 2).

Table 2. Pro-resolving lipids. * Antagonist.

Name	Formula	Receptor	Anti-EMT	Source	Yields (pg/mg)
	AA-derived lipoxins				
LxA4	5S,6R,15S-trihydroxy-7E,9E,11Z,13E-ETE	FPR2, GPR32	+	*Toxoplasma gondii*	~0.43 ng
LxB4	5S,14R,15S-trihydroxy-6E,8Z,10E,12E-ETE				
AT-LxA4	5S,6R,15R-trihydroxy-7E,9E,11Z,13E-eicosatetraenoic acid	FPR2			
AT-LxB4	5S,14R,15R-trihydroxy-6E,8Z,10E,12E-eicosatrienoic acid				
	EPA-derived resolvins				
RvE1	5S,12R,18R-trihydroxy-6Z,8E,10E,14Z,16E-EPA	CMKLR1, BLT *, TRPV1 *,	+ (?)	*Candida albicans*	
18S-RvE1	5S,12R,18S-trihydroxy-6Z,8E,10E,14Z,16E-EPA	CMKLR1, BLT *			
RvE2	5S,18R-dihydroxy-6E,8Z,11Z,14Z,16E-EPA	CMKLR1, BLT *		*Trypanosoma cruzi*	9.5–23.6
RvE3	17R,18R/S-dihydroxy-5Z,8Z,11Z,13E,15E-EPA				
	DHA-derived resolvins				
RvD1	7S,8R,17S-trihydroxy-4Z,9E,11E,13Z,15E,19Z-DHA	GPR32, FPR2, TRPV3 *,	+	*Trypanosoma cruzi*, *Oncorhynchus mykiss*	1.8–7.0
RvD2	7S,16R,17S-trihydroxy-4Z,8E,10Z,12E,14E,19Z-DHA	GPR32, GPR18, FPR2,	+	*Oncorhynchus mykiss*	
RvD3	4S,11R,17S-trihydroxy-5Z,7E,9E,13Z,15E,19Z-DHA	GPR32			
RvD4	4S,5R,17S-trihydroxy-6E,8E,10Z,13Z,15E,19Z-DHA				

Table 2. *Cont.*

Name	Formula	Receptor	Anti-EMT	Source	Yields (pg/mg)
RvD5	7S,17S-dihydroxy-4Z,8E,10Z,13Z,15E,19Z-DHA	GPR32		*Trypanosoma cruzi; Oncorhynchus mykiss*	
RvD6	4S,17S-dihydroxy-5E,7Z,10Z,13Z,15E,19Z-DHA				
AT-RvD1	7S,8R,17R-trihydroxy-4Z,9E,11E,13Z,15E,19Z-DHA	FPR2, GPR32, TRPV3 *,	+		
AT-RvD2	7S,16R,17R-trihydroxy-4Z,8E,10Z,12E,14E,19Z-DHA				
AT-RvD3	4S,11R,17R-trihydroxy-5Z,7E,9E,13Z,15E,19Z-DHA	GPR32			
AT-RvD4	4S,5R,17R-trihydroxy-6E,8E,10Z,13Z,15E,19ZDHA				
AT-RvD5	7S,17R-dihydroxy-4Z,8E,10Z,13Z,15E,19Z-DHA				
AT-RvD6	4S,17R-dihydroxy-5E,7Z,10Z,13Z,15E,19Z-DHA				
	n-3 DPA-derived resolvins				
RvT1	7,13R,20-trihydroxy-8E,10Z,14E,16Z,18E-DPA				
RvT2	7,8,13R-trihydroxy-9E,11E,14E,16Z,19Z-DPA				
RvT3	7,12,13R-trihydroxy-8Z,10E,14E,16Z,19Z-DPA				
RvT4	7,13R-dihydroxy-8E,10Z,14E,16Z,19Z-DPA				
RvD1n-3	7,8,17-trihydroxy-8,10,13,15,19-DPA				
RvD2n-3	7,16,17-trihydroxy-8,10,12,14,19-DPA				
RvD5n-3	7,17-dihydroxy-8,10,13,15,19-DPA				
	DHA-derived protectins/neuroprotectins				
PD1(NPD1)	10R,17S-dihydroxy-4Z,7Z,11E,13E,15Z,19Z-DHA			*Oncorhynchus mykiss*	
PDX	10S,17S-dihydroxy-4Z,7Z,11E,13Z,15E,19Z-DHA		+ (?)		
22-hydroxy-PD1	10R,17S,22-trihydroxy-4Z,7Z,11E,13E,15Z,19Z-DHA				
AT-PD1	10R,17R-dihydroxy-4Z,7Z,11E,13E,15Z,19Z-DHA				
Ent-AT-NPD1	10S,17S-Dihydroxy-4Z,7Z,11E,13E,15Z,19Z-DHA				
	n-3 DPA-derived protectins/neuroprotectins				
PD1n-3	10,17-dihydroxy-7,11,13,15,19-DPA				
PD2n-3	16,17-dihydroxy-7,10,12,14,19-DPA				
	DHA-derived maresins				
MaR1	7R,14S-dihydroxy-4Z,8E,10E,12Z,16Z,19Z-DHA	TRPV1 *, TRPA1 *	+ (?)		
MaR2	13R,14S-dihydroxy-4Z,7Z,9E,11E,16Z,19Z-DHA				
7-epi-MaR1	7S,14S-dihydroxy-4Z,8E,10Z,12E,16Z,19Z-DHA				
MaR-L1	14S,22-dihydroxy-4Z,7Z,10Z,12E,16Z,19Z-DHA				
MaR-L2	14R,22-dihydroxy-4Z,7Z,10Z,12E,16Z,19Z-DHA				
	n-3 DPA-derived maresins				
MaR1n-3	7S,14S-dihydroxy-8E,10E,12Z,16Z,19Z-DPA				
MaR2n-3	13,14-dihydroxy-7Z,9,11,16Z,19Z-DPA				
MaR3n-3	7,14-dihydroxy-8,10,12,16Z,19Z-DPA				

* antagonist; +(?) confirmed in non-cancer cells.

Pro-resolving lipids are generally known to act via GPCR receptors such as BLT1, CMKLR1, FPRL1 (ALX/FPR2), GPR18, and GPR3 [41,281]. CMKLR1 is a receptor with high affinity for RvE1, as measured via radioligand-binding assay (Kd = 11.3–5.4 nM) [282,283]. BLT1 is also a receptor with low affinity for RvE1, as measured via radioligand-binding assay (Kd = 45 nM) [284]. CMKLR1 has not been reported to be associated with EMT in cancer cells but seems to be associated with EMT in diabetic nephropathy [285]. FPRL1 (ALX/FPR2) not only possesses an annexin A1 [286] protein but also LxA4 (Kd = 1.7 nM, radioligand binding) [287], AT-RvD1 (EC50 = 1.8×10^{-10}, -arrestin receptor system [288], RvD1 (Kd = 0.17 ± 0.06 nM, radioligand assay; EC50 = 4.5 × 10–11, -arrestin receptor system) [288], RvD3 (activation of FPRL1 at 100 nM) [289] and other similar proteins as a ligand. FPRL1 mediates EMT inhibition by LxA4 and RvD1 [290,291].

GRP18 has been reported as a receptor for RvD2 on performing GPCR–arrestin-based screening (Kd of 9.6 ± 0.9 nM, radioligand binding) [292]. GPR32 is known as the receptor of RvD1 (EC50 = 3.6×10^{-12}, -arrestin receptor system) [293] and RvD5 (activation of GPR32 at the range of 10^{-13}–10^{-9} M) [294]. GPR32 is also activated by RvD3 and AT-RvD3 (0.1 pM–10 nM) [295]; it mediates EMT Inhibition by RvD1 [291]. RvD1 also promotes wound healing in pulmonary epithelial cells and mediates EMT inhibition [296].

LxA4 (0.1–0.8 μM) can suppress TGF-1 signaling in pancreatic cancer cells, reverse mesenchymal features and block invasion and migration via a FPR2 [297]. LxA4 (0.01–0.1 μM) can also suppress estrogen-induced EMT via LxA4 receptor-dependent manner in endometriosis [298]. LxA4 (0.2 μM) and its analogue can suppress EMT, migration and metastasis of HCC by regulating integrin-linked kinase axis [299].

100 nM of RvD1 and RvD2 can inhibit TGF-1-induced EMT of A549 lung cancer cells through FPR2/ALXR and GPR32 [291]. GPR32 can recognize RvD1 (EC_{50} = 8.8 pM from -arrestin receptor system) as an endogenous ligand [293]. Aspirin-triggered RvD1 (10 ng/mL) can block TGF-1-induced EMT of A549 lung cancer cells via suppression of the mTOR pathway by reducing the expression of pyruvate kinase M2 [300]. RvD1 (400 nM) prevents EMT of HCC cells by inhibiting paracrine of cancer-associated fibroblast-derived cartilage oligomeric matrix protein [301].

MaR1 (0.1–10 ng/mice), a docosahexaenoic acid-derived pro-resolution lipid, can protect skin from inflammation and oxidative stress caused by UVB irradiation [302]. Receptor for MaR1 is yet unknown. The effect of MaR1 on EMT of cancer cells has not been reported yet except one study has shown that incubating DHA with A549 lung cancer cells can produce MaR1 (1.58 ng/mL) and PD1 (1.67 ng/mL) [303]. MaR1 (10 nM) can inhibit TGF-1-induced proliferation, migration, and differentiation in human lung fibroblasts [304].

PD1 has anti-inflammatory and survival effects on neuronal diseases such as Alzheimer's disease and retinal degenerations [305]. PDX (1–100 nM), one of PD1 derivatives, can suppress bleomycin-induced lung fibrosis through blocking EMT [6,306]. However, it is hard to find reports about the effect of PDs on the EMT of cancer cells [6,306]. However, it is hard to find reports about the effect of PDs on the EMT of cancer cells.

Pro-resolving lipid classes are mainly originated from ω-3 fatty acids which constitute a group of essential fats that humans cannot synthesize endogenously [307]. Several pro-resolving lipids can be obtained by total synthesis. Studies on derivatives are in progress.

It is possible to produce pro-resolving lipids by biological methods. EPA and DHA are the major long chain ω-3 fatty acids in the diet. Algae are the major producers of EPA and DHA in the ecosystem. Therefore, fish that consumes algae contains a lot of EPA and DHA [307]. Accordingly, algae or fish might be a starting point for isolating or producing pro-resolving lipids in industrial scale. Brain cells of rainbow trout (*Oncorhynchus mykiss*) can produce novel DHA-derived Rvs and PDs (Table 2) [308]. However, baking reduces proportions of PG, Rv, and hydroxy-fatty acid in farm-raised Atlantic salmon (*Salmo salar*; Table 2) [309].

Infectious organisms can produce pro-resolving lipids to control host inflammation. Thus, supraphysiological levels of LxA4 are generated during infection by *Toxoplasma gondii*, which in turn reduces IL-12 production by dendritic cells, thus dampening Th1-type cell-mediated immune responses (Table 2) [310].

Candida albicans can modulate host defense by biosynthesizing the pro-resolving lipid RvE1 [311]. *C. albicans* can biosynthesize nanogram quantities of RvE1 from EPA without collaboration of other cellular partners. It can also biosynthesize PDs (Table 2) [311]. *Trypanosoma cruzi* is a protozoan parasite that causes Chagas disease and produces the RvD1, RvD5, and RvE2 (Table 2) [312]. These reports suggested that algae, fish, and some infectious organisms might be applied to produce pro-resolving lipids.

5. Perspectives

Several natural compounds have anti-inflammatory activities and/or anti-EMT activities. However, few reports have explained their anti-EMT activity by anti-inflammatory or pro-resolving mechanisms. Therefore, examining anti-EMT activities of natural compounds based on their anti-inflammatory or pro-resolving activities and assuring their anti-EMT activities in vivo might be important in the future. Especially, it might be a reasonable way to study the anti-EMT activity of natural products through interaction with the tumor microenvironment [255].

The flux of inflammatory or pro-resolving lipids from arachidonic acid related pathway (Figure 2) could change if one pathway is blocked by natural anti-inflammatory or pro-resolving compounds. Therefore, prevention of the production of these lipids might influence levels of other inflammatory and pro-resolving lipids. Thus, further studies on effects of anti-inflammatory compounds or pro-resolving lipids on EMT of cancer might require entire profiles of lipid metabolites affecting the EMT.

6. Conclusions

Expansion of studies about other pro-resolving lipids and nature-derived pro-resolving compounds to inhibit EMT are needed. To do this, various types of pro-resolving lipids and new compounds from natural sources should be procured and diffused to researcher without barrier. Thus, collaboration of natural chemistry researchers with other fields is also required.

Funding: This study was supported by grants (NRF-2017R1A2A1A05000878 and NRF-2018R1A5A2023127) of the Basic Science Research Program, through the National Research Foundation (NRF) of Korea.

Acknowledgments: The author apologizes to researchers in the EMT field whose work we are not able to cite here due to space limits.

Conflicts of Interest: The author declares no conflicts of interest.

References

1. Brabletz, T.; Kalluri, R.; Nieto, M.A.; Weinberg, R.A. EMT in cancer. *Nat. Rev. Cancer* **2018**, *18*, 128–134. [CrossRef] [PubMed]
2. Marcucci, F.; Stassi, G.; De Maria, R. Epithelial-mesenchymal transition: A new target in anticancer drug discovery. *Nat. Rev. Drug Discov.* **2016**, *15*, 311–325. [CrossRef] [PubMed]
3. Reka, A.K.; Kuick, R.; Kurapati, H.; Standiford, T.J.; Omenn, G.S.; Keshamouni, V.G. Identifying inhibitors of epithelial-mesenchymal transition by connectivity map-based systems approach. *J. Thorac. Oncol.* **2011**, *6*, 1784–1792. [CrossRef] [PubMed]
4. David, J.M.; Dominguez, C.; Palena, C. Pharmacological and immunological targeting of tumor mesenchymalization. *Pharmacol. Ther.* **2017**, *170*, 212–225. [CrossRef]
5. Hanahan, D.; Weinberg, R.A. Hallmarks of cancer: The next generation. *Cell* **2011**, *144*, 646–674. [CrossRef]
6. Lee, C.H. Epithelial-mesenchymal transition: Initiation by cues from chronic inflammatory tumor microenvironment and termination by anti-inflammatory compounds and specialized pro-resolving lipids. *Biochem. Pharmacol.* **2018**, *158*, 261–273. [CrossRef]
7. Xu, J.; Lamouille, S.; Derynck, R. TGF-beta-induced epithelial to mesenchymal transition. *Cell Res.* **2009**, *19*, 156–172. [CrossRef]
8. Smith, C.L.; Baek, S.T.; Sung, C.Y.; Tallquist, M.D. Epicardial-derived cell epithelial-to-mesenchymal transition and fate specification require PDGF receptor signaling. *Circ. Res.* **2011**, *108*, e15–e26. [CrossRef]
9. Strutz, F.; Zeisberg, M.; Ziyadeh, F.N.; Yang, C.Q.; Kalluri, R.; Muller, G.A.; Neilson, E.G. Role of basic fibroblast growth factor-2 in epithelial-mesenchymal transformation. *Kidney Int.* **2002**, *61*, 1714–1728. [CrossRef]
10. Farrell, J.; Kelly, C.; Rauch, J.; Kida, K.; Garcia-Munoz, A.; Monsefi, N.; Turriziani, B.; Doherty, C.; Mehta, J.P.; Matallanas, D.; et al. HGF induces epithelial-to-mesenchymal transition by modulating the mammalian hippo/MST2 and ISG15 pathways. *J. Proteome Res.* **2014**, *13*, 2874–2886. [CrossRef]
11. Liao, G.; Wang, M.; Ou, Y.; Zhao, Y. IGF-1-induced epithelial-mesenchymal transition in MCF-7 cells is mediated by MUC1. *Cell Signal* **2014**, *26*, 2131–2137. [CrossRef] [PubMed]

12. Janse van Rensburg, H.J.; Yang, X. The roles of the Hippo pathway in cancer metastasis. *Cell Signal* **2016**, *28*, 1761–1772. [CrossRef] [PubMed]
13. Hollier, B.G.; Tinnirello, A.A.; Werden, S.J.; Evans, K.W.; Taube, J.H.; Sarkar, T.R.; Sphyris, N.; Shariati, M.; Kumar, S.V.; Battula, V.L.; et al. FOXC2 expression links epithelial-mesenchymal transition and stem cell properties in breast cancer. *Cancer Res.* **2013**, *73*, 1981–1992. [CrossRef] [PubMed]
14. Tang, Y.; Weiss, S.J. Snail/Slug-YAP/TAZ complexes cooperatively regulate mesenchymal stem cell function and bone formation. *Cell Cycle* **2017**, *16*, 399–405. [CrossRef] [PubMed]
15. Lehmann, W.; Mossmann, D.; Kleemann, J.; Mock, K.; Meisinger, C.; Brummer, T.; Herr, R.; Brabletz, S.; Stemmler, M.P.; Brabletz, T. ZEB1 turns into a transcriptional activator by interacting with YAP1 in aggressive cancer types. *Nat. Commun.* **2016**, *7*, 10498. [CrossRef]
16. Roselli, M.; Fernando, R.I.; Guadagni, F.; Spila, A.; Alessandroni, J.; Palmirotta, R.; Costarelli, L.; Litzinger, M.; Hamilton, D.; Huang, B.; et al. Brachyury, a driver of the epithelial-mesenchymal transition, is overexpressed in human lung tumors: An opportunity for novel interventions against lung cancer. *Clin. Cancer Res.* **2012**, *18*, 3868–3879. [CrossRef]
17. Arnold, S.J.; Stappert, J.; Bauer, A.; Kispert, A.; Herrmann, B.G.; Kemler, R. Brachyury is a target gene of the Wnt/beta-catenin signaling pathway. *Mech. Dev.* **2000**, *91*, 249–258. [CrossRef]
18. Zhang, J.; Liu, Y.; Zhang, J.; Cui, X.; Li, G.; Wang, J.; Ren, H.; Zhang, Y. FOXQ1 promotes gastric cancer metastasis through upregulation of Snail. *Oncol. Rep.* **2016**, *35*, 3607–3613. [CrossRef]
19. Fujita, T.; Azuma, Y.; Fukuyama, R.; Hattori, Y.; Yoshida, C.; Koida, M.; Ogita, K.; Komori, T. Runx2 induces osteoblast and chondrocyte differentiation and enhances their migration by coupling with PI3K-Akt signaling. *J. Cell Biol.* **2004**, *166*, 85–95. [CrossRef]
20. Cao, Z.; Sun, B.; Zhao, X.; Zhang, Y.; Gu, Q.; Liang, X.; Dong, X.; Zhao, N. The Expression and Functional Significance of Runx2 in Hepatocellular Carcinoma: Its Role in Vasculogenic Mimicry and Epithelial-Mesenchymal Transition. *Int. J. Mol. Sci.* **2017**, *18*, 500. [CrossRef]
21. Lamouille, S.; Xu, J.; Derynck, R. Molecular mechanisms of epithelial–mesenchymal transition. *Nat. Rev. Mol. Cell Biol.* **2014**, *15*, 178–196. [CrossRef] [PubMed]
22. Li, Y.; Ke, Q.; Shao, Y.; Zhu, G.; Li, Y.; Geng, N.; Jin, F.; Li, F. GATA1 induces epithelial-mesenchymal transition in breast cancer cells through PAK5 oncogenic signaling. *Oncotarget* **2015**, *6*, 4345–4356. [CrossRef] [PubMed]
23. Wu, X.; Xin, Z.; Zou, Z.; Lu, C.; Yu, Z.; Feng, S.; Pan, P.; Hao, G.; Dong, Y.; Yang, Y. SRY-related high-mobility-group box 4: Crucial regulators of the EMT in cancer. *Semin. Cancer Biol.* **2019**. [CrossRef] [PubMed]
24. Tiwari, N.; Tiwari, V.K.; Waldmeier, L.; Balwierz, P.J.; Arnold, P.; Pachkov, M.; Meyer-Schaller, N.; Schubeler, D.; van Nimwegen, E.; Christofori, G. Sox4 is a master regulator of epithelial-mesenchymal transition by controlling Ezh2 expression and epigenetic reprogramming. *Cancer Cell* **2013**, *23*, 768–783. [CrossRef]
25. Jolly, M.K.; Boareto, M.; Huang, B.; Jia, D.; Lu, M.; Ben-Jacob, E.; Onuchic, J.N.; Levine, H. Implications of the Hybrid Epithelial/Mesenchymal Phenotype in Metastasis. *Front. Oncol.* **2015**, *5*, 155. [CrossRef]
26. Elisha, Y.; Kalchenko, V.; Kuznetsov, Y.; Geiger, B. Dual role of E-cadherin in the regulation of invasive collective migration of mammary carcinoma cells. *Sci. Rep.* **2018**, *8*, 4986. [CrossRef]
27. Park, S.Y.; Choi, J.H.; Nam, J.S. Targeting Cancer Stem Cells in Triple-Negative Breast Cancer. *Cancers (Basel)* **2019**, *11*, 965. [CrossRef]
28. Batlle, E.; Clevers, H. Cancer stem cells revisited. *Nat. Med.* **2017**, *23*, 1124–1134. [CrossRef]
29. Chaffer, C.L.; Marjanovic, N.D.; Lee, T.; Bell, G.; Kleer, C.G.; Reinhardt, F.; D'Alessio, A.C.; Young, R.A.; Weinberg, R.A. Poised chromatin at the ZEB1 promoter enables breast cancer cell plasticity and enhances tumorigenicity. *Cell* **2013**, *154*, 61–74. [CrossRef]
30. Nieto, M.A.; Huang, R.Y.-J.; Jackson, R.A.; Thiery, J.P. EMT: 2016. *Cell* **2016**, *166*, 21–45. [CrossRef]
31. Du, B.; Shim, J.S. Targeting Epithelial-Mesenchymal Transition (EMT) to Overcome Drug Resistance in Cancer. *Molecules* **2016**, *21*, 965. [CrossRef] [PubMed]
32. Fischer, K.R.; Durrans, A.; Lee, S.; Sheng, J.; Li, F.; Wong, S.T.; Choi, H.; El Rayes, T.; Ryu, S.; Troeger, J.; et al. Epithelial-to-mesenchymal transition is not required for lung metastasis but contributes to chemoresistance. *Nature* **2015**, *527*, 472–476. [CrossRef] [PubMed]
33. Zheng, X.; Carstens, J.L.; Kim, J.; Scheible, M.; Kaye, J.; Sugimoto, H.; Wu, C.C.; LeBleu, V.S.; Kalluri, R. Epithelial-to-mesenchymal transition is dispensable for metastasis but induces chemoresistance in pancreatic cancer. *Nature* **2015**, *527*, 525–530. [CrossRef]

34. Hu, B.; Cheng, J.W.; Hu, J.W.; Li, H.; Ma, X.L.; Tang, W.G.; Sun, Y.F.; Guo, W.; Huang, A.; Zhou, K.Q.; et al. KPNA3 Confers Sorafenib Resistance to Advanced Hepatocellular Carcinoma via TWIST Regulated Epithelial-Mesenchymal Transition. *J. Cancer* **2019**, *10*, 3914–3925. [CrossRef] [PubMed]
35. Akalay, I.; Janji, B.; Hasmim, M.; Noman, M.Z.; Andre, F.; De Cremoux, P.; Bertheau, P.; Badoual, C.; Vielh, P.; Larsen, A.K.; et al. Epithelial-to-mesenchymal transition and autophagy induction in breast carcinoma promote escape from T-cell-mediated lysis. *Cancer Res.* **2013**, *73*, 2418–2427. [CrossRef] [PubMed]
36. Kudo-Saito, C.; Shirako, H.; Takeuchi, T.; Kawakami, Y. Cancer metastasis is accelerated through immunosuppression during Snail-induced EMT of cancer cells. *Cancer Cell* **2009**, *15*, 195–206. [CrossRef] [PubMed]
37. Ricciardi, M.; Zanotto, M.; Malpeli, G.; Bassi, G.; Perbellini, O.; Chilosi, M.; Bifari, F.; Krampera, M. Epithelial-to-mesenchymal transition (EMT) induced by inflammatory priming elicits mesenchymal stromal cell-like immune-modulatory properties in cancer cells. *Br. J. Cancer* **2015**, *112*, 1067–1075. [CrossRef]
38. Chen, L.; Gibbons, D.L.; Goswami, S.; Cortez, M.A.; Ahn, Y.H.; Byers, L.A.; Zhang, X.; Yi, X.; Dwyer, D.; Lin, W.; et al. Metastasis is regulated via microRNA-200/ZEB1 axis control of tumour cell PD-L1 expression and intratumoral immunosuppression. *Nat. Commun.* **2014**, *5*, 5241. [CrossRef]
39. Reiman, J.M.; Knutson, K.L.; Radisky, D.C. Immune promotion of epithelial-mesenchymal transition and generation of breast cancer stem cells. *Cancer Res.* **2010**, *70*, 3005–3008. [CrossRef]
40. Fullerton, J.N.; Gilroy, D.W. Resolution of inflammation: A new therapeutic frontier. *Nat. Rev. Drug Discov.* **2016**, *15*, 551–567. [CrossRef]
41. Lee, C.H. Resolvins as new fascinating drug candidates for inflammatory diseases. *Arch. Pharm. Res.* **2012**, *35*, 3–7. [CrossRef] [PubMed]
42. Mariathasan, S.; Newton, K.; Monack, D.M.; Vucic, D.; French, D.M.; Lee, W.P.; Roose-Girma, M.; Erickson, S.; Dixit, V.M. Differential activation of the inflammasome by caspase-1 adaptors ASC and Ipaf. *Nature* **2004**, *430*, 213–218. [CrossRef] [PubMed]
43. Guo, H.; Callaway, J.B.; Ting, J.P. Inflammasomes: Mechanism of action, role in disease, and therapeutics. *Nat. Med.* **2015**, *21*, 677–687. [CrossRef] [PubMed]
44. Martinon, F.; Burns, K.; Tschopp, J. The inflammasome: A molecular platform triggering activation of inflammatory caspases and processing of proIL-beta. *Mol. Cell* **2002**, *10*, 417–426. [CrossRef]
45. Song, S.; Qiu, D.; Luo, F.; Wei, J.; Wu, M.; Wu, H.; Du, C.; Du, Y.; Ren, Y.; Chen, N.; et al. Knockdown of NLRP3 alleviates high glucose or TGFB1-induced EMT in human renal tubular cells. *J. Mol. Endocrinol.* **2018**, *61*, 101–113. [CrossRef] [PubMed]
46. Yin, X.F.; Zhang, Q.; Chen, Z.Y.; Wang, H.F.; Li, X.; Wang, H.X.; Li, H.X.; Kang, C.M.; Chu, S.; Li, K.F.; et al. NLRP3 in human glioma is correlated with increased WHO grade, and regulates cellular proliferation, apoptosis and metastasis via epithelial-mesenchymal transition and the PTEN/AKT signaling pathway. *Int. J. Oncol.* **2018**, *53*, 973–986. [CrossRef] [PubMed]
47. Li, X.; Yan, X.; Wang, Y.; Wang, J.; Zhou, F.; Wang, H.; Xie, W.; Kong, H. NLRP3 inflammasome inhibition attenuates silica-induced epithelial to mesenchymal transition (EMT) in human bronchial epithelial cells. *Exp. Cell Res.* **2018**, *362*, 489–497. [CrossRef]
48. Tian, R.; Zhu, Y.; Yao, J.; Meng, X.; Wang, J.; Xie, H.; Wang, R. NLRP3 participates in the regulation of EMT in bleomycin-induced pulmonary fibrosis. *Exp. Cell Res.* **2017**, *357*, 328–334. [CrossRef]
49. Romero, C.A.; Remor, A.; Latini, A.; De Paul, A.L.; Torres, A.I.; Mukdsi, J.H. Uric acid activates NRLP3 inflammasome in an in-vivo model of epithelial to mesenchymal transition in the kidney. *J. Mol. Histol.* **2017**, *48*, 209–218. [CrossRef]
50. Wang, H.; Wang, Y.; Du, Q.; Lu, P.; Fan, H.; Lu, J.; Hu, R. Inflammasome-independent NLRP3 is required for epithelial-mesenchymal transition in colon cancer cells. *Exp. Cell Res.* **2016**, *342*, 184–192. [CrossRef]
51. Serhan, C.N.; Chiang, N.; Dalli, J. The resolution code of acute inflammation: Novel pro-resolving lipid mediators in resolution. *Semin. Immunol.* **2015**, *27*, 200–215. [CrossRef] [PubMed]
52. Serhan, C.N.; Chiang, N.; Dalli, J. New pro-resolving n-3 mediators bridge resolution of infectious inflammation to tissue regeneration. *Mol. Aspects Med.* **2018**, *64*, 1–17. [CrossRef] [PubMed]
53. Dalli, J.; Chiang, N.; Serhan, C.N. Elucidation of novel 13-series resolvins that increase with atorvastatin and clear infections. *Nat. Med.* **2015**, *21*, 1071–1075. [CrossRef] [PubMed]
54. Dominguez, C.; David, J.M.; Palena, C. Epithelial-mesenchymal transition and inflammation at the site of the primary tumor. *Semin. Cancer Biol.* **2017**, *47*, 177–184. [CrossRef] [PubMed]

55. Lou, Y.; Diao, L.; Cuentas, E.R.; Denning, W.L.; Chen, L.; Fan, Y.H.; Byers, L.A.; Wang, J.; Papadimitrakopoulou, V.A.; Behrens, C.; et al. Epithelial-Mesenchymal Transition Is Associated with a Distinct Tumor Microenvironment Including Elevation of Inflammatory Signals and Multiple Immune Checkpoints in Lung Adenocarcinoma. *Clin. Cancer Res.* **2016**, *22*, 3630–3642. [CrossRef] [PubMed]
56. Vanneman, M.; Dranoff, G. Combining immunotherapy and targeted therapies in cancer treatment. *Nat. Rev. Cancer* **2012**, *12*, 237–251. [CrossRef] [PubMed]
57. Bradley, J.R. TNF-mediated inflammatory disease. *J. Pathol.* **2008**, *214*, 149–160. [CrossRef]
58. Sheen-Chen, S.M.; Chen, W.J.; Eng, H.L.; Chou, F.F. Serum concentration of tumor necrosis factor in patients with breast cancer. *Breast Cancer Res. Treat.* **1997**, *43*, 211–215. [CrossRef]
59. Bates, R.C.; Mercurio, A.M. Tumor necrosis factor-alpha stimulates the epithelial-to-mesenchymal transition of human colonic organoids. *Mol. Biol. Cell* **2003**, *14*, 1790–1800. [CrossRef]
60. Kozlowski, L.; Zakrzewska, I.; Tokajuk, P.; Wojtukiewicz, M.Z. Concentration of interleukin-6 (IL-6), interleukin-8 (IL-8) and interleukin-10 (IL-10) in blood serum of breast cancer patients. *Rocz. Akad. Med. Bialymst.* **2003**, *48*, 82–84.
61. Sullivan, N.J.; Sasser, A.K.; Axel, A.E.; Vesuna, F.; Raman, V.; Ramirez, N.; Oberyszyn, T.M.; Hall, B.M. Interleukin-6 induces an epithelial-mesenchymal transition phenotype in human breast cancer cells. *Oncogene* **2009**, *28*, 2940–2947. [CrossRef] [PubMed]
62. Wu, J.; Zhang, J.; Shen, B.; Yin, K.; Xu, J.; Gao, W.; Zhang, L. Long noncoding RNA lncTCF7, induced by IL-6/STAT3 transactivation, promotes hepatocellular carcinoma aggressiveness through epithelial-mesenchymal transition. *J. Exp. Clin. Cancer Res.* **2015**, *34*, 116. [CrossRef] [PubMed]
63. Yamada, D.; Kobayashi, S.; Wada, H.; Kawamoto, K.; Marubashi, S.; Eguchi, H.; Ishii, H.; Nagano, H.; Doki, Y.; Mori, M. Role of crosstalk between interleukin-6 and transforming growth factor-beta 1 in epithelial-mesenchymal transition and chemoresistance in biliary tract cancer. *Eur. J. Cancer* **2013**, *49*, 1725–1740. [CrossRef] [PubMed]
64. Bates, R.C.; DeLeo, M.J., 3rd; Mercurio, A.M. The epithelial-mesenchymal transition of colon carcinoma involves expression of IL-8 and CXCR-1-mediated chemotaxis. *Exp. Cell Res.* **2004**, *299*, 315–324. [CrossRef] [PubMed]
65. Li, X.-J.; Peng, L.-X.; Shao, J.-Y.; Lu, W.-H.; Zhang, J.-X.; Chen, S.; Chen, Z.-Y.; Xiang, Y.-Q.; Bao, Y.-N.; Zheng, F.-J. As an independent unfavorable prognostic factor, IL-8 promotes metastasis of nasopharyngeal carcinoma through induction of epithelial–mesenchymal transition and activation of AKT signaling. *Carcinogenesis* **2012**, *33*, 1302–1309. [CrossRef] [PubMed]
66. Visciano, C.; Liotti, F.; Prevete, N.; Franco, R.; Collina, F.; De Paulis, A.; Marone, G.; Santoro, M.; Melillo, R. Mast cells induce epithelial-to-mesenchymal transition and stem cell features in human thyroid cancer cells through an IL-8-Akt-Slug pathway. *Oncogene* **2015**, *34*, 5175–5186. [CrossRef]
67. Qu, J.; Cheng, T.; Liu, L.; Heng, J.; Liu, X.; Sun, Z.; Wang, W.; Li, K.; Yang, N. Mast cells induce epithelial-to-mesenchymal transition and migration in non-small cell lung cancer through IL-8/Wnt/beta-catenin pathway. *J. Cancer* **2019**, *10*, 3830–3841. [CrossRef]
68. Fu, X.T.; Dai, Z.; Song, K.; Zhang, Z.J.; Zhou, Z.J.; Zhou, S.L.; Zhao, Y.M.; Xiao, Y.S.; Sun, Q.M.; Ding, Z.B.; et al. Macrophage-secreted IL-8 induces epithelial-mesenchymal transition in hepatocellular carcinoma cells by activating the JAK2/STAT3/Snail pathway. *Int. J. Oncol.* **2015**, *46*, 587–596. [CrossRef]
69. Palena, C.; Hamilton, D.H.; Fernando, R.I. Influence of IL-8 on the epithelial-mesenchymal transition and the tumor microenvironment. *Future Oncol.* **2012**, *8*, 713–722. [CrossRef]
70. Rouvier, E.; Luciani, M.F.; Mattei, M.G.; Denizot, F.; Golstein, P. CTLA-8, cloned from an activated T cell, bearing AU-rich messenger RNA instability sequences, and homologous to a herpesvirus saimiri gene. *J. Immunol.* **1993**, *150*, 5445–5456.
71. Zhang, Q.; Liu, S.; Parajuli, K.; Zhang, W.; Zhang, K.; Mo, Z.; Liu, J.; Chen, Z.; Yang, S.; Wang, A. Interleukin-17 promotes prostate cancer via MMP7-induced epithelial-to-mesenchymal transition. *Oncogene* **2017**, *36*, 687–699. [CrossRef] [PubMed]
72. Huang, Q.; Han, J.; Fan, J.; Duan, L.; Guo, M.; Lv, Z.; Hu, G.; Chen, L.; Wu, F.; Tao, X.; et al. IL-17 induces EMT via Stat3 in lung adenocarcinoma. *Am. J. Cancer Res.* **2016**, *6*, 440–451. [PubMed]
73. Xiang, T.; Long, H.; He, L.; Han, X.; Lin, K.; Liang, Z.; Zhuo, W.; Xie, R.; Zhu, B. Interleukin-17 produced by tumor microenvironment promotes self-renewal of CD133+ cancer stem-like cells in ovarian cancer. *Oncogene* **2015**, *34*, 165–176. [CrossRef] [PubMed]

74. Paudel, Y.N.; Angelopoulou, E.; Piperi, C.; Balasubramaniam, V.; Othman, I.; Shaikh, M.F. Enlightening the role of high mobility group box 1 (HMGB1) in inflammation: Updates on receptor signalling. *Eur. J. Pharmacol.* **2019**, *858*, 172487. [CrossRef] [PubMed]
75. Sun, S.; Zhang, W.; Cui, Z.; Chen, Q.; Xie, P.; Zhou, C.; Liu, B.; Peng, X.; Zhang, Y. High mobility group box-1 and its clinical value in breast cancer. *OncoTargets Ther.* **2015**, *8*, 413–419. [CrossRef]
76. Zhang, J.; Shao, S.; Han, D.; Xu, Y.; Jiao, D.; Wu, J.; Yang, F.; Ge, Y.; Shi, S.; Li, Y.; et al. High mobility group box 1 promotes the epithelial-to-mesenchymal transition in prostate cancer PC3 cells via the RAGE/NF-kappaB signaling pathway. *Int. J. Oncol.* **2018**, *53*, 659–671. [CrossRef]
77. Zhu, L.; Li, X.; Chen, Y.; Fang, J.; Ge, Z. High-mobility group box 1: A novel inducer of the epithelial-mesenchymal transition in colorectal carcinoma. *Cancer Lett.* **2015**, *357*, 527–534. [CrossRef]
78. Moorre, K.; OíGarra, A.; de Waal Malefyt, R.; Viera, P.; Mosmann, T. Interleukin-10. *Annu. Rev. Imunol.* **1993**, *11*, 165–190. [CrossRef]
79. Liu, C.Y.; Xu, J.Y.; Shi, X.Y.; Huang, W.; Ruan, T.Y.; Xie, P.; Ding, J.L. M2-polarized tumor-associated macrophages promoted epithelial-mesenchymal transition in pancreatic cancer cells, partially through TLR4/IL-10 signaling pathway. *Lab. Invest.* **2013**, *93*, 844–854. [CrossRef]
80. Miyazono, K. Transforming growth factor-beta signaling in epithelial-mesenchymal transition and progression of cancer. *Proc. Jpn. Acad. Ser. B Phys. Biol. Sci.* **2009**, *85*, 314–323. [CrossRef]
81. Ivanovic, V.; Todorovic-Rakovic, N.; Demajo, M.; Neskovic-Konstantinovic, Z.; Subota, V.; Ivanisevic-Milovanovic, O.; Nikolic-Vukosavljevic, D. Elevated plasma levels of transforming growth factor-beta 1 (TGF-beta 1) in patients with advanced breast cancer: Association with disease progression. *Eur. J. Cancer* **2003**, *39*, 454–461. [CrossRef]
82. Ahmadi, A.; Najafi, M.; Farhood, B.; Mortezaee, K. Transforming growth factor-beta signaling: Tumorigenesis and targeting for cancer therapy. *J. Cell Physiol.* **2019**, *234*, 12173–12187. [CrossRef] [PubMed]
83. Maeda, H.; Kuwahara, H.; Ichimura, Y.; Ohtsuki, M.; Kurakata, S.; Shiraishi, A. TGF-beta enhances macrophage ability to produce IL-10 in normal and tumor-bearing mice. *J. Immunol.* **1995**, *155*, 4926–4932. [PubMed]
84. Li, Y.; Wang, P.; Zhao, J.; Li, H.; Liu, D.; Zhu, W. HMGB1 attenuates TGF-beta-induced epithelial-mesenchymal transition of FaDu hypopharyngeal carcinoma cells through regulation of RAGE expression. *Mol. Cell Biochem.* **2017**, *431*, 1–10. [CrossRef] [PubMed]
85. Loh, J.K.; Hwang, S.L.; Lieu, A.S.; Huang, T.Y.; Howng, S.L. The alteration of prostaglandin E2 levels in patients with brain tumors before and after tumor removal. *J. Neurooncol.* **2002**, *57*, 147–150. [CrossRef] [PubMed]
86. O'Callaghan, G.; Houston, A. Prostaglandin E2 and the EP receptors in malignancy: Possible therapeutic targets? *Br. J. Pharmacol.* **2015**, *172*, 5239–5250. [CrossRef]
87. Dohadwala, M.; Yang, S.C.; Luo, J.; Sharma, S.; Batra, R.K.; Huang, M.; Lin, Y.; Goodglick, L.; Krysan, K.; Fishbein, M.C.; et al. Cyclooxygenase-2-dependent regulation of E-cadherin: Prostaglandin E(2) induces transcriptional repressors ZEB1 and snail in non-small cell lung cancer. *Cancer Res.* **2006**, *66*, 5338–5345. [CrossRef]
88. Mann, J.R.; Backlund, M.G.; Buchanan, F.G.; Daikoku, T.; Holla, V.R.; Rosenberg, D.W.; Dey, S.K.; DuBois, R.N. Repression of prostaglandin dehydrogenase by epidermal growth factor and snail increases prostaglandin E2 and promotes cancer progression. *Cancer Res.* **2006**, *66*, 6649–6656. [CrossRef]
89. Ha, B.; Lee, E.B.; Cui, J.; Kim, Y.; Jang, H.H. YB-1 overexpression promotes a TGF-beta1-induced epithelial-mesenchymal transition via Akt activation. *Biochem. Biophys. Res. Commun.* **2015**, *458*, 347–351. [CrossRef]
90. Zhang, A.; Wang, M.-H.; Dong, Z.; Yang, T. Prostaglandin E2 is a potent inhibitor of epithelial-to-mesenchymal transition: Interaction with hepatocyte growth factor. *Am. J. Physiol. Renal. Physiol.* **2006**, *291*, F1323–F1331. [CrossRef]
91. Takai, E.; Tsukimoto, M.; Kojima, S. TGF-1 downregulates COX-2 expression leading to decrease of PGE2 production in human lung cancer A549 cells, which is involved in fibrotic response to TGF-1. *PLoS ONE* **2013**, *8*, e76346. [CrossRef]
92. Carpagnano, G.E.; Palladino, G.P.; Lacedonia, D.; Koutelou, A.; Orlando, S.; Foschino-Barbaro, M.P. Neutrophilic airways inflammation in lung cancer: The role of exhaled LTB-4 and IL-8. *BMC Cancer* **2011**, *11*, 226. [CrossRef] [PubMed]
93. Park, M.K.; Park, Y.; Shim, J.; Lee, H.J.; Kim, S.; Lee, C.H. Novel involvement of leukotriene B4 receptor 2 through ERK activation by PP2A down-regulation in leukotriene B4-induced keratin phosphorylation and reorganization of pancreatic cancer cells. *Biochim. Biophys. Acta (BBA) Mol. Cell Res.* **2012**, *1823*, 2120–2129. [CrossRef]

94. Kim, H.; Choi, J.A.; Kim, J.H. Ras promotes transforming growth factor-beta (TGF-beta)-induced epithelial-mesenchymal transition via a leukotriene B4 receptor-2-linked cascade in mammary epithelial cells. *J. Biol. Chem.* **2014**, *289*, 22151–22160. [CrossRef]
95. Kim, Y.R.; Park, M.K.; Kang, G.J.; Kim, H.J.; Kim, E.J.; Byun, H.J.; Lee, M.Y.; Lee, C.H. Leukotriene B4 induces EMT and vimentin expression in PANC-1 pancreatic cancer cells: Involvement of BLT2 via ERK2 activation. *Prostaglandins Leukot. Essent. Fatty Acids* **2016**, *115*, 67–76. [CrossRef] [PubMed]
96. Burke, L.; Butler, C.T.; Murphy, A.; Moran, B.; Gallagher, W.M.; O'Sullivan, J.; Kennedy, B.N. Evaluation of Cysteinyl Leukotriene Signaling as a Therapeutic Target for Colorectal Cancer. *Front. Cell Dev. Biol.* **2016**, *4*, 103. [CrossRef]
97. Zhou, Y.; Guo, D.; Li, H.; Jie, S. Circulating LTD4 in patients with hepatocellular carcinoma. *Tumor Biol.* **2011**, *32*, 139–144. [CrossRef] [PubMed]
98. Kanaoka, Y.; Boyce, J.A. Cysteinyl leukotrienes and their receptors; emerging concepts. *Allergy Asthma Immunol. Res.* **2014**, *6*, 288–295. [CrossRef]
99. Hosoki, K.; Kainuma, K.; Toda, M.; Harada, E.; Chelakkot-Govindalayathila, A.L.; Roeen, Z.; Nagao, M.; D'Alessandro-Gabazza, C.N.; Fujisawa, T.; Gabazza, E.C. Montelukast suppresses epithelial to mesenchymal transition of bronchial epithelial cells induced by eosinophils. *Biochem. Biophys. Res. Commun.* **2014**, *449*, 351–356. [CrossRef]
100. Lukic, A.; Wahlund, C.J.E.; Gomez, C.; Brodin, D.; Samuelsson, B.; Wheelock, C.E.; Gabrielsson, S.; Radmark, O. Exosomes and cells from lung cancer pleural exudates transform LTC4 to LTD4, promoting cell migration and survival via CysLT1. *Cancer Lett.* **2019**, *444*, 1–8. [CrossRef]
101. Kang, H.S.; Lee, J.Y.; Kim, C.J. Anti-inflammatory activity of arctigenin from Forsythiae Fructus. *J. Ethnopharmacol.* **2008**, *116*, 305–312. [CrossRef] [PubMed]
102. Xu, Y.; Lou, Z.; Lee, S.H. Arctigenin represses TGF-beta-induced epithelial mesenchymal transition in human lung cancer cells. *Biochem. Biophys. Res. Commun.* **2017**, *493*, 934–939. [CrossRef] [PubMed]
103. Li, A.; Wang, J.; Zhu, D.; Zhang, X.; Pan, R.; Wang, R. Arctigenin suppresses transforming growth factor-beta1-induced expression of monocyte chemoattractant protein-1 and the subsequent epithelial-mesenchymal transition through reactive oxygen species-dependent ERK/NF-kappaB signaling pathway in renal tubular epithelial cells. *Free Radic. Res.* **2015**, *49*, 1095–1113. [CrossRef] [PubMed]
104. Yang, X.; Yang, J.; Zou, H. Baicalin inhibits IL-17-mediated joint inflammation in murine adjuvant-induced arthritis. *Clin. Dev. Immunol.* **2013**, *2013*, 268065. [CrossRef] [PubMed]
105. Chung, H.; Choi, H.S.; Seo, E.K.; Kang, D.H.; Oh, E.S. Baicalin and baicalein inhibit transforming growth factor-beta1-mediated epithelial-mesenchymal transition in human breast epithelial cells. *Biochem. Biophys. Res. Commun.* **2015**, *458*, 707–713. [CrossRef] [PubMed]
106. Wang, Y.; Wang, H.; Zhou, R.; Zhong, W.; Lu, S.; Ma, Z.; Chai, Y. Baicalin inhibits human osteosarcoma cells invasion, metastasis, and anoikis resistance by suppressing the transforming growth factor-beta1-induced epithelial-to-mesenchymal transition. *Anticancer Drugs* **2017**, *28*, 581–587. [CrossRef]
107. Zhou, T.; Zhang, A.; Kuang, G.; Gong, X.; Jiang, R.; Lin, D.; Li, J.; Li, H.; Zhang, X.; Wan, J.; et al. Baicalin inhibits the metastasis of highly aggressive breast cancer cells by reversing epithelial-to-mesenchymal transition by targeting beta-catenin signaling. *Oncol. Rep.* **2017**, *38*, 3599–3607. [CrossRef]
108. Jin, Y.; Khadka, D.B.; Cho, W.J. Pharmacological effects of berberine and its derivatives: A patent update. *Expert Opin. Ther. Pat.* **2016**, *26*, 229–243. [CrossRef]
109. Zhang, Y.; Li, X.; Zhang, Q.; Li, J.; Ju, J.; Du, N.; Liu, X.; Chen, X.; Cheng, F.; Yang, L.; et al. Berberine hydrochloride prevents postsurgery intestinal adhesion and inflammation in rats. *J. Pharmacol. Exp. Ther.* **2014**, *349*, 417–426. [CrossRef]
110. Chu, S.C.; Yu, C.C.; Hsu, L.S.; Chen, K.S.; Su, M.Y.; Chen, P.N. Berberine reverses epithelial-to-mesenchymal transition and inhibits metastasis and tumor-induced angiogenesis in human cervical cancer cells. *Mol. Pharmacol.* **2014**, *86*, 609–623. [CrossRef]
111. Liu, C.H.; Tang, W.C.; Sia, P.; Huang, C.C.; Yang, P.M.; Wu, M.H.; Lai, I.L.; Lee, K.H. Berberine inhibits the metastatic ability of prostate cancer cells by suppressing epithelial-to-mesenchymal transition (EMT)-associated genes with predictive and prognostic relevance. *Int. J. Med. Sci.* **2015**, *12*, 63–71. [CrossRef] [PubMed]

112. Wang, J.; Kang, M.; Wen, Q.; Qin, Y.T.; Wei, Z.X.; Xiao, J.J.; Wang, R.S. Berberine sensitizes nasopharyngeal carcinoma cells to radiation through inhibition of Sp1 and EMT. *Oncol. Rep.* **2017**, *37*, 2425–2432. [CrossRef] [PubMed]
113. Harmer, R. Occurrence, chemistry and application of betanin. *Food Chem.* **1980**, *5*, 81–90. [CrossRef]
114. Martinez, R.M.; Longhi-Balbinot, D.T.; Zarpelon, A.C.; Staurengo-Ferrari, L.; Baracat, M.M.; Georgetti, S.R.; Sassonia, R.C.; Verri, W.A., Jr.; Casagrande, R. Anti-inflammatory activity of betalain-rich dye of Beta vulgaris: Effect on edema, leukocyte recruitment, superoxide anion and cytokine production. *Arch. Pharm. Res.* **2015**, *38*, 494–504. [CrossRef] [PubMed]
115. Sutariya, B.; Taneja, N.; Badgujar, L.; Saraf, M. Modulatory effect of betanin on high glucose induced epithelial to mesenchymal transition in renal proximal tubular cells. *Biomed. Pharmacother.* **2017**, *89*, 18–28. [CrossRef] [PubMed]
116. Turpaev, K.; Welsh, N. Brusatol inhibits the response of cultured beta-cells to pro-inflammatory cytokines in vitro. *Biochem. Biophys. Res. Commun.* **2015**, *460*, 868–872. [CrossRef] [PubMed]
117. Lu, Z.; Lai, Z.Q.; Leung, A.W.N.; Leung, P.S.; Li, Z.S.; Lin, Z.X. Exploring brusatol as a new anti-pancreatic cancer adjuvant: Biological evaluation and mechanistic studies. *Oncotarget* **2017**, *8*, 84974–84985. [CrossRef]
118. Park, M.K.; Lee, H.J.; Choi, J.K.; Kim, H.J.; Kang, J.H.; Lee, E.J.; Kim, Y.R.; Kang, J.H.; Yoo, J.K.; Cho, H.Y.; et al. Novel anti-nociceptive effects of cardamonin via blocking expression of cyclooxygenase-2 and transglutaminase-2. *Pharmacol. Biochem. Behav.* **2014**, *118*, 10–15. [CrossRef]
119. Park, M.K.; Choi, J.K.; Kim, H.J.; Nakahata, N.; Lim, K.M.; Kim, S.Y.; Lee, C.H. Novel inhibitory effects of cardamonin on thromboxane A2-induced scratching response: Blocking of Gh/transglutaminase-2 binding to thromboxane A2 receptor. *Pharmacol. Biochem. Behav.* **2014**, *126*, 131–135. [CrossRef]
120. Kim, E.J.; Kim, H.J.; Park, M.K.; Kang, G.J.; Byun, H.J.; Lee, H.; Lee, C.H. Cardamonin Suppresses TGF-beta1-Induced Epithelial Mesenchymal Transition via Restoring Protein Phosphatase 2A Expression. *Biomol. Ther. (Seoul)* **2015**, *23*, 141–148. [CrossRef]
121. Shrivastava, S.; Jeengar, M.K.; Thummuri, D.; Koval, A.; Katanaev, V.L.; Marepally, S.; Naidu, V.G.M. Cardamonin, a chalcone, inhibits human triple negative breast cancer cell invasiveness by downregulation of Wnt/beta-catenin signaling cascades and reversal of epithelial-mesenchymal transition. *Biofactors* **2017**, *43*, 152–169. [CrossRef] [PubMed]
122. Park, M.K.; Jo, S.H.; Lee, H.J.; Kang, J.H.; Kim, Y.R.; Kim, H.J.; Lee, E.J.; Koh, J.Y.; Ahn, K.O.; Jung, K.C.; et al. Novel suppressive effects of cardamonin on the activity and expression of transglutaminase-2 lead to blocking the migration and invasion of cancer cells. *Life Sci.* **2013**, *92*, 154–160. [CrossRef] [PubMed]
123. Lo, A.H.; Liang, Y.C.; Lin-Shiau, S.Y.; Ho, C.T.; Lin, J.K. Carnosol, an antioxidant in rosemary, suppresses inducible nitric oxide synthase through down-regulating nuclear factor-kappaB in mouse macrophages. *Carcinogenesis* **2002**, *23*, 983–991. [CrossRef] [PubMed]
124. Guerrero, I.C.; Andres, L.S.; Leon, L.G.; Machin, R.P.; Padron, J.M.; Luis, J.G.; Delgadillo, J. Abietane diterpenoids from Salvia pachyphylla and S. clevelandii with cytotoxic activity against human cancer cell lines. *J. Nat. Prod.* **2006**, *69*, 1803–1805. [CrossRef] [PubMed]
125. Yeo, I.J.; Park, J.H.; Jang, J.S.; Lee, D.Y.; Park, J.E.; Choi, Y.E.; Joo, J.H.; Song, J.K.; Jeon, H.O.; Hong, J.T. Inhibitory effect of Carnosol on UVB-induced inflammation via inhibition of STAT3. *Arch. Pharm Res.* **2019**, *42*, 274–283. [CrossRef]
126. Kashyap, D.; Kumar, G.; Sharma, A.; Sak, K.; Tuli, H.S.; Mukherjee, T.K. Mechanistic insight into carnosol-mediated pharmacological effects: Recent trends and advancements. *Life Sci.* **2017**, *169*, 27–36. [CrossRef]
127. Giacomelli, C.; Daniele, S.; Natali, L.; Iofrida, C.; Flamini, G.; Braca, A.; Trincavelli, M.L.; Martini, C. Carnosol controls the human glioblastoma stemness features through the epithelial-mesenchymal transition modulation and the induction of cancer stem cell apoptosis. *Sci. Rep.* **2017**, *7*, 15174. [CrossRef]
128. Zhao, H.; Wang, Z.; Tang, F.; Zhao, Y.; Feng, D.; Li, Y.; Hu, Y.; Wang, C.; Zhou, J.; Tian, X.; et al. Carnosol-mediated Sirtuin 1 activation inhibits Enhancer of Zeste Homolog 2 to attenuate liver fibrosis. *Pharmacol. Res.* **2018**, *128*, 327–337. [CrossRef]
129. Venkatesha, S.H.; Moudgil, K.D. Celastrol suppresses experimental autoimmune encephalomyelitis via MAPK/SGK1-regulated mediators of autoimmune pathology. *Inflamm. Res.* **2019**, *68*, 285–296. [CrossRef]
130. Kang, H.; Lee, M.; Jang, S.W. Celastrol inhibits TGF-beta1-induced epithelial-mesenchymal transition by inhibiting Snail and regulating E-cadherin expression. *Biochem. Biophys. Res. Commun.* **2013**, *437*, 550–556. [CrossRef]

131. Kashyap, D.; Sharma, A.; Tuli, H.S.; Sak, K.; Mukherjee, T.; Bishayee, A. Molecular targets of celastrol in cancer: Recent trends and advancements. *Crit. Rev. Oncol. Hematol.* **2018**, *128*, 70–81. [CrossRef] [PubMed]
132. Wang, S.; Cai, R.; Ma, J.; Liu, T.; Ke, X.; Lu, H.; Fu, J. The natural compound codonolactone impairs tumor induced angiogenesis by downregulating BMP signaling in endothelial cells. *Phytomedicine* **2015**, *22*, 1017–1026. [CrossRef] [PubMed]
133. Liu, C.; Zhao, H.; Ji, Z.H.; Yu, X.Y. Neuroprotection of atractylenolide III from Atractylodis macrocephalae against glutamate-induced neuronal apoptosis via inhibiting caspase signaling pathway. *Neurochem. Res.* **2014**, *39*, 1753–1758. [CrossRef] [PubMed]
134. Wang, W.; Chen, B.; Zou, R.; Tu, X.; Tan, S.; Lu, H.; Liu, Z.; Fu, J. Codonolactone, a sesquiterpene lactone isolated from Chloranthus henryi Hemsl, inhibits breast cancer cell invasion, migration and metastasis by downregulating the transcriptional activity of Runx2. *Int. J. Oncol.* **2014**, *45*, 1891–1900. [CrossRef] [PubMed]
135. Cunningham, K.G.; Manson, W.; Spring, F.S.; Hutchinson, S.A. Cordycepin, a metabolic product isolated from cultures of Cordyceps militaris (Linn.) Link. *Nature* **1950**, *166*, 949. [CrossRef]
136. Qing, R.; Huang, Z.; Tang, Y.; Xiang, Q.; Yang, F. Cordycepin negatively modulates lipopolysaccharide-induced cytokine production by up-regulation of heme oxygenase-1. *Int. Immunopharmacol.* **2017**, *47*, 20–27. [CrossRef]
137. Wang, C.W.; Lee, B.H.; Tai, C.J. The inhibition of cordycepin on cancer stemness in TGF-beta induced chemo-resistant ovarian cancer cell. *Oncotarget* **2017**, *8*, 111912–111921. [CrossRef]
138. Su, N.W.; Wu, S.H.; Chi, C.W.; Liu, C.J.; Tsai, T.H.; Chen, Y.J. Metronomic Cordycepin Therapy Prolongs Survival of Oral Cancer-Bearing Mice and Inhibits Epithelial-Mesenchymal Transition. *Molecules* **2017**, *22*, 629. [CrossRef]
139. Yao, W.L.; Ko, B.S.; Liu, T.A.; Liang, S.M.; Liu, C.C.; Lu, Y.J.; Tzean, S.S.; Shen, T.L.; Liou, J.Y. Cordycepin suppresses integrin/FAK signaling and epithelial-mesenchymal transition in hepatocellular carcinoma. *Anticancer Agents Med. Chem.* **2014**, *14*, 29–34. [CrossRef]
140. Feng, Z.; Zheng, W.; Li, X.; Lin, J.; Xie, C.; Li, H.; Cheng, L.; Wu, A.; Ni, W. Cryptotanshinone protects against IL-1beta-induced inflammation in human osteoarthritis chondrocytes and ameliorates the progression of osteoarthritis in mice. *Int. Immunopharmacol.* **2017**, *50*, 161–167. [CrossRef]
141. Zhang, Y.; Cabarcas, S.M.; Zheng, J.I.; Sun, L.; Mathews, L.A.; Zhang, X.; Lin, H.; Farrar, W.L. Cryptotanshinone targets tumor-initiating cells through down-regulation of stemness genes expression. *Oncol. Lett.* **2016**, *11*, 3803–3812. [CrossRef] [PubMed]
142. Zhang, L.; Cheng, X.; Gao, Y.; Zhang, C.; Bao, J.; Guan, H.; Yu, H.; Lu, R.; Xu, Q.; Sun, Y. Curcumin inhibits metastasis in human papillary thyroid carcinoma BCPAP cells via down-regulation of the TGF-beta/Smad2/3 signaling pathway. *Exp. Cell Res.* **2016**, *341*, 157–165. [CrossRef]
143. Bahrami, A.; Majeed, M.; Sahebkar, A. Curcumin: A potent agent to reverse epithelial-to-mesenchymal transition. *Cell Oncol. (Dordr)* **2019**, *42*, 405–421. [CrossRef] [PubMed]
144. Jiang, G.-M.; Xie, W.-Y.; Wang, H.-S.; Du, J.; Wu, B.-P.; Xu, W.; Liu, H.-F.; Xiao, P.; Liu, Z.-G.; Li, H.-Y. Curcumin combined with FAPαc vaccine elicits effective antitumor response by targeting indolamine-2, 3-dioxygenase and inhibiting EMT induced by TNF-α in melanoma. *Oncotarget* **2015**, *6*, 25932–25942.
145. Aedo-Aguilera, V.; Carrillo-Beltran, D.; Calaf, G.M.; Munoz, J.P.; Guerrero, N.; Osorio, J.C.; Tapia, J.C.; Leon, O.; Contreras, H.R.; Aguayo, F. Curcumin decreases epithelialmesenchymal transition by a Pirindependent mechanism in cervical cancer cells. *Oncol. Rep.* **2019**. [CrossRef]
146. Huber, M.A.; Beug, H.; Wirth, T. Epithelial-mesenchymal transition: NF-kappaB takes center stage. *Cell Cycle* **2004**, *3*, 1477–1480. [CrossRef]
147. Tyagi, N.; Singh, D.K.; Dash, D.; Singh, R. Curcumin Modulates Paraquat-Induced Epithelial to Mesenchymal Transition by Regulating Transforming Growth Factor-beta (TGF-beta) in A549 Cells. *Inflammation* **2019**, *42*, 1441–1455. [CrossRef]
148. Yin, J.; Wang, L.; Wang, Y.; Shen, H.; Wang, X.; Wu, L. Curcumin reverses oxaliplatin resistance in human colorectal cancer via regulation of TGF-beta/Smad2/3 signaling pathway. *OncoTargets Ther.* **2019**, *12*, 3893–3903. [CrossRef]
149. Lim, W.C.; Kim, H.; Kim, Y.J.; Choi, K.C.; Lee, I.H.; Lee, K.H.; Kim, M.K.; Ko, H. Dioscin suppresses TGF-beta1-induced epithelial-mesenchymal transition and suppresses A549 lung cancer migration and invasion. *Bioorg. Med. Chem. Lett.* **2017**, *27*, 3342–3348. [CrossRef]

150. Wu, S.; Xu, H.; Peng, J.; Wang, C.; Jin, Y.; Liu, K.; Sun, H.; Qin, J. Potent anti-inflammatory effect of dioscin mediated by suppression of TNF-α-induced VCAM-1, ICAM-1and EL expression via the NF-κB pathway. *Biochimie* **2015**, *110*, 62–72. [CrossRef]
151. Chen, B.; Zhou, S.; Zhan, Y.; Ke, J.; Wang, K.; Liang, Q.; Hou, Y.; Zhu, P.; Ao, W.; Wei, X.; et al. Dioscin Inhibits the Invasion and Migration of Hepatocellular Carcinoma HepG2 Cells by Reversing TGF-beta1-Induced Epithelial-Mesenchymal Transition. *Molecules* **2019**, *24*, 2222. [CrossRef]
152. Chang, H.Y.; Kao, M.C.; Way, T.D.; Ho, C.T.; Fu, E. Diosgenin suppresses hepatocyte growth factor (HGF)-induced epithelial-mesenchymal transition by down-regulation of Mdm2 and vimentin. *J. Agric. Food Chem.* **2011**, *59*, 5357–5363. [CrossRef] [PubMed]
153. Bhuvanalakshmi, G.; Rangappa, K.S.; Dharmarajan, A.; Sethi, G.; Kumar, A.P.; Warrier, S. Breast Cancer Stem-Like Cells Are Inhibited by Diosgenin, a Steroidal Saponin, by the Attenuation of the Wnt beta-Catenin Signaling via the Wnt Antagonist Secreted Frizzled Related Protein-4. *Front. Pharmacol.* **2017**, *8*, 124. [CrossRef] [PubMed]
154. Sogo, T.; Terahara, N.; Hisanaga, A.; Kumamoto, T.; Yamashiro, T.; Wu, S.; Sakao, K.; Hou, D.X. Anti-inflammatory activity and molecular mechanism of delphinidin 3-sambubioside, a Hibiscus anthocyanin. *BioFactors* **2015**, *41*, 58–65. [CrossRef] [PubMed]
155. Ouanouki, A.; Lamy, S.; Annabi, B. Anthocyanidins inhibit epithelial–mesenchymal transition through a TGF/Smad2 signaling pathway in glioblastoma cells. *Mol. Carcinog.* **2017**, *56*, 1088–1099. [CrossRef] [PubMed]
156. Lim, W.C.; Kim, H.; Ko, H. Delphinidin inhibits epidermal growth factor-induced epithelial-to-mesenchymal transition in hepatocellular carcinoma cells. *J. Cell Biochem.* **2019**, *120*, 9887–9899. [CrossRef]
157. Byun, E.B.; Kim, W.S.; Sung, N.Y.; Byun, E.H. Epigallocatechin-3-Gallate Regulates Anti-Inflammatory Action Through 67-kDa Laminin Receptor-Mediated Tollip Signaling Induction in Lipopolysaccharide-Stimulated Human Intestinal Epithelial Cells. *Cell Physiol. Biochem.* **2018**, *46*, 2072–2081. [CrossRef]
158. Shi, J.; Liu, F.; Zhang, W.; Liu, X.; Lin, B.; Tang, X. Epigallocatechin-3-gallate inhibits nicotine-induced migration and invasion by the suppression of angiogenesis and epithelial-mesenchymal transition in non-small cell lung cancer cells. *Oncol. Rep.* **2015**, *33*, 2972–2980. [CrossRef]
159. Li, T.; Zhao, N.; Lu, J.; Zhu, Q.; Liu, X.; Hao, F.; Jiao, X. Epigallocatechin gallate (EGCG) suppresses epithelial-Mesenchymal transition (EMT) and invasion in anaplastic thyroid carcinoma cells through blocking of TGF-beta1/Smad signaling pathways. *Bioengineered* **2019**, *10*, 282–291. [CrossRef]
160. Boldbaatar, A.; Lee, S.; Han, S.; Jeong, A.L.; Ka, H.I.; Buyanravjikh, S.; Lee, J.H.; Lim, J.S.; Lee, M.S.; Yang, Y. Eupatolide inhibits the TGF-beta1-induced migration of breast cancer cells via downregulation of SMAD3 phosphorylation and transcriptional repression of ALK5. *Oncol. Lett.* **2017**, *14*, 6031–6039. [CrossRef]
161. Lee, J.; Tae, N.; Lee, J.J.; Kim, T.; Lee, J.H. Eupatolide inhibits lipopolysaccharide-induced COX-2 and iNOS expression in RAW264.7 cells by inducing proteasomal degradation of TRAF6. *Eur. J. Pharmacol.* **2010**, *636*, 173–180. [CrossRef] [PubMed]
162. Fan, Y.; Piao, C.H.; Hyeon, E.; Jung, S.Y.; Eom, J.E.; Shin, H.S.; Song, C.H.; Chai, O.H. Gallic acid alleviates nasal inflammation via activation of Th1 and inhibition of Th2 and Th17 in a mouse model of allergic rhinitis. *Int. Immunopharmacol.* **2019**, *70*, 512–519. [CrossRef] [PubMed]
163. BenSaad, L.A.; Kim, K.H.; Quah, C.C.; Kim, W.R.; Shahimi, M. Anti-inflammatory potential of ellagic acid, gallic acid and punicalagin A&B isolated from Punica granatum. *BMC Complement. Altern. Med.* **2017**, *17*, 47. [CrossRef]
164. Sunil Gowda, S.N.; Rajasowmiya, S.; Vadivel, V.; Banu Devi, S.; Celestin Jerald, A.; Marimuthu, S.; Devipriya, N. Gallic acid-coated sliver nanoparticle alters the expression of radiation-induced epithelial-mesenchymal transition in non-small lung cancer cells. *Toxicol. In Vitro* **2018**, *52*, 170–177. [CrossRef]
165. Chang, Y.C.; Chen, P.N.; Chu, S.C.; Lin, C.Y.; Kuo, W.H.; Hsieh, Y.S. Black tea polyphenols reverse epithelial-to-mesenchymal transition and suppress cancer invasion and proteases in human oral cancer cells. *J. Agric. Food Chem.* **2012**, *60*, 8395–8403. [CrossRef]
166. Ren, J.; Li, L.; Wang, Y.; Zhai, J.; Chen, G.; Hu, K. Gambogic acid induces heme oxygenase-1 through Nrf2 signaling pathway and inhibits NF-kappaB and MAPK activation to reduce inflammation in LPS-activated RAW264.7 cells. *Biomed. Pharmacother.* **2019**, *109*, 555–562. [CrossRef]
167. Zhao, K.; Zhang, S.; Song, X.; Yao, Y.; Zhou, Y.; You, Q.; Guo, Q.; Lu, N. Gambogic acid suppresses cancer invasion and migration by inhibiting TGFbeta1-induced epithelial-to-mesenchymal transition. *Oncotarget* **2017**, *8*, 27120–27136. [CrossRef]

168. Yue, Q.; Feng, L.; Cao, B.; Liu, M.; Zhang, D.; Wu, W.; Jiang, B.; Yang, M.; Liu, X.; Guo, D. Proteomic Analysis Revealed the Important Role of Vimentin in Human Cervical Carcinoma HeLa Cells Treated With Gambogic Acid. *Mol. Cell. Proteom.* **2016**, *15*, 26–44. [CrossRef]
169. Tharmarajah, L.; Samarakoon, S.R.; Ediriweera, M.K.; Piyathilaka, P.; Tennekoon, K.H.; Senathilake, K.S.; Rajagopalan, U.; Galhena, P.B.; Thabrew, I. In Vitro Anticancer Effect of Gedunin on Human Teratocarcinomal (NTERA-2) Cancer Stem-Like Cells. *Biomed. Res. Int.* **2017**, *2017*, 2413197. [CrossRef]
170. Borges, P.V.; Moret, K.H.; Raghavendra, N.M.; Maramaldo Costa, T.E.; Monteiro, A.P.; Carneiro, A.B.; Pacheco, P.; Temerozo, J.R.; Bou-Habib, D.C.; das Gracas Henriques, M.; et al. Protective effect of gedunin on TLR-mediated inflammation by modulation of inflammasome activation and cytokine production: Evidence of a multitarget compound. *Pharmacol. Res.* **2017**, *115*, 65–77. [CrossRef]
171. Subramani, R.; Gonzalez, E.; Nandy, S.B.; Arumugam, A.; Camacho, F.; Medel, J.; Alabi, D.; Lakshmanaswamy, R. Gedunin inhibits pancreatic cancer by altering sonic hedgehog signaling pathway. *Oncotarget* **2017**, *8*, 10891–10904. [CrossRef] [PubMed]
172. Bhattarai, G.; Poudel, S.B.; Kook, S.H.; Lee, J.C. Anti-inflammatory, anti-osteoclastic, and antioxidant activities of genistein protect against alveolar bone loss and periodontal tissue degradation in a mouse model of periodontitis. *J. Biomed. Mater. Res. A* **2017**, *105*, 2510–2521. [CrossRef] [PubMed]
173. Wang, A.; Wei, J.; Lu, C.; Chen, H.; Zhong, X.; Lu, Y.; Li, L.; Huang, H.; Dai, Z.; Han, L. Genistein suppresses psoriasis-related inflammation through a STAT3-NF-kappaB-dependent mechanism in keratinocytes. *Int. Immunopharmacol.* **2019**, *69*, 270–278. [CrossRef] [PubMed]
174. Spagnuolo, C.; Moccia, S.; Russo, G.L. Anti-inflammatory effects of flavonoids in neurodegenerative disorders. *Eur. J. Med. Chem.* **2018**, *153*, 105–115. [CrossRef]
175. Lee, G.A.; Hwang, K.A.; Choi, K.C. Roles of Dietary Phytoestrogens on the Regulation of Epithelial-Mesenchymal Transition in Diverse Cancer Metastasis. *Toxins (Basel)* **2016**, *8*, 162. [CrossRef]
176. Zhou, P.; Wang, C.; Hu, Z.; Chen, W.; Qi, W.; Li, A. Genistein induces apoptosis of colon cancer cells by reversal of epithelial-to-mesenchymal via a Notch1/NF-kappaB/slug/E-cadherin pathway. *BMC Cancer* **2017**, *17*, 813. [CrossRef]
177. Ma, J.; Zeng, F.; Ma, C.; Pang, H.; Fang, B.; Lian, C.; Yin, B.; Zhang, X.; Wang, Z.; Xia, J. Synergistic reversal effect of epithelial-to-mesenchymal transition by miR-223 inhibitor and genistein in gemcitabine-resistant pancreatic cancer cells. *Am. J. Cancer Res.* **2016**, *6*, 1384–1395.
178. Nam, H.H.; Nan, L.; Park, J.C.; Choo, B.K. Geraniin ameliorate experimental acute reflux esophagitis via NF-κB regulated anti-inflammatory activities in rats. *App. Biol. Chem.* **2019**, *62*, 13. [CrossRef]
179. Lee, J.C.; Tsai, C.Y.; Kao, J.Y.; Kao, M.C.; Tsai, S.C.; Chang, C.S.; Huang, L.J.; Kuo, S.C.; Lin, J.K.; Way, T.D. Geraniin-mediated apoptosis by cleavage of focal adhesion kinase through up-regulation of Fas ligand expression in human melanoma cells. *Mol. Nutr. Food Res.* **2008**, *52*, 655–663. [CrossRef]
180. Ko, H. Geraniin inhibits TGF-beta1-induced epithelial-mesenchymal transition and suppresses A549 lung cancer migration, invasion and anoikis resistance. *Bioorg. Med. Chem. Lett.* **2015**, *25*, 3529–3534. [CrossRef]
181. Won, J.H.; Kim, J.Y.; Yun, K.J.; Lee, J.H.; Back, N.I.; Chung, H.G.; Chung, S.A.; Jeong, T.S.; Choi, M.S.; Lee, K.T. Gigantol isolated from the whole plants of Cymbidium goeringii inhibits the LPS-induced iNOS and COX-2 expression via NF-kappaB inactivation in RAW 264.7 macrophages cells. *Planta Med.* **2006**, *72*, 1181–1187. [CrossRef] [PubMed]
182. Unahabhokha, T.; Chanvorachote, P.; Pongrakhananon, V. The attenuation of epithelial to mesenchymal transition and induction of anoikis by gigantol in human lung cancer H460 cells. *Tumour Biol.* **2016**, *37*, 8633–8641. [CrossRef] [PubMed]
183. Bhummaphan, N.; Chanvorachote, P. Gigantol Suppresses Cancer Stem Cell-Like Phenotypes in Lung Cancer Cells. *Evid. Based Complement. Alternat. Med.* **2015**, *2015*, 836564. [CrossRef] [PubMed]
184. Kotake, D.; Hirasawa, N. Activation of a retinoic acid receptor pathway by thiazolidinediones induces production of vascular endothelial growth factor/vascular permeability factor in OP9 adipocytes. *Eur. J. Pharmacol.* **2013**, *707*, 95–103. [CrossRef]
185. Baek, S.H.; Ko, J.H.; Lee, J.H.; Kim, C.; Lee, H.; Nam, D.; Lee, J.; Lee, S.G.; Yang, W.M.; Um, J.Y.; et al. Ginkgolic Acid Inhibits Invasion and Migration and TGF-beta-Induced EMT of Lung Cancer Cells Through PI3K/Akt/mTOR Inactivation. *J. Cell Physiol.* **2017**, *232*, 346–354. [CrossRef]
186. Yi, Y.S. Roles of ginsenosides in inflammasome activation. *J. Ginseng Res.* **2019**, *43*, 172–178. [CrossRef]

187. Tian, L.; Shen, D.; Li, X.; Shan, X.; Wang, X.; Yan, Q.; Liu, J. Ginsenoside Rg3 inhibits epithelial-mesenchymal transition (EMT) and invasion of lung cancer by down-regulating FUT4. *Oncotarget* **2016**, *7*, 1619–1632. [CrossRef]
188. Wang, J.; Tian, L.; Khan, M.N.; Zhang, L.; Chen, Q.; Zhao, Y.; Yan, Q.; Fu, L.; Liu, J. Ginsenoside Rg3 sensitizes hypoxic lung cancer cells to cisplatin via blocking of NF-kappaB mediated epithelial-mesenchymal transition and stemness. *Cancer Lett.* **2018**, *415*, 73–85. [CrossRef]
189. Liu, D.; Liu, T.; Teng, Y.; Chen, W.; Zhao, L.; Li, X. Ginsenoside Rb1 inhibits hypoxia-induced epithelial-mesenchymal transition in ovarian cancer cells by regulating microRNA-25. *Exp. Ther. Med.* **2017**, *14*, 2895–2902. [CrossRef]
190. Zhang, L.; Shan, X.; Chen, Q.; Xu, D.; Fan, X.; Yu, M.; Yan, Q.; Liu, J. Downregulation of HDAC3 by ginsenoside Rg3 inhibits epithelial-mesenchymal transition of cutaneous squamous cell carcinoma through c-Jun acetylation. *J. Cell. Physiol.* **2019**, *234*, 22207–22219. [CrossRef]
191. Cheng, Z.; Xing, D. Ginsenoside Rg3 inhibits growth and epithelial-mesenchymal transition of human oral squamous carcinoma cells by down-regulating miR-221. *Eur. J. Pharmacol.* **2019**, *853*, 353–363. [CrossRef] [PubMed]
192. Chang, H.Y.; Chen, S.Y.; Wu, C.H.; Lu, C.C.; Yen, G.C. Glycyrrhizin Attenuates the Process of Epithelial-to-Mesenchymal Transition by Modulating HMGB1 Initiated Novel Signaling Pathway in Prostate Cancer Cells. *J. Agric. Food Chem.* **2019**, *67*, 3323–3332. [CrossRef] [PubMed]
193. Woodbury, A.; Yu, S.P.; Wei, L.; Garcia, P. Neuro-modulating effects of honokiol: A review. *Front. Neurol.* **2013**, *4*, 130. [CrossRef] [PubMed]
194. Lee, J.; Jung, E.; Park, J.; Jung, K.; Lee, S.; Hong, S.; Park, J.; Park, E.; Kim, J.; Park, S.; et al. Anti-inflammatory effects of magnolol and honokiol are mediated through inhibition of the downstream pathway of MEKK-1 in NF-kappaB activation signaling. *Planta Med.* **2005**, *71*, 338–343. [CrossRef]
195. Lv, X.Q.; Qiao, X.R.; Su, L.; Chen, S.Z. Honokiol inhibits EMT-mediated motility and migration of human non-small cell lung cancer cells in vitro by targeting c-FLIP. *Acta Pharmacol. Sin.* **2016**, *37*, 1574–1586. [CrossRef]
196. Avtanski, D.B.; Nagalingam, A.; Bonner, M.Y.; Arbiser, J.L.; Saxena, N.K.; Sharma, D. Honokiol inhibits epithelial-mesenchymal transition in breast cancer cells by targeting signal transducer and activator of transcription 3/Zeb1/E-cadherin axis. *Mol. Oncol.* **2014**, *8*, 565–580. [CrossRef]
197. Li, W.; Wang, Q.; Su, Q.; Ma, D.; An, C.; Ma, L.; Liang, H. Honokiol suppresses renal cancer cells' metastasis via dual-blocking epithelial-mesenchymal transition and cancer stem cell properties through modulating miR-141/ZEB2 signaling. *Mol. Cells* **2014**, *37*, 383–388. [CrossRef]
198. Joo, Y.N.; Eun, S.Y.; Park, S.W.; Lee, J.H.; Chang, K.C.; Kim, H.J. Honokiol inhibits U87MG human glioblastoma cell invasion through endothelial cells by regulating membrane permeability and the epithelial-mesenchymal transition. *Int. J. Oncol.* **2014**, *44*, 187–194. [CrossRef]
199. Nesterova, Y.V.; Povetieva, T.N.; Suslov, N.I.; Zyuz'kov, G.N.; Aksinenko, S.G.; Pushkarskii, S.V.; Krapivin, A.V. Anti-inflammatory activity of diterpene alkaloids from Aconitum baikalense. *Bull. Exp. Biol. Med.* **2014**, *156*, 665–668. [CrossRef]
200. Feng, H.T.; Zhao, W.W.; Lu, J.J.; Wang, Y.T.; Chen, X.P. Hypaconitine inhibits TGF-beta1-induced epithelial-mesenchymal transition and suppresses adhesion, migration, and invasion of lung cancer A549 cells. *Chin. J. Nat. Med.* **2017**, *15*, 427–435. [CrossRef]
201. Frohlich, J.K.; Stein, T.; da Silva, L.A.; Biavatti, M.W.; Tonussi, C.R.; Lemos-Senna, E. Antinociceptive and anti-inflammatory activities of the Jatropha isabellei dichloromethane fraction and isolation and quantitative determination of jatrophone by UFLC-DAD. *Pharm. Biol.* **2017**, *55*, 1215–1222. [CrossRef] [PubMed]
202. Fatima, I.; El-Ayachi, I.; Taotao, L.; Lillo, M.A.; Krutilina, R.I.; Seagroves, T.N.; Radaszkiewicz, T.W.; Hutnan, M.; Bryja, V.; Krum, S.A.; et al. The natural compound Jatrophone interferes with Wnt/beta-catenin signaling and inhibits proliferation and EMT in human triple-negative breast cancer. *PLoS ONE* **2017**, *12*, e0189864. [CrossRef] [PubMed]
203. Wei, Y.; Liu, J.; Zhang, H.; Du, X.; Luo, Q.; Sun, J.; Liu, F.; Li, M.; Xu, F.; Wei, K.; et al. Ligustrazine attenuates inflammation and the associated chemokines and receptors in ovalbumine-induced mouse asthma model. *Environ. Toxicol. Pharmacol.* **2016**, *46*, 55–61. [CrossRef] [PubMed]

204. Hashim, P.; Selamat, J.; Syed Muhammad, S.K.; Ali, A. Effect of mass and turning time on free amino acid, peptide-N, sugar and pyrazine concentration during cocoa fermentation. *J. Sci. Food Agric.* **1998**, *78*, 543–550. [CrossRef]
205. Donkor, P.O.; Chen, Y.; Ding, L.; Qiu, F. Locally and traditionally used Ligusticum species—A review of their phytochemistry, pharmacology and pharmacokinetics. *J. Ethnopharmacol.* **2016**, *194*, 530–548. [CrossRef]
206. Luan, Y.; Liu, J.; Liu, X.; Xue, X.; Kong, F.; Sun, C.; Wang, J.; Liu, L.; Jia, H. Tetramethypyrazine inhibits renal cell carcinoma cells through inhibition of NKG2D signaling pathways. *Int. J. Oncol.* **2016**, *49*, 1704–1712. [CrossRef]
207. Chen, K.C.; Chen, C.Y.; Lin, C.R.; Yang, T.Y.; Chen, T.H.; Wu, L.C.; Wu, C.C. Luteolin attenuates TGF-beta1-induced epithelial-mesenchymal transition of lung cancer cells by interfering in the PI3K/Akt-NF-kappaB-Snail pathway. *Life Sci.* **2013**, *93*, 924–933. [CrossRef]
208. Gao, G.; Ge, R.; Li, Y.; Liu, S. Luteolin exhibits anti-breast cancer property through up-regulating miR-203. *Artif. Cells Nanomed. Biotechnol.* **2019**, *47*, 3265–3271. [CrossRef]
209. Lin, D.; Kuang, G.; Wan, J.; Zhang, X.; Li, H.; Gong, X.; Li, H. Luteolin suppresses the metastasis of triple-negative breast cancer by reversing epithelial-to-mesenchymal transition via downregulation of beta-catenin expression. *Oncol. Rep.* **2017**, *37*, 895–902. [CrossRef]
210. Li, C.; Wang, Q.; Shen, S.; Wei, X.; Li, G. HIF-1alpha/VEGF signaling-mediated epithelial-mesenchymal transition and angiogenesis is critically involved in anti-metastasis effect of luteolin in melanoma cells. *Phytother. Res.* **2019**, *33*, 798–807. [CrossRef]
211. Liu, Y.; Lang, T.; Jin, B.; Chen, F.; Zhang, Y.; Beuerman, R.W.; Zhou, L.; Zhang, Z. Luteolin inhibits colorectal cancer cell epithelial-to-mesenchymal transition by suppressing CREB1 expression revealed by comparative proteomics study. *J. Proteom.* **2017**, *161*, 1–10. [CrossRef] [PubMed]
212. Lin, T.H.; Hsu, W.H.; Tsai, P.H.; Huang, Y.T.; Lin, C.W.; Chen, K.C.; Tsai, I.H.; Kandaswami, C.C.; Huang, C.J.; Chang, G.D.; et al. Dietary flavonoids, luteolin and quercetin, inhibit invasion of cervical cancer by reduction of UBE2S through epithelial-mesenchymal transition signaling. *Food Funct.* **2017**, *8*, 1558–1568. [CrossRef] [PubMed]
213. Zang, M.D.; Hu, L.; Fan, Z.Y.; Wang, H.X.; Zhu, Z.L.; Cao, S.; Wu, X.Y.; Li, J.F.; Su, L.P.; Li, C.; et al. Luteolin suppresses gastric cancer progression by reversing epithelial-mesenchymal transition via suppression of the Notch signaling pathway. *J. Transl. Med.* **2017**, *15*, 52. [CrossRef] [PubMed]
214. Dia, V.P.; Pangloli, P. Epithelial-to-Mesenchymal Transition in Paclitaxel-Resistant Ovarian Cancer Cells Is Downregulated by Luteolin. *J. Cell. Physiol.* **2017**, *232*, 391–401. [CrossRef]
215. Huang, X.; Dai, S.; Dai, J.; Xiao, Y.; Bai, Y.; Chen, B.; Zhou, M. Luteolin decreases invasiveness, deactivates STAT3 signaling, and reverses interleukin-6 induced epithelial-mesenchymal transition and matrix metalloproteinase secretion of pancreatic cancer cells. *OncoTargets Ther.* **2015**, *8*, 2989–3001. [CrossRef]
216. Seo, J.Y.; Lee, C.; Hwang, S.W.; Chun, J.; Im, J.P.; Kim, J.S. Nimbolide Inhibits Nuclear Factor-capital KA, CyrillicB Pathway in Intestinal Epithelial Cells and Macrophages and Alleviates Experimental Colitis in Mice. *Phytother. Res.* **2016**, *30*, 1605–1614. [CrossRef]
217. Subramani, R.; Gonzalez, E.; Arumugam, A.; Nandy, S.; Gonzalez, V.; Medel, J.; Camacho, F.; Ortega, A.; Bonkoungou, S.; Narayan, M.; et al. Nimbolide inhibits pancreatic cancer growth and metastasis through ROS-mediated apoptosis and inhibition of epithelial-to-mesenchymal transition. *Sci. Rep.* **2016**, *6*, 19819. [CrossRef]
218. Lin, H.; Qiu, S.; Xie, L.; Liu, C.; Sun, S. Nimbolide suppresses non-small cell lung cancer cell invasion and migration via manipulation of DUSP4 expression and ERK1/2 signaling. *Biomed. Pharmacother.* **2017**, *92*, 340–346. [CrossRef]
219. Wang, Z.; Jiang, W.; Zhang, Z.; Qian, M.; Du, B. Nitidine chloride inhibits LPS-induced inflammatory cytokines production via MAPK and NF-kappaB pathway in RAW 264.7 cells. *J. Ethnopharmacol.* **2012**, *144*, 145–150. [CrossRef]
220. Sun, M.; Zhang, N.; Wang, X.; Li, Y.; Qi, W.; Zhang, H.; Li, Z.; Yang, Q. Hedgehog pathway is involved in nitidine chloride induced inhibition of epithelial-mesenchymal transition and cancer stem cells-like properties in breast cancer cells. *Cell Biosci.* **2016**, *6*, 44. [CrossRef]
221. Cheng, Z.; Guo, Y.; Yang, Y.; Kan, J.; Dai, S.; Helian, M.; Li, B.; Xu, J.; Liu, C. Nitidine chloride suppresses epithelial-to-mesenchymal transition in osteosarcoma cell migration and invasion through Akt/GSK-3beta/Snail signaling pathway. *Oncol. Rep.* **2016**, *36*, 1023–1029. [CrossRef] [PubMed]

222. Fan, H.; Gao, Z.; Ji, K.; Li, X.; Wu, J.; Liu, Y.; Wang, X.; Liang, H.; Liu, Y.; Li, X.; et al. The in vitro and in vivo anti-inflammatory effect of osthole, the major natural coumarin from Cnidium monnieri (L.) Cuss, via the blocking of the activation of the NF-kappaB and MAPK/p38 pathways. *Phytomedicine* **2019**, *58*, 152864. [CrossRef] [PubMed]
223. Hung, C.M.; Kuo, D.H.; Chou, C.H.; Su, Y.C.; Ho, C.T.; Way, T.D. Osthole suppresses hepatocyte growth factor (HGF)-induced epithelial-mesenchymal transition via repression of the c-Met/Akt/mTOR pathway in human breast cancer cells. *J. Agric. Food Chem.* **2011**, *59*, 9683–9690. [CrossRef] [PubMed]
224. Lin, Y.C.; Lin, J.C.; Hung, C.M.; Chen, Y.; Liu, L.C.; Chang, T.C.; Kao, J.Y.; Ho, C.T.; Way, T.D. Osthole inhibits insulin-like growth factor-1-induced epithelial to mesenchymal transition via the inhibition of PI3K/Akt signaling pathway in human brain cancer cells. *J. Agric. Food Chem.* **2014**, *62*, 5061–5071. [CrossRef]
225. Wen, Y.C.; Lee, W.J.; Tan, P.; Yang, S.F.; Hsiao, M.; Lee, L.M.; Chien, M.H. By inhibiting snail signaling and miR-23a-3p, osthole suppresses the EMT-mediated metastatic ability in prostate cancer. *Oncotarget* **2015**, *6*, 21120–21136. [CrossRef]
226. Feng, H.; Lu, J.J.; Wang, Y.; Pei, L.; Chen, X. Osthole inhibited TGF beta-induced epithelial-mesenchymal transition (EMT) by suppressing NF-kappaB mediated Snail activation in lung cancer A549 cells. *Cell Adh. Migr.* **2017**, *11*, 464–475. [CrossRef]
227. Chen, Y.; Qi, Z.; Qiao, B.; Lv, Z.; Hao, Y.; Li, H. Oxymatrine can attenuate pathological deficits of Alzheimer's disease mice through regulation of neuroinflammation. *J. Neuroimmunol.* **2019**, *334*, 576978. [CrossRef]
228. Chen, Y.; Chen, L.; Zhang, J.Y.; Chen, Z.Y.; Liu, T.T.; Zhang, Y.Y.; Fu, L.Y.; Fan, S.Q.; Zhang, M.Q.; Gan, S.Q.; et al. Oxymatrine reverses epithelial-mesenchymal transition in breast cancer cells by depressing alphabeta3 integrin/FAK/PI3K/Akt signaling activation. *OncoTargets Ther.* **2019**, *12*, 6253–6265. [CrossRef]
229. Liang, L.; Huang, J. Oxymatrine inhibits epithelial-mesenchymal transition through regulation of NF-kappaB signaling in colorectal cancer cells. *Oncol. Rep.* **2016**, *36*, 1333–1338. [CrossRef]
230. Xiong, Y.; Wang, J.; Zhu, H.; Liu, L.; Jiang, Y. Chronic oxymatrine treatment induces resistance and epithelialmesenchymal transition through targeting the long non-coding RNA MALAT1 in colorectal cancer cells. *Oncol. Rep.* **2018**, *39*, 967–976. [CrossRef]
231. Yan, D.; Saito, K.; Ohmi, Y.; Fujie, N.; Ohtsuka, K. Paeoniflorin, a novel heat shock protein-inducing compound. *Cell Stress Chaperones* **2004**, *9*, 378–389. [CrossRef] [PubMed]
232. Tu, J.; Guo, Y.; Hong, W.; Fang, Y.; Han, D.; Zhang, P.; Wang, X.; Korner, H.; Wei, W. The Regulatory Effects of Paeoniflorin and Its Derivative Paeoniflorin-6′-O-Benzene Sulfonate CP-25 on Inflammation and Immune Diseases. *Front. Pharmacol.* **2019**, *10*, 57. [CrossRef] [PubMed]
233. Zhang, J.W.; Li, L.X.; Wu, W.Z.; Pan, T.J.; Yang, Z.S.; Yang, Y.K. Anti-Tumor Effects of Paeoniflorin on Epithelial-To-Mesenchymal Transition in Human Colorectal Cancer Cells. *Med. Sci. Monit.* **2018**, *24*, 6405–6413. [CrossRef] [PubMed]
234. Wang, Z.; Liu, Z.; Yu, G.; Nie, X.; Jia, W.; Liu, R.E.; Xu, R. Paeoniflorin Inhibits Migration and Invasion of Human Glioblastoma Cells via Suppression Transforming Growth Factor beta-Induced Epithelial-Mesenchymal Transition. *Neurochem. Res.* **2018**, *43*, 760–774. [CrossRef]
235. Fukuhara, Y.; Yoshida, D. Paeonol: A bio-antimutagen isolated from a crude drug, Moutan cortex. *Agric. Biol. Chem.* **1987**, *51*, 1441–1442.
236. Lei, H.; Wen, Q.; Li, H.; Du, S.; Wu, J.J.; Chen, J.; Huang, H.; Chen, D.; Li, Y.; Zhang, S.; et al. Paeonol Inhibits Lipopolysaccharide-Induced HMGB1 Translocation from the Nucleus to the Cytoplasm in RAW264.7 Cells. *Inflammation* **2016**, *39*, 1177–1187. [CrossRef]
237. Yang, L.; Xing, S.; Wang, K.; Yi, H.; Du, B. Paeonol attenuates aging MRC-5 cells and inhibits epithelial-mesenchymal transition of premalignant HaCaT cells induced by aging MRC-5 cell-conditioned medium. *Mol. Cell Biochem.* **2018**, *439*, 117–129. [CrossRef]
238. Wang, M.; Li, Q. Parthenolide could become a promising and stable drug with anti-inflammatory effects. *Nat. Prod. Res.* **2015**, *29*, 1092–1101. [CrossRef]
239. Zhu, S.M.; Park, Y.R.; Seo, S.Y.; Kim, I.H.; Lee, S.T.; Kim, S.W. Parthenolide inhibits transforming growth factor beta1-induced epithelial-mesenchymal transition in colorectal cancer cells. *Intest. Res.* **2019**. [CrossRef]
240. Kim, S.L.; Park, Y.R.; Lee, S.T.; Kim, S.W. Parthenolide suppresses hypoxia-inducible factor-1alpha signaling and hypoxia induced epithelial-mesenchymal transition in colorectal cancer. *Int. J. Oncol.* **2017**, *51*, 1809–1820. [CrossRef]

241. Ai, X.Y.; Zhang, H.; Gao, S.Y.; Qin, Y.; Zhong, W.L.; Gu, J.; Li, M.; Qiao, K.L.; Tian, Q.; Cui, Z.H.; et al. Sesquiterpene binding Gly-Leu-Ser/Lys-"co-adaptation pocket" to inhibit lung cancer cell epithelial-mesenchymal transition. *Oncotarget* **2017**, *8*, 70192–70203. [CrossRef] [PubMed]
242. Akhtar, N.; Jafri, L.; Green, B.D.; Kalsoom, S.; Mirza, B. A Multi-Mode Bioactive Agent Isolated From Ficus microcarpa L. Fill. With Therapeutic Potential for Type 2 Diabetes Mellitus. *Front. Pharmacol.* **2018**, *9*, 1376. [CrossRef] [PubMed]
243. Akhtar, N.; Syed, D.N.; Khan, M.I.; Adhami, V.M.; Mirza, B.; Mukhtar, H. The pentacyclic triterpenoid, plectranthoic acid, a novel activator of AMPK induces apoptotic death in prostate cancer cells. *Oncotarget* **2016**, *7*, 3819–3831. [CrossRef] [PubMed]
244. Akhtar, N.; Syed, D.N.; Lall, R.K.; Mirza, B.; Mukhtar, H. Targeting epithelial to mesenchymal transition in prostate cancer by a novel compound, plectranthoic acid, isolated from Ficus microcarpa. *Mol. Carcinog.* **2018**, *57*, 653–663. [CrossRef]
245. Chatterjee, A.; Dutta, C. Alkaloids of Piper longum Linn—I: Structure and synthesis of piperlongumine and piperlonguminine. *Tetrahedron* **1967**, *23*, 1769–1781. [CrossRef]
246. Kim, N.; Do, J.; Bae, J.S.; Jin, H.K.; Kim, J.H.; Inn, K.S.; Oh, M.S.; Lee, J.K. Piperlongumine inhibits neuroinflammation via regulating NF-kappaB signaling pathways in lipopolysaccharide-stimulated BV2 microglia cells. *J. Pharmacol. Sci.* **2018**, *137*, 195–201. [CrossRef]
247. Park, M.J.; Lee, D.E.; Shim, M.K.; Jang, E.H.; Lee, J.K.; Jeong, S.Y.; Kim, J.H. Piperlongumine inhibits TGF-beta-induced epithelial-to-mesenchymal transition by modulating the expression of E-cadherin, Snail1, and Twist1. *Eur. J. Pharmacol.* **2017**, *812*, 243–249. [CrossRef]
248. Messeha, S.S.; Zarmouh, N.O.; Mendonca, P.; Kolta, M.G.; Soliman, K.F.A. The attenuating effects of plumbagin on pro-inflammatory cytokine expression in LPS-activated BV-2 microglial cells. *J. Neuroimmunol.* **2017**, *313*, 129–137. [CrossRef]
249. Sakunrangsit, N.; Kalpongnukul, N.; Pisitkun, T.; Ketchart, W. Plumbagin Enhances Tamoxifen Sensitivity and Inhibits Tumor Invasion in Endocrine Resistant Breast Cancer through EMT Regulation. *Phytother. Res.* **2016**, *30*, 1968–1977. [CrossRef]
250. Pan, S.T.; Qin, Y.; Zhou, Z.W.; He, Z.X.; Zhang, X.; Yang, T.; Yang, Y.X.; Wang, D.; Zhou, S.F.; Qiu, J.X. Plumbagin suppresses epithelial to mesenchymal transition and stemness via inhibiting Nrf2-mediated signaling pathway in human tongue squamous cell carcinoma cells. *Drug Des. Devel. Ther.* **2015**, *9*, 5511–5551. [CrossRef]
251. Wang, F.; Wang, Q.; Zhou, Z.W.; Yu, S.N.; Pan, S.T.; He, Z.X.; Zhang, X.; Wang, D.; Yang, Y.X.; Yang, T.; et al. Plumbagin induces cell cycle arrest and autophagy and suppresses epithelial to mesenchymal transition involving PI3K/Akt/mTOR-mediated pathway in human pancreatic cancer cells. *Drug Des. Devel. Ther.* **2015**, *9*, 537–560. [CrossRef] [PubMed]
252. Qiu, J.X.; Zhou, Z.W.; He, Z.X.; Zhao, R.J.; Zhang, X.; Yang, L.; Zhou, S.F.; Mao, Z.F. Plumbagin elicits differential proteomic responses mainly involving cell cycle, apoptosis, autophagy, and epithelial-to-mesenchymal transition pathways in human prostate cancer PC-3 and DU145 cells. *Drug Des. Devel. Ther.* **2015**, *9*, 349–417. [CrossRef] [PubMed]
253. Wang, Q.; Zhou, X.; Zhao, Y.; Xiao, J.; Lu, Y.; Shi, Q.; Wang, Y.; Wang, H.; Liang, Q. Polyphyllin I Ameliorates Collagen-Induced Arthritis by Suppressing the Inflammation Response in Macrophages Through the NF-kappaB Pathway. *Front. Immunol.* **2018**, *9*, 2091. [CrossRef] [PubMed]
254. Lou, W.; Chen, Y.; Zhu, K.Y.; Deng, H.; Wu, T.; Wang, J. Polyphyllin I Overcomes EMT-Associated Resistance to Erlotinib in Lung Cancer Cells via IL-6/STAT3 Pathway Inhibition. *Biol. Pharm. Bull.* **2017**, *40*, 1306–1313. [CrossRef] [PubMed]
255. Mak, K.K.; Wu, A.T.; Lee, W.H.; Chang, T.C.; Chiou, J.F.; Wang, L.S.; Wu, C.H.; Huang, C.Y.F.; Shieh, Y.S.; Chao, T.Y. Pterostilbene, a bioactive component of blueberries, suppresses the generation of breast cancer stem cells within tumor microenvironment and metastasis via modulating NF-κB/microRNA 448 circuit. *Mol. Nutr. Food Res.* **2013**, *57*, 1123–1134. [CrossRef] [PubMed]
256. Su, C.-M.; Lee, W.-H.; Wu, A.T.; Lin, Y.-K.; Wang, L.-S.; Wu, C.-H.; Yeh, C.-T. Pterostilbene inhibits triple-negative breast cancer metastasis via inducing microRNA-205 expression and negatively modulates epithelial-to-mesenchymal transition. *J. Nutr. Biochem.* **2015**, *26*, 675–685. [CrossRef]
257. Huang, Y.; Du, J.; Mi, Y.; Li, T.; Gong, Y.; Ouyang, H.; Hou, Y. Long Non-coding RNAs Contribute to the Inhibition of Proliferation and EMT by Pterostilbene in Human Breast Cancer. *Front. Oncol.* **2018**, *8*, 629. [CrossRef]

258. Inoue, H.; Nakata, R. Resveratrol Targets in Inflammation. *Endocr. Metab. Immune Disord. Drug Targets* **2015**, *15*, 186–195. [CrossRef]
259. Ji, Q.; Liu, X.; Han, Z.; Zhou, L.; Sui, H.; Yan, L.; Jiang, H.; Ren, J.; Cai, J.; Li, Q. Resveratrol suppresses epithelial-to-mesenchymal transition in colorectal cancer through TGF-1/Smads signaling pathway mediated Snail/E-cadherin expression. *BMC Cancer* **2015**, *15*, 97. [CrossRef]
260. Wang, H.; Zhang, H.; Tang, L.; Chen, H.; Wu, C.; Zhao, M.; Yang, Y.; Chen, X.; Liu, G. Resveratrol inhibits TGF-1-induced epithelial-to-mesenchymal transition and suppresses lung cancer invasion and metastasis. *Toxicology* **2013**, *303*, 139–146. [CrossRef]
261. Li, W.; Ma, J.; Ma, Q.; Li, B.; Han, L.; Liu, J.; Xu, Q.; Duan, W.; Yu, S.; Wang, F.; et al. Resveratrol inhibits the epithelial-mesenchymal transition of pancreatic cancer cells via suppression of the PI-3K/Akt/NF-kappaB pathway. *Curr. Med. Chem.* **2013**, *20*, 4185–4194. [CrossRef] [PubMed]
262. Simic, P.; Williams, E.O.; Bell, E.L.; Gong, J.J.; Bonkowski, M.; Guarente, L. SIRT1 suppresses the epithelial-to-mesenchymal transition in cancer metastasis and organ fibrosis. *Cell Rep.* **2013**, *3*, 1175–1186. [CrossRef] [PubMed]
263. Byles, V.; Zhu, L.; Lovaas, J.D.; Chmilewski, L.K.; Wang, J.; Faller, D.V.; Dai, Y. SIRT1 induces EMT by cooperating with EMT transcription factors and enhances prostate cancer cell migration and metastasis. *Oncogene* **2012**, *31*, 4619–4629. [CrossRef] [PubMed]
264. Zhang, X.C.; Chen, J.Q.; Li, B. Salvianolic acid A suppresses CCL-20 expression in TNF-alpha-treated macrophages and ApoE-deficient mice. *J. Cardiovasc. Pharmacol.* **2014**, *64*, 318–325. [CrossRef] [PubMed]
265. Tang, Q.; Zhong, H.; Xie, F.; Xie, J.; Chen, H.; Yao, G. Expression of miR-106b-25 induced by salvianolic acid B inhibits epithelial-to-mesenchymal transition in HK-2 cells. *Eur. J. Pharmacol.* **2014**, *741*, 97–103. [CrossRef]
266. Wang, Q.L.; Tao, Y.Y.; Yuan, J.L.; Shen, L.; Liu, C.H. Salvianolic acid B prevents epithelial-to-mesenchymal transition through the TGF-beta1 signal transduction pathway in vivo and in vitro. *BMC Cell Biol.* **2010**, *11*, 31. [CrossRef]
267. Drobnik, J.; Drobnik, E. Timeline and bibliography of early isolations of plant metabolites (1770–1820) and their impact to pharmacy: A critical study. *Fitoterapia* **2016**, *115*, 155–164. [CrossRef]
268. Kang, H.; Jeong, H.D.; Choi, H.Y. The chloroform fraction of Solanum nigrum suppresses nitric oxide and tumor necrosis factor-alpha in LPS-stimulated mouse peritoneal macrophages through inhibition of p38, JNK and ERK1/2. *Am. J. Chin. Med.* **2011**, *39*, 1261–1273. [CrossRef]
269. Shen, K.H.; Liao, A.C.; Hung, J.H.; Lee, W.J.; Hu, K.C.; Lin, P.T.; Liao, R.F.; Chen, P.S. alpha-Solanine inhibits invasion of human prostate cancer cell by suppressing epithelial-mesenchymal transition and MMPs expression. *Molecules* **2014**, *19*, 11896–11914. [CrossRef]
270. Li, Y.; Liu, Y.; Yan, X.; Liu, Q.; Zhao, Y.H.; Wang, D.W. Pharmacological Effects and Mechanisms of Chinese Medicines Modulating NLRP3 Inflammasomes in Ischemic Cardio/Cerebral Vascular Disease. *Am. J. Chin. Med.* **2018**, *46*, 1727–1741. [CrossRef]
271. Wang, D.X.; Zou, Y.J.; Zhuang, X.B.; Chen, S.X.; Lin, Y.; Li, W.L.; Lin, J.J.; Lin, Z.Q. Sulforaphane suppresses EMT and metastasis in human lung cancer through miR-616-5p-mediated GSK3beta/beta-catenin signaling pathways. *Acta Pharmacol. Sin.* **2017**, *38*, 241–251. [CrossRef] [PubMed]
272. Wu, J.; Han, J.; Hou, B.; Deng, C.; Wu, H.; Shen, L. Sulforaphane inhibits TGF-beta-induced epithelial-mesenchymal transition of hepatocellular carcinoma cells via the reactive oxygen species-dependent pathway. *Oncol. Rep.* **2016**, *35*, 2977–2983. [CrossRef] [PubMed]
273. Shan, Y.; Zhang, L.; Bao, Y.; Li, B.; He, C.; Gao, M.; Feng, X.; Xu, W.; Zhang, X.; Wang, S. Epithelial-mesenchymal transition, a novel target of sulforaphane via COX-2/MMP2, 9/Snail, ZEB1 and miR-200c/ZEB1 pathways in human bladder cancer cells. *J. Nutr. Biochem.* **2013**, *24*, 1062–1069. [CrossRef] [PubMed]
274. Song, D.; Zhao, J.; Deng, W.; Liao, Y.; Hong, X.; Hou, J. Tannic acid inhibits NLRP3 inflammasome-mediated IL-1beta production via blocking NF-kappaB signaling in macrophages. *Biochem. Biophys. Res. Commun.* **2018**, *503*, 3078–3085. [CrossRef] [PubMed]
275. Pattarayan, D.; Sivanantham, A.; Krishnaswami, V.; Loganathan, L.; Palanichamy, R.; Natesan, S.; Muthusamy, K.; Rajasekaran, S. Tannic acid attenuates TGF-beta1-induced epithelial-to-mesenchymal transition by effectively intervening TGF-beta signaling in lung epithelial cells. *J. Cell. Physiol.* **2018**, *233*, 2513–2525. [CrossRef] [PubMed]

276. Bale, S.; Pulivendala, G.; Godugu, C. Withaferin A attenuates bleomycin-induced scleroderma by targeting FoxO3a and NF-kappabeta signaling: Connecting fibrosis and inflammation. *Biofactors* **2018**, *44*, 507–517. [CrossRef]
277. Kyakulaga, A.H.; Aqil, F.; Munagala, R.; Gupta, R.C. Withaferin A inhibits Epithelial to Mesenchymal Transition in Non-Small Cell Lung Cancer Cells. *Sci. Rep.* **2018**, *8*, 15737. [CrossRef]
278. Lee, J.; Hahm, E.R.; Marcus, A.I.; Singh, S.V. Withaferin A inhibits experimental epithelial-mesenchymal transition in MCF-10A cells and suppresses vimentin protein level in vivo in breast tumors. *Mol. Carcinog.* **2015**, *54*, 417–429. [CrossRef]
279. Yang, Z.; Garcia, A.; Xu, S.; Powell, D.R.; Vertino, P.M.; Singh, S.; Marcus, A.I. Withania somnifera root extract inhibits mammary cancer metastasis and epithelial to mesenchymal transition. *PLoS ONE* **2013**, *8*, e75069. [CrossRef]
280. Zhang, Q.; Zhu, B.; Li, Y. Resolution of Cancer-Promoting Inflammation: A New Approach for Anticancer Therapy. *Front. Immunol.* **2017**, *8*, 71. [CrossRef]
281. Pirault, J.; Back, M. Lipoxin and Resolvin Receptors Transducing the Resolution of Inflammation in Cardiovascular Disease. *Front. Pharmacol.* **2018**, *9*, 1273. [CrossRef] [PubMed]
282. Ohira, T.; Arita, M.; Omori, K.; Recchiuti, A.; Van Dyke, T.E.; Serhan, C.N. Resolvin E1 receptor activation signals phosphorylation and phagocytosis. *J. Biol. Chem.* **2010**, *285*, 3451–3461. [CrossRef] [PubMed]
283. Arita, M.; Bianchini, F.; Aliberti, J.; Sher, A.; Chiang, N.; Hong, S.; Yang, R.; Petasis, N.A.; Serhan, C.N. Stereochemical assignment, antiinflammatory properties, and receptor for the omega-3 lipid mediator resolvin E1. *J. Exp. Med.* **2005**, *201*, 713–722. [CrossRef] [PubMed]
284. Arita, M.; Ohira, T.; Sun, Y.P.; Elangovan, S.; Chiang, N.; Serhan, C.N. Resolvin E1 selectively interacts with leukotriene B4 receptor BLT1 and ChemR23 to regulate inflammation. *J. Immunol.* **2007**, *178*, 3912–3917. [CrossRef]
285. Wang, X.; Guo, J.; Wu, Q.; Niu, C.; Cheng, G.; Liu, D.; Liu, Z.; Zhao, Z.; Xiao, J. Chemerin/chemR23 association with endothelial-mesenchymal transition in diabetic nephropathy. *Int. J. Clin. Exp. Pathol.* **2017**, *10*, 7408–7416.
286. Hayhoe, R.P.; Kamal, A.M.; Solito, E.; Flower, R.J.; Cooper, D.; Perretti, M. Annexin 1 and its bioactive peptide inhibit neutrophil-endothelium interactions under flow: Indication of distinct receptor involvement. *Blood* **2006**, *107*, 2123–2130. [CrossRef]
287. Fiore, S.; Maddox, J.F.; Perez, H.D.; Serhan, C.N. Identification of a human cDNA encoding a functional high affinity lipoxin A4 receptor. *J. Exp. Med.* **1994**, *180*, 253–260. [CrossRef]
288. Krishnamoorthy, S.; Recchiuti, A.; Chiang, N.; Fredman, G.; Serhan, C.N. Resolvin D1 receptor stereoselectivity and regulation of inflammation and proresolving microRNAs. *Am. J. Pathol.* **2012**, *180*, 2018–2027. [CrossRef]
289. Arnardottir, H.H.; Dalli, J.; Norling, L.V.; Colas, R.A.; Perretti, M.; Serhan, C.N. Resolvin D3 Is Dysregulated in Arthritis and Reduces Arthritic Inflammation. *J. Immunol.* **2016**, *197*, 2362–2368. [CrossRef]
290. Zugazagoitia, J.; Guedes, C.; Ponce, S.; Ferrer, I.; Molina-Pinelo, S.; Paz-Ares, L. Current challenges in cancer treatment. *Clin. Ther.* **2016**, *38*, 1551–1566. [CrossRef]
291. Lee, H.J.; Park, M.K.; Lee, E.J.; Lee, C.H. Resolvin D1 inhibits TGF-beta1-induced epithelial mesenchymal transition of A549 lung cancer cells via lipoxin A4 receptor/formyl peptide receptor 2 and GPR32. *Int. J. Biochem. Cell Biol.* **2013**, *45*, 2801–2807. [CrossRef] [PubMed]
292. Chiang, N.; Dalli, J.; Colas, R.A.; Serhan, C.N. Identification of resolvin D2 receptor mediating resolution of infections and organ protection. *J. Exp. Med.* **2015**, *212*, 1203–1217. [CrossRef] [PubMed]
293. Krishnamoorthy, S.; Recchiuti, A.; Chiang, N.; Yacoubian, S.; Lee, C.H.; Yang, R.; Petasis, N.A.; Serhan, C.N. Resolvin D1 binds human phagocytes with evidence for proresolving receptors. *Proc. Natl. Acad. Sci. USA* **2010**, *107*, 1660–1665. [CrossRef] [PubMed]
294. Chiang, N.; Fredman, G.; Backhed, F.; Oh, S.F.; Vickery, T.; Schmidt, B.A.; Serhan, C.N. Infection regulates pro-resolving mediators that lower antibiotic requirements. *Nature* **2012**, *484*, 524–528. [CrossRef] [PubMed]
295. Dalli, J.; Winkler, J.W.; Colas, R.A.; Arnardottir, H.; Cheng, C.Y.; Chiang, N.; Petasis, N.A.; Serhan, C.N. Resolvin D3 and aspirin-triggered resolvin D3 are potent immunoresolvents. *Chem. Biol.* **2013**, *20*, 188–201. [CrossRef] [PubMed]
296. Zheng, S.; Wang, Q.; D'Souza, V.; Bartis, D.; Dancer, R.; Parekh, D.; Gao, F.; Lian, Q.; Jin, S.; Thickett, D.R. ResolvinD1 stimulates epithelial wound repair and inhibits TGF-beta-induced EMT whilst reducing fibroproliferation and collagen production. *Lab. Invest.* **2018**, *98*, 130–140. [CrossRef] [PubMed]

297. Zong, L.; Chen, K.; Jiang, Z.; Chen, X.; Sun, L.; Ma, J.; Zhou, C.; Xu, Q.; Duan, W.; Han, L.; et al. Lipoxin A4 reverses mesenchymal phenotypes to attenuate invasion and metastasis via the inhibition of autocrine TGF-beta1 signaling in pancreatic cancer. *J. Exp. Clin. Cancer Res.* **2017**, *36*, 181. [CrossRef]
298. Wu, R.F.; Huang, Z.X.; Ran, J.; Dai, S.J.; Lin, D.C.; Ng, T.W.; Chen, Q.X.; Chen, Q.H. Lipoxin A4 Suppresses Estrogen-Induced Epithelial-Mesenchymal Transition via ALXR-Dependent Manner in Endometriosis. *Reprod. Sci.* **2018**, *25*, 566–578. [CrossRef]
299. Xu, F.; Zhou, X.; Hao, J.; Dai, H.; Zhang, J.; He, Y.; Hao, H. Lipoxin A4 and its analog suppress hepatocarcinoma cell epithelial-mesenchymal transition, migration and metastasis via regulating integrin-linked kinase axis. *Prostaglandins Other Lipid Mediat.* **2018**, *137*, 9–19. [CrossRef]
300. Liu, Y.; Yuan, X.; Li, W.; Cao, Q.; Shu, Y. Aspirin-triggered resolvin D1 inhibits TGF-beta1-induced EMT through the inhibition of the mTOR pathway by reducing the expression of PKM2 and is closely linked to oxidative stress. *Int. J. Mol. Med.* **2016**, *38*, 1235–1242. [CrossRef]
301. Sun, L.; Wang, Y.; Wang, L.; Yao, B.; Chen, T.; Li, Q.; Liu, Z.; Liu, R.; Niu, Y.; Song, T.; et al. Resolvin D1 prevents epithelial-mesenchymal transition and reduces the stemness features of hepatocellular carcinoma by inhibiting paracrine of cancer-associated fibroblast-derived COMP. *J. Exp. Clin. Cancer Res.* **2019**, *38*, 170. [CrossRef] [PubMed]
302. Cezar, T.L.C.; Martinez, R.M.; Rocha, C.D.; Melo, C.P.B.; Vale, D.L.; Borghi, S.M.; Fattori, V.; Vignoli, J.A.; Camilios-Neto, D.; Baracat, M.M.; et al. Treatment with maresin 1, a docosahexaenoic acid-derived pro-resolution lipid, protects skin from inflammation and oxidative stress caused by UVB irradiation. *Sci. Rep.* **2019**, *9*, 3062. [CrossRef] [PubMed]
303. Gdula-Argasinska, J.; Czepiel, J.; Wozniakiewicz, A.; Wojton, K.; Grzywacz, A.; Wozniakiewicz, M.; Jurczyszyn, A.; Perucki, W.; Librowski, T. n-3 Fatty acids as resolvents of inflammation in the A549 cells. *Pharmacol. Rep.* **2015**, *67*, 610–615. [CrossRef] [PubMed]
304. Sun, Q.; Wu, Y.; Zhao, F.; Wang, J. Maresin 1 inhibits transforming growth factor-beta1-induced proliferation, migration and differentiation in human lung fibroblasts. *Mol. Med. Rep.* **2017**, *16*, 1523–1529. [CrossRef]
305. Bazan, N.G. Neuroprotectin D1-mediated anti-inflammatory and survival signaling in stroke, retinal degenerations, and Alzheimer's disease. *J. Lipid Res.* **2009**, *50*, S400–S405. [CrossRef]
306. Li, H.; Hao, Y.; Zhang, H.; Ying, W.; Li, D.; Ge, Y.; Ying, B.; Cheng, B.; Lian, Q.; Jin, S. Posttreatment with Protectin DX ameliorates bleomycin-induced pulmonary fibrosis and lung dysfunction in mice. *Sci. Rep.* **2017**, *7*, 46754. [CrossRef]
307. Balta, M.G.; Loos, B.G.; Nicu, E.A. Emerging Concepts in the Resolution of Periodontal Inflammation: A Role for Resolvin E1. *Front. Immunol.* **2017**, *8*, 1682. [CrossRef]
308. Hong, S.; Tjonahen, E.; Morgan, E.L.; Lu, Y.; Serhan, C.N.; Rowley, A.F. Rainbow trout (Oncorhynchus mykiss) brain cells biosynthesize novel docosahexaenoic acid-derived resolvins and protectins-Mediator lipidomic analysis. *Prostaglandins Other Lipid Mediat.* **2005**, *78*, 107–116. [CrossRef]
309. Raatz, S.K.; Golovko, M.Y.; Brose, S.A.; Rosenberger, T.A.; Burr, G.S.; Wolters, W.R.; Picklo, M.J., Sr. Baking reduces prostaglandin, resolvin, and hydroxy-fatty acid content of farm-raised Atlantic salmon (Salmo salar). *J. Agric. Food Chem.* **2011**, *59*, 11278–11286. [CrossRef]
310. Bannenberg, G.L.; Aliberti, J.; Hong, S.; Sher, A.; Serhan, C. Exogenous pathogen and plant 15-lipoxygenase initiate endogenous lipoxin A4 biosynthesis. *J. Exp. Med.* **2004**, *199*, 515–523. [CrossRef]
311. Haas-Stapleton, E.J.; Lu, Y.; Hong, S.; Arita, M.; Favoreto, S.; Nigam, S.; Serhan, C.N.; Agabian, N. Candida albicans modulates host defense by biosynthesizing the pro-resolving mediator resolvin E1. *PLoS ONE* **2007**, *2*, e1316. [CrossRef] [PubMed]
312. Colas, R.A.; Ashton, A.W.; Mukherjee, S.; Dalli, J.; Akide-Ndunge, O.B.; Huang, H.; Desruisseaux, M.S.; Guan, F.; Jelicks, L.A.; Matos Dos Santos, F.; et al. Trypanosoma cruzi Produces the Specialized Proresolving Mediators Resolvin D1, Resolvin D5, and Resolvin E2. *Infect. Immun.* **2018**, *86*. [CrossRef] [PubMed]

cancers

MDPI

Review

Persistent Organic Pollutants and Breast Cancer: A Systematic Review and Critical Appraisal of the Literature

Kaoutar Ennour-Idrissi [1,2,3,4], Pierre Ayotte [3,5,6] and Caroline Diorio [1,2,3,4,*]

1 Axe Oncologie, Centre de Recherche du CHU de Québec-Université Laval, Quebec City, QC G1E 6W2, Canada
2 Centre de Recherche sur le Cancer, Université Laval, Quebec City, QC G1R 3S3, Canada
3 Département de Médecine Sociale et Préventive, Faculté de Médecine, Université Laval, Quebec City, QC G1V 0A6, Canada
4 Centre des Maladies du Sein Deschênes-Fabia, Hôpital du Saint-Sacrement, Quebec City, QC G1S 4L8, Canada
5 Axe santé des Populations et Pratiques Optimales en Santé, Centre de Recherche du CHU de Québec, Université Laval, Quebec City, QC G1E 6W2, Canada
6 Centre de Toxicologie du Québec (CTQ), INSPQ, Quebec City, QC G1V 5B3, Canada
* Correspondence: Caroline.Diorio@crchudequebec.ulaval.ca; Tel.: +1-418-682-7511 (ext. 84726)

Received: 4 July 2019; Accepted: 24 July 2019; Published: 27 July 2019

Abstract: Persistent organic pollutants (POPs) bioaccumulate in the food chain and have been detected in human blood and adipose tissue. Experimental studies demonstrated that POPs can cause and promote growth of breast cancer. However, inconsistent results from epidemiological studies do not support a causal relationship between POPs and breast cancer in women. To identify individual POPs that are repeatedly found to be associated with both breast cancer incidence and progression, and to demystify the observed inconsistencies between epidemiological studies, we conducted a systematic review of 95 studies retrieved from three main electronic databases. While no clear pattern of associations between blood POPs and breast cancer incidence could be drawn, POPs measured in breast adipose tissue were more clearly associated with higher breast cancer incidence. POPs were more consistently associated with worse breast cancer prognosis whether measured in blood or breast adipose tissue. In contrast, POPs measured in adipose tissue other than breast were inversely associated with both breast cancer incidence and prognosis. Differences in biological tissues used for POPs measurement and methodological biases explain the discrepancies between studies results. Some individual compounds associated with both breast cancer incidence and progression, deserve further investigation.

Keywords: breast cancer; persistent organic pollutants; breast cancer risk; breast cancer prognostic; systematic review

---

1. Introduction

Persistent organic pollutants (POPs) are a group of chemical substances of synthetic origin used for industrial, agricultural or domestic purposes, that persist in the environment and bioaccumulate in the food chain due to their lipophilic properties [1,2]. POPs have been detected in human blood, adipose tissue and human milk and have been linked to the increase in the incidence of hormone-dependent breast cancers [1,3–8].

Given the abundance of adipose tissue in the human breast, mammary epithelial cells exposure to POPs sequestered in breast adipose tissue may promote carcinogenesis and progression of mammary cancers [9]. In fact, numerous in vitro studies have demonstrated that some POPs stimulate the growth of estrogen receptor (ER)-positive breast cancer cells [10–12]. In animal studies, exposure to some POPs, particularly during the perinatal period, impairs breast tissue development and increases its susceptibility to carcinogens and the incidence of precancerous and cancerous breast lesions [13]. In addition to their endocrine disrupting effect either as agonists or as antagonists of endogenous hormones [14], POPs can interfere with estrogen synthesis by disrupting adipose tissue functioning [15,16], interact with transcription factors [17], induce genotoxic enzymes [17] and cytochrome 450 leading to increased levels of reactive oxygen species [18], and induce trans-generational phenotypic changes by altering the epigenome [19].

Although experimental studies demonstrate that POPs can cause and promote growth of breast cancer, several observational studies conducted in humans yielded inconsistent results regarding the implication of POPs in women breast cancers [20–26]. Observational studies are known to be prone to different biases that vary according to studies designs [27]. To draw meaningful conclusions about a causal relationship between POPs and breast cancer in women, a systematic comparison of the strengths and weaknesses of studies should be performed to triangulate their findings to provide assurance that the observed findings are actually real [27]. Thus, the objective of the present systematic review of the literature was to evaluate the observed associations between POPs and breast cancer risk and prognosis to identify individual POPs that are repeatedly found to be associated with both breast cancer incidence and progression, and to provide an explanation to the observed inconsistencies between studies.

2. Materials and Methods

A systematic review was conducted following a pre-established protocol and according to the general methodology of Cochrane reviews [28]. Considering the expected methodological diversity and heterogeneity between eligible studies, the great susceptibility of observational designs to selection bias and the variability in methods used to control for confounding, no quantitative synthesis was planned [28].

2.1. *Search Methods for Identification of Studies*

An electronic search of the following databases was performed, from inception to December 2018: MEDLINE (via PubMed), EMBASE and CENTRAL (Cochrane Central Register of Controlled Trials). Search strategies were developed for each of these databases with text words and index terms referring to POPs, breast cancer risk and breast cancer prognosis, and excluding animal studies (Table S1). No language or publication date restrictions were applied. Reference lists of relevant reviews and of included studies were scanned for any additional relevant studies not otherwise identified.

2.2. *Criteria for Considering Studies for This Review*

2.2.1. Types of Studies

Any observational or intervention study that evaluated the association between POPs and breast cancer risk, survival or a meaningful breast cancer prognostic factor, whatever the design was eligible for inclusion. No restrictions were applied regarding language or type (articles, short reports and abstracts) of publication.

2.2.2. Types of Participants

Women included in the studies before or after breast cancer diagnosis, regardless of age, menopausal status, breast cancer type, disease stage and treatment regimen, were eligible. No participants were excluded based on ethnicity.

2.2.3. Types of Exposures

Studies that measured exposure to any lipophilic POP, in a lipid rich biological human sample (peripheral blood and adipose tissue), whatever the method of measurement, were eligible.

2.2.4. Types of Outcomes

Breast cancer risk, measured by breast cancer incidence, prevalence or breast mammographic density (a recognized breast cancer risk factor) and breast cancer survival, including overall survival (all-cause mortality), breast cancer-specific survival (breast cancer-specific mortality), and breast cancer-free survival (breast cancer recurrence), were the primary outcomes. Studies that assessed the association of POPs with meaningful breast cancer prognostic factors (age, stage, tumor size, lymph node involvement, histological type, grade and molecular subtype) were also eligible.

2.3. Data Collection and Analysis

2.3.1. Selection of Studies

The references identified by the search strategy were reviewed by one author (K.E-I.) in a two-step process. First, the title and abstract of each study were screened to exclude obviously non-eligible studies and second, the full text of retained articles was examined and subjected to evaluation using the predefined eligibility criteria. Whenever required, a second review author (C.D.) was consulted. When required, further information was sought from the authors by email.

2.3.2. Data Extraction

Data extraction was performed using an exhaustive standardized form designed for this review. Information about study design (inclusion criteria, sample size and methodology), participants and tumors characteristics at diagnosis (age, menopausal status, tumor invasiveness, tumor ER status), exposure assessment (timing, tissue sample, method of measurement, lipid-adjustment, list of all contaminants evaluated, treatment of non-detectable values), measured outcomes and reported results (any reported measure of association, adjustment variables, and statistical model selection procedure) were collected. For observational studies, special attention was paid to distinguishing between adjusted and unadjusted results, and to the variable selection method used in multivariate analyses. Studies definition of each characteristic or variable retained was recorded. In the case of multiple publications related to the same study, the publication reporting the outcomes of interest to the present review or the one with the longest follow-up of these outcomes was considered as the reference, and information was supplemented by secondary publications as required. Abstracts with insufficient information and data to permit inclusion were excluded from the qualitative synthesis (Table S2). Data were extracted twice over the course of several days to ensure their consistency.

2.3.3. Assessment of Risk of Bias in Retained Studies

Based on the "STrengthening the Reporting of OBservational studies in Epidemiology." (STROBE) statements [28], and the rating approach of the "Risk Of Bias in Non-randomized Studies-of Interventions" (ROBINS-I) tool [27], the following domains were evaluated for risk of bias of included studies: selection of participants into the study, exposure measurement, outcome measurement, potential confounding accounted for, missing data, and selective reporting.

Assessment of the risk of bias was performed twice by a review author (K.E-I.), both for the risk of bias in each study and for the overall risk of bias across studies. When required, a second reviewer (C.D.) was consulted.

2.3.4. Assessment of Heterogeneity

Differences between studies, including study design, participant characteristics (age and menopausal status), tumor characteristics (invasiveness, ER status, and treatment received), exposure measurement (timing, type of tissue sample) and different levels of risk of bias were considered for exploring possible sources of heterogeneity.

2.3.5. Data Synthesis

Given that high heterogeneity between studies was expected, quantitative synthesis of data was not considered appropriate. A systematic qualitative synthesis of study characteristics and results was performed for risk, mortality, and prognostic factors associations with POPs exposure, and separately for each type of tissue sample. The results were considered adjusted only when all important confounders were considered into the models. For breast cancer risk, authors should have considered at minimum age, body mass index or any other estimation of body fat, and breastfeeding or parity as potential confounders. For breast cancer mortality, authors should have adjusted at minimum for age. In addition, studies of breast adipose POPs should have considered breastfeeding or parity as potential confounders. A positive association was defined as an observed higher risk or mortality with higher POPs exposure whereas a negative association was defined as an observed inverse association.

3. Results

3.1. *Results of the Search*

Of the 11,015 references retrieved by electronic search, 95 met eligibility criteria (Figure 1), of which 85 reported breast cancer incidence or prevalence outcomes [29–113], six reported mortality outcomes [41,45,114–117] and nine reported breast cancer prognostic factors [66,90,116,118–123]. The majority of studies of breast cancer risk were case-control studies (n = 81) whereas studies of breast cancer prognosis included five cohort studies for mortality and nine cross-sectional studies for breast cancer prognostic factors. Overall, POPs were measured in peripheral blood in 63 studies, in breast adipose tissue in 32 studies, in adipose tissue other than breast in five studies and in breast tumors in four studies (Figure 1).

3.2. *Description of Studies*

The 95 included studies were published between 1976 and 2018, and involved between five and 902 breast cancer patients (median = 113).

3.2.1. Studies of Breast Cancer Risk

Characteristics of the 61 studies that examined associations between peripheral blood POPs and breast cancer risk are summarized in Table 1. These studies included breast cancer patients between 40 and 66 years of mean age with varying proportions of premenopausal and postmenopausal patients. Ten studies included at least 80% of postmenopausal patients, of which three studies included exclusively postmenopausal patients. The proportion of invasive breast cancers was not reported in 35 studies and varied in the remaining 26 studies between 62 and 100%. Twenty studies included at least 80% of invasive breast cancers of which 13 studies included exclusively invasive breast cancers. The proportion of estrogen receptor (ER) positive breast cancers was not reported in 40 studies and varied in the remaining 19 studies between 32% and 87%. Two studies included at least 80% of ER-positive breast cancers (Table 1 and Table S3).

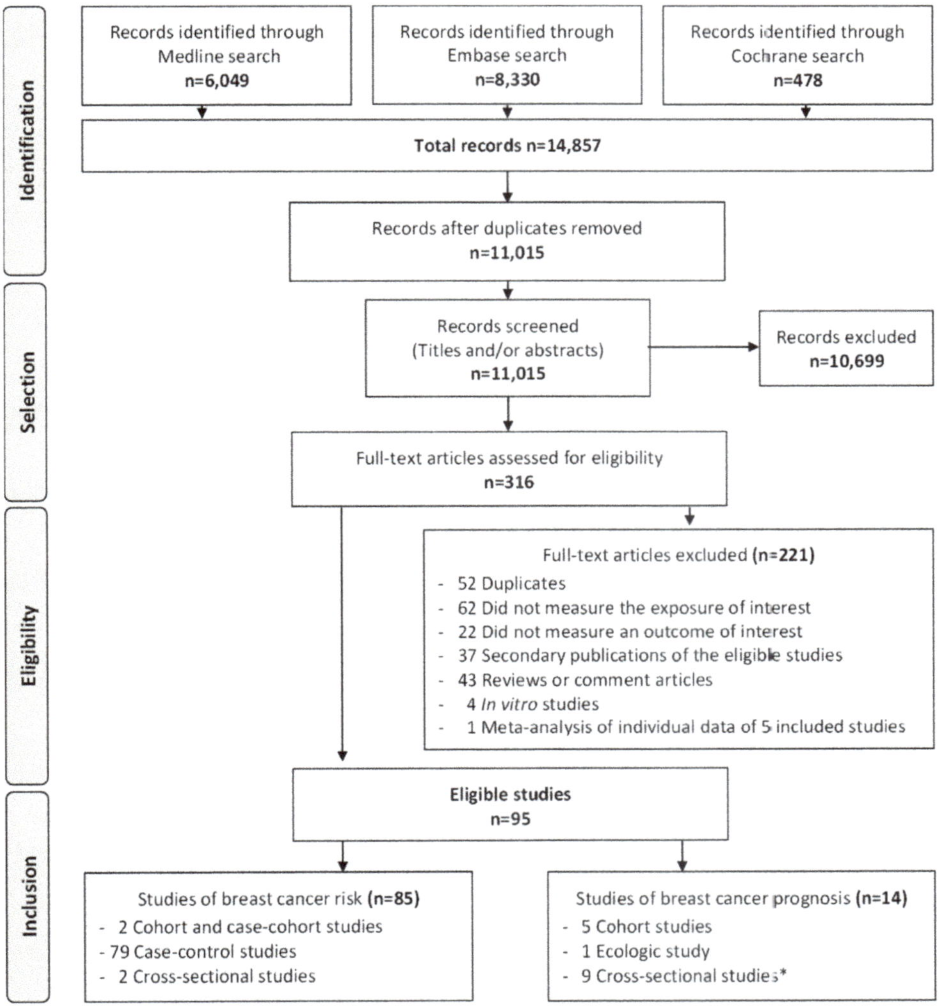

Figure 1. Flow Diagram according to PRISMA (Preferred Reporting Items of Systematic Reviews and Meta-Analyses) [PRISMA], with modifications. *One cohort study on breast cancer mortality also reported cross-sectional analyses of prognostic factors.

Characteristics of the 26 studies that examined associations between breast adipose tissue POPs and breast cancer risk are summarized in Table 2. All these studies were hospital-based case-control studies with hospital-controls and included breast cancer patients between 40 and 63 years of mean age with varying proportions of premenopausal and postmenopausal patients. Only one study included at least 80% of postmenopausal patients. The proportion of invasive breast cancers was not reported in 11 studies and varied in the remaining 15 studies between 76% and 100%, with eight studies including exclusively invasive breast cancers. The proportion of ER-positive breast cancers was not reported in 15 studies and varied in the remaining 11 studies between 45% and 90%. No study included at least 80% of ER-positive breast cancers (Table 2 and Supplementary Table S4).

Table 1. Summary characteristics of studies of peripheral blood POPs and breast cancer risk ($n = 61$).

Design	Cohort studies: $n = 1$ Case-cohort studies: $n = 1$ Cohort nested case-control studies with incidence density (risk-set) sampling: $n = 13$ Cohort nested case-control studies with cumulative density (exclusive) sampling: $n = 3$ Population-based case-control studies not nested in a defined cohort: $n = 9$ Hospital-based case-control studies with community controls: $n = 8$ Hospital-based case-control studies with both community-controls and hospital-controls: $n = 4$ Hospital-based case-control studies with hospital-controls: $n = 19$ Case-control studies (unclassified): $n = 1$ Cross-sectional studies: $n = 2$	
Breast cancer patients	Number of breast cancer patients: 20 to 902 Mean age: 40 to 66 years; NR in 13 studies. Postmenopausal patients: 0% to 100%, median = 61%; two studies exclusively in premenopausal patients; NR in 19 studies. Invasive breast cancers: 62% to 100%, median = 99.5%; NR in 35 studies Estrogen receptor positive breast cancers: 32% to 87%, median 64%; NR in 40 studies	
POPs	*Measurement method:* GC-ECD: $n = 43$ HPLC-MS-MS: $n = 5$ GC-MS: $n = 4$ GC-ID-HRMS: $n = 3$ HPLC/FD: $n = 2$ LC-MS-MS: $n = 1$ GC-IDMS: $n = 1$ GC: $n = 3$ MS: $n = 1$ NR: $n = 2$	*Lipid adjustment:* $n = 39$; NR in 18 studies *Measured POPs:* PCBs: $n = 38$ Organochlorines: $n = 48$ Dioxins: $n = 2$ PFAS: $n = 3$ Phthalates: $n = 1$ Parabens: $n = 1$ BPA: $n = 2$ PBBs: $n = 1$

n: number of studies; NR: not reported; POPs: persistent organic pollutants; MS: mass spectrometry; GC-ECD: gas chromatography with electron Capture Detector; HLPC-MS-MS: high-performance liquid chromatography-tandem mass spectrometry; HPLC/FD: high-performance liquid chromatography with fluorescence detection; LC-MS-MS: liquid chromatography–tandem mass spectrometry; GC-MS-MS: gas chromatography-tandem mass spectrometry; GC-IDMS: gas-chromatography isotope-dilution mass-spectrometry; GC-ID-HRMS: gas chromatography-isotope dilution high-resolution mass spectrometry; HR-GC-ECD: high-resolution gas chromatography with micro-electron capture detection; GC: gas chromatography; PCBs:. polychlorinated biphenyls; PFAS: perfluoroalkyl substances; BPA: bisphenol A; PBBs: polybrominated biphenyls.

Table 2. Summary characteristics of studies of breast adipose tissue POPs and breast cancer risk ($n = 26$).

Design	Hospital-based case-control studies with hospital-controls: $n = 26$	
Breast cancer patients	Number of breast cancer patients: 5 to 304 Mean age: 40 to 63 years; NR in three studies. Postmenopausal patients: 49% to 82%, median = 56%; NR in 16 studies. Invasive breast cancers: 76% to 100%, median = 100%; NR in 11 studies Estrogen receptor positive breast cancers: 45% to 79%, median 64%; NR in 15 studies	
POPs	*Measurement method:* GC-ECD: $n = 14$ GC-MS: $n = 9$ GC: $n = 4$ *Lipid adjustment:* $n = 23$; NR in three studies	*Measured POPs:* PCBs: $n = 16$ Organochlorines: $n = 21$ Dioxins: $n = 2$ PBDEs: $n = 1$

n: number of studies; NR: not reported; POPs: persistent organic pollutants; GC-ECD: gas chromatography with electron Capture Detector; GC-MS: gas-chromatography mass-spectrometry; GC: gas chromatography; PCBs: polychlorinated biphenyls; PFAS: perfluoroalkyl substances; PBDE: polybrominated diphenyl ethers.

Four studies examined associations between POPs in breast tumors and breast cancer risk, of which three included breast tissue surrounding malignant tumors for cases. Two studies compared malignant tumors of cases to benign tumors of controls whereas the two other studies compared malignant tumors of cases to normal tissue of controls. All these four studies were hospital-bases case-control studies with hospital controls, in which cases were on average 50 to 60 years old in the two studies that reported age at diagnosis. One study included exclusively invasive cancers, whereas the three other studies did not report proportion of invasive cancers. No study reported proportion of menopausal patients or the proportion of ER-positive breast cancers (Table S4).

Two studies examined associations between POPs in buttock adipose tissue and breast cancer risk. The recent one was a cohort-nested case-control study with incidence density (risk-set) sampling and included 409 postmenopausal breast cancer patients aged 58 years old in average, of which 78% had ER-positive breast cancers. The proportion of invasive tumors was not reported. The other study was a hospital-based case-control study with community controls including 265 breast cancer patients aged 62 years old in average. The proportions of postmenopausal patients, invasive tumors and ER-positive tumors were not reported (Table S5).

3.2.2. Studies of Breast Cancer Prognosis

Characteristics of the 14 studies that examined the association between POPs and breast cancer prognosis are summarized in Table 3.

Table 3. Summary characteristics of studies of POPs and breast cancer prognosis (n = 14).

Design	Studies of mortality: n = 6 Cohort studies n = 5 Ecologic study: n = 1 Studies of prognostic factors: n = 10 Cross-sectional studies: n = 10	
Breast cancer patients	Studies of mortality: n = 6 Number of breast cancer patients: 161 to 633 Mean age: 58 to 66 years; NR in one study. Postmenopausal patients: 66% to 100% (in two studies); NR in three studies. Invasive breast cancers: 71% to 86% (in two studies); NR in three studies Estrogen receptor positive breast cancers: 72% to 78% (in three studies); NR in three studies Studies of prognostic factors: n = 10 Number of breast cancer patients: 40 to 409 Mean age: 52 to 65 years Postmenopausal patients: 41 % to 100%, median = 63%; NR in four studies. Invasive breast cancers: 85% to 100%, median = 100%; NR in three studies Estrogen receptor positive breast cancers: 50% to 86%, median 68%	
POPs	Studies of mortality: n = 6 *Type of sample:* Peripheral blood: n = 3 Breast adipose tissue: n =1 Adipose tissue other than breast: n = 2 *Measurement method:* GC-ECD: n = 4 GC-MS: n = 1 NR: n = 1 *Lipid adjustment:* n = 5; NR in 1 study *Measured POPs:* PCBs: n = 4 Organochlorines: n = 5	Studies of prognostic factors: n = 10 *Type of sample:* Peripheral blood: n = 3 Breast adipose tissue: n = 6 Adipose tissue other than breast: n = 1 *Measurement method:* GC-ECD: n = 6 GC-MS: n = 3 HLPC-MS-MS: n = 1 *Lipid adjustment:* n = 9; NR in 1 study *Measured POPs:* PCBs: n = 4 Organochlorines: n = 6

n: number of studies; NR: not reported; POPs: persistent organic pollutants; GC-MS: gas-chromatography mass-spectrometry; GC-ECD: gas chromatography with electron Capture Detector; HLPC-MS-MS: high-performance liquid chromatography-tandem mass spectrometric; PCBs: polychlorinated biphenyls.

Six studies examined mortality outcomes, with three measuring POPs in peripheral blood, one measuring POPs in breast adipose tissue and two measuring POPs in adipose tissue other than breast. Patients were aged between 58 and 66 years old in average with only two studies reporting the proportion of postmenopausal women (66% and 100% respectively), and two studies reporting proportion of invasive cancers (71% and 86% respectively). The proportion of ER-positive breast cancers varied between 72% and 78% in the three studies that have reported patients ER status (Tables S5 and S6).

Ten studies examined breast cancer prognostic factors, with three measuring POPs in peripheral blood, six measuring POPs in breast adipose tissue and one measuring POPs in adipose tissue other than breast. Patients were aged between 52 and 65 years old in average with varying proportions of premenopausal and postmenopausal women and none with at least 80% of postmenopausal patients. The proportion of invasive cancers varied between 85% and 100% and the proportion of ER-positive breast cancers varied between 50% and 86% with only two studies including at least 80% of ER-positive breast cancers (Table S8).

3.3. *Risk of Bias in Retained Studies*

Overall, studies reporting associations between peripheral blood POPs and breast cancer risk ranged from moderate to critical risk of bias, whereas studies reporting associations between adipose tissue POPs and breast cancer risk were more likely to be at serious or critical risk of bias.

Overall, studies reporting associations between POPs, measured in peripheral blood or in adipose tissue, and mortality outcomes were at serious risk of bias, whereas studies reporting associations with prognostic factors ranged from serious to critical risk of bias.

3.4. *Systematic Data Synthesis*

3.4.1. Studies of Breast Cancer Risk

Among the 61 studies that examined associations between peripheral blood POPs and breast cancer risk, 30 reported a positive association with at least one POP, six reported a negative association with at least one POP, and 20 reported no association. Cohort-nested case-control studies with cumulative density (exclusive) sampling, population-based case-control studies not nested in a defined cohort and hospital-based case-control studies with both community- and hospital-controls were more likely to observe an association. Studies that included at least 80% of postmenopausal patients were more likely to observe an association whereas studies that included less than 80% of postmenopausal patients were more likely to observe no association. Studies that included at least 80% of invasive cancers and those with non-reported proportions of invasive breast cancers were more likely to observe an association. Studies with non-reported proportions of ER-positive breast cancers were more likely to observe an association whereas studies including less than 80% ER-positive breast cancers were slightly more likely to observe no association (Table S3).

Among the 26 studies that examined associations between breast adipose tissue POPs and breast cancer risk, 10 reported a positive association with at least one POP, two reported a negative association with at least one POP, and seven reported no association. Studies reporting the proportion of invasive breast cancers were slightly more likely to observe an association. Studies with non-reported proportions of ER-positive breast cancers were more likely to observe a positive association. The four studies of POPs in breast tumors did not report minimally adjusted estimates of risk (Table S4).

The two studies that examined associations between POPs in buttock adipose tissue and breast cancer risk reported negative associations (Table S5).

3.4.2. Studies of Breast Cancer Prognosis

All three studies that examined peripheral blood POPs reported positive associations with both breast cancer all-cause and specific mortality, of which one study also reported a negative association with all-cause mortality (Table S6). The only study of breast adipose tissue POPs and breast cancer mortality reported a positive association with breast cancer recurrence. The two studies of POPs in adipose tissue other than breast reported negative associations with all-cause and breast cancer specific mortality respectively, of which one study reported a positive association with breast cancer specific mortality (Table S7).

Among the three studies that examined peripheral blood POPs and breast cancer prognostic factors, one study reported a positive association with tumor size and lymph-node involvement. Among the six studies of breast adipose tissue POPs, three studies examined associations with meaningful breast cancer prognostic factors but none of them reported minimally adjusted estimates. The only study of buttock adipose tissue POPs and breast cancer prognostic factors did not report minimally adjusted estimates (Table S8).

3.4.3. Individual POPs and Breast Cancer Risk and Prognosis

One to 71 individual POPs were measured in studies of breast cancer risk (median = 9). Organochlorines were measured in 69 studies, of which 43 in blood, 21 in breast adipose tissue, two in adipose tissue other than breast and three in breast tumor. Polychlorinated biphenyls (PCBs) were measured in 57 studies, of which 38 were in blood, 16 in breast adipose tissue, one in adipose tissue other than breast and two in breast tumors. Dioxins were measured in four studies, of which two in blood and two in breast adipose tissue. Perfluoroalkyl substances (PFAS) were measured in three studies in blood. Bisphenol A (BPA) was measured in two studies in blood, polybrominated flame retardants (PBBs and PBDEs) in one study in blood and one study in breast adipose tissue whereas mono-ethyl phthalate (MEP) and parabens were measured in blood in one study respectively (Table S9).

One to 35 individual POPs were measured in studies of breast cancer prognosis (median = 25). Organochlorines were measured in five studies of breast cancer mortality, of which three were in blood, one in breast adipose tissue and two in adipose tissue other than breast, whereas six studies measured organochlorines in relation to prognostic factors, of which three were in blood, four in breast adipose tissue and one in adipose tissue other than breast. PCBs were measured in four studies of breast cancer mortality, of which three were in blood, one in breast adipose tissue and one in adipose tissue other than breast, whereas four studies measured PCBs in relation to prognostic factors, of which two in blood, four in breast adipose tissue and one in adipose tissue other than breast (Tables S9 and S10). Parabens were measured in breast adipose tissue in one study in relation to prognostic factors.

The magnitude of the reported associations between POPs and breast cancer risk and mortality are summarized in Table 4.

Table 4. Main results summary of studies reporting positive * associations between POPs and breast cancer risk and mortality.

Type of Tissue Sample	Studies with Positive * Associations/Total Studies	Range of Associations ** Estimate [95% CI]
	Breast cancer risk	
Blood	29/61	From OR = 1.1 [1.0–1.2] (*Heptachlor, continuous*) to OR = 7.6 [1.1–51.4] (*PCBs group 1a, variable form NR*)
Breast adipose tissue	10/26	From OR = 1.1 [1.0–1.3] (*PCB 180, quart 4 vs. quart 1*) to OR = 10.5 [2.0–55.3] (*β-HCH, >0.1 mg/kg vs. <0.1 mg/kg*)
Adipose tissue other than breast	0/2	NA
	Breast cancer mortality	
Blood	3/3	From HR = 1.9 [1.1–3.4] (*15-year breast cancer mortality PCB 174, tert 3 vs. tert 1*) to HR = 5.8 [1.6–20.5] (*breast cancer recurrence and/or death, Dieldrin, quart 4 vs. quart 1*)
Breast adipose tissue	1/1	From HR = 2.6 [1.0–7.1] (*Breast cancer recurrence, PCB 153, tert 3 vs. tert1*) to HR = 4.0 [1.3–4.9] (*PCB 118, Breast cancer recurrence, tert 3 vs. tert1*)
Adipose tissue other than breast	1/2	MRR = 1.3 [1.1–1.5] (*Breast-cancer specific mortality, Dieldrin, linear estimates per inter-quartile range*)

POPs: persistent organic pollutants; CI: Confidence interval; NA: Not applicable; NR: not reported; * positive association: an observed higher risk or mortality with higher POPs exposure; ** Adjusted for all important confounders; OR: odds ratio; MRR: mortality rate ratio; PCB: Polychlorinated biphenyls; HR: hazard ratio, β-HCH: β-Hexachlorocyclohexane.

When considering POPs positively associated with breast cancer risk in at least 10% of studies and at least two studies and no reported negative associations, eight individual POPs were consistently positively associated with breast cancer risk in blood: *p,p′*-Dichlorodiphenyldichloroethylene (*p,p′*-DDE), total or not specified DDE, β- Hexachlorocyclohexane (β-HCH), Dieldrin, PCB 118, PCB 138, PCB 170, PCB 180. Three individual POPs were consistently positively associated with breast cancer risk in breast adipose tissue: *p,p′*-DDE, total or not specified DDE, PCB 105. When considering POPs positively associated with breast cancer risk in at least one study and no reported negative associations, six individual POPs were positively associated with breast cancer risk in both blood and breast adipose tissue: *p,p′*-DDE, total or not specified DDE, Hexachlorobenzene (HCB), β-HCH, PCB 118 and PCB 180 (Tables S11 and S12).

When considering POPs positively associated with breast cancer mortality in at least 10% of studies and no reported negative associations, total PCBs and four individual POPs were positively associated with breast cancer mortality in blood: *p,p′*-Dichlorodiphenyltrichloroethane (*p,p′*-DDT), Dieldrin, PCB 174, PCB 177. Total PCBs and three individual POPs were positively associated with breast cancer mortality in breast adipose tissue: PCB 118, PCB 153, PCB 167 (Tables S13 and S14). Six individual POPs were positively associated with breast cancer prognostic factors in blood, in at least one study and with no reported negative associations: *p,p′*-DDE, Oxychlordane, *trans*-Nonachlor, β-HCH, PCB 138, PCB 153 (Table S16).

Three individual POPs were positively associated with both breast cancer risk and prognosis either in blood or in breast adipose tissue: *p,p′*-DDE, β-HCH and PCB 118 (Tables S11–S15).

4. Discussion

The present systematic review of POPs and breast cancer indicates that studies of blood POPs and breast cancer risk accounted for much of the observed inconsistencies of epidemiological studies results. POPs measured in breast adipose tissue were more clearly associated with higher breast cancer risk. POPs were more consistently associated with worse breast cancer prognosis, whether measured in blood or breast adipose tissue, whereas POPs measured in adipose tissue other than breast were inversely associated with both breast cancer risk and prognosis. Some individual POPs measured in blood and breast adipose tissue were consistently associated with higher breast cancer risk and worse prognosis. However, the overall strength of evidence is weak, since few studies contributed to estimations of associations and the overall risk of bias in these studies ranged from moderate to critical.

The inconsistencies between studies of blood POPs and breast cancer risk could be explained by methodological biases. In fact, more than half of these studies have measured POPs at the time of diagnosis which does not necessarily reflect the cumulative lifetime exposure to POPs and early-life exposures during critical windows of vulnerability [124]. Even though the majority of population-nested case-control studies and the only cohort study have measured POPs several months to many years before breast cancer occurrence, a point measurement of blood concentration of POPs is more likely to reflect recent dietary intakes and liver function [125,126] and can be affected by various events over time, such as weight loss or gain, pregnancies and breastfeeding [124–126]. The complex misclassification of POPs exposure resulting from blood measurements could have biased the observed associations toward the null, i.e., toward the observation of weaker associations or no associations at all.

In this regard, adipose tissue, as a storage compartment for lipophilic POPs [127], is a more appropriate medium for estimating lifetime exposure to POPs. The observation of consistently positive associations with breast cancer risk in studies of breast adipose tissue POPs but consistently negative associations with POPs measured in adipose tissue other than breast is in line with the existent evidence of a protective function of adipose tissue in the wildlife and points toward the metabolic and toxicokinetic differences between different types of adipose tissue [127]. By accumulating POPs, adipose tissue away from breast decreases their availability to other tissues, thereby limiting their toxicity to the breast, whereas accumulation of POPs in breast adipose tissue exposes breast epithelial cells to their chronic local release. In fact, ultrastructural methods revealed regional differences in morphology of human subcutaneous tissues [128]. Abdominal adipose tissue, classified as deposit white adipose tissue, having large adipose cells and a poor collagenic component whereas adipocytes of breast adipose tissue, classified as structural white adipose tissue, are covered by a relatively dense connective capsule [128]. These regional differences in morphology explain the known regional differences in the metabolism of subcutaneous fatty depots that are related to their various functions. Thus, our results suggest that differences between adipose tissue subtypes may also have a toxicokinetic impact on POPs.

Moreover, more than half of studies of blood POPs and breast cancer risk included more premenopausal than postmenopausal breast cancer patients. Although environmental exposures may be involved in premenopausal breast cancer occurrence, these cancers are primarily driven by a strong genetic susceptibility and are more often ER-negative cancers [129]. Furthermore, the increase in breast cancer incidence over the last decades reflects the increase in the incidence of postmenopausal breast cancers, which are more often ER-positive breast cancers [130], and thus more susceptible to the hormone-disrupting effects of POPs [8]. The selection bias created by inclusion of large proportions of premenopausal breast cancers, which are less likely to be related to POPs exposure, could have biased the observed associations toward the null. In fact, we observed that studies that included less than 80% of postmenopausal patients and those including fewer than 80% ER-positive breast cancers were more likely to observe no association.

Another important issue was related to statistical methods used for selecting potential confounders. If the majority of studies of blood POPs and breast cancer risk have considered important confounders in their statistical models, methods used for selecting potential confounders were not always appropriate. In particular, the majority of case-control studies used the change in estimate method, which is not appropriate for accurate estimations of associations. Changes in estimates may be observed when adjusting for colliders (i.e., non-confounders that introduce a selection bias) and when non-collapsible effect measures such as odds ratios are used [131]. The bias introduced by this method can be difficult to predict when numerous variables are tested for confounding and can lead to discrepant studies results.

The strengths of the present systematic review include the use of the Cochrane Reviews rigorous methodology, the extensive and highly sensitive search strategy to retrieve as many relevant studies as possible, the use of a pre-established protocol, the assessment of the risk of bias, and the systematic analysis of results, in addition to considering sources of heterogeneity between studies results. Limitations include the lack of high-quality evidence inherent in observational study designs and the overall critical risk of bias in included studies.

Finally, although the present systematic review has identified some individual POPs associated with both breast cancer risk and prognosis that deserve further investigation, it should be emphasized that different POPs have different metabolic profiles and can have synergistic or antagonistic effects, and that proportions of different POPs may vary from one person to another. Thus, approaches considering the simultaneous exposure to different POPs may be more relevant than the isolated analysis of individual POPs.

5. Conclusions

Over the past three decades, numerous epidemiological studies have attempted to assess the association between exposure to POPs and breast cancer. Despite the apparent inconsistencies between studies, which were mainly due to methodological biases and to differences in the biological sample used for exposure measurement, when considering all studies (peripheral blood and adipose tissue) and all outcomes together (risk and prognosis), there was a trend toward a positive association between exposure to POPs and breast cancer that deserves further investigation. Future studies need to use rigorous methodology by including the relevant study population, using an appropriate biological sample for POPs measurement, controlling properly for confounding and assessing combined effects of POPs.

Supplementary Materials: The following are available online at http://www.mdpi.com/2072-6694/11/8/1063/s1, Table S1: Search strategy for Medline via PubMed, Table S2: Studies not included in the qualitative synthesis, Table S3: Studies of POPs measured in peripheral blood and breast cancer risk, Table S4: Studies of POPs measured in breast adipose tissue and breast cancer risk, Table S5: Studies of POPs measured in adipose tissue other than breast and breast cancer risk, Table S6: Studies of POPs measured in peripheral blood and mortality among breast cancer patients, Table S7: Studies of POPs measured in adipose tissue and mortality among breast cancer patients, Table S8: Studies of POPs and breast cancer prognostic factors, Table S9: Results of studies of POPs and breast cancer risk, Table S10: Results of studies of POPs and mortality in breast cancer patients, Table S11: Results of studies of POPs and prognostic factors in breast cancer patients, Table S12: Summary results of POPs associated positively with breast cancer risk, Table S13: Summary results of POPs associated negatively with breast cancer risk, Table S14: Summary results of POPs associated positively with mortality among breast cancer patient, Table S15: Summary results of POPs associated negatively with mortality among breast cancer patients, Table S16: Summary results of POPs associated positively with breast cancer prognostic factors.

Author Contributions: Conceptualization, K.E.-I. and C.D.; methodology, K.E.-I. and C.D.; formal analysis, K.E.-I. and C.D.; resources, C.D.; writing—original draft preparation, K.E.-I.; writing—review and editing, C.D. and P.A.; supervision, C.D.

Funding: This research received no external funding. K.E.-I. is a recipient of the Vanier Canada Graduate Scholarship. C.D. was a recipient of the Canadian Breast Cancer Foundation-Canadian Cancer Society Capacity Development award (award #703003) and holds a Senior Investigator Award from the Fonds de recherche du Québec-Santé.

Conflicts of Interest: The authors declare no conflict of interest. The funders had no role in the design of the study; in the collection, analyses, or interpretation of data; in the writing of the manuscript, or in the decision to publish the results.

References

1. Connell, D.W.; Miller, G.J.; Mortimer, M.R.; Shaw, G.R.; Anderson, S.M. Persistent Lipophilic Contaminants and Other Chemical Residues in the Southern Hemisphere. *Crit. Rev. Environ. Sci. Technol.* **1999**, *29*, 47–82. [CrossRef]
2. Kelly, B.C.; Ikonomou, M.G.; Blair, J.D.; Morin, A.E.; Gobas, F.A.P.C. Food Web-Specific Biomagnification of Persistent Organic Pollutants. *Science* **2007**, *317*, 236–239. [CrossRef] [PubMed]
3. Davis, D.L.; Bradlow, H.L.; Wolff, M.; Woodruff, T.; Hoel, D.G.; Anton-Culver, H. Medical hypothesis: Xenoestrogens as preventable causes of breast cancer. *Environ. Health Perspect.* **1993**, *101*, 372–377. [CrossRef] [PubMed]
4. Hunter, D.J.; Kelsey, K.T. Pesticide Residues and Breast Cancer: The Harvest of a Silent Spring? *J. Natl. Cancer Inst.* **1993**, *85*, 598–599. [CrossRef] [PubMed]
5. Ahlborg, U.G.; Lipworth, L.; Titus-Ernstoff, L.; Hsieh, C.-C.; Hanberg, A.; Baron, J.; Trichopoulos, D.; Adami, H.-O. Organochlorine Compounds in Relation to Breast Cancer, Endometrial Cancer, and Endometriosis: An Assessment of the Biological and Epidemiological Evidence. *Crit. Rev. Toxicol.* **1995**, *25*, 463–531. [CrossRef] [PubMed]
6. Calafat, A.M.; Needham, L.L. Human exposures and body burdens of endocrine-disrupting chemicals. In *Endocrine-Disrupting Chemicals: From Basic Research to Clinical Practice*; Gore, A.C., Ed.; Humana Press: Totowa, NJ, USA, 2007; pp. 253–268.
7. Diamanti-Kandarakis, E.; Bourguignon, J.-P.; Giudice, L.C.; Hauser, R.; Prins, G.S.; Soto, A.M.; Zoeller, R.T.; Gore, A.C. Endocrine-Disrupting Chemicals: An Endocrine Society Scientific Statement. *Endocr. Rev.* **2009**, *30*, 293–342. [CrossRef] [PubMed]
8. Dey, S.; Soliman, A.S.; Merajver, S.D. Xenoestrogens may be the cause of high and increasing rates of hormone receptor positive breast cancer in the world. *Med. Hypotheses* **2009**, *72*, 652–656. [CrossRef]
9. Phillips, K.P.; Foster, W.G. Key Developments in Endocrine Disrupter Research and Human Health. *J. Toxicol. Environ. Health Part B* **2008**, *11*, 322–344. [CrossRef] [PubMed]
10. Soto, A.M.; Chung, K.L.; Sonnenschein, C. The pesticides endosulfan, toxaphene, and dieldrin have estrogenic effects on human estrogen-sensitive cells. *Environ. Health Perspect.* **1994**, *102*, 380–383. [CrossRef]
11. Shekhar, P.V.M.; Werdell, J.; Basrur, V.S. Environmental Estrogen Stimulation of Growth and Estrogen Receptor Function in Preneoplastic and Cancerous Human Breast Cell Lines. *J. Natl. Cancer Inst.* **1997**, *89*, 1774–1782. [CrossRef]

12. Steinmetz, R.; Young, P.C.; Caperell-Grant, A.; Gize, E.A.; Madhukar, B.V.; Ben-Jonathan, N.; Bigsby, R.M. Novel estrogenic action of the pesticide residue beta-hexachlorocyclohexane in human breast cancer cells. *Cancer Res.* **1996**, *56*, 5403–5409. [PubMed]
13. Reaves, D.K.; Ginsburg, E.; Bang, J.J.; Fleming, J.M. Persistent organic pollutants and obesity: Are they potential mechanisms for breast cancer promotion? *Endocr. Relat. Cancer* **2015**, *22*, R69–R86. [CrossRef] [PubMed]
14. Tabb, M.M.; Blumberg, B. New Modes of Action for Endocrine-Disrupting Chemicals. *Mol. Endocrinol.* **2006**, *20*, 475–482. [CrossRef] [PubMed]
15. Letcher, R.J.; Van Holsteijn, I.; Drenth, H.-J.; Norstrom, R.J.; Bergman, Å.; Safe, S.; Pieters, R.; Berg, M.V.D. Cytotoxicity and Aromatase (CYP19) Activity Modulation by Organochlorines in Human Placental JEG-3 and JAR Choriocarcinoma Cells. *Toxicol. Appl. Pharmacol.* **1999**, *160*, 10–20. [CrossRef] [PubMed]
16. Wojtowicz, A.K.; Milewicz, T.; Gregoraszczuk, E.Ł. DDT and its metabolite DDE alter steroid hormone secretion in human term placental explants by regulation of aromatase activity. *Toxicol Lett.* **2007**, *173*, 24–30. [CrossRef] [PubMed]
17. Yanez, L.; Borja-Aburto, V.H.; Rojas, E.; de la Fuente, H.; Gonzalez-Amaro, R.; Gomez, H.; Jongitud, A.A.; Diaz-Barriga, F. DDT induces DNA damage in blood cells. Studies in vitro and in women chronically exposed to this insecticide. *Environ. Res.* **2004**, *94*, 18–24. [CrossRef]
18. Karami-Mohajeri, S.; Abdollahi, M. Toxic influence of organophosphate, carbamate, and organochlorine pesticides on cellular metabolism of lipids, proteins, and carbohydrates: A systematic review. *Hum. Exp. Toxicol.* **2011**, *30*, 1119–1140. [CrossRef]
19. Knower, K.C.; To, S.Q.G.; Leung, Y.-K.; Ho, S.-M.; Clyne, C.D. Endocrine disruption of the epigenome: A breast cancer link. *Endocr. Relat. Cancer* **2014**, *21*, T33–T55. [CrossRef]
20. Falkner, K.; Moysich, K.; Menezes, R.; Baker, J. Environmental Exposure to Polychlorinated Biphenyls and Breast Cancer Risk. *Rev. Environ. Health* **2002**, *17*, 263–278.
21. Khanjani, N.; Hoving, J.L.; Forbes, A.B.; Sim, M.R. Systematic Review and Meta-analysis of Cyclodiene Insecticides and Breast Cancer. *J. Environ. Sci. Health Part C* **2007**, *25*, 23–52. [CrossRef]
22. Gray, J.; Evans, N.; Taylor, B.; Rizzo, J.; Walker, M. State of the Evidence: The Connection between Breast Cancer and the Environment. *Int. J. Occup. Environ. Health* **2009**, *15*, 43–78. [CrossRef]
23. Ingber, S.Z.; Buser, M.C.; Pohl, H.R.; Abadin, H.G.; Murray, H.E.; Scinicariello, F. DDT/DDE and breast cancer: A meta-analysis. *Regul. Toxicol. Pharmacol.* **2013**, *67*, 421–433. [CrossRef] [PubMed]
24. Leng, L.; Li, J.; Luo, X.-M.; Kim, J.-Y.; Li, Y.-M.; Guo, X.-M.; Chen, X.; Yang, Q.-Y.; Li, G.; Tang, N.-J. Polychlorinated biphenyls and breast cancer: A congener-specific meta-analysis. *Environ. Int.* **2016**, *88*, 133–141. [CrossRef] [PubMed]
25. Gray, J.M.; Rasanayagam, S.; Engel, C.; Rizzo, J. State of the evidence 2017: An update on the connection between breast cancer and the environment. *Environ. Health* **2017**, *16*, 94. [CrossRef] [PubMed]
26. Rodgers, K.M.; Udesky, J.O.; Rudel, R.A.; Brody, J.G. Environmental chemicals and breast cancer: An updated review of epidemiological literature informed by biological mechanisms. *Environ. Res.* **2018**, *160*, 152–182. [CrossRef]
27. Sterne, J.A.C.; Higgins, J.P.T.; Elbers, R.G.; Reeves, B.C. The Development Group for ROBINS-I. Risk of Bias in Non-Randomized Studies of Interventions (ROBINS-I): Detailed Guidance. Updated 12 October 2016. Available online: http://www.riskofbias.info (accessed on 9 June 2019).
28. Higgins, J.P.T.; Green, S. Cochrane Handbook for Systematic Reviews of Interventions Version 5.1.0. Updated March 2011. The Cochrane Collaboration. 2011. Available online: www.handbook.cochrane.org (accessed on 31 December 2018).
29. Warner, M.; Mocarelli, P.; Samuels, S.; Needham, L.; Brambilla, P.; Eskenazi, B. Dioxin Exposure and Cancer Risk in the Seveso Women's Health Study. *Environ. Health Perspect.* **2011**, *119*, 1700–1705. [CrossRef]
30. Bonefeld-Jørgensen, E.C.; Long, M.; Fredslund, S.O.; Bossi, R.; Olsen, J. Breast cancer risk after exposure to perfluorinated compounds in Danish women: A case-control study nested in the Danish National Birth Cohort. *Cancer Causes Control* **2014**, *25*, 1439–1448. [CrossRef]
31. Hurley, S.; Goldberg, D.; Wang, M.; Park, J.-S.; Petreas, M.; Bernstein, L.; Anton-Culver, H.; Nelson, D.O.; Reynolds, P. Breast cancer risk and serum levels of per- and poly-fluoroalkyl substances: A case-control study nested in the California Teachers Study. *Environ. Health* **2018**, *17*, 83. [CrossRef]

32. Cohn, B.A.; La Merrill, M.; Krigbaum, N.Y.; Yeh, G.; Park, J.-S.; Zimmermann, L.; Cirillo, P.M. DDT Exposure in Utero and Breast Cancer. *J. Clin. Endocrinol. Metab.* **2015**, *100*, 2865–2872. [CrossRef]
33. Cohn, B.A.; Terry, M.B.; Plumb, M.; Cirillo, P.M. Exposure to polychlorinated biphenyl (PCB) congeners measured shortly after giving birth and subsequent risk of maternal breast cancer before age 50. *Breast Cancer Res. Treat.* **2012**, *136*, 267–275. [CrossRef]
34. Iwasaki, M.; Inoue, M.; Sasazuki, S.; Kurahashi, N.; Itoh, H.; Usuda, M.; Tsugane, S. Plasma organochlorine levels and subsequent risk of breast cancer among Japanese women: A nested case-control study. *Sci. Total Environ.* **2008**, *402*, 176–183. [CrossRef] [PubMed]
35. Rubin, C.H.; Lanier, A.; Kieszak, S.; Brock, J.W.; Koller, K.R.; Strosnider, H.; Needham, L.; Zahm, S.; Harpster, A. Breast cancer among Alaska Native women potentially exposed to environmental organochlorine chemicals. *Int. J. Circumpolar Health* **2006**, *65*, 18–27. [CrossRef] [PubMed]
36. Wolff, M.S.; Zeleniuch-Jacquotte, A.; Dubin, N.; Toniolo, P. Risk of breast cancer and organochlorine exposure. *Cancer Epidemiol. Biomark. Prev.* **2000**, *9*, 271–277.
37. Terrell, M.L.; Rosenblatt, K.A.; Wirth, J.; Cameron, L.L.; Marcus, M. Breast cancer among women in Michigan following exposure to brominated flame retardants. *Occup. Environ. Med.* **2016**, *73*, 564–567. [CrossRef] [PubMed]
38. Ward, E.M.; Schulte, P.; Grajewski, B.; Andersen, A.; Patterson, D.G., Jr.; Turner, W.; Jellum, E.; Deddens, J.A.; Friedland, J.; Roeleveld, N.; et al. Serum organochlorine levels and breast cancer: A nested case-control study of Norwegian women. *Cancer Epidemiol. Biomark. Prev.* **2000**, *9*, 1357–1367.
39. Helzlsouer, K.J.; Alberg, A.J.; Huang, H.Y.; Hoffman, S.C.; Strickland, P.T.; Brock, J.W.; Burse, V.W.; Needham, L.L.; Bell, D.A.; Lavigne, J.A.; et al. Serum concentrations of organochlorine compounds and the subsequent development of breast cancer. *Cancer Epidemiol. Biomark. Prev.* **1999**, *8*, 525–532.
40. Dorgan, J.F.; Brock, J.W.; Rothman, N.; Needham, L.L.; Miller, R.; Stephenson, H.E.; Schussler, N.; Taylor, P.R. Serum organochlorine pesticides and PCBs and breast cancer risk: Results from a prospective analysis (USA). *Cancer Causes Control* **1999**, *10*, 1–11. [CrossRef] [PubMed]
41. Høyer, A.P.; Jørgensen, T.; Brock, J.W.; Grandjean, P. Organochlorine exposure and breast cancer survival. *J. Clin. Epidemiol.* **2000**, *53*, 323–330. [CrossRef]
42. Krieger, N.; Wolff, M.S.; Hiatt, R.A.; Rivera, M.; Vogelman, J.; Orentreich, N. Breast Cancer and Serum Organochlorines: A Prospective Study among White, Black, and Asian Women. *J. Natl. Cancer Inst.* **1994**, *86*, 589–599. [CrossRef] [PubMed]
43. Wolff, M.S.; Toniolo, P.G.; Lee, E.W.; Rivera, M.; Dubin, N. Blood Levels of Organochlorine Residues and Risk of Breast Cancer. *J. Natl. Cancer Inst.* **1993**, *85*, 648–652. [CrossRef]
44. Laden, F.; Ishibe, N.; Hankinson, S.E.; Wolff, M.S.; Gertig, D.M.; Hunter, D.J.; Kelsey, K.T. Polychlorinated biphenyls, cytochrome P450 1A1, and breast cancer risk in the Nurses' Health Study. *Cancer Epidemiol. Biomark. Prev.* **2002**, *11*, 1560–1565.
45. Høyer, A.P.; Jørgensen, T.; Rank, F.; Grandjean, P. Organochlorine exposures influence on breast cancer risk and survival according to estrogen receptor status: A Danish cohort-nested case-control study. *BMC Cancer* **2001**, *1*, 8. [CrossRef]
46. Høyer, A.P.; Jørgensen, T.; Grandjean, P.; Hartvig, H.B. Repeated measurements of organochlorine exposure and breast cancer risk (Denmark). *Cancer Causes Control* **2000**, *11*, 177–184. [CrossRef] [PubMed]
47. Morgan, M.; Deoraj, A.; Felty, Q.; Roy, D. Environmental estrogen-like endocrine disrupting chemicals and breast cancer. *Mol. Cell. Endocrinol.* **2017**, *457*, 89–102. [CrossRef] [PubMed]
48. Pastor-Barriuso, R.; Fernández, M.F.; Castaño-Vinyals, G.; Whelan, D.; Pérez-Gómez, B.; Llorca, J.; Villanueva, C.M.; Guevara, M.; Molina, J.-M.M.; Artacho-Cordon, F.; et al. Total Effective Xenoestrogen Burden in Serum Samples and Risk for Breast Cancer in a Population-Based Multicase-Control Study in Spain. *Environ. Health Perspect.* **2016**, *124*, 1575–1582. [CrossRef] [PubMed]
49. Tang, M.; Zhao, M.; Shanshan, Z.; Chen, K.; Zhang, C.; Liu, W. Assessing the underlying breast cancer risk of Chinese females contributed by dietary intake of residual DDT from agricultural soils. *Environ. Int.* **2014**, *73*, 208–215. [CrossRef]
50. Gatto, N.M.; Longnecker, M.P.; Press, M.F.; Sullivan-Halley, J.; McKean-Cowdin, R.; Bernstein, L. Serum Organochlorines and Breast Cancer: A Case-Control Study among African-American Women. *Cancer Causes Control* **2007**, *18*, 29–39. [CrossRef]

51. Li, Y.; Millikan, R.C.; Bell, D.A.; Cui, L.; Tse, C.K.; Newman, B.; Conway, K. Polychlorinated biphenyls, cytochrome P450 1A1 (CYP1A1) polymorphisms, and breast cancer risk among African American women and white women in North Carolina: A population-based case-control study. *Breast Cancer Res.* **2005**, *7*, R12–R18. [CrossRef]
52. Millikan, R.; DeVoto, E.; Duell, E.J.; Tse, C.K.; Savitz, D.A.; Beach, J.; Edmiston, S.; Jackson, S.; Newman, B. Dichlorodiphenyldichloroethene, polychlorinated biphenyls, and breast cancer among African-American and white women in North Carolina. *Cancer Epidemiol. Biomark. Prev.* **2000**, *9*, 1233–1240.
53. Moysich, K.B.; Shields, P.G.; Freudenheim, J.L.; Schisterman, E.F.; Vena, J.E.; Kostyniak, P.; Greizerstein, H.; Marshall, J.R.; Graham, S.; Ambrosone, C.B. Polychlorinated biphenyls, cytochrome P4501A1 polymorphism, and postmenopausal breast cancer risk. *Cancer Epidemiol. Biomark. Prev.* **1999**, *8*, 41–44. [CrossRef]
54. Gammon, M.D.; Wolff, M.S.; Neugut, A.I.; Eng, S.M.; Teitelbaum, S.L.; Britton, J.A.; Terry, M.B.; Levin, B.; Stellman, S.D.; Kabat, G.C.; et al. Environmental toxins and breast cancer on Long Island. II. Organochlorine compound levels in blood. *Cancer Epidemiol. Biomark. Prev.* **2002**, *11*, 686–697.
55. Weber, J.P.; Romieu, I.; Hernandez-Avila, M.; Lazcano-Ponce, E.; Dewailly, E. Breast Cancer, Lactation History, and Serum Organochlorines. *Am. J. Epidemiol.* **2000**, *152*, 363–370.
56. Arrebola, J.P.; Belhassen, H.; Artacho-Cordon, F.; Ghali, R.; Ghorbel, H.; Boussen, H.; Perez-Carrascosa, F.M.; Expósito, J.; Hedhili, A.; Olea, N. Risk of female breast cancer and serum concentrations of organochlorine pesticides and polychlorinated biphenyls: A case–control study in Tunisia. *Sci. Total Environ.* **2015**, *520*, 106–113. [CrossRef] [PubMed]
57. Boada, L.D.; Zumbado, M.; Henriquez-Hernandez, L.A.; Almeida-Gonzalez, M.; Alvarez-Leon, E.E.; Serra-Majem, L.; Luzardo, O.P. Complex organochlorine pesticide mixtures as determinant factor for breast cancer risk: A population-based case–control study in the Canary Islands (Spain). *Environ. Health* **2012**, *11*, 28. [CrossRef] [PubMed]
58. Bonefeld-Jorgensen, E.C.; Long, M.; Bossi, R.; Ayotte, P.; Asmund, G.; Krüger, T.; Ghisari, M.; Mulvad, G.; Kern, P.; Nzulumiki, P.; et al. Perfluorinated compounds are related to breast cancer risk in greenlandic inuit: A case control study. *Environ. Health* **2011**, *10*, 88. [CrossRef] [PubMed]
59. Li, J.-Y.; Wu, D.-S.; Yang, F.; Zeng, H.-Y.; Lei, F.-M.; Zhou, W.-D.; Li, H.; Tao, P. Study on serum organochlorines pesticides (DDTs) level, CYP1A1 genetic polymorphism and risk of breast cancer: A case control study. *Zhonghua Liuxingbingxue Zazhi* **2006**, *27*, 217–222. [PubMed]
60. Pavuk, M.; Cerhan, J.R.; Lynch, C.F.; Kocan, A.; Petrik, J.; Chovancova, J. Case-control study of PCBs, other organochlorines and breast cancer in Eastern Slovakia. *J. Expo. Sci. Environ. Epidemiol.* **2003**, *13*, 267–275. [CrossRef]
61. Soliman, A. Serum organochlorine levels and history of lactation in Egypt. *Environ. Res.* **2003**, *92*, 110–117. [CrossRef]
62. Dello Iacovo, R.; Celentano, E.; Strollo, A.M.; Iazzetta, G.; Capasso, I.; Randazzo, G. Organochlorines and breast cancer. A study on Neapolitan women. *Adv. Exp. Med. Biol.* **1999**, *472*, 57–66.
63. Andrada-Serpa, M.J.; Carmo, P.A.; Barreto, H.H.; Inomata, O.N.; Kussumi, T.A.; Mendonça, G.A.; Eluf-Neto, J.; Andrada-Serpa, M.J. Organochlorines and breast cancer: A case-control study in Brazil. *Int. J. Cancer* **1999**, *83*, 596–600.
64. Wielsøe, M.; Kern, P.; Bonefeld-Jørgensen, E.C. Serum levels of environmental pollutants is a risk factor for breast cancer in Inuit: A case control study. *Environ. Health* **2017**, *16*, 56. [CrossRef] [PubMed]
65. Zhang, Y.; Wise, J.P.; Holford, T.R.; Xie, H.; Boyle, P.; Zahm, S.H.; Rusiecki, J.; Zou, K.; Zhang, B.; Zhu, Y.; et al. Serum Polychlorinated Biphenyls, Cytochrome P-450 1A1 Polymorphisms, and Risk of Breast Cancer in Connecticut Women. *Am. J. Epidemiol.* **2004**, *160*, 1177–1183. [CrossRef] [PubMed]
66. Demers, A.; Ayotte, P.; Brisson, J.; Dodin, S.; Robert, J.; Dewailly, E. Risk and aggressiveness of breast cancer in relation to plasma organochlorine concentrations. *Cancer Epidemiol. Biomark. Prev.* **2000**, *9*, 161–166.
67. Zheng, T.; Holford, T.R.; Mayne, S.T.; Tessari, J.; Ward, B.; Carter, D.; Owens, P.H.; Boyle, P.; Dubrow, R.; Archibeque-Engle, S.; et al. Risk of female breast cancer associated with serum polychlorinated biphenyls and 1,1-dichloro-2,2′-bis(p-chlorophenyl)ethylene. *Cancer Epidemiol. Biomark. Prev.* **2000**, *9*, 167–174.
68. Holmes, A.K.; Koller, K.R.; Kieszak, S.M.; Sjodin, A.; Calafat, A.M.; Sacco, F.D.; Varner, D.W.; Lanier, A.P.; Rubin, C.H. Case-control study of breast cancer and exposure to synthetic environmental chemicals among Alaska Native women. *Int. J. Circumpolar Health* **2014**, *73*, 25760. [CrossRef] [PubMed]

69. Zhang, H.; Liu, L.; Zhang, P.; Zhao, Y.; Wu, X.; Ni, W. A case-control study on the relationship between organochlorine and female breast cancer. *J. Hyg. Res.* **2013**, *42*, 44–48.
70. Recio-Vega, R.; Velazco-Rodriguez, V.; Ocampo-Gómez, G.; Hernandez-Gonzalez, S.; Ruiz-Flores, P.; Lopez-Marquez, F.; Recio-Vega, R.; Velazco-Rodriguez, V.; Ocampo-Gómez, G.; Hernandez-Gonzalez, S.; et al. Serum levels of polychlorinated biphenyls in Mexican women and breast cancer risk. *J. Appl. Toxicol.* **2011**, *31*, 270–278. [CrossRef] [PubMed]
71. Itoh, H.; Iwasaki, M.; Hanaoka, T.; Kasuga, Y.; Yokoyama, S.; Onuma, H.; Nishimura, H.; Kusama, R.; Tsugane, S. Serum organochlorines and breast cancer risk in Japanese women: A case-control study. *Cancer Causes Control* **2009**, *20*, 567–580. [CrossRef]
72. Yang, M.; Ryu, J.H.; Jeon, R.; Kang, D.; Yoo, K.Y. Effects of bisphenol A on breast cancer and its risk factors. *Arch. Toxicol.* **2009**, *83*, 281–285. [CrossRef]
73. Siddiqui, M.; Anand, M.; Mehrotra, P.; Sarangi, R.; Mathur, N. Biomonitoring of organochlorines in women with benign and malignant breast disease. *Environ. Res.* **2005**, *98*, 250–257. [CrossRef]
74. Charlier, C.J.; Albert, A.I.; Zhang, L.; Dubois, N.G.; Plomteux, G.J. Polychlorinated biphenyls contamination in women with breast cancer. *Clin. Chim. Acta* **2004**, *347*, 177–181. [CrossRef] [PubMed]
75. Charlier, C.; Foidart, J.-M.; Pitance, F.; Herman, P.; Gaspard, U.; Meurisse, M.; Plomteux, G. Environmental dichlorodiphenyltrichlorethane or hexachlorobenzene exposure and breast cancer: Is there a risk? *Clin. Chem. Lab. Med.* **2004**, *42*, 222–227. [CrossRef] [PubMed]
76. Rusiecki, J.A.; Holford, T.R.; Zahm, S.H.; Zheng, T. Polychlorinated biphenyls and breast cancer risk by combined estrogen and progesterone receptor status. *Eur. J. Epidemiol.* **2004**, *19*, 793–801. [CrossRef] [PubMed]
77. Charlier, C.; Albert, A.; Herman, P.; Hamoir, E.; Gaspard, U.; Meurisse, M.; Plomteux, G. Breast cancer and serum organochlorine residues. *Occup. Environ. Med.* **2003**, *60*, 348–351. [CrossRef] [PubMed]
78. Lopez-Carrillo, L.; Lopez-Cervantes, M.; Torres-Sanchez, L.; Blair, A.; Cebrian, M.E.; Garcia, R.M. Serum levels of beta-hexachlorocyclohexane, hexachlorobenzene and polychlorinated biphenyls and breast cancer in Mexican women. *Eur. J. Cancer Prev.* **2002**, *11*, 129–135. [CrossRef]
79. Mathur, V.; Bhatnagar, P.; Sharma, R.G.; Acharya, V.; Sexana, R. Breast cancer incidence and exposure to pesticides among women originating from Jaipur. *Environ. Int.* **2002**, *28*, 331–336. [CrossRef]
80. Ahmed, M.T.; Loutfy, N.; El Shiekh, E. Residue levels of DDE and PCBs in the blood serum of women in the Port Said region of Egypt. *J. Hazard. Mater.* **2002**, *89*, 41–48. [CrossRef]
81. Wolff, M.S.; Berkowitz, G.S.; Brower, S.; Senie, R.; Bleiweiss, I.J.; Tartter, P.; Pace, B.; Roy, N.; Wallenstein, S.; Weston, A. Organochlorine Exposures and Breast Cancer Risk in New York City Women. *Environ. Res.* **2000**, *84*, 151–161. [CrossRef]
82. Burger, M.; Mate, M.; Lavina, R.; Carzoglio, J.; Antonaz, R.; Rampoldi, O. Role of the organochlorine pesticides in breast cancer. *Rev. Toxicol.* **2000**, *17*, 79–82.
83. Olaya-Contreras, P.; Rodríguez-Villamil, J.; Posso-Valencia, H.J.; Cortez, J.E. Organochlorine exposure and breast cancer risk in Colombian women. *Cad. Saúde Públ.* **1998**, *14*, S125–S132. [CrossRef]
84. López-Carrillo, L.; Blair, A.; López-Cervantes, M.; Cebrián, M.; Rueda, C.; Reyes, R.; Mohar, A.; Bravo, J. Dichlorodiphenyltrichloroethane serum levels and breast cancer risk: A case-control study from Mexico. *Cancer Res.* **1997**, *57*, 3728–3732. [PubMed]
85. Schecter, A.; Toniolo, P.; Dai, L.C.; Thuy, L.T.B.; Wolff, M.S. Blood Levels of DDT and Breast Cancer Risk Among Women Living in the North of Vietnam. *Arch. Environ. Contam. Toxicol.* **1997**, *33*, 453–456. [CrossRef] [PubMed]
86. Dewailly, É.; Dodin, S.; Verreault, R.; Ayotte, P.; Sauvé, L.; Morin, J.; Brisson, J. High Organochlorine Body Burden in Women with Estrogen Receptor-Positive Breast Cancer. *J. Natl. Cancer Inst.* **1994**, *86*, 232–234. [CrossRef] [PubMed]
87. Ye, H.Z. *A Case-Control Study on the Relationship between Organochlorine and Breast Cancer*; Zhejiang University: Hangzhou, China, 2009.
88. Sprague, B.L.; Trentham-Dietz, A.; Hedman, C.J.; Wang, J.; Hemming, J.D.; Hampton, J.M.; Buist, D.S.; Bowles, E.J.A.; Sisney, G.S.; Burnside, E.S. Circulating serum xenoestrogens and mammographic breast density. *Breast Cancer Res.* **2013**, *15*, R45. [CrossRef] [PubMed]
89. Diorio, C.; Dumas, I.; Sandanger, T.M.; Ayotte, P. Levels of circulating polychlorinated biphenyls and mammographic breast density. *Anticancer Res.* **2013**, *33*, 5483–5489. [PubMed]

90. He, Y.; Peng, L.; Zhang, W.; Liu, C.; Yang, Q.; Zheng, S.; Bao, M.; Huang, Y.; Wu, K. Adipose tissue levels of polybrominated diphenyl ethers and breast cancer risk in Chinese women: A case-control study. *Environ. Res.* **2018**, *167*, 160–168. [CrossRef] [PubMed]
91. He, T.-T.; Zuo, A.-J.; Wang, J.-G.; Zhao, P. Organochlorine pesticides accumulation and breast cancer: A hospital-based case-control study. *Tumor Boil.* **2017**, *39*. [CrossRef] [PubMed]
92. Ociepa-Zawal, M.; Rubis, B.; Wawrzynczak, D.; Wachowiak, R.; Trzeciak, W.H. Accumulation of environmental estrogens in adipose tissue of breast cancer patients. *J. Environ. Sci. Health Part A* **2010**, *45*, 305–312. [CrossRef]
93. Hurley, S.; Reynolds, P.; Goldberg, D.; Nelson, D.O.; Jeffrey, S.S.; Petreas, M. Adipose levels of polybrominated diphenyl ethers and risk of breast cancer. *Breast Cancer Res. Treat.* **2011**, *129*, 505–511. [CrossRef]
94. Cassidy, R.A.; Natarajan, S.; Vaughan, G.M. The link between the insecticide heptachlor epoxide, estradiol, and breast cancer. *Breast Cancer Res. Treat.* **2005**, *90*, 55–64. [CrossRef]
95. Waliszewski, S.M.; Bermudez, M.T.; Infanzon, R.M.; Silva, C.S.; Carvajal, O.; Trujillo, P.; Arroyo, S.G.; Pietrini, R.V.; Saldanña, V.A.; Melo, G.; et al. Persistent Organochlorine Pesticide Levels in Breast Adipose Tissue in Women with Malignant and Benign Breast Tumors. *Bull. Environ. Contam. Toxicol.* **2005**, *75*, 752–759. [CrossRef] [PubMed]
96. McCready, D.; Aronson, K.J.; Chu, W.; Fan, W.; Vesprini, D.; Narod, S.A. Breast Tissue Organochlorine Levels and Metabolic Genotypes in Relation to Breast Cancer Risk Canada. *Cancer Causes Control* **2004**, *15*, 399–418. [CrossRef] [PubMed]
97. Ibarluzea, J.J.; Fernandez, M.F.; Santa-Marina, L.; Olea-Serrano, M.F.; Rivas, A.M.; Aurrekoetxea, J.J.; Exposito, J.; Lorenzo, M.; Torne, P.; Villalobos, M.; et al. Breast cancer risk and the combined effect of environmental estrogens. *Cancer Causes Control* **2004**, *15*, 591–600. [CrossRef] [PubMed]
98. Lucena, R.A.; Allam, M.F.; Costabeber, I.H.; Villarejo, M.L.; Navajas, R.F. Breast cancer risk factors: PCB congeners. *Eur. J. Cancer Prev.* **2001**, *10*, 117–119. [CrossRef] [PubMed]
99. Charles, M.J.; Schell, M.J.; Willman, E.; Gross, H.B.; Lin, Y.; Sonnenberg, S.; Graham, M.L. Organochlorines and 8-Hydroxy-2′-Deoxyguanosine (8-OHdG) in Cancerous and Noncancerous Breast Tissue: Do the Data Support the Hypothesis That Oxidative DNA Damage Caused by Organochlorines Affects Breast Cancer? *Arch. Environ. Contam. Toxicol.* **2001**, *41*, 386–395. [PubMed]
100. Woolcott, C.G.; Aronson, K.J.; Hanna, W.M.; Sengupta, S.K.; McCready, D.R.; Sterns, E.E.; Miller, A.B. Organochlorines and breast cancer risk by receptor status, tumor size, and grade (Canada). *Cancer Causes Control* **2001**, *12*, 395–404. [CrossRef] [PubMed]
101. Zheng, T.; Holford, T.R.; Tessari, J.; Mayne, S.T.; Zahm, S.H.; Owens, P.H.; Zhang, B.; Ward, B.; Carter, D.; Zhang, Y.; et al. Oxychlordane and trans-nonachlor in breast adipose tissue and risk of female breast cancer. *J. Epidemiol. Biostat.* **2000**, *5*, 153–160.
102. Stellman, S.D.; Djordjevic, M.V.; Britton, J.A.; Muscat, J.E.; Citron, M.L.; Kemeny, M.; Busch, E.; Gong, L. Breast cancer risk in relation to adipose concentrations of organochlorine pesticides and polychlorinated biphenyls in Long Island, New York. *Cancer Epidemiol. Biomark. Prev.* **2000**, *9*, 1241–1249.
103. Bagga, D.; Anders, K.H.; Wang, H.-J.; Roberts, E.; Glaspy, J.A. Organochlorine Pesticide Content of Breast Adipose Tissue from Women with Breast Cancer and Control Subjects. *J. Natl. Cancer Inst.* **2000**, *92*, 750–753. [CrossRef]
104. Liljegren, G.; Hardell, L.; Lindström, G.; Dahl, P.; Magnuson, A. Case-control study on breast cancer and adipose tissue concentrations of congener specific polychlorinated biphenyls, DDE and hexachlorobenzene. *Eur. J. Cancer Prev.* **1998**, *7*, 135–140.
105. Guttes, S.; Failing, K.; Neumann, K.; Kleinstein, J.; Georgii, S.; Brunn, H. Chlororganic Pesticides and Polychlorinated Biphenyls in Breast Tissue of Women with Benign and Malignant Breast Disease. *Arch. Environ. Contam. Toxicol.* **1998**, *35*, 140–147. [CrossRef] [PubMed]
106. Hardell, L.; Lindström, G.; Liljegren, G.; Dahl, P.; Magnuson, A. Increased concentrations of octachlorodibenzo-p-dioxin in cases with breast cancer—Results from a case—Control study. *Eur. J. Cancer Prev.* **1996**, *5*, 351–357. [CrossRef] [PubMed]
107. Djordjevic, M.V.; Hoffmann, D.; Fan, J.; Prokopczyk, B.; Citron, M.L.; Stellman, S.D. Assessment of chlorinated pesticides and polychlorinated biphenyls in adipose breast tissue using a supercritical fluid extraction method. *Carcinogenesis* **1994**, *15*, 2581–2585. [CrossRef] [PubMed]

108. Falck, F.; Ricci, A.; Wolff, M.S.; Godbold, J.; Deckers, P. Pesticides and polychlorinated biphenyl residues in human breast lipids and their relation to breast cancer. *Arch. Environ. Health Int. J.* **1992**, *47*, 143–146.
109. Mussalo-Rauhamaa, H.; Häsänen, E.; Pyysalo, H.; Antervo, K.; Kauppila, R.; Pantzar, P. Occurrence of beta-hexachlorocyclohexane in breast cancer patients. *Cancer* **1990**, *66*, 2124–2128. [CrossRef]
110. Unge, M.; Kiær, H.; Blichert-Toft, M.; Olsen, J.; Clausen, J. Organochlorine compounds in human breast fat from deceased with and without breast cancer and in a biopsy material from newly diagnosed patients undergoing breast surgery. *Environ. Res.* **1984**, *34*, 24–28. [CrossRef]
111. Wassermann, M.; Nogueira, D.P.; Tomatis, L.; Mirra, A.P.; Shibata, H.; Arie, G.; Cucos, S.; Wassermann, D. Organochlorine compounds in neoplastic and adjacent apparently normal breast tissue. *Bull. Environ. Contam. Toxicol.* **1976**, *15*, 478–484. [CrossRef]
112. Brauner, E.V.; Loft, S.; Wellejus, A.; Autrup, H.; Tjonneland, A.; Raaschou-Nielsen, O. Adipose tissue PCB levels and CYP1B1 and COMT genotypes in relation to breast cancer risk in postmenopausal Danish women. *Int. J. Environ. Health Res.* **2014**, *24*, 256–268. [CrossRef]
113. Van't Veer, P.; Lobbezoo, I.E.; Martin-Moreno, J.M.; Guallar, E.; Gomez-Aracena, J.; Kardinaal, A.F.; Kohlmeier, L.; Martin, B.C.; Strain, J.J.; Thamm, M.; et al. DDT (dicophane) and postmenopausal breast cancer in Europe: Case-control study. *BMJ* **1997**, *315*, 81–85. [CrossRef]
114. Parada, H., Jr.; Wolff, M.S.; Engel, L.S.; White, A.J.; Eng, S.M.; Cleveland, R.J.; Khankari, N.K.; Teitelbaum, S.L.; Neugut, A.I.; Gammon, M.D. Organochlorine insecticides DDT and chlordane in relation to survival following breast cancer. *Int. J. Cancer* **2016**, *138*, 565–575. [CrossRef]
115. Muscat, J.E.; Britton, J.A.; Djordjevic, M.V.; Citron, M.L.; Kemeny, M.; Busch-Devereaux, E.; Pittman, B.; Stellman, S.D. Adipose concentrations of organochlorine compounds and breast cancer recurrence in Long Island, New York. *Cancer Epidemiol. Biomark. Prev.* **2003**, *12*, 1474–1478.
116. Roswall, N.; Sørensen, M.; Tjønneland, A.; Raaschou-Nielsen, O. Organochlorine concentrations in adipose tissue and survival in postmenopausal, Danish breast cancer patients. *Environ. Res.* **2018**, *163*, 237–248. [CrossRef] [PubMed]
117. Cocco, P.; Kazerouni, N.; Zahm, S.H. Cancer Mortality and Environmental Exposure to DDE in the United States. *Environ. Health Perspect.* **2000**, *108*, 1. [PubMed]
118. Arrebola, J.P.; Fernández-Rodríguez, M.; Artacho-Cordón, F.; Garde, C.; Perez-Carrascosa, F.; Linares, I.; Tovar, I.; González-Alzaga, B.; Expósito, J.; Torne, P.; et al. Associations of persistent organic pollutants in serum and adipose tissue with breast cancer prognostic markers. *Sci. Total Environ.* **2016**, *566*, 41–49. [CrossRef] [PubMed]
119. Charlier, C.J.; Dejardin, M.-T.C. Increased Risk of Relapse after Breast Cancer with Exposure to Organochlorine Pollutants. *Bull. Environ. Contam. Toxicol.* **2007**, *78*, 1–4. [CrossRef]
120. Ellsworth, R.E.; Kostyniak, P.J.; Chi, L.-H.; Shriver, C.D.; Costantino, N.S.; Ellsworth, D.L. Organochlorine pesticide residues in human breast tissue and their relationships with clinical and pathological characteristics of breast cancer. *Environ. Toxicol.* **2018**, *33*, 876–884. [CrossRef] [PubMed]
121. Eldakroory, S.; El Morsi, D.; Abdel-Rahman, R.; Roshdy, S.; Gouida, M.; Khashaba, E. Correlation between toxic organochlorine pesticides and breast cancer. *Hum. Exp. Toxicol.* **2017**, *39*, 1326–1334. [CrossRef]
122. Barr, L.; Metaxas, G.; Harbach, C.A.J.; Savoy, L.A.; Darbre, P.D. Measurement of paraben concentrations in human breast tissue at serial locations across the breast from axilla to sternum. *J. Appl. Toxicol.* **2012**, *32*, 219–232. [CrossRef]
123. Munoz-de-Toro, M.; Durando, M.; Beldomenico, P.M.; Beldomenico, H.R.; Kass, L.; Garcia, S.R.; Luque, E.H. Estrogenic microenvironment generated by organochlorine residues in adipose mammary tissue modulates biomarker expression in ERalpha-positive breast carcinomas. *Breast Cancer Res.* **2006**, *8*, R47. [CrossRef]
124. Verner, M.A.; Bachelet, D.; McDougall, R.; Charbonneau, M.; Guenel, P.; Haddad, S. A case study addressing the reliability of polychlorinated biphenyl levels measured at the time of breast cancer diagnosis in representing early-life exposure. *Cancer Epidemiol. Biomark. Prev.* **2011**, *20*, 281–286. [CrossRef]
125. Moser, G.A.; McLachlan, M.S. The influence of dietary concentration on the absorption and excretion of persistent lipophilic organic pollutants in the human intestinal tract. *Chemosphere* **2001**, *45*, 201–211. [CrossRef]
126. Alcock, R.E.; Sweetman, A.J.; Juan, C.Y.; Jones, K.C. A generic model of human lifetime exposure to persistent organic contaminants: Development and application to PCB-101. *Environ. Pollut.* **2000**, *110*, 253–265. [CrossRef]

127. La Merrill, M.; Emond, C.; Kim, M.J.; Antignac, J.P.; Le Bizec, B.; Clement, K.; Birnbaum, L.S.; Barouki, R. Toxicological function of adipose tissue: Focus on persistent organic pollutants. *Environ. Health Perspect.* **2013**, *121*, 162–169. [CrossRef] [PubMed]
128. Sbarbati, A.; Accorsi, D.; Benati, D.; Marchetti, L.; Orsini, G.; Rigotti, G.; Panettiere, P. Subcutaneous adipose tissue classification. *Eur. J. Histochem.* **2010**, *54*, 48. [CrossRef]
129. Strobel, E.-S.; Fritschka, E. Hereditary Premenopausal Breast Cancer. *Oncol. Res. Treat.* **2002**, *25*, 24–27. [CrossRef] [PubMed]
130. Bray, F.; McCarron, P.; Parkin, D.M. The changing global patterns of female breast cancer incidence and mortality. *Breast Cancer Res.* **2004**, *6*, 229–239. [CrossRef] [PubMed]
131. Hernán, M.A.; Robins, J.M. *Causal Inference*; Chapman & Hall/CRC: Boca Raton, FL, USA, 2019.

cancers

MDPI

Review

Regulation of Cell Signaling Pathways by Berberine in Different Cancers: Searching for Missing Pieces of an Incomplete Jig-Saw Puzzle for an Effective Cancer Therapy

Ammad Ahmad Farooqi [1], Muhammad Zahid Qureshi [2], Sumbul Khalid [3], Rukset Attar [4], Chiara Martinelli [5], Uteuliyev Yerzhan Sabitaliyevich [6], Sadykov Bolat Nurmurzayevich [7], Simona Taverna [8], Palmiro Poltronieri [9] and Baojun Xu [10,*]

1. Department of Molecular Oncology, Institute of Biomedical and Genetic Engineering (IBGE), Islamabad 44000, Pakistan; ammadfarooqi@rlmclahore.com
2. Department of Chemistry, GCU, 54000 Lahore, Pakistan; qureshienv@yahoo.com
3. Department of Bioinformatics and Biotechnology, International Islamic University, Islamabad 44000, Pakistan; sumbul.khalid@iiu.edu.pk
4. Department of Obstetrics and Gynecology, Yeditepe University Hospital, 34755 Istanbul, Turkey; ruksetattar@hotmail.com
5. Istituto Italiano di Tecnologia, Smart Bio-Interfaces, Pontedera, 56025 Pisa, Italy; chiara.martinelli@iit.it
6. Department of Postgraduate Education and Research, Kazakhstan Medical University KSPH, Almaty 050004, Kazakhstan; e.uteuliyev@ksph.kz
7. Department of General Surgery, Kazakhstan Medical University KSPH, Almaty 050004, Kazakhstan; gkb5gkb5@mail.ru
8. Department of Biomedical Science, Institute of Biomedicine and Molecular Immunology "A. Monroy", National Research Council, 90146 Palermo, Italy; simonataverna.unipa@gmail.com
9. Department of Agrifood, National Research Council Italy Institute of Sciences of Food Productions (CNR-ISPA) Via Lecce-Monteroni km 7, 73100 Lecce, Italy; palmiro.poltronieri@ispa.cnr.it
10. Food Science and Technology Program, Division of Science and Technology, Beijing Normal University-Hong Kong Baptist University United International College, Zhuhai 519087, China

* Correspondence: baojunxu@uic.edu.hk; Tel.: +86-0756-3620636

Received: 7 March 2019; Accepted: 25 March 2019; Published: 4 April 2019

Abstract: There has been a renewed interest in the identification of natural products having premium pharmacological properties and minimum off-target effects. In accordance with this approach, natural product research has experienced an exponential growth in the past two decades and has yielded a stream of preclinical and clinical insights which have deeply improved our knowledge related to the multifaceted nature of cancer and strategies to therapeutically target deregulated signaling pathways in different cancers. In this review, we have set the spotlight on the scientifically proven ability of berberine to effectively target a myriad of deregulated pathways.

Keywords: berberine; signaling pathways; oncogenic cascades; TRAIL; microRNAs; cancer therapy

1. Introduction

Berberine, a natural alkaloid compound, is found in several medicinal plants. Typically, berberine is commercially produced from a Chinese medicinal plant *Coptis chinensis*. Berberine has captivated a substantial proportion of appreciation because of its remarkable pharmacological properties. Recent advancements in high-throughput techniques have helped us to demystify various hierarchically organized signaling complexes which play an instrumental role in cancer development and progression. Basic reviews related to the ability of berberine to improve worsening conditions in different diseases have previously been published, so our aim is not to summarize pharmacological importance of berberine in different diseases, but we have restricted our discussion specifically to berberine mediated targeting of signaling cascades in different cancers. In this review, we present current views related to how berberine effectively targets different deregulated oncogenic cascades and highlight key practical and conceptual questions that will be helpful to shape the next dimension of investigation into the ability of berberine to efficiently target different signaling cascades. We will start our overview with one of the most widely investigated cancer killing molecules: TNF-related apoptosis-inducing ligand (TRAIL).

2. Berberine Mediated Restoration of TRAIL-Mediated Apoptosis

There has always been a quest to identify the molecules having significant cancer killing activity and minimal off-target effects. In accordance with this view, the discovery of TRAIL revolutionized the field of molecular oncology [1,2]. However, initial claims were partially challenged by contemporary researchers because TRAIL was ineffective against different cancers. In-depth studies revealed that TRAIL transduced the signals intracellularly through death receptors (DR4, DR5) [3,4]. However, loss of cell surface appearance of death receptors was a frequently noted mechanism in TRAIL-resistance cancers. Moreover, imbalance of pro- and anti-apoptotic proteins was also reported in TRAIL resistant cancers. The advent of high-throughput technologies has helped us to uncover the highly orchestrated nature of TRAIL mediated signaling, which is initialized through extrinsic and intrinsic pathways. In this section, we will summarize recent developments and discuss unresolved and outstanding research questions.

Berberine has been shown to potently induce AMP-activated protein kinase (AMPK) in cancer cells [5]. Expectedly, berberine mediated apoptosis inducing effects were severely impaired in AMPKα-dominant negative (DN) expressing or AMPKα knockdown cancer cells. Knockdown of DR5 significantly abrogated TRAIL-berberine-induced apoptosis [5]. TRAIL and berberine combinatorially enhanced p38-MAPK phosphorylation [6]. p38-MAPK inhibition enhanced apoptosis in EGFR (epidermal growth factor receptor)-overexpressing MDA-MB-468 TNBC cells [6].

TRAIL and berberine significantly activated caspase-3 and cleavage of PARP in TRAIL-resistant MDA-MB-468 BCa cells [7]. In a murine 4T1 BCa model, berberine potentiated the efficaciousness of the anti-DR5 antibody and effectively blocked tumor growth and lung metastases [7]. Mcl-1 and c-FLIP have been shown to negatively regulate TRAIL-mediated apoptosis. Berberine dose-dependently induced degradation of Mcl-1 and c-FLIP [8]. However, treatment with a proteasome inhibitor MG132 interfered with berberine-mediated downregulation of Mcl-1 and c-FLIP [8].

While great efforts over the past few years have advanced our understanding of the berberine mediated regulation of the TRAIL-mediated pathway, much of this work has focused on preliminary information about the ability of berberine to improve TRAIL-induced killing activity. There are still many questions which need detailed research, for example, how berberine mechanistically regulates expression of death receptors in different cancers. Does it inhibit receptor degradation, or does it interfere with epigenetic silencing to restore expression of death receptors? How does berberine improve formation of death inducing signaling complex (DISC) while simultaneously targeting negative regulators which inhibit DISC formation? In the following section, we will discuss how berberine modulates the WNT/β-catenin pathway to inhibit cancer.

3. Regulation of WNT Pathway by Berberine

Levels of cytoplasmic β-catenin are controlled by a multi-proteins destruction complex which induces β-catenin phosphorylation, which is required for β-catenin ubiquitination and its subsequent degradation by proteasomes [9]. Berberine efficiently inhibited nuclear accumulation of β-catenin. Co-immunoprecipitation studies revealed that berberine increased the interaction between APC (adenomatous polyposis coli) and β-catenin [9] (shown in Figure 1). These findings shed light on the ability of berberine to stimulate the expression of APC and negatively regulate β-catenin by increasing physical interaction between these proteins.

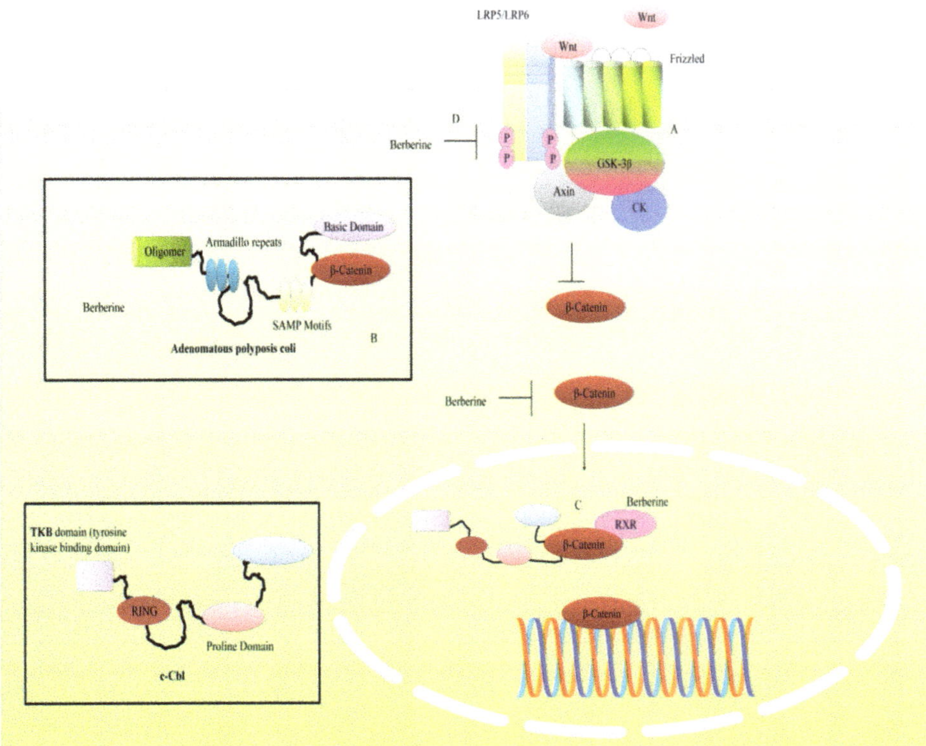

Figure 1. (**A**) WNT/ β-catenin mediated intracellular signaling. β-catenin moved into the nucleus to stimulate expression of target genes. (**B**) Berberine promoted interaction of β-catenin and APC to enhance degradation of β-catenin. (**C**) Berberine also promoted interaction of β-catenin with c-Cbl in nucleus that also induced degradation of β-catenin. (**D**) Berberine inhibited phosphorylation of LRP5/6 and GSK-3β. Abbreviations: c-Cbl (CASITAS B-lineage lymphoma protooncogene), LRP5/6 (Low density lipoprotein receptor-related protein), CK (Casein Kinase), GSK-3β (Glycogen synthase kinase), RXR (Retinoid X-receptor), WNT (Wingless/Integrase), SAMP (Ser-Ala-Met-Pro motif), RING (Really interesting new gene).

There is evidence of mechanistic regulation of the WNT pathway by berberine in hepatocellular carcinoma cell line (SMMC-7721). However, these findings should be tested in other cancers. In a report, researchers demonstrated that berberine worked synergistically with HMQ1611, a taspine derivative and suppressed the phosphorylation of LRP5/6 and GSK3β [10]. Berberine and HMQ1611 combinatorially downregulated WNT5A, Frizzled8, CK1 (casein kinase 1) and APC. Overall this study uncovered distinct steps of β-catenin phosphorylation and degradation by berberine and HMQ1611 [10]. Importantly, berberine was unable to significantly inhibit tumor growth individually in mice xenografted with SMMC-7721 cells.

Berberine decreased levels of WNT5A and cytoplasmic β-catenin in both SGC7901 and AGS cells [11]. Treatment with berberine and galangin decreased β-catenin and WNT3A in esophageal carcinoma cells [12]. Berberine concentration-dependently downregulated mRNA expression of β-catenin in colon cancer cells [13]. RXRα (retinoid X receptor α) agonists have been shown to promote RXRα binding to β-catenin to induce ubiquitination and degradation of β-catenin [14]. Berberine dose-dependently enhanced physical association of RXRα with β-catenin. Berberine induced nuclear translocation of E3 ubiquitin ligase c-Cbl which modulated degradation of β-catenin [14] (shown in Figure 1). Overall, these findings clearly suggested that berberine enhanced degradation of β-catenin by promoting its interaction with c-Cbl.

It seems clear that berberine has potential to regulate the WNT pathway in different cancers, but it needs to be tested tactically in xenografted mice. Future studies must converge on identification of the modes opted by berberine to inhibit the WNT pathway in different cancers. Does it interfere with importin and exportin proteins to inhibit nuclear accumulation? Are there further previously unexplored ubiquitin ligases which can target β-catenin? Can berberine effectively target LRP5/6 and Frizzled receptors in other cancers to also efficiently inhibit cancer proliferation? In the upcoming section we will highlight how berberine regulates the Janus kinases-signal transducer and activator of transcription proteins (JAK-STAT) pathway in different cancers.

4. Targeting of JAK-STAT Pathway

Kinases of the Janus kinase (JAK) family and transcriptional factors of the STAT (signal transducer and activator of transcription) family form a highly dynamic and orchestrated membrane-to-nucleus signaling module that has been extensively investigated, and an overwhelmingly increasing list of scientific reports have provided evidence of natural products mediated targeting of the JAK-STAT pathway in different cancers. More importantly and excitingly, coupling of massively parallel DNA sequencing with chromatin-immunoprecipitation has enabled researchers to capture thousands of STAT-binding sites.

Berberine reduced protein levels of STAT3 and inhibited the phosphorylation at 705th tyrosine and 727th serine in cholangiocarcinoma cell lines [15] (shown in Figure 2). Berberine exerted inhibitory effects on constitutive and IL-6-triggered activation of STAT3 in NPC (nasopharyngeal carcinoma) cells [16]. TAFs (tumor-associated fibroblasts) secreted IL-6 and the conditioned media harvested from the fibroblasts induced STAT3 activation in NPC cells. Activation of STAT3 by conditioned media of TAFs was blocked by berberine [16].

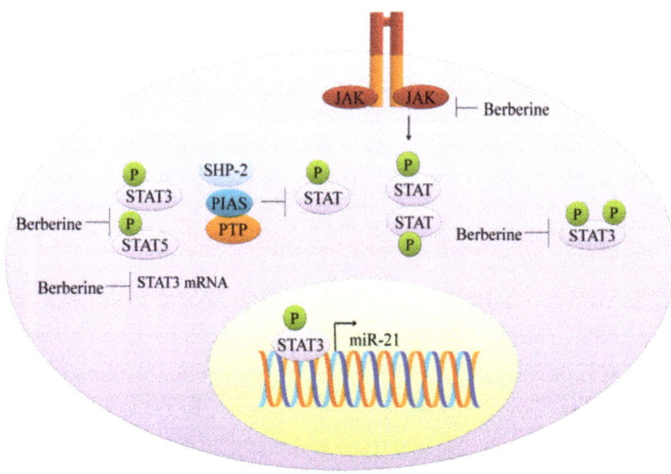

Figure 2. JAK-STAT signaling. Berberine has been shown to inhibit phosphorylation of STAT3 and STAT5. Berberine markedly reduced mRNA levels of STAT3. Functionally active STAT3 moved into the nucleus and stimulated expression of miR-21. Abbreviations: STAT (signal transducer and activator of transcription), JAK (Janus kinase), PIAS (Protein inhibitors of activated STATs), PTP (Protein-tyrosine phosphatase), SHP-2 (Src homology phosphatase-2).

Berberine significantly decreased the phosphorylated levels of JAK2 and STAT3 in colorectal cancer cells [17]. Interestingly, p-JAK2 and p-STAT3 were found to be remarkably enhanced in COX2 (cyclooxygenase-2) overexpressing colorectal cancer cells. COX2 overexpression induced activation of JAK-STAT signaling further upregulated matrix metalloproteinases (MMP)-2 and MMP-9 in colorectal cancer cells. However, berberine effectively interrupted COX2/JAK/STAT signaling [17].

Colonization of *Fusobacterium nucleatum* in the intestine may contribute to colorectal cancer (18). Levels of p-STAT3 and p-STAT5 were found to be enhanced after inoculation of *F. nucleatum* in C57BL/6-APC$^{Min/+}$ mice and wild-type C57BL/6 mice. Moreover, *F. nucleatum*-induced increases in quantities of p-STAT3 and p-STAT5 were found to be considerably reduced in mice treated with berberine [18] (shown in Figure 2).

Doxorubicin, a widely used chemotherapeutic drug has been shown to induce activation of STAT3 in lung cancer cells [19]. However, berberine markedly inhibited doxorubicin-triggered STAT3 activation. Besides, berberine promoted degradation of STAT3 by enhancing ubiquitination [19]. Berberine also enhanced killing effects of 5-fluorouracil by STAT3 inactivation and repressing the expression of survivin in gastric cancer cells [20].

5. Targeting of the mTOR Pathway: Could Berberine Modify Extracellular Vesicle-Composition?

The mammalian target of rapamycin (mTOR), a serine/threonine kinase of the PI3K (phosphoinositide 3-kinases)-related kinase family, is conserved on an evolutionary scale that coordinates different cellular processes. mTOR forms two structurally and functionally active complexes: mTOR complex 1 (mTORC1) and 2 (mTORC2). These two multi-component complexes are involved in physiological and pathological functions, such as macromolecules synthesis, homeostasis maintenance, cytoskeleton remodeling, angiogenesis, survival, response to stress and autophagy [21]. Considering the key role of mTOR in cell proliferation and differentiation, its deregulation contributes to cancer onset and progression [22].

The cellular metabolism mediated by mTOR is involved in the connections between cancer cells and tumor microenvironment during cancer advance and drug resistance acquisition, indicating the potential benefits of PI3K-Akt (protein kinase B)-mTOR pathway blockage. This inhibition contributes to reduce proliferation, migration, and survival of cancer cells, and increase tumor immunosurveillance through down-regulation of immunosuppression and anti-tumor immune stimulation [23]. mTORC1 is a downstream component in several pathways frequently altered in cancer, including the PI3K/Akt and MAPK (mitogen-activated protein kinases) pathways, that induces mTORC1 hyperactivation in many human cancers. Besides, mTORC2 signaling has a key role in tumors for its role in Akt activation, that induces tumor growth mechanisms such as glucose metabolism and apoptosis inhibition [24]. Recent reports show that mTOR is involved in lipid metabolism [25]; the critical step of this signaling cascade is the activation of proteins by phosphorylation at different sites. An important upstream target of mTOR is ERK (extracellular signal-regulated kinases), that regulates mTOR negatively and in turn enhances autophagy. ERK, a protein of MAPK cascade, is a central integrator of extracellular signals which are transduced by single cytokines or hormones or activated by cellular mechanical stresses that influence lipid metabolism [26,27].

The mTOR inhibitors, called "rapalogs," used as anti-cancer drugs belonged to a class of rapamycin derivatives. The first rapalog, approved for advanced renal carcinoma management was temsirolimus, followed by everolimus. These rapalogs did not show significant results in clinical practice as compared to the results obtained in pre-clinical studies. The rapalogs and catalytic mTOR inhibitors were useful in immunosuppression in a small number of cancers [28,29].

Several natural compounds such as berberine, resveratrol, curcumin, quercetin and others can modulate the mTOR pathway [30–32]. Recent studies have revealed that berberine has anti-tumor effects, through inhibition of the mTOR-signaling pathway. Berberine, anisoquinoline alkaloid isolated from *Berberis vulgaris* L, has anti-diarrheic, anti-inflammation, and anti-microbial activities [33].

Nowadays, several studies have shown that berberine is effective against glioma, colorectal, lung, prostate and ovarian cancer [26].

Berberine can modulate different pathways, such as cellular glucose metabolism and the HIF-1α (hypoxia-inducible factor 1α)-mTOR axis. In this context, Wang et al. [26] indicated that berberine modulated the metabolism of glioblastoma multiforme cells, induced autophagy and reduced glucose metabolism. These changes reduced tumor growth and invasiveness, induced apoptosis, by AMPK/mTOR/ULK1 (Unc-51 like autophagy activating kinase) pathway inhibition. Berberine reduced cancer progression in vivo, which clearly indicated the potential clinical benefits of alkaloids extract from plants in cancer therapy [26].

Mao et al. provided evidence that berberine played a central role in regulation of cellular glucose metabolism in colon cancer cells [34]. They studied the effects of berberine in colon cancer cell lines and findings revealed that berberine inhibited glucose uptake and reduced the transcription of genes, such as *GLUT1* (glucose transporter 1), *LDHA* (lactate dehydrogenase A) and *HK2* (hexokinase 2), involved in glucose metabolism of colon cancer cells. This mechanism is mediated by HIF-1α protein synthesis inhibition through mTOR pathway suppression. The molecular study indicated that HIF-1α protein expression, a well-known transcription factor critical for dysregulated cancer cell glucose metabolism, was considerably inhibited in berberine-treated colon cancer cell lines [34]. It was reported that berberine activated AMPK that in turn inhibited mTOR, in in vitro studies and in mouse models of colon carcinogenesis in early stages of tumorigenesis. Berberine also interfered with the NF-κB (nuclear factor kappa-light-chain-enhancer of activated B cells) pathway and effectively inhibited colon cancer progression [33].

Berberine may also induce autophagy in human liver carcinoma cell lines, through activation of Beclin-1 and inhibition of mTOR signaling by suppressing the activity of Akt and up-regulating P38 MAPK signaling [35,36]. The role of berberine in mTOR pathway modulation has been also demonstrated in hematological malignancies. Ma and collaborators showed the synergism of TPD7 and berberine in leukemia Jurkat cell growth inhibition through ephrin-B2 signaling modulation [37]. There are direct pieces of evidence which shed light on synergistic antitumor activities of rapamycin and berberine treatment in hepato-carcinoma cell lines. There was a marked decrease in phosphorylated p70S6 kinase 1 protein levels, a downstream effector of mTOR in cells combinatorially treated with rapamycin and berberine as compared to cells treated with rapamycin or berberine alone [38]. It was also demonstrated that berberine and cinnamaldehyde reduced the susceptibility of mice to lung carcinogenesis induced by urethane, and reversed the urethane-induced AMPK, mTOR, AQP-1 (aquaporin 1) and NF-κB expression patterns [39]. Overall these reports advocated the role of berberine as a new compound for cancer therapy.

Recent findings indicate that extracellular vesicles (EVs) play a key role in different steps of cancer progression, transporting oncogenic proteins and nucleic acids [40–42]. EVs are named in different ways based on their origin, diameter and mechanism of release. The two population of EVs better characterized are exosomes and micro-vesicles [43,44]. Hypoxia induces wide changes in the tumor microenvironment, and several reports show EVs central role in this mechanism [45]. Besides HIF-1, other pathways such as PI3K/Akt/mTOR are induced in tumor cells under hypoxia. It was demonstrated that hypoxia promoted prostate cancer progression and hypoxis-induced-exosomes remodeled the cancer microenvironment [46]. Moreover, exosomes released by mesenchymal stem cells (MSC) dose-dependently reduced VEGF (vascular endothelial growth factor) expression and secretion mainly through interfering with the mTOR/HIF-1α axis in breast cancer cells. MSC-derived exosomes, enriched in miR-100, were taken up by breast cancer cells [47]. microRNA-100 efficiently downregulated VEGF in breast cancer cells.

This evidence suggests the possible role of berberine in modifying EV-composition, as has already been demonstrated for other natural compounds such as curcumin. It is exciting to note that exosomes released by curcumin-treated CML cells contained considerably higher levels of miR-21. Consequently, these miR-21 loaded exosomes were taken up by HUVECs (human umbilical vein endothelial cells) and it was mechanistically shown that miR-21 directly targeted MARCKS (myristoylated alanine-rich C-kinase substrate) and inhibited angiogenic phenotypes [48]. Curcumin also induced selective packaging of miR-21 in exosomes and played a central role in reshaping post-transcriptional network in recipient cells [49]. Furthermore, in CML cells, curcumin modulates other molecular pathways thus altering the metabolism of glucose that in myeloproliferative disease is a consequence of non-hypoxic activation of HIF-1α [50]. It was demonstrated that curcumin promoted miR-22 mediated targeting of importin 7 that resulted in a significant reduction in nuclear accumulation of HIF-1α [51].

It will be interesting to see if berberine demonstrated potent activity to promote release of exosomes loaded with tumor suppressor microRNAs and proteins.

6. Regulation of Epigenetic Modulators by Berberine

Histone marks are motives enabling the recruitment of chromatin complexes that activate or repress transcription. Histone modifications such as acetylation and methylation at specific positions are signals recognized by these complexes. Berberine was shown to upregulate some histone deacetylases (HDAC) of class II, such as sirtuin SIRT1 (sirtuin 1), producing an antiatherogenic effect, and suppression of foam cell formation in THP-1-derived macrophages treated with oxidized low-density lipoprotein [52]. RNA silencing of SIRT1 or AMPK blocked the berberine action.

Rel proteins have emerged as complex modulators of carcinogenesis and we still have to explore their functionalities in malignancies and response to cancer therapeutics [53]. Set9 (lysine methyltransferase) induced methylation of the RelA/p65 subunit, which inhibited nuclear accumulation of NF-κB and repressed transcriptional upregulation of miR-21. Berberine dose-dependently induced generation of ROS, arrested cancer cells in G(2)/M phase and induced apoptosis in U266 cells [53]. Overall, the findings clearly suggested that berberine promoted Set9 mediated methylation of p65 to limit shuttling of NF-κB into the nucleus. Berberine mediated inhibition of translocation of NF-κB into the nucleus resulted in inhibition of miR-21 and B-cell lymphoma 2 (Bcl-2).

The protective effects of berberine against metabolic syndrome might rely on increasing mitochondrial SIRT3 activity and stimulating glycolysis, independent of AMPK activation [54,55].

The growth arrest and DNA damage-inducible protein GADD45α (growth arrest and DNA damage 45α) is a DNA demethylation regulator recruited by TCF21 antisense RNA inducing demethylation (TARID) lncRNA to enable transcription of the TCF21 gene coding for tumor-suppressor gene transcription factor 21 [56]. As detailed in the next paragraphs, administration of *Coptidis rhyzoma* aqueous extract, resulted in a higher expression of miR-23a, and in up-regulation of GADD45a, a chromatin relaxer, decreasing DNA methylation through recruitment of the 5-hydroxymethylcytosine (5hmC) hyperproducing enzyme TET, and thymine DNA glycosidase (TDG) [57–59].

Berberine induced a decrease in activity of two DNA methylases, DNMT1 (DNA (cytosine-5)-methyltransferase 1) and DNMT3, that, through DNA hypomethylation, induced an increase of p53 [60]; p53 activation was observed following the increase in miR-23a. through repression of Nek6, in HCC cells in response to Rhizoma Coptis aqueous extract [60].

Various long noncoding RNA (lncRNAs), through their tertiary structure, work as scaffolds to recruit protein complexes such as chromatin modifiers, polycomb repressive complex (PRC), and transcriptional regulators to euchromatin regions: many lncRNAs have been associated with pharmacological drugs as well as to cisplatin treatment [61]. In various studies, berberine has been able to induce or repress some lncRNA [62].

7. Regulation of microRNAs by Berberine

Many bioactive compounds have been proposed for their regulatory effects on non-coding RNAs, either long ncRNAs (lncRNA), or small RNAs such as microRNAs [63–65]. These miRNAs have added new layers of complexity in the context of post-transcriptional regulation, controlling the availability of mRNAs, and selection of the mRNAs that they recognize as targets by complementary seed sequences to initialize the degradation process of mRNAs. miRNAs exert a regulatory role in the post-translational process.

OncomiRs are oncogenic due to their ability to support cell proliferation, apoptosis inhibition, cell stemness, while tumor suppressor miRNAs are involved in nodes or networks leading to differentiation, cell cycle inhibition and growth arrest and apoptosis.

There are feedback loops and feedforward loops, sustaining the oncogenic activity of ncRNAs: for instance, a feedback loop has been described in lymphoma between miR-17-92, MYC, the protein kinase Chk2 and hu antigen R (HUR) [66]. Transcription factors can induce the expression of protein-coding genes as well as of miRNAs, that can target the mRNAs of the induced genes, in a feedforward loop. In lymphoma cells, miR-17-92 regulates MYC mRNA levels through the inhibition of Chk2, causing the depression of RNA-binding protein HUR, and its binding to MYC mRNA, preventing MYC translation. Thus, berberine suppresses the growth of multiple myelomas, either by down-regulation of three oncogenic miRNA clusters and other mRNAs, or by involvement of p53 and MAP kinases [67].

Several researchers pointed out to a correlation between berberine treatment and expression of non-coding RNAs, either lncRNAs or microRNAs. In cancer studies, treatment of multiple myeloma cells with berberine, significantly suppressed three oncogenic miRNA clusters, miR-17-92, miR-106-25, and miR-99a-125b. Berberine mediated downregulation of miR-99a-125b was found to be correlated with the regulation of p53, MAP Kinases and ErbB oncogene, leading to cell cycle arrest in the G2-phase and to apoptosis [63].

In colon cancers, berberine was effective in downregulating miR-429, with increase in its target, SOX-2; berberine up-regulated miR-296-5p and effectively interfered with the Pin1–β-catenin–cyclin D1 signal transduction cascade. Berberine inhibited the growth of HepG2 cells and induced the upregulation of miR-22-3p. miR-22-3p directly targeted SP1 and suppressed expression of its target genes, BCL2 (B-cell lymphoma 2) and CCND1 (cyclin D1) [68] (shown in Figure 3).

It was shown that berberine suppresses interleukin 6 (IL6), a factor required for cell growth in multiple myeloma cells (U266), through negative regulation of the signal transducer and activator of transcription 3 (STAT3), and this induces inhibition of miR-21 expression [69]. STAT3 regulates miR-21 expression through binding to STAT3 binding sites in the promoter (shown in Figure 2). Additionally, in ovarian cancer cells (SKOV3), berberine sensitized to cisplatin treatment through inhibition of miR-21, and subsequent activation of PDCD4, a tumor suppressor.

It is not straightforward to determine the role of miRNAs in tumor promotion or suppression, that depends on the context-specific, cell-type specific dual role of certain miRNAs [66]. This depends on the varieties of targets of miRNAs. There are miRNAs, such as miR-25 and miR-125b, that act as tumor suppressors in some tumor types and as oncogenes in others. In particular, in stem cells, miRNAs that in other cell types are regulated and respond to the bioactive supplements, may not decrease in levels, for their role in cell stemness. Berberine, in the form of a Rhizoma Coptis aqueous extract, resulted not effective in decreasing miR-21 levels, while it supported a higher expression of miR-23a, by up-regulating p21/GADD45a tumor suppressor gene (shown in Figure 3), causing HCC cells to arrest the growth in G2/M phase [19,70]. Furthermore, miR-23a was shown to repress Nek6 and to regulate p53 transcriptional activity.

Autophagy influences glucose and lipid metabolism in adipocytes. Berberine was shown to decrease miR-30a and miR-376b, preventing basal autophagy in 3T3-L1 adipocytes. MiR-30 interacts with the 3′-untranslated region of Beclin 1 (BECN1), thus the reduction in miR-30a levels increased BECN1 to form BECN1 complexes that induce autophagy (beclin homolog in yeast, Atg6, is known as autophagy promoting factor) [71]. This leads to reduced fat deposits and an increase in brown fat tissue.

Berberine was shown to exert an anti-apoptotic role in development of preimplantation embryos in vitro, by maintaining high levels of miR-21. Berberine up-regulated Bcl-2, in 2- and 4-cell embryos and blastocysts, and down-regulated caspase-3 and PTEN.

When the pre-miRNA is processed by the RISC complex, the guide strand is considered the mature, active miRNA, while the complementary passenger strand (termed miRNA*) is thought to be devoid of function. However, some miRNA star has been related to a function, some have a feedback role to regulate the RISC processing, and some have been shown to have anti-oncogene activity. This seems the case for miR-21*, or miR21-3p, affecting cancer cell growth [72]. In HepG2 hepatocellular carcinoma cells, berberine increased the levels of miR-21-3p, with a role in tumor growth inhibition and induction of apoptosis. In hepatic cancers, miR-21* antitumor activity relies on inhibition of MAT2A and MAT2B methionine adenosyltransferases mRNAs, with consequent increased levels of S-adenosyl-methionine (SAM). Methionine adenosyltransferase (MAT) played a central role in growth of hepatoma cells. miR-21-3p had previously been shown to directly target MAT2A and MAT2B in HepG2 cells. Berberine induced apoptosis in HepG2 cells mainly through miR-21-3p mediated targeting of MAT2A and MAT2B [72].

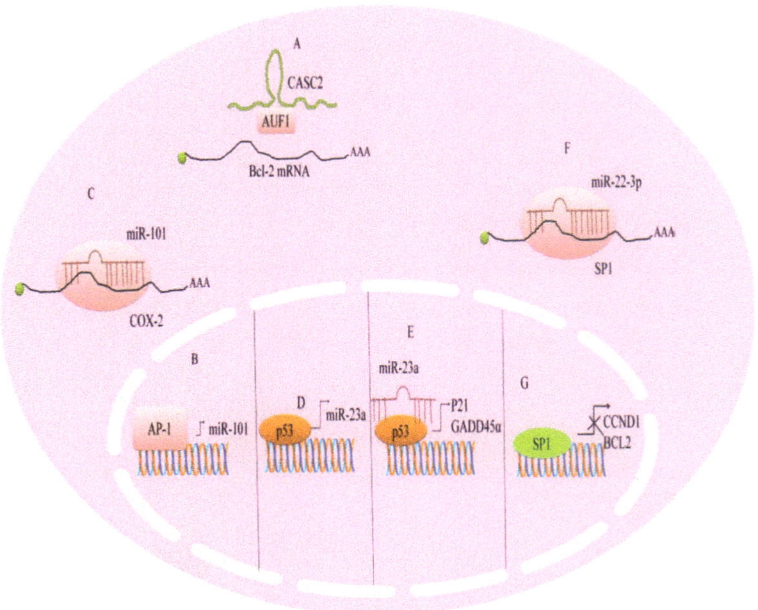

Figure 3. Regulation of non-coding RNAs by berberine. (**A**) CASC2 interacted with AUF1 and prevented its binding to AU rich sequences present within mRNA of Bcl-2. (**B**,**C**) AP-1 transcriptionally upregulated miR-101. miR-101 directly targeted COX-2. (**D**–**G**) P53 induced transcriptional upregulation of miR-23a. Additionally, miR-23a worked synchronously and stimulated the expression of GADD45a and p21. miR-22-3p directly targeted SP1. SP1 mediated upregulation of CCND1 and BCL2. Abbreviations: AUF1 (AU-rich element RNA-binding protein-1), BCL2 (B-cell CLL/lymphoma-2), *GADD45α* (Growth arrest- and DNA damage-inducible gene), CCND1 (Cyclin D1), CASC2 (long non-coding RNA), SP1 (specificity protein-1), AP-1 (Activator protein-1).

Ovarian cancer cells resistant to cisplatin, such as A2780 and A2780/DDP lines, were incubated with berberine combined with cisplatin, showing a significantly lower survival rate [73]. This effect was found to be related to inhibition of miR-93 expression, that translated in re-expression of PTEN tumor suppressor and recovery of AKT signaling [73]. Bcl-w has been shown to be targeted by miRNAs. In gastric cancers [60], berberine upregulated miR-203 and restored cisplatin-sensitivity in gastric cancer cells [74].

In liver cancers, berberine inhibited cell proliferation and viability in HepG2, Hep3B, and SNU-182 lines. Berberine treatment increased the expression of tumor suppressor such as Kruppel-like factor 6 (KLF6), activating transcription factor 3 (ATF3) and p21, a cell cycle inhibitor, and down-regulated the oncogene E2F transcription factor 1 (E2F1). A possible mechanism of the upregulation of protein coding genes may be hypothesized through downregulation of the respective miRNAs.

Berberine supplementation led to the miR29-b suppression, increasing insulin-like growth factor-binding protein (IGFBP1) expression in the liver; miR29-b suppression caused an increase in AMPK activity and a reduction of lipid storage in diabetic and obese patients [60]. Activation of AKT positively affected glucose uptake, reducing the glucose levels in blood.

LncRNAs are deregulated after berberine treatment in hepatocellular cancer, similarly to the effects seen after curcumin treatment on lncRNAs and on epigenetic changes seen in hepatocellular cancer [75]. A lncRNA, ANRIL, expressed at increased levels in type 2 diabetic (T2D) patients, causing an increase of CREB (cAMP response element-binding protein) expression, may be affected by berberine. Berberine down-regulated miR-122 and this caused a decrease in SREBP-1 levels in palmitic acid-treated HepG2 cells. Berberine was able to reduce the levels of glucose in T2D patients and in nonalcoholic fatty liver disease, by down-regulating lncRNA052686 and miR-122 [61,62,71,76,77]. The hyperlipidemic effect of berberine was linked to the inhibition of C/EBPα and PPARγ2 expression through inhibition of phosphorylated CREB binding to C/EBPβ promoter. MRAK052686, a lncRNA downregulated in diabetes, was induced by berberine. MRAK052686 co-localize at the 3′UTR of Zbtb20, coding for a protein regulating glucose homeostasis. The co-expression of MRAK052686 and Zbtb20 increases the level of the protein, improving glucose homeostasis [62].

8. Nanotechnological Strategies to Improve the Delivery of Berberine

Due to its outstanding antitumoral properties, many efforts have been devoted in designing carriers for berberine delivery as an anti-cancer therapeutic agent. Both inorganic and organic nanomaterials have been exploited for this purpose (shown in Figure 4).

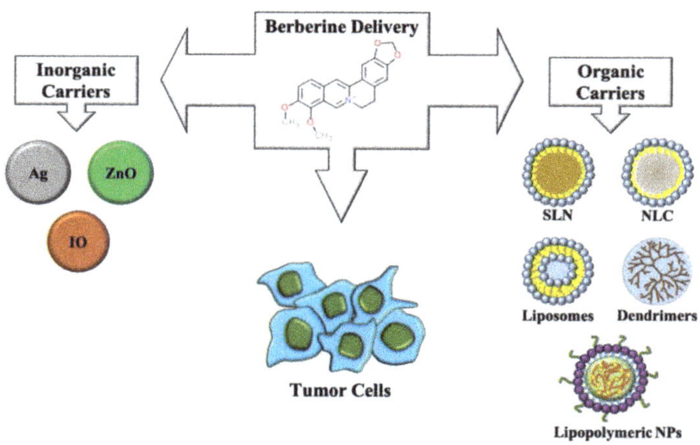

Figure 4. Berberine delivery strategies. On the left, inorganic nanocarriers are shown: Ag, Silver nanoparticles; ZnO, zinc oxide nanoparticles; IO, Iron oxide nanoparticles. On the right, organic nanocarriers are shown: SLN, solid lipid nanoparticles; NLC, nanostructured lipid carriers; Liposomes; Dendrimers; Lipopolymeric nanoparticles.

Silver nanoparticles proved successful in delivering berberine to human tongue squamous carcinoma SCC-25 cells, blocking cell cycle and increasing Bax/Bcl-2 ratio gene expression, thus indicating activation of pro-apoptotic pathways triggered by mitochondrial dysfunction [78]. Interestingly, silver nanoparticles carrying berberine displayed elevated cytotoxicity in different breast cancer cell lines and inhibition of tumor growth in vivo [79]. The same research group fabricated citrate-capped silver nanoparticles loaded with berberine and conjugated to polyethylene glycol-functionalized folic acid and demonstrated that they were able to induce apoptosis and variations in gene expression in breast cancer cells. In MDA-MB-231 athymic nude mice models, a significative reduction of tumor progression was observed [80].

Zinc oxide nanoparticles carrying berberine were recently synthesized for lung cancer therapy. They displayed antiproliferative activity in A549 cells and no significant toxicity in vivo. Moreover, due to the intrinsic properties of ZnO, these materials have been exploited as photothermal therapy (PTT) agents, leading to thermal ablation of cancer cells [81].

An innovative approach has been recently adopted by creating iron oxide nanoparticles complexed with hypoxic cell sensitizer sanazole together with berberine, able to target hypoxic tumors in vivo. Hypoxia is known to induce expression of HIF-1-alpha and consequent activation of angiogenesis related genes. Interestingly, after administration of nanoparticles, transcriptional downregulation of these and many genes linked to cell proliferation and metastasis was observed and clearly correlated with a reduction of tumor volume in vivo [82].

Many studies have demonstrated the feasibility of designing organic nanoparticles for berberine delivery. Solid lipid nanoparticles have been synthesized with good stability, high berberine loading and huge entrapment efficiency, essential parameters for successful clinical evaluation. These nanomaterials inhibited cell proliferation of MCF-7, HepG2, and A549 cancer cells inhibited cell cycle progression and apoptosis in MCF-7 cells [83].

Nanostructured lipid carriers efficiently delivered berberine to H22 hepatocarcinoma cells with high antitumor efficacy [84]. Lipid nanoparticles covered with the self-tumor targeting polymers lactoferrin and hyaluronic acid were fabricated for berberine and rapamycin delivery to lung cancer cells and displayed improved internalization and selective cytotoxicity. Detectable reduction in the number of lung foci and vascular endothelial growth factor levels were further observed [85]. The same approach was adopted by administering inhalable nanoparticles to in vivo models of lung cancer, achieving a significant decrease in lung weight, reduction in lung adenomatous foci number and diameter and in angiogenic markers expression [86].

Amine terminated G4 PAMAM have been developed with conjugated berberine and demonstrated specific cytotoxicity in different human breast cancer cell lines [87]. Lipopolymeric micelles have been developed that greatly improved berberine water solubility up to 300% with low toxicity and induced apoptosis in treated monolayer and spheroid cultures of human prostate carcinomas [88]. Finally, berberine loaded folate acid modified chitosan nanoparticles demonstrated effective in inhibiting proliferation and migration and inducing apoptosis and necrosis in human nasopharyngeal carcinoma cells CNE-1 [89].

9. Conclusions

Berberine has emerged as an excellent natural product having significant biological activity [90]. It has demonstrated premium activity against different cancers. Increasingly sophisticated cutting-edge research has uncovered tremendous chemopreventive ability of berberine to modulate signaling pathways [91,92].

In the last few years, many studies have been focused on unraveling intrinsic properties of berberine and its ability to interfere with intracellular pathways. In particular, there is an increasing interest in exploiting this natural compound as an effective anticancer drug. Although some issues remain to be solved, such as its poor water solubility/stability and low bioavailability, many nanotechnological approaches have allowed the design of ad hoc delivery systems, making berberine application for cancer treatment feasible. Many different nanocarriers, based both on inorganic and organic materials, have been developed and proven to be effective in overcoming some of the above-mentioned issues and in delivering berberine in an extremely efficient manner in many different cancer experimental models. In the near future, further research will provide crucial answers needed to pave the way to berberine clinical application.

Funding: This research received no external funding.

Conflicts of Interest: The authors declare no conflict of interest.

Abbreviations

Breast cancer cells	MDA-MB-231, MCF-7, MDA-MB-468
hepatocellular carcinoma cells	SNU-182, HepG2, Hep3B, H22 SMMC-7721
gastric cancer cells	SGC7901, AGS
lung carcinoma	A549
myeloma	U266 cell line
ovarian cancer cell line	SKOV3
Akt	protein kinase B
AMPK	AMP-activated protein kinase
AQP-1	aquaporin 1
APC	adenomatous polyposis coli
Bcl-2	B-cell lymphoma 2
CCND1	cyclin D1
c-FLIP	cellular FADD like interdeukin-1-p-converting enzyme inhibitory protein
CK1	casein kinase 1
COX2	cyclooxygenase-2
CREB	cAMP response element-binding protein
DISC	death inducing signaling complex
DNMT1	DNA (cytosine-5)-methyltransferase 1
DR	death receptor
EGFR	epidermal growth factor receptor
ERK	extracellular signal-regulated kinases
EVs	extracellular vesicles
GADD45α	growth arrest and DNA damage 45α
GLUT1	glucose transporter 1
HIF-1α	hypoxia-inducible factor 1α
HK2	hexokinase 2
HUVECs	human umbilical vein endothelial cells
IO	iron oxide
JAK-STAT	Janus kinases-signal transducer and activator of transcription proteins
LDHA	lactate dehydrogenase A
lncRNAs	long noncoding RNA
LRP	low-density lipoprotein receptor-related protein
MAPK	mitogen-activated protein kinases
MARCKS	myristoylated alanine-rich C-kinase substrate
MMP	matrix metalloproteinases
mTOR	mammalian target of rapamycin
NF-κB	nuclear factor kappa-light-chain-enhancer of activated B cells
NLC	nanostructured lipid carriers
NP	nanoparticles
NPC	nasopharyngeal carcinoma
PI3K	phosphoinositide 3-kinases
RXRα	retinoid X receptor α
SIRT1	sirtuin 1
SLN	solid lipid nanoparticles
T2D	type 2 diabetes
TAFs	tumor-associated fibroblasts
TARID	TCF21 antisense RNA inducing demethylation
TNF	tumor necrosis factor
TRAIL	TNF-related apoptosis-inducing ligand
ULK1	Unc-51 like autophagy activating kinase
VEGF	vascular endothelial growth factor
ZnO	zinc oxide.

References

1. De Miguel, D.; Lemke, J.; Anel, A.; Walczak, H.; Martinez-Lostao, L. Onto better TRAILs for cancer treatment. *Cell Death Differ.* **2016**, *23*, 733–747. [CrossRef]
2. Lemke, J.; von Karstedt, S.; Zinngrebe, J.; Walczak, H. Getting TRAIL back on track for cancer therapy. *Cell Death Differ.* **2014**, *21*, 1350–1364. [CrossRef]
3. Von Karstedt, S.; Montinaro, A.; Walczak, H. Exploring the TRAILs less travelled: TRAIL in cancer biology and therapy. *Nat. Rev. Cancer* **2017**, *17*, 352–366. [CrossRef]
4. Stuckey, D.W.; Shah, K. TRAIL on trial: Preclinical advances in cancer therapy. *Trends Mol. Med.* **2013**, *19*, 685–694. [CrossRef]
5. Ke, R.; Vishnoi, K.; Viswakarma, N.; Santha, S.; Das, S.; Rana, A.; Rana, B. Involvement of AMP-activated protein kinase and death receptor 5 in TRAIL-berberine-induced apoptosis of cancer cells. *Sci. Rep.* **2018**, *8*, 5521. [CrossRef] [PubMed]
6. Refaat, A.; Abdelhamed, S.; Saiki, I.; Sakurai, H. Inhibition of p38 mitogen-activated protein kinase potentiates the apoptotic effect of berberine/tumor necrosis factor-related apoptosis-inducing ligand combination therapy. *Oncol. Lett.* **2015**, *10*, 1907–1911. [CrossRef] [PubMed]
7. Refaat, A.; Abdelhamed, S.; Yagita, H.; Inoue, H.; Yokoyama, S.; Hayakawa, Y.; Saiki, I. Berberine enhances tumor necrosis factor-related apoptosis-inducing ligand-mediated apoptosis in breast cancer. *Oncol. Lett.* **2013**, *6*, 840–844. [CrossRef]
8. Lee, S.J.; Noh, H.; Sung, E.G.; Song, I.H.; Kim, J.Y.; Kwon, T.K.; Lee, T.J. Berberine sensitizes TRAIL-induced apoptosis through proteasome-mediated downregulation of c-FLIP and Mcl-1 proteins. *Int. J. Oncol.* **2011**, *38*, 485–492. [CrossRef] [PubMed]
9. Zhang, J.; Cao, H.; Zhang, B.; Cao, H.; Xu, X.; Ruan, H.; Yi, T.; Tan, L.; Qu, R.; Song, G.; et al. Berberine potently attenuates intestinal polyps growth in ApcMin mice and familial adenomatous polyposis patients through inhibition of Wnt signaling. *J. Cell. Mol. Med.* **2013**, *17*, 1484–1493. [CrossRef] [PubMed]
10. Dai, B.; Ma, Y.; Yang, T.; Fan, M.; Yu, R.; Su, Q.; Wang, H.; Liu, F.; Yang, C.; Zhang, Y. Synergistic effect of berberine and HMQ1611 impairs cell proliferation and migration by regulating Wnt signaling pathway in hepatocellular carcinoma. *Phytother. Res.* **2019**, *18*. [CrossRef]
11. Hu, Q.; Li, L.; Zou, X.; Xu, L.; Yi, P. Berberine attenuated proliferation, invasion and migration by targeting the AMPK/HNF4α/WNT5A pathway in gastric carcinoma. *Front. Pharmacol.* **2018**, *9*, 1150. [CrossRef] [PubMed]
12. Ren, K.; Zhang, W.; Wu, G.; Ren, J.; Lu, H.; Li, Z.; Han, X. Synergistic anti-cancer effects of galangin and berberine through apoptosis induction and proliferation inhibition in oesophageal carcinoma cells. *Biomed. Pharmacother.* **2016**, *84*, 1748–1759. [CrossRef] [PubMed]
13. Wu, K.; Yang, Q.; Mu, Y.; Zhou, L.; Liu, Y.; Zhou, Q.; He, B. Berberine inhibits the proliferation of colon cancer cells by inactivating Wnt/β-catenin signaling. *Int. J. Oncol.* **2012**, *41*, 292–298. [CrossRef] [PubMed]
14. Ruan, H.; Zhan, Y.Y.; Hou, J.; Xu, B.; Chen, B.; Tian, Y.; Wu, D.; Zhao, Y.; Zhang, Y.; Chen, X.; et al. Berberine binds RXRα to suppress β-catenin signaling in colon cancer cells. *Oncogene* **2017**, *36*, 6906–6918. [CrossRef] [PubMed]
15. Puthdee, N.; Seubwai, W.; Vaeteewoottacharn, K.; Boonmars, T.; Cha'on, U.; Phoomak, C.; Wongkham, S. Berberine induces cell cycle arrest in cholangiocarcinoma cell lines via inhibition of NF-κB and STAT3 pathways. *Biol. Pharm. Bull.* **2017**, *40*, 751–757. [CrossRef] [PubMed]
16. Tsang, C.M.; Cheung, Y.C.; Lui, V.W.; Yip, Y.L.; Zhang, G.; Lin, V.W.; Cheung, K.C.; Feng, Y.; Tsao, S.W. Berberine suppresses tumorigenicity and growth of nasopharyngeal carcinoma cells by inhibiting STAT3 activation induced by tumor associated fibroblasts. *BMC Cancer* **2013**, *13*, 619. [CrossRef]
17. Liu, X.; Ji, Q.; Ye, N.; Sui, H.; Zhou, L.; Zhu, H.; Fan, Z.; Cai, J.; Li, Q. Berberine inhibits invasion and metastasis of colorectal cancer cells via COX-2/PGE2 mediated JAK2/STAT3 signaling pathway. *PLoS ONE* **2015**, *8*, e0123478. [CrossRef]
18. Yu, Y.; Yu, T.; Zhao, H.J.; Sun, T.T.; Chen, H.M.; Chen, H.Y.; An, H.F.; Weng, Y.R.; Yu, J.; Li, M.; et al. Berberine may rescue *Fusobacterium nucleatum*-induced colorectal tumorigenesis by modulating the tumor microenvironment. *Oncotarget* **2015**, *6*, 32013–32026. [CrossRef]
19. Zhu, T.; Li, L.L.; Xiao, G.F.; Luo, Q.Z.; Liu, Q.Z.; Yao, K.T.; Xiao, G.H. Berberine increases doxorubicin sensitivity by suppressing STAT3 in lung cancer. *Am. J. Chin. Med.* **2015**, *43*, 1487–1502. [CrossRef] [PubMed]

20. Pandey, A.; Vishnoi, K.; Mahata, S.; Tripathi, S.C.; Misra, S.P.; Misra, V.; Mehrotra, R.; Dwivedi, M.; Bharti, A.C. Berberine and curcumin target Survivin and STAT3 in gastric cancer cells and synergize actions of standard chemotherapeutic 5-Fluorouracil. *Nutr. Cancer* **2015**, *67*, 1293–1304. [CrossRef] [PubMed]
21. Saxton, R.A.; Sabatini, D.M. mTOR signaling in growth, metabolism, and disease. *Cell* **2017**, *169*, 361–371. [CrossRef] [PubMed]
22. Conciatori, F.; Bazzichetto, C.; Falcone, I.; Pilotto, S.; Bria, E.; Cognetti, F.; Milella, M.; Ciuffreda, L. Role of mTOR signaling in tumor microenvironment: An overview. *Int. J. Mol. Sci.* **2018**, *19*, 2453. [CrossRef] [PubMed]
23. Weichhart, T.; Hengstschlager, M.; Linke, M. Regulation of innate immune cell function by mTOR. *Nat. Rev. Immunol.* **2015**, *15*, 599–614. [CrossRef] [PubMed]
24. Zhang, H.; Bajraszewski, N.; Wu, E.; Wang, H.; Moseman, A.P.; Dabora, S.L.; Griffin, J.D.; Kwiatkowski, D.J. PDGFRs are critical for PI3K/Akt activation and negatively regulated by mTOR. *J. Clin. Investig.* **2007**, *117*, 730–738. [CrossRef] [PubMed]
25. Han, J.; Liang, H.; Tian, D.; Du, J.; Wang, Q.; Xi, P.; Wang, H.; Li, Y. mTOR remains unchanged in diet-resistant (DR) rats despite impaired LKB1/AMPK cascade in adipose tissue. *Biochem. Biophys. Res. Commun.* **2016**, *476*, 333–339. [CrossRef] [PubMed]
26. Wang, J.; Qi, Q.; Feng, Z.; Zhang, X.; Huang, B.; Chen, A.; Prestegarden, L.; Li, X.; Wang, J. Berberine induces autophagy in glioblastoma by targeting the AMPK/mTOR/ULK1-pathway. *Oncotarget* **2016**, *7*, 66944–66958. [CrossRef] [PubMed]
27. Knebel, B.; Lehr, S.; Hartwig, S.; Haas, J.; Kaber, G.; Dicken, H.D.; Susanto, F.; Bohne, L.; Jacob, S.; Nitzgen, U.; et al. Phosphorylation of sterol regulatory element-binding protein (SREBP)-1c by p38 kinases, ERK and JNK influences lipid metabolism and the secretome of human liver cell line HepG2. *Arch. Physiol. Biochem.* **2014**, *120*, 216–227. [CrossRef] [PubMed]
28. Saran, U.; Foti, M.; Dufour, J.F. Cellular and molecular effects of the mTOR inhibitor everolimus. *Clin. Sci. (Lond.)* **2015**, *129*, 895–914. [CrossRef]
29. Bukowski, R.M. Temsirolimus: A safety and efficacy review. *Expert Opin. Drug Saf.* **2012**, *11*, 861–879. [CrossRef] [PubMed]
30. Hambright, H.G.; Batth, I.S.; Xie, J.; Ghosh, R.; Kumar, A.P. Palmatine inhibits growth and invasion in prostate cancer cell: Potential role for rpS6/NFkappaB/FLIP. *Mol. Carcinog.* **2015**, *54*, 1227–1234. [CrossRef]
31. Giovannini, L.; Bianchi, S. Role of nutraceutical SIRT1 modulators in AMPK and mTOR pathway: Evidence of a synergistic effect. *Nutrition* **2017**, *34*, 82–96. [CrossRef] [PubMed]
32. Park, D.; Jeong, H.; Lee, M.N.; Koh, A.; Kwon, O.; Yang, Y.R.; Noh, J.; Suh, P.-G.; Park, H.; Ryu, S.H. Resveratrol induces autophagy by directly inhibiting mTOR through ATP competition. *Sci. Rep.* **2016**, *6*, 21772. [CrossRef] [PubMed]
33. Li, W.; Hua, B.; Saud, S.M.; Lin, H.; Hou, W.; Matter, M.S.; Jia, L.; Colburn, N.H.; Young, M.R. Berberine regulates AMP-activated protein kinase signaling pathways and inhibits colon tumorigenesis in mice. *Mol. Carcinog.* **2015**, *54*, 1096–1109. [CrossRef] [PubMed]
34. Mao, L.; Chen, Q.; Gong, K.; Xu, X.; Xie, Y.; Zhang, W.; Cao, H.; Hu, T.; Hong, X.; Zhan, Y.-Y. Berberine decelerates glucose metabolism via suppression of mTORdependent HIF1alpha protein synthesis in colon cancer cells. *Oncol. Rep.* **2018**, *39*, 2436–2442. [PubMed]
35. Wang, N.; Feng, Y.; Zhu, M.; Tsang, C.M.; Man, K.; Tong, Y.; Tsao, S.-W. Berberine induces autophagic cell death and mitochondrial apoptosis in liver cancer cells: The cellular mechanism. *J. Cell. Biochem.* **2010**, *111*, 1426–1436. [CrossRef] [PubMed]
36. Paquette, M.; El-Houjeiri, L.; Pause, A. mTOR pathways in cancer and autophagy. *Cancers* **2018**, *10*, 18. [CrossRef] [PubMed]
37. Ono, M.; Kosaka, N.; Tominaga, N.; Yoshioka, Y.; Takeshita, F.; Takahashi, R.; Yoshida, M.; Tsuda, H.; Tamura, K.; Ochiya, T. Exosomes from bone marrow mesenchymal stem cells contain a microRNA that promotes dormancy in metastatic breast cancer cells. *Sci. Signal.* **2014**, *7*, ra63. [CrossRef]
38. Guo, N.; Yan, A.; Gao, X.; Chen, Y.; He, X.; Hu, Z.; Mi, M.; Tang, X.; Gou, X. Berberine sensitizes rapamycinmediated human hepatoma cell death in vitro. *Mol. Med. Rep.* **2014**, *10*, 3132–3138. [CrossRef]
39. Meng, M.; Geng, S.; Du, Z.; Yao, J.; Zheng, Y.; Li, Z.; Zhang, Z.; Li, J.; Duan, Y.; Du, G. Berberine and cinnamaldehyde together prevent lung carcinogenesis. *Oncotarget* **2017**, *8*, 76385–76397. [CrossRef]

40. Pucci, M.; Reclusa Asiain, P.; Durendez Saez, E.; Jantus-Lewintre, E.; Malarani, M.; Khan, S.; Fontana, S.; Naing, A.; Passiglia, F.; Raez, LE.; et al. Extracellular vesicles as miRNA nano-shuttles: Dual role in tumor progression. *Target. Oncol.* **2018**, *13*, 175–187. [CrossRef]
41. Reclusa, P.; Taverna, S.; Pucci, M.; Durendez, E.; Calabuig, S.; Manca, P.; Serrano, M.J.; Sober, L.; Pauwels, P.; Russo, A.; et al. Exosomes as diagnostic and predictive biomarkers in lung cancer. *J. Thorac. Dis.* **2017**, *9*, S1373–S1382. [CrossRef] [PubMed]
42. Ruivo, C.F.; Adem, B.; Silva, M.; Melo, S.A. The biology of cancer exosomes: Insights and new perspectives. *Cancer Res.* **2017**, *77*, 6480–6488. [CrossRef] [PubMed]
43. Raposo, G.; Stoorvogel, W. Extracellular vesicles: Exosomes, microvesicles, and friends. *J. Cell Biol.* **2013**, *200*, 373–383. [CrossRef] [PubMed]
44. Tkach, M.; Thery, C. Communication by extracellular vesicles: Where we are and where we need to go. *Cell* **2016**, *164*, 1226–1232. [CrossRef]
45. Endzelins, E.; Abols, A.; Buss, A.; Zandberga, E.; Palviainen, M.; Siljander, P.; Linē, A. Extracellular vesicles derived from hypoxic colorectal cancer cells confer metastatic phenotype to non-metastatic cancer cells. *Anticancer Res.* **2018**, *38*, 5139–5147. [CrossRef]
46. Deep, G.; Panigrahi, G.K. Hypoxia-induced signaling promotes prostate cancer progression: Exosomes role as messenger of hypoxic response in tumor microenvironment. *Crit. Rev. Oncog.* **2015**, *20*, 419–434. [CrossRef] [PubMed]
47. Pakravan, K.; Babashah, S.; Sadeghizadeh, M.; Mowla, S.J.; Mossahebi-Mohammadi, M.; Ataei, F.; Dana, N.; Javan, M. MicroRNA-100 shuttled by mesenchymal stem cell-derived exosomes suppresses in vitro angiogenesis through modulating the mTOR/HIF-1alpha/VEGF signaling axis in breast cancer cells. *Cell. Oncol. (Dordr.)* **2017**, *40*, 457–470. [CrossRef] [PubMed]
48. Taverna, S.; Fontana, S.; Monteleone, F.; Pucci, M.; Saieva, L.; De Caro, V.; Cardinale, V.G.; Giallombardo, M.; Vicario, E.; Rolfo, C.; et al. Curcumin modulates chronic myelogenous leukemia exosomes composition and affects angiogenic phenotype via exosomal miR-21. *Oncotarget* **2016**, *7*, 30420–30439. [CrossRef] [PubMed]
49. Taverna, S.; Giallombardo, M.; Pucci, M.; Flugy, A.; Manno, M.; Raccosta, S.; Rolfo, C.; De Leo, G.; Alessandro1, R. Curcumin inhibits in vitro and in vivo chronic myelogenous leukemia cells growth: A possible role for exosomal disposal of miR-21. *Oncotarget* **2015**, *6*, 21918–21933. [CrossRef] [PubMed]
50. Zhao, F.; Mancuso, A.; Bui, T.V.; Tong, X.; Gruber, J.J.; Swider, C.R.; Sanchez, P.V.; Lum, J.J.; Sayed, N.; Melo, J.V.; et al. Imatinib resistance associated with BCR-ABL upregulation is dependent on HIF-1alpha-induced metabolic reprograming. *Oncogene* **2010**, *29*, 2962–2972. [CrossRef] [PubMed]
51. Monteleone, F.; Taverna, S.; Alessandro, R.; Fontana, S. SWATH-MS based quantitative proteomics analysis reveals that curcumin alters the metabolic enzyme profile of CML cells by affecting the activity of miR-22/IPO7/HIF-1alpha axis. *J. Exp. Clin. Cancer Res.* **2018**, *37*, 170. [CrossRef] [PubMed]
52. Chi, L.; Peng, L.; Pan, N.; Hu, X.; Zhang, Y. The anti-atherogenic effects of berberine on foam cell formation are mediated through the upregulation of sirtuin 1. *Int. J. Mol. Med.* **2014**, *34*, 1087–1093. [CrossRef] [PubMed]
53. Hu, H.Y.; Li, K.P.; Wang, X.Y.; Liu, Y.; Lu, Z.G.; Dong, R.H.; Guo, H.B.; Zhang, M.X. Set9, NF-κB, and microRNA-21 mediate berberine-induced apoptosis of human multiple myeloma cells. *Acta Pharmacol. Sin.* **2013**, *34*, 157–166. [CrossRef] [PubMed]
54. Xu, M.; Xiao, Y.; Yin, J.; Hou, W.; Yu, X.; Shen, L.; Liu, F.; Wei, L.; Jia, W. Berberine promotes glucose consumption independently of AMP-activated protein kinase activation. *PLoS ONE* **2014**, *9*, e103702. [CrossRef] [PubMed]
55. Teodoro, J.S.; Duarte, F.V.; Gomes, A.P.; Varela, A.T.; Peixoto, F.M.; Rolo, A.P.; Palmeira, C.M. Berberine reverts hepatic mitochondrial dysfunction in high-fat fed rats: A possible role for SirT3 activation. *Mitochondrion* **2013**, *13*, 637–646. [CrossRef]
56. Arab, K.; Park, Y.J.; Lindroth, A.M.; Schäfer, A.; Oakes, C.; Weichenhan, D.; Lukanova, A.; Lundin, E.; Risch, A.; Meister, M.; et al. Long noncoding RNA *TARID* directs demethylation and activation of the tumor suppressor *TCF21* via *GADD45A*. *Mol. Cell* **2014**, *55*, 604–614. [CrossRef] [PubMed]
57. Li, Z.; Gu, T.P.; Weber, A.R.; Shen, J.Z.; Li, B.Z.; Xie, Z.G.; Yin, R.; Guo, F.; Liu, X.; Tang, F.; et al. Gadd45a promotes DNA demethylation through TDG. *Nucleic Acids Res.* **2015**, *43*, 3986–3997. [CrossRef] [PubMed]
58. Kienhöfer, S.; Musheev, M.U.; Stapf, U.; Helm, M.; Schomacher, L.; Niehrs, C.; Schäfer, A. GADD45a physically and functionally interacts with TET1. *Differentiation* **2015**, *90*, 59–68. [CrossRef] [PubMed]

59. Chen, K.; Long, Q.; Wang, T.; Zhao, D.; Zhou, Y.; Qi, J.; Wu, Y.; Li, S.; Chen, C.; Zeng, X.; et al. Gadd45a is a heterochromatin relaxer that enhances iPS cell generation. *EMBO Rep.* **2016**, *17*, 1641–1656. [CrossRef] [PubMed]
60. Ayati, S.H.; Fazeli, B.; Momtazi-Borojeni, A.A.; Cicero, A.F.G.; Pirro, M.; Sahebkar, A. Regulatory effects of berberine on microRNome in cancer and other conditions. *Crit. Rev. Oncol. Hematol.* **2017**, *116*, 147–158. [CrossRef]
61. Wang, Y.; Wang, Z.; Xu, J.; Li, J.; Li, S.; Zhang, M.; Yang, D. Systematic identification of non-coding pharmacogenomic landscape in cancer. *Nat. Commun.* **2018**, *9*, 3192. [CrossRef]
62. Yuan, X.; Wang, J.; Tang, X.; Li, Y.; Xia, P.; Gao, X. Berberine ameliorates nonalcoholic fatty liver disease by a global modulation of hepatic mRNA and lncRNA expression profiles. *J. Transl. Med.* **2015**, *13*, 24. [CrossRef] [PubMed]
63. Biersack, B. Non-coding RNA/microRNA-modulatory dietary factors and natural products for improved cancer therapy and prevention: Alkaloids, organosulfur compounds, aliphatic carboxylic acids and water-soluble vitamins. *Non-Coding RNA Res.* **2016**, *1*, 51–63. [CrossRef] [PubMed]
64. McCubrey, J.A.; Abrams, S.L.; Lertpiriyapong, K.; Cocco, L.; Ratti, S.; Martelli, A.M.; Candido, S.; Libra, M.; Murata, R.M.; Rosalen, P.L.; et al. Effects of berberine, curcumin, resveratrol alone and in combination with chemotherapeutic drugs and signal transduction inhibitors on cancer cells—Power of nutraceuticals. *Adv. Biol. Regul.* **2018**, *67*, 190–211. [CrossRef] [PubMed]
65. McCubrey, J.A.; Lertpiriyapong, K.; Steelman, L.S.; Abrams, S.L.; Yang, L.V.; Murata, R.M.; Rosalen, P.L.; Scalisi, A.; Neri, L.M.; Cocco, L.; et al. Effects of resveratrol, curcumin, berberine and other nutraceuticals on aging, cancer development, cancer stem cells and microRNAs. *Aging (Albany, NY)* **2017**, *9*, 1477–1536. [CrossRef] [PubMed]
66. Anastasiadou, E.; Jacob, L.S.; Slack, F.J. Noncoding RNA networks in cancer. *Nat. Rev. Cancer* **2018**, *18*, 5–18. [CrossRef] [PubMed]
67. Feng, M.; Luo, X.; Gu, C.; Li, Y.; Zhu, X.; Fei, J. Systematic analysis of berberine-induced signaling pathway between miRNA clusters and mRNAs and identification of mir-99a-125b cluster function by seed-targeting inhibitors in multiple myeloma cells. *RNA Biol.* **2015**, *12*, 82–91. [CrossRef]
68. Chen, J.; Wu, F.X.; Luo, H.L.; Liu, J.J.; Luo, T.; Bai, T.; Li, L.Q.; Fan, X.H. Berberine upregulates miR-22-3p to suppress hepatocellular carcinoma cell proliferation by targeting Sp1. *Am. J. Transl. Res.* **2016**, *8*, 4932–4941.
69. Luo, X.; Gu, J.; Zhu, R.; Feng, M.; Zhu, X.; Li, Y.; Fei, J. Integrative analysis of differential miRNA and functional study of miR-21 by seed-targeting inhibition in multiple myeloma cells in response to berberine. *BMC Syst. Biol.* **2014**, *8*, 82. [CrossRef] [PubMed]
70. Hong, M.; Wang, N.; Tan, H.Y.; Tsao, S.W.; Feng, Y. MicroRNAs and Chinese medicinal herbs: New possibilities in cancer therapy. *Cancers* **2015**, *7*, 1643–1657. [CrossRef] [PubMed]
71. Tillhon, M.; Guamán Ortiz, L.M.; Lombardi, P.; Scovassi, A.I. Berberine: New perspectives for old remedies. *Biochem. Pharmacol.* **2012**, *84*, 1260–1267. [CrossRef] [PubMed]
72. Lo, T.F.; Tsai, W.C.; Chen, S.T. MicroRNA-21-3p, a berberine-induced miRNA, directly down-regulates human methionine adenosyltransferases 2A and 2B and inhibits hepatoma cell growth. *PLoS ONE* **2013**, *8*, e75628. [CrossRef]
73. Chen, Q.; Qin, R.; Fang, Y.; Li, H. Berberine sensitizes human ovarian cancer cells to cisplatin through miR-93/PTEN/Akt signaling pathway. *Cell. Physiol. Biochem.* **2015**, *36*, 956–965. [CrossRef] [PubMed]
74. You, H.Y.; Xie, X.M.; Zhang, W.J.; Zhu, H.L.; Jiang, F.Z. Berberine modulates cisplatin sensitivity of human gastric cancer cells by upregulation of miR-203. *In Vitro Cell Dev. Biol. Anim.* **2016**, *52*, 857–863. [CrossRef] [PubMed]
75. Zamani, M.; Sadeghizadeh, M.; Behmanesh, M.; Najafi, F. Dendrosomal curcumin increases expression of the long non-coding RNA gene MEG3 via up-regulation of epi-miRs in hepatocellular cancer. *Phytomedicine* **2015**, *22*, 961–967. [CrossRef] [PubMed]
76. Chang, W. Non-coding RNAs and berberine: A new mechanism of its anti-diabetic activities. *Eur. J. Pharmacol.* **2016**, *795*, 8–12. [CrossRef] [PubMed]
77. Wei, S.; Zhang, M.; Yu, Y.; Lan, X.; Yao, F.; Yan, X.; Chen, L.; Hatch, G.M. Berberine attenuates development of the hepatic gluconeogenesis and lipid metabolism disorder in type 2 diabetic mice and in palmitate-incubated HepG2 cells through suppression of the HNF-4alpha miR122 pathway. *PLoS ONE* **2016**, *11*, e0152097. [CrossRef] [PubMed]

78. Dziedzic, A.; Kubina, R.; Bułdak, R.J.; Skonieczna, M.; Cholewa, K. Silver nanoparticles exhibit the dose-dependent anti-proliferative effect against human squamous carcinoma cells attenuated in the presence of berberine. *Molecules* **2016**, *21*, 365. [CrossRef] [PubMed]
79. Bhanumathi, R.; Vimala, K.; Shanthi, K.; Thangaraj, R.; Kannan, S. Bioformulation of silver nanoparticles as berberine carrier cum anticancer agent against breast cancer. *New J. Chem.* **2017**, *41*, 14466–14477. [CrossRef]
80. Bhanumathi, R.; Manivannan, M.; Thangaraj, R.; Kannan, S. Drug-carrying capacity and anticancer effect of the folic acid- and berberine-loaded silver nanomaterial to regulate the AKT-ERK pathway in breast cancer. *ACS Omega* **2018**, *3*, 8317–8328. [CrossRef] [PubMed]
81. Kim, S.; Lee, S.Y.; Cho, H.J. Berberine and zinc oxide-based nanoparticles for the chemo-photothermal therapy of lung adenocarcinoma. *Biochem. Biophys. Res. Commun.* **2018**, *501*, 765–770. [CrossRef]
82. Sreeja, S.; Krishnan Nair, C.K. Tumor control by hypoxia-specific chemotargeting of iron-oxide nanoparticle—Berberine complexes in a mouse model. *Life Sci.* **2018**, *195*, 71–80. [CrossRef] [PubMed]
83. Wang, L.; Li, H.; Wang, S.; Liu, R.; Wu, Z.; Wang, C.; Wang, Y.; Chen, M. Enhancing the antitumor activity of berberine hydrochloride by solid lipid nanoparticle encapsulation. *AAPS PharmSciTech* **2014**, *15*, 834–844. [CrossRef] [PubMed]
84. Wang, Z.P.; Wu, J.B.; Chen, T.S.; Zhou, Q.; Wang, Y.F. In vitro and in vivo antitumor efficacy of berberine-nanostructured lipid carriers against H22 tumor. In *Biophotonics and Immune Responses X*; SIPE: Bellingham, WA, USA, 2015.
85. Kabary, D.M.; Helmy, M.W.; Elkhodairy, K.A.; Fang, J.Y.; Elzoghby, A.O. Hyaluronate/lactoferrin layer-by-layer-coated lipid nanocarriers for targeted co-delivery of rapamycin and berberine to lung carcinoma. *Colloids Surf. B Biointerfaces* **2018**, *169*, 183–194. [CrossRef]
86. Kabary, D.M.; Helmy, M.W.; Abdelfattah, E.Z.A.; Fang, J.Y.; Elkhodairy, K.A.; Elzoghby, A.O. Inhalable multi-compartmental phospholipid enveloped lipid core nanocomposites for localized mTOR inhibitor/herbal combined therapy of lung carcinoma. *Eur. J. Pharm. Biopharm.* **2018**, *130*, 152–164. [CrossRef] [PubMed]
87. Gupta, L.; Sharma, A.K.; Gothwal, A.; Khan, M.S.; Khinchi, M.P.; Qayum, A.; Singh, S.K.; Gupta, U. Dendrimer encapsulated and conjugated delivery of berberine: A novel approach mitigating toxicity and improving in vivo pharmacokinetics. *Int. J. Pharm.* **2017**, *528*, 88–99. [CrossRef] [PubMed]
88. Shen, R.; Kim, J.J.; Yao, M.; Elbayoumi, T.A. Development and evaluation of vitamin E D-α-tocopheryl polyethylene glycol 1000 succinate-mixed polymeric phospholipid micelles of berberine as an anticancer nanopharmaceutical. *Int. J. Nanomed.* **2016**, *11*, 1687–1700.
89. Wang, Y.; Wen, B.; Yu, H.; Ding, D.; Zhang, J.J.; Zhang, Y.; Zhao, L.; Zhang, W. Berberine hydrochloride-loaded chitosan nanoparticles effectively targets and suppresses human nasopharyngeal carcinoma. *J. Biomed Nanotechnol.* **2018**, *14*, 1486–1495. [CrossRef]
90. Agnarelli, A.; Natali, M.; Garcia-Gil, M.; Pesi, R.; Tozzi, M.G.; Ippolito, C.; Bernardini, N.; Vignali, R.; Batistoni, R.; Bianucci, A.M.; et al. Cell-specific pattern of berberine pleiotropic effects on different human cell lines. *Sci. Rep.* **2018**, *8*, 10599. [CrossRef]
91. Mohammadinejad, R.; Ahmadi, Z.; Tavakol, S.; Ashrafizadeh, M. Berberine as a potential autophagy modulator. *J. Cell. Physiol.* **2019**. [CrossRef]
92. Bianchi, S.; Giovannini, L. Inhibition of mTOR/S6K1/4E-BP1 signaling by nutraceutical SIRT1 modulators. *Nutr. Cancer* **2018**, *70*, 490–501. [CrossRef] [PubMed]

cancers

MDPI

Article

Lambda-Carrageenan Enhances the Effects of Radiation Therapy in Cancer Treatment by Suppressing Cancer Cell Invasion and Metastasis through Racgap1 Inhibition

Ping-Hsiu Wu [1,2], Yasuhito Onodera [1,3], Frances C. Recuenco [1,†], Amato J. Giaccia [4], Quynh-Thu Le [4], Shinichi Shimizu [1,5], Hiroki Shirato [1,6] and Jin-Min Nam [1,6,*]

1 Global Station for Quantum Medical Science and Engineering, Global Institution for Collaborative Research and Education (GI-CoRE), Hokkaido University, Sapporo, Hokkaido 060-8638, Japan
2 Department of Radiation Oncology, Graduate School of Medicine, Hokkaido University, Sapporo, Hokkaido 060-8638, Japan
3 Department of Molecular Biology, Faculty of Medicine, Hokkaido University, Sapporo, Hokkaido 060-8638, Japan
4 Department of Radiation Oncology, Stanford University School of Medicine, Stanford, CA 94305, USA
5 Department of Radiation Oncology, Faculty of Medicine, Hokkaido University, Sapporo, Hokkaido 060-8638, Japan
6 Research Center for Cooperative Project, Faculty of Medicine, Hokkaido University, Sapporo, Hokkaido 060-8638, Japan
* Correspondence: jinmini@med.hokudai.ac.jp; Tel.: +81-11-706-5076; Fax: +81-11-706-7005
† Current address: Cavite State University, Cavite 4122, Philippines.

Received: 25 June 2019; Accepted: 12 August 2019; Published: 16 August 2019

Abstract: Radiotherapy is used extensively in cancer treatment, but radioresistance and the metastatic potential of cancer cells that survive radiation remain critical issues. There is a need for novel treatments to improve radiotherapy. Here, we evaluated the therapeutic benefit of λ-carrageenan (CGN) to enhance the efficacy of radiation treatment and investigated the underlying molecular mechanism. CGN treatment decreased viability in irradiated cancer cells and enhanced reactive oxygen species accumulation, apoptosis, and polyploid formation. Additionally, CGN suppressed radiation-induced chemoinvasion and invasive growth in 3D lrECM culture. We also screened target molecules using a gene expression microarray analysis and focused on Rac GTPase-activating protein 1 (RacGAP1). Protein expression of RacGAP1 was upregulated in several cancer cell lines after radiation, which was significantly suppressed by CGN treatment. Knockdown of RacGAP1 decreased cell viability and invasiveness after radiation. Overexpression of RacGAP1 partially rescued CGN cytotoxicity. In a mouse xenograft model, local irradiation followed by CGN treatment significantly decreased tumor growth and lung metastasis compared to either treatment alone. Taken together, these results suggest that CGN may enhance the effectiveness of radiation in cancer therapy by decreasing cancer cell viability and suppressing both radiation-induced invasive activity and distal metastasis through downregulating RacGAP1 expression.

Keywords: carrageenan; invasion; metastasis; RacGAP1; radiotherapy

1. Introduction

Radiotherapy is a standard treatment to eliminate cancer cells by inducing a variety of cellular events, including the accumulation of reactive oxygen species (ROS) [1] and apoptotic cell death [2]. In clinical practice, radiotherapy is used as a post-operative treatment after resection, treatment for inoperable tumors, and may even replace surgery where organ preservation is desired [3,4]. Although radiotherapy is an effective anticancer therapy, differences in radiation response among different types of cancers [5,6], along with the side effects of high dose or large field IR [7], limit its efficacy. To increase its clinical impact, various chemotherapeutic drugs are administered with radiotherapy. However, these drugs can have their toxicity to normal tissues [8]. Therefore, novel therapies that are free from side effects but at the same time can improve radiotherapy by increasing cancer cell death and reducing distal metastasis are highly desirable.

Lambda-carrageenan (CGN), a family of linear sulfated polysaccharides, has diverse biological activities, which include anti-coagulant [9], anti-viral [10], and anti-tumor effects [11]. In addition, carrageenans are safely used as a food additive under FDA regulations in the United States [12]. Recently, several studies have reported the anti-tumor effects of CGN by stimulating an immune response in mice [11,13]. However, the effect of CGN in combination with IR on cancer treatment and its molecular mechanisms are not known. Considering the safety profile of this agent, we evaluated the effect of CGN as an adjuvant treatment to radiotherapy.

In several types of cancer, recent studies have identified upregulated expression of RacGAP1 (also known as MgcRacGAP or hCYK-4), a member of the guanine triphosphatase (GTPase) activation protein (GAP) family and suggested its potential role in promoting tumor progression [14–16]. RacGAP1 regulates the activation of Rho GTPase, which is reported to drive tumor growth [17] and to act as an oncogene in basal-like breast cancers [18]. Moreover, RacGAP1 is required for integrin-related invasive cell migration in the three-dimensional extracellular matrix (3D ECM) [19]. In clinical studies, RacGAP1 has attracted increasing attention as a predictive biomarker for metastasis and prognosis in several types of cancer [15,20].

In this study, we investigated the effect of CGN combined with ionizing radiation (IR) on cancer treatment and determined underlying molecular mechanisms. We found that CGN treatment after IR decreased cancer cell survival and invasiveness. Gene expression analysis showed that RacGAP1 was upregulated after IR treatment, and significantly suppressed by CGN treatment. Furthermore, CGN treatment after IR significantly suppressed tumor growth and lung metastasis in an in vivo model. These results indicate that CGN is a potential therapeutic adjuvant to radiotherapy, improving its therapeutic effect by suppressing RacGAP1.

2. Results

2.1. CGN Treatment Decreases the Cell Viability in Irradiated Cancer Cell Lines

The possibility of using polysaccharides as an adjuvant to improve the anti-tumor efficacy of traditional chemotherapy has been raised [21,22]. In this study, we determined the anti-tumor effects of CGN combined with IR in the MDA-MB-231 breast cancer, FaDu head and neck cancer, PANC-1 pancreatic cancer, and 4T1 murine breast cancer cell lines. In each case, when compared to IR alone, cell number decreased significantly when IR and CGN were combined. The result also showed that CGN alone caused damage to cancer cells (Figure 1A). To further analyze the decrease in cell number in each condition, we next stained the cells with propidium iodide (PI) and Annexin V. CGN treatment following IR led to a significant increase in PI-positive cells (Figure 1B) and Annexin V-positive cells (Figure 1C) compared to the IR alone group. Annexin V/PI double staining analysis showed that late apoptosis was increased in cells treated with IR and CGN compared to those treated with IR alone (Figure S1). These results suggest that CGN combined with IR decreases viability in several cancer cell lines by induction of apoptotic cell death. We evaluated the effect of CGN and IR treatment in non-malignant epithelial cell line MCF10A (Figure S2A–C). The CGN-induced toxicity following IR treatment in MCF10A cells was not higher than that in other malignant cells.

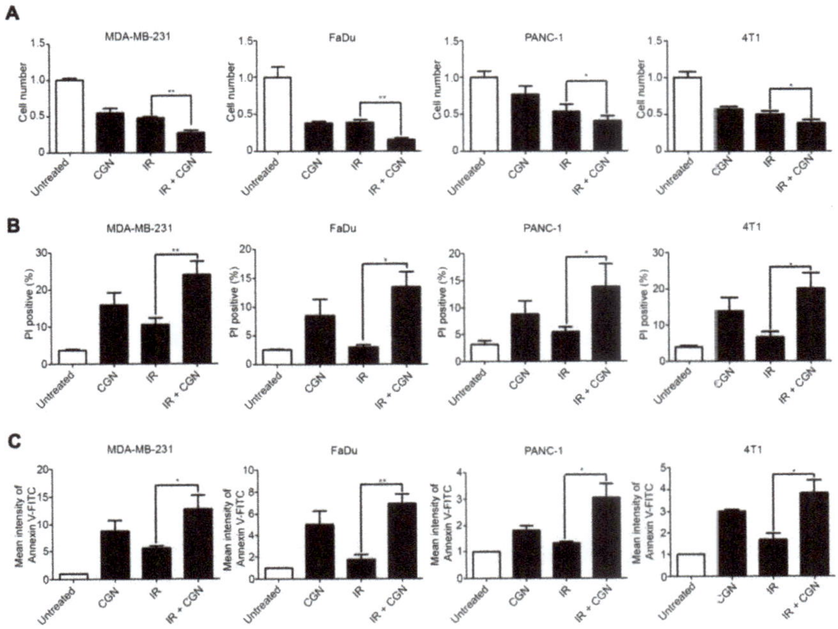

Figure 1. λ-carrageenan (CGN) treatment decreases cell viability and increases apoptosis in irradiated cancer cell lines. Cells were treated with 4 Gy IR, followed by CGN on the next day, and then analyzed 72 h after IR. (**A**) Cell viability was quantified by cell counting in MDA-MB-231, FaDu, PANC-1, and 4T1 cell lines. (**B**) The percentage of dead cells was measured by PI staining, followed by flow cytometry. The proportion of dead cells was quantified by gating the population of PI-positive cells. (p-values: MDA-MB-231, 0.0056; FaDu, 0.0129; PANC-1, 0.0489; 4T1, 0.0468.) (**C**) Apoptotic cells were measured by Annexin V-FITC staining, followed by flow cytometry. Mean fluorescence intensity of FITC was calculated and normalized to the untreated group. Columns, mean ($n \geq 3$); bars, SE. *, $p < 0.05$; **, $p < 0.01$.

2.2. IR Combined with CGN Treatment Increases ROS Accumulation in MDA-MB-231 Breast Cancer Cells

Elevation of ROS is an important factor in the control of cancer cell death in radiotherapy [23]. It is known that IR induces ROS, which mediate apoptotic cell death and mitotic failure. Additionally, CGN has been reported to increase the production of ROS in human colonic epithelial cells [24]. We analyzed cellular ROS levels using DCFDA, which fluoresces when oxidized by ROS. Increased ROS levels were observed in the IR and CGN treated cells, compared to IR alone (Figure 2A). CGN or IR alone also showed an increase in ROS accumulation. High levels of ROS are known to activate caspase-3 and caspase-8, which are the key proteins of apoptosis [25]. The activities of caspase-3 and caspase-8, but not caspase-9, were elevated after IR followed by CGN in comparison to IR alone in MDA-MB-231 cells (Figure 2B). Consistent with these results, an increase in cleaved caspase-3 level in the IR and CGN treated cells, compared to the other groups, was also confirmed by western blot (Figure S3). These results indicate that apoptosis-related cell death is efficiently induced by CGN following IR, which is consistent with the PI and Annexin V staining (Figure 1).

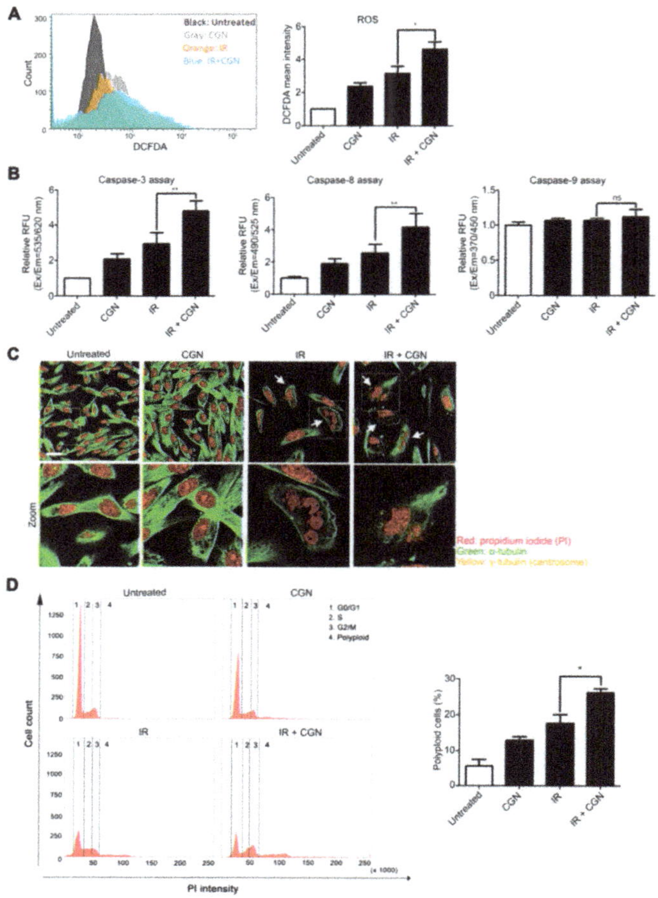

Figure 2. IR exposure in combination with CGN increases ROS accumulation in MDA-MB-231 cells. Cells were treated with 4 Gy IR, followed by CGN on the next day, and then analyzed 72 h after IR. (**A**) ROS was measured by DCFDA. Columns, mean ($n = 5$); bars, SE. *, $p < 0.05$. (**B**) Caspase-3, caspase-8, and caspase-9 activities were detected by microplate reader at specific wavelengths: caspase-3 excitation (Ex)/emission (Em) = 535/620 nm; caspase-8 Ex/Em = 490/525 nm; caspase-9 Ex/Em = 370/450 nm. Columns, mean ($n = 5$); bars, SE. **, $p < 0.01$; ns, not significant. (**C**) Cells stained with α-tubulin (green) and PI (red) after treatments. Bar, 25 μm. (**D**) To measure polyploid populations, cells were treated with staining solution and PI and analyzed by flow cytometry. Columns, mean ($n = 3$); bars, SE. *, $p < 0.05$.

Besides apoptotic cell death, IR is known to cause mitotic catastrophe [26,27], a mechanism of mitosis-linked cell death resulting in polyploid cell formation [28]. Generation of ROS is also reported to permit inappropriate entry into mitosis and induce mitotic catastrophe [29]. To determine whether mitotic catastrophe was induced by CGN combined with IR, we analyzed polyploid formation in the cells by immunofluorescence. Under confocal fluorescence microscopy, abnormal polyploid giant cells were observed in both the IR alone and combined treatment groups (Figure 2C). The proportion of polyploid cells was significantly increased by combined treatment with CGN and IR compared to IR alone (Figure 2D). These data suggest that CGN can increase ROS accumulation in irradiated cells, which may further enhance caspase-mediated apoptosis and mitosis-related cell death.

2.3. CGN Inhibits the Radiation-Induced Invasiveness of Breast Cancer Cell Lines

Cancer cells with high invasive capacity are correlated with poor prognosis [30,31]. Several groups have reported that failure of tumor control by IR could be associated with cancer invasiveness and subsequent distal metastasis [32,33], highlighting a potentially undesirable effect of radiotherapy. Our previous studies showed that cancer cell invasiveness could be increased in the surviving population after IR treatment through integrin-mediated pathways [34,35]. We, therefore, investigated CGN's effect on the invasiveness of surviving cells after IR. The invasive activity was increased in the breast cancer cell lines after IR treatment, as we have reported previously [35].

Interestingly, the invasive ability of MDA-MB-231 (Figure 3A) and 4T1 (Figure 3B) breast cancer cell lines was significantly lower in the combined treatment with IR and CGN compared to IR alone. Cell viability was not affected during the invasion assay (Figure S4). These data indicate that CGN suppresses the IR-related invasiveness of these cells. Compared to the results of cytotoxicity depicted in Figures 1 and 2, CGN showed a higher anti-invasive effect which is specific in post-IR cells. This effect was more obvious in MDA-MB-231 cells, which leads to a more significant reduction in invasive ability than that of CGN alone, suggesting the possibility that CGN suppresses specific mechanisms, which induced the increase in invasiveness in post-IR cancer cells. To further confirm the effect of CGN in the invasive growth of cancer cells, we performed 3D laminin-rich ECM (lrECM) culture. Culturing cells in 3D lrECM is a common method to assess the physiologically relevant morphogenesis and oncogenic properties of non-malignant or cancerous mammary epithelial cells [36]. As shown in Figure 3C, untreated MDA-MB-231 cells under 3D culture displayed aggressive invasive growth with stellate protrusions extending into the lrECM. The formation of protrusions was reduced in cancer cells that received the combination of CGN and IR treatment, indicating suppression of their invasive capacity in 3D lrECM culture.

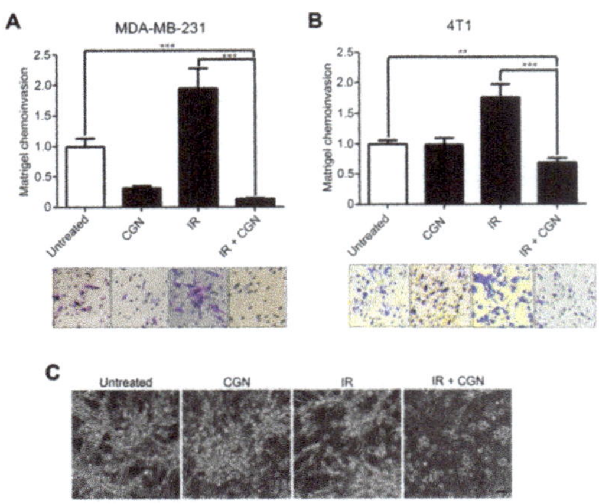

Figure 3. CGN inhibits the IR-induced invasive activity and 3D lrECM growth in breast cancer cells. (**A**, **B**) The invasive activity was measured by Matrigel chemoinvasion assay after IR and/or CGN treatments in MDA-MB-231 (**A**) and 4T1 (**B**) cells. Columns, mean (n = 4); bars, SE. **, $p < 0.01$; ***, $p < 0.001$. (**C**) MDA-MB-231 cells were cultured in 3D lrECM. Bar, 50 μm.

2.4. Upregulation of RacGAP1 Is Involved in Cancer Cell Survival and Invasion after IR

To determine the effects on molecular pathways, differential gene expression in each treatment group was assessed by cDNA microarray. We selected the genes that were both upregulated by IR treatment and suppressed by the following CGN treatment (Figure S5). From these genes, we focused on RacGAP1 as a potential target (Figure 4A) because this protein is intimately connected with integrin signaling, which we have been mainly working on [34,35,37]. Moreover, RacGAP1 is recently reported to have important roles in oncogenic activity [19], tumor progression [15,16], and cancer invasion [38,39].

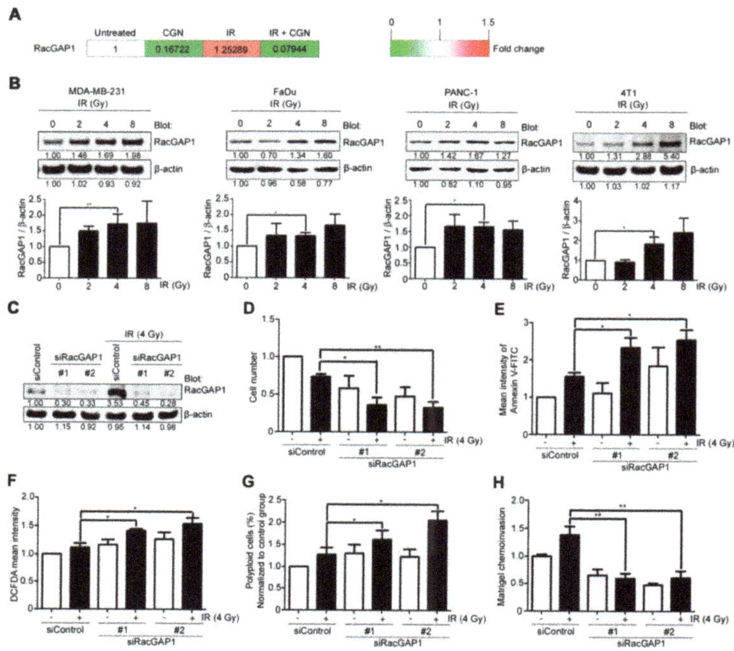

Figure 4. Upregulation of RacGAP1 is involved in cancer cell survival and invasion after IR. (**A**) Gene expression was analyzed by cDNA microarray. RacGAP1 expression level is shown by the heat map. The values were normalized to the untreated group. (**B**) RacGAP1 protein expressions with different doses of IR were analyzed in MDA-MB-231, FaDu, PANC-1, and 4T1 cell lines. (**C**) MDA-MB-231 cells were transfected with siRNA duplexes targeting RacGAP1 (# 1 or # 2) or the control sequence, as indicated. Cell lysates were subjected to western blot two days after transfection. (**D–H**) MDA-MB-231 cells were transfected with siRNAs and incubated for two days, and then treated with 4 Gy IR. Each experiment was performed 24 h after IR treatment. Cell viability was quantified by cell counting (**D**). Apoptotic cells were measured by Annexin V-FITC staining and flow cytometry (**E**). ROS was measured by DCFDA (**F**). Cells were treated with staining solution and PI, and then the cell cycle was analyzed by flow cytometry. The percentage of polyploid cells in each group was normalized with control group (**G**). The invasive activity was measured by Matrigel chemoinvasion assay (**H**). Columns, mean ($n \geq 3$); bars, SE. *, $p < 0.05$; **, $p < 0.01$. Data shown in (**D–H**) were normalized to that of MDA-MB-231 cells transfected with siRNA control. For the continuous full length-image of Western blot signals, please refer to Supplementary Materials Figure S7.

To investigate its role in irradiated cells, protein levels of RacGAP1 were determined after IR treatment in different types of cancer cell lines (Figure 4B). RacGAP1 was found to be upregulated after IR treatment, suggesting that it plays a role in the cellular response to radiation. Compared to the control group, knockdown of RacGAP1 by siRNA (Figure 4C and Figure S6) resulted in a significant decrease in cell viability following IR treatment (Figure 4D). Knockdown of RacGAP1 also increased the apoptosis, ROS accumulation, and polyploid formation after IR treatment in MDA-MB-231 cells (Figure 4E–G). Furthermore, knockdown of RacGAP1 effectively suppressed the IR-induced invasion activity of cancer cells (Figure 4H). These results indicate that RacGAP1 inhibition could increase the effectiveness of IR by reducing both cell viability and IR-induced invasiveness, which was also achieved by adding CGN after IR.

2.5. RacGAP1 Expression Is Suppressed by CGN in MDA-MB-231 Cells

The cDNA microarray data suggest that CGN treatment suppresses RacGAP1 (Figure 4A). Consistent with the gene expression data, we also confirmed that the protein expression of RacGAP1 was significantly downregulated by CGN (Figure 5A). Moreover, elevation of RacGAP1 protein levels following IR, and its suppression after CGN treatment were also confirmed by western blot (Figure 5B). Immunofluorescent staining showed that RacGAP1 mainly localized in the nucleus and that its nuclear level was increased after IR and suppressed by CGN treatment (Figure 5C).

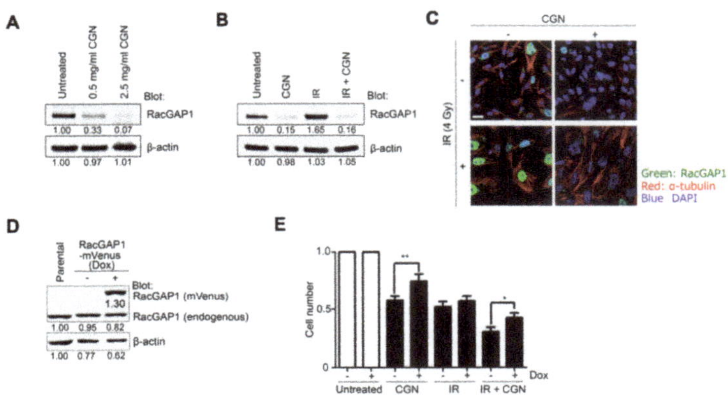

Figure 5. RacGAP1 expression is suppressed by CGN in MDA-MB-231 cells. (**A**) Protein expression of RacGAP1 was analyzed after treatment with different concentrations of CGN in MDA-MB-231 cells. Total cell lysates were subjected to western blot. (**B**) RacGAP1 protein expression was analyzed using cell lysates after IR and/or CGN treatment in MDA-MB-231 cells. (**C**) Immunofluorescence images show RacGAP1 (green), α-tubulin (red) and nuclei (blue). Bar, 25 μm. (**D**) RacGAP1-mVenus expression was induced by doxycycline in MDA-MB-231 cells. RacGAP1 expression was analyzed by western blot using anti-RacGAP1 antibody. Dox, doxycycline. (**E**) Cell number was quantified by cell counting. Columns, mean ($n = 3$); bars, SE. *, $p < 0.05$; **, $p < 0.01$. For the continuous full length-image of Western blot signals, please refer to Supplementary Materials Figure S7.

To further confirm that the downregulation of RacGAP1 is involved in the cytotoxic effect of CGN, we generated a doxycycline-inducible overexpression system of RacGAP1 (Figure 5D). RacGAP1 overexpression partially reduced the cytotoxicity caused by CGN with or without IR treatment (Figure 5E), which suggests that RacGAP1 is indeed an important molecular target in CGN treatment.

2.6. CGN in Combination with IR Decreases Tumor Size and Metastasis in a Mouse Xenograft Model

To determine the in vivo effect of adjuvant CGN treatment after IR, we used a 4T1 xenograft animal model with the experimental schedule shown in Figure 6A. Tumor growth was significantly decreased in the group treated with CGN after IR compared to either CGN or IR treatment alone (Figure 6B). Similarly, the terminal tumor size at day 25 was smallest in the CGN and IR group (Figure 6C). Local invasion is the initial step in the spread of cancer cells from a local site to distant metastasis sites [30]. Therefore, we next determined the therapeutic effect of CGN in combination with IR on distant metastasis. Compared to the other groups, combined treatment with IR and CGN significantly inhibited lung metastasis (Figure 6D). We then confirmed RacGAP1 expression within tumors. Consistent with the in vitro data shown in Figure 5, RacGAP1 was suppressed in 4T1 tumors treated by either CGN alone or CGN in combination with IR (Figure 6E). Taken together, these results indicate that adjuvant CGN treatment effectively suppresses primary tumor growth and reduces metastatic potential, which may be the result of RacGAP1 suppression.

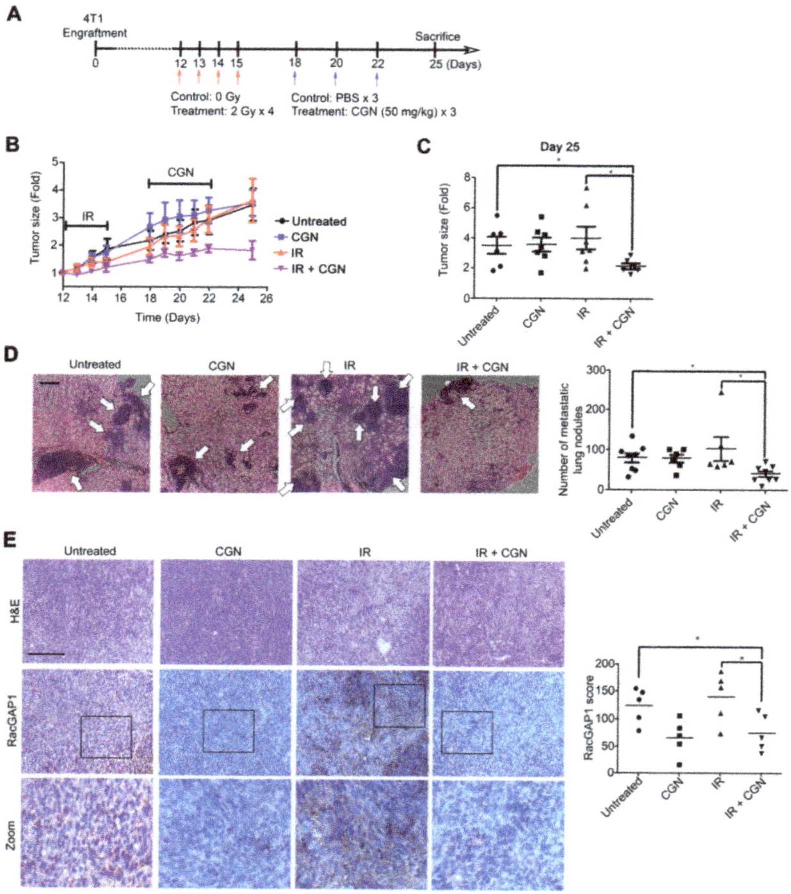

Figure 6. Radiation followed by CGN treatment decreases tumor size and metastasis in a 4T1 mouse xenograft model. (**A**) Treatment schedule of each group. Mouse 4T1 cells were injected into Balb/c mice (n = 6–7 in each group). (**B**) 4T1 tumor sizes were measured and normalized to size at day 12 in each group. (**C**) Relative tumor size for each group was measured on Day 25. Scatter plot; mean ± SE. *, $p < 0.05$. (**D**) Representative H&E images of lung sections. Bar, 500 μm. Arrow, metastatic lung nodules. The number of metastatic lung nodules was counted (right panel). Scatter plot; mean ± SE. *, $p < 0.05$. (**E**) Sections from 4T1 tumors were subjected to IHC staining with antibodies against RacGAP1. Bar, 100 μm. RacGAP1 expression level was determined by scoring, as described in the methods (right panel). Scatter plot; bars, mean (n = 5). *, $p < 0.05$.

3. Discussion

In this study, we tested CGN as an adjuvant therapy to improve the effectiveness of radiotherapy. Administration of CGN to IR treatment increased cancer cell death. Furthermore, CGN treatment resulted in notable suppression of IR-induced cancer cell invasion. RacGAP1 signaling is a possible molecular mechanism of CGN effect after IR treatment on cancer cells. In the 4T1 xenograft model, combined treatment of IR and CGN significantly suppressed the tumor growth and lung metastasis.

Although local radiotherapy improves cancer treatment outcomes, recurrence or distant metastases following local treatment remain major therapeutic challenges. Local recurrences or distant metastases could be partially due to the enhancement of invasive properties in surviving cancer cells after IR [33]. We and others have shown that integrins are involved in the acquisition of cancer cell invasion after IR [34,35,40]. Several studies have reported that α5β1-integrin trafficking regulates RacGAP1 activation, which is essential to promote pseudopod extension and cancer cell invasion [19,38]. Here, we show that upregulation of RacGAP1 after IR is accompanied by increased invasion activity, and depletion of RacGAP1 significantly suppressed IR-induced cancer cell invasion (Figure 4). Moreover, CGN suppresses IR-induced invasiveness, which is partially restored by overexpression of RacGAP1 (Figure 5). These findings suggest that upregulation of RacGAP1 in cancer cells after IR treatment may be one of the pivotal mechanisms contributing to IR-induced invasiveness. The detailed regulatory mechanisms of RacGAP1 on IR-induced invasion should be investigated in future studies.

Previous studies suggest that RacGAP1 is a potential therapeutic target for the treatment of highly aggressive cancers. The clinical significance of RacGAP1 has been widely reported, and its expression in tumors is associated with more aggressive phenotypes in many cancers, including high-grade breast cancer, epithelial ovarian cancer, gastric cancer, colorectal cancer, and hepatocellular carcinoma in the transition from low- to high-invasive disease [15,20,39,41,42]. Besides, RacGAP1 is implicated in the resistance to doxorubicin treatment in squamous cell carcinoma [43]. To our knowledge, this is the first report to connect RacGAP1 to radiation resistance. In addition, we reveal a novel method for targeting RacGAP1 by administering CGN after radiation therapy, which may lead to future clinical application.

Besides RacGAP1, other genes were also found to be upregulated by IR and suppressed by the following CGN treatment (Figure S5). Within these genes, *AKAP9*, *CENPE*, *PRKCI*, *RDX*, *RECQL*, and *USO1* have also been reported to be associated with cancer progression [44–55]. *AKAP9* (encodes A-kinase anchor proteins-9) is involved in the development of metastasis of several cancers, including colorectal cancer [44], breast cancer [45], lung cancer [46], melanomas [47], thyroid carcinomas [48]. Inhibition of *CENPE*-encoded protein CENPE (Centromere-associated protein E), a kinetochore-associated mitotic kinesin, has been shown to induce cancer cell apoptosis and tumor regression [49]. *PRKCI* (encodes protein kinase C, iota) is overexpressed in ovarian cancer [50] and was suggested to promote immune suppression [51]. *RDX* (encodes radixin) is overexpressed in many tumor tissues and was suggested to enhance colon cancer cell invasion [52]. *RECQL*-encoded RecQ helicase-like protein is a DNA helicase which plays a vital role in the DNA damage response, and the mutation of *RECQL* has been suggested as a plausible candidate breast cancer susceptibility gene [53]. Knockdown of *USO1* (encodes general vesicular transport factor p115) was shown to inhibit cell proliferation and induce cell apoptosis in multiple myeloma cells [54] and colon cancer cells [55]. Although these molecules have various functions in cancer cells, they may be involved in radioresistance or radiation-induced invasion that can be targeted by CGN adjuvant treatment. Roles of these molecules and related mechanisms could be investigated in the future.

In our study, an increase in the proportion of polyploid cells was noted after the combined treatment with IR and CGN. Polyploid cells are considered the result of enhanced mitotic catastrophe [27,56]. Eriksson et al. revealed that IR treatment leads to a dose-dependent induction of morphological mitotic catastrophes and polyploid formation [28], which is accompanied by delayed DNA damage from IR [26]. The proportion of polyploid cells in HeLa Hep2 cells increases from 2.8 ± 1.3% to 17.6 ± 2.1% following treatment with 10 Gy [28]. Our data show that the proportion of polyploid cells in MDA-MB-231 cells was 5.6 ± 2.7% in the untreated group, 12.9 ± 1.7% in the CGN group, 17.6 ± 3.3% in the 4 Gy IR group, and 26.1 ± 1.9% in the IR and CGN group. Our results show that IR followed by CGN led to a higher proportion of polyploid cells, indicating that CGN could enhance the induction of cellular mitotic catastrophe following IR treatment. Aside from CGN, a variety of anticancer drugs are also known to induce mitotic catastrophe by influencing the stability of microtubule spindles or defective cell cycle checkpoints [57]. Mansilla et al. reported the result of mitotic catastrophe caused by the chemotherapeutic agent doxorubicin and the anthracycline antibiotic WP631 in MDA-MB-231 cells and MCF-7/VP cells [58]. In their study, treating cells with both agents resulted in increased polyploid formation, followed by increased cell death. The activation of the caspase-3-related apoptotic pathway was only observed in MDA-MB-231 cells treated with doxorubicin, indicating that caspase is not mandatory for cell death induced by mitotic catastrophe. In our study, cells cultured with CGN increased caspase-3 activity as well, suggesting that different patterns of cell death are also caused by CGN.

In Figure 6, single IR or CGN failed to control the tumor growth and metastasis in vivo. In this mouse model, we chose a low dose of IR treatment on a big tumor (about 500 mm^3) to elucidate the adjuvant effect of CGN with IR. IR was treated from day 12 after injection of 4T1 cells. In fact, on day 12, the tumor size is already too big to be affected by 2 Gy $\times$ 4 dosages. In our preliminary experiments, when we treat the mice with higher doses (2 Gy $\times$ 5) on the smaller tumor (less than 100 mm^3), we observed the effect by IR alone as compared with the untreated group. Several studies have reported the utilization of CGN in cancer immunotherapy. CGN promotes dendritic cell maturation through toll-like receptor 4 signaling [13]. CGN-treated dendritic cells significantly inhibited the tumor growth of murine lung tumor TC-1 compared with the control group [13]. In addition, a previous report showed that intratumoral injection of CGN decreases tumor growth in the B16-F10 or 4T1 tumor model in vivo by stimulating immune responses [11]. They treated with CGN every two days in the early stage of the subcutaneous 4T1 tumor growth, which decreased significantly 25 days after tumor inoculation in the CGN-treated group compared to the control group. In contrast, we showed that CGN treatment alone did not inhibit in vivo 4T1 tumor growth (Figure 6B), while it decreased cell viability in vitro (Figure 1A). We started CGN injection in the late stage of tumor growth and limited the treatment to three injections to better assess the adjuvant effect of CGN for radiation therapy. Therefore, the different in vivo results for CGN alone may be due to the differences in starting time and a total number of CGN injections. It has also been reported that radiation affects immune cells surrounding the irradiated tumors [59], but its role in either stimulating or dampening anti-tumor responses is not fully understood. To effectively induce the immune response by CGN combined with IR, the treatment condition should be optimized. Further investigation into the alteration of infiltrating immune cell subsets following treatment with IR and CGN may lead to a better understanding of their combined cytotoxic effects.

Carrageenan has been considered to cause inflammatory events of the gastrointestinal tract, which may limit the utilization to use in clinical treatment. Early studies reported that CGN administered in drinking water or diet could cause intestinal inflammation and ulcers in animals [60,61]. This phenomenon has been exacerbated in animal studies where CGN injected into confined spaces in an animal's body, such as the hind paw, pleural space, or peritoneal cavity [62,63]. However, safety studies conducted over the last 15–20 years in which CGN was administered to test animals through the diet have not shown any adverse effects [64]. These conflicting results are thought to be due to the difference in the purity of "carrageenan" [65]. McKim et al. found that commercial CGN can be diluted with sugars (dextrose or sucrose) and, thus, the potential to inadvertently add contaminants, such as bacteria, is high. In our study, the purity of CGN we used is 88.72 ± 2.21%, which is higher than the previous study used [65]. On the other hands, the amount of glucose was 4.23 ± 0.62% of the initial weight. These results suggest that high purity CGN which we used would be difficult to cause a pro-inflammatory effect.

In our study, the expression of RacGAP1 was found to be significantly suppressed by CGN treatment. Although the detailed mechanism is still unclear, other studies suggest possible pathways. For instance, Signal transducer and activator of transcription 3 (STAT3) is a transcriptional factor which has been reported to activate the transcription of RacGAP1 in hepatocellular carcinoma cells [42]. Several studies reported that some polysaccharides from plants could induce biological effects in cells via suppression of STAT3-related pathways [66,67]. Therefore, it is reasonable to the hypothesis that CGN, as a natural polysaccharide, regulates the expression of RacGAP1 via STAT3-related pathway. These studies may provide us the clues about the mechanism of how CGN regulates RacGAP1, which are worth further investigation.

Taken together, our results suggest a possible therapeutic strategy involving CGN treatment as an adjuvant to radiotherapy for the suppression of tumor growth and the reduction of distant metastasis.

4. Materials and Methods

4.1. Cell Culture

MDA-MD-231 human breast cancer cell line (ATCC® HTB-26™), PANC-1 human pancreatic cancer cell line (ATCC® CRL-1469™), FaDu human head and neck cancer cell line (ATCC® HTB-43™) and 4T1 mouse mammary carcinoma cell line (ATCC® CRL-2539™) were purchased from American Type Culture Collection (ATCC; Manassas, VA, USA). MDA-MB-231 and PANC-1 cells were cultured in Dulbecco's modified Eagle's medium (DMEM; Nacalai Tesque, Kyoto, Japan) containing 10% fetal bovine serum (FBS; HyClone, GE Healthcare Life Sciences, Logan, UT, USA). FaDu cells were cultured in minimum essential medium eagle (Sigma-Aldrich, St Louis, MO, USA) containing 10% FBS. 4T1 cells were cultured in RPMI-1640 (Sigma-Aldrich) containing 10% FBS.

4.2. Carrageenan

Lambda-carrageenan plant mucopolysaccharide (Sigma-Aldrich, Lot number BCBP8978V) was dissolved in Milli-Q water at a concentration of 10 mg/mL. The typical molecular weight of λ-carrageenan was reported as 1054 kDa [68]. The purity of carrageenan used in this study was 88.72 ± 2.21%, which was determined by EDTA/2-propanol recovery method [65]. The amount of glucose/dextrose dissolved in the wash solution was also determined by Picoprobe Glucose Assay Kit (Abcam, Cambridge, UK), and the results suggest that 4.23 ± 0.62% of the initial weight is accounted by glucose. To dissolve CGN in water, the solution of CGN was gently shaking for 24 h in 37 °C. And then, CGN was filtered through 0.45 μm filters (Advantec, Tokyo, Japan). Cells were treated with 2.5 mg/mL CGN or Milli-Q water for 24 h after 4 Gy IR treatment.

4.3. Irradiations

Cells were irradiated with 4 Gy 130 kV X-rays using a CellRad X-ray generator (CellRad; Faxitron, Tucson, AZ, USA). Mice were irradiated with a daily fraction of 2 Gy 125 kV X-rays for four days (HITACHI).

4.4. Cell Viability Assay

Cells were seeded and treated with 4 Gy IR at 50–60% confluency. Twenty-four hours after IR, cells were treated with 2.5 mg/mL CGN (approximately 2.2 mg/mL CGN is contained considering purity mentioned above) or Milli-Q water for 48 h and then subjected to each experiment. Cell viability and cytotoxicity were examined by cell counting using the trypan blue exclusion method and flow cytometry after PI staining. The proportion of PI-positive cells was quantified by placing polygon gate.

4.5. Apoptosis Analysis

Annexin V staining was performed using an Annexin V-FITC Apoptosis Detection Kit (Abcam). Cells more than 1×10^5 were harvested, resuspended in 500 μL binding buffer, and incubated with Annexin V-FITC and PI for 5 min at room temperature in the dark. Fluorescence was analyzed using a FACSAria III flow cytometer (BD Biosciences, Franklin Lakes, NJ, USA). Mean fluorescence intensity of FITC was calculated and normalized to the untreated group.

To determine caspase activity, caspase-3, caspase-8, and caspase-9 multiplex activity assay kit (Abcam) was used. Briefly, cells were treated and seeded in 96-well plates at 2×10^4 cells/100 μL FACS buffer (2% FBS in PBS). After incubation at 37 °C, 5% CO_2 for 1 h, fluorescence was monitored using a microplate reader (CLARIOstar; BMG LABTECH, Ortenberg, Germany) with the following wavelengths: caspase-3 excitation (Ex)/emission (Em) = 535/620 nm; caspase-8 Ex/Em = 490/525 nm; caspase-9 Ex/Em = 370/450 nm.

4.6. ROS Detection Assay

To measure ROS levels in cells, treated cells were stained with 20 μM dichlorofluorescein diacetate (DCFDA) for 30 min at 37 °C using a Cellular ROS detection assay kit (Abcam). Cells were then analyzed using a FACSAria III flow cytometer.

4.7. DNA Content and Polyploidy Analysis

For polyploidy analysis, 70% of ethanol was added slowly to cell pellets. Cells were stored at -80 °C overnight, and then cells were centrifuged and washed with cold PBS two times. Cells were then resuspended in 300 μL staining solution (0.1% (v/v) Triton X-100, 2 mg RNase A (NIPPON GENE, Tokyo, Japan) and 400 μL of 500 μg/mL PI (Setareh Biotech, Eugene, OR, USA) in 10 mL PBS. After incubation at 37 °C for 15 min, samples were analyzed by a FACSAria III flow cytometer.

4.8. Immunofluorescence

Cells were fixed in 4% paraformaldehyde (PFA), permeabilized with 0.2% Triton X-100/PBS, and then washed with PBS. For examination of polyploidy, cells were incubated with an α-tubulin (Cell Signaling Technology, Danvers, MA, USA) antibody after blocking, and then washed with PBS, followed by incubation with an Alexa Fluor secondary antibody. Cell nuclei were counterstained with PI. For images of RacGAP1 localization, cells were stained with an anti-RacGAP1 antibody (Abcam), an α-tubulin antibody and DAPI. Images were acquired by Leica True Confocal Scanning (TCS) SP8 microscope system (Leica Microsystems, Wetzlar, Germany).

4.9. Matrigel Invasion Assay

The Matrigel chemoinvasion assay was performed using Biocoat Matrigel invasion chambers (Corning Inc., Corning, NY, USA) or 24-well hanging inserts 8.0 μm PET Millicell cell culture inserts (Merck Millipore, Darmstadt, Germany) coated with Matrigel growth factor reduced (GFR) basement membrane matrix (Corning Inc.). For coating Millicell inserts, 100 μL serum-free medium containing 400 μg/mL Matrigel was evenly distributed on the membrane of the Millicell insert chamber, followed by incubation at 37 °C for at least 2 h. During the invasion assay, cells suspended in the DMEM with 0.1% BSA were seeded on the upper chambers, and the lower wells were filled with DMEM with 10% FBS. After incubation for 8 h, cells that migrated out onto the lower surface of the membranes were fixed in 4% PFA and stained with 1% crystal violet. Data were collected from four independent experiments and normalized to the results of the untreated group.

4.10. Western Blotting

Western blotting was performed as described previously [35]. Briefly, cell lysates were separated by SDS-PAGE or Nu-PAGE Bis-Tris protein gels (Thermo Fisher Scientific, Waltham, MA, USA), and then transferred onto a polyvinylidene fluoride (PVDF) membrane (Merck Millipore), and then blocked with Odyssey® blocking buffer (LI-COR Biosciences, Lincoln, NE, USA). Membranes were probed with primary antibodies, anti-RacGAP1 (Proteintech, Rosemont, IL, USA or Abcam) or anti-β-actin (Sigma-Aldrich, St Louis, MO, USA), and then washed with Tris-buffered saline Tween-20 (TBST). The membranes were incubated with secondary antibodies and then washed with TBST. The signals were detected with an Odyssey CLx Imager (LI-COR Biosciences).

4.11. Microarray Analysis

After treatment with CGN and IR, total RNA of MDA-MB-231 cells was isolated using a NucleoSpin® RNA kit (MACHEREY-NAGEL, Düren, Germany). For the microarray analysis, the high sensitivity 3D-Gene® Human oligo chip 25k version 2.10 (Toray Industries, Tokyo, Japan) was used. The data were normalized and analyzed by Toray Industries. Gene expression values lower than 100 after global normalization were excluded. To focus on the effects of CGN, genes were selected by the following order (Figure S1). First, genes increased 1.25-fold or higher in the IR group compared to the untreated group were selected. Second, genes reduced to one-eighth or less in IR and CGN treatment groups, compared to untreated, were selected. The genes that meet both of these criteria were selected as candidates for further analyses.

4.12. siRNA and Transfection

To knockdown RacGAP1, siRNAs with the following sequences were used: negative control: 5′-GUUUAUUGACAAGUUAAGAdTdT-3′ (sense), 5′-UCUUAACUUGUCAAUAAACdTdT-3′ (antisense); siRacGAP1 # 1: 5′-CAGGUGGAUGUAGAGAUCAAAdTdT-3′ (sense), 5′-UUUGAUCUCUACAUCCACCUGdTdT-3′ (antisense); siRacGAP1 # 2 5′-CUAGGACGACAAGGCAACUUUdTdT-3′ (sense), 5′-AAAGUUGCCUUGUCGUCCUAGdTdT-3′ (antisense). The siRNA duplexes were synthesized by Hokkaido System Science. Cells were transfected with siRNA duplexes using Lipofectamine RNAiMAX (Thermo Fisher Scientific).

4.13. Overexpression

For RacGAP1 overexpression, complementary DNA (cDNA) of RacGAP1 was obtained by PCR from the first-strand cDNA of MDA-MB-231 cells. The RacGAP1 cDNA was subcloned into the mVenus N1 vector and followed by subcloning into a PiggyBac transposon-based doxycycline-inducible vector, pPB-TRE3G-MCS-CEH-rtTA3-IP [69]. Transfection of the resulting RacGAP1-mVenus plasmid, together with a hyperactive PiggyBac transposase vector [69] to MDA-MB-231 cells was performed by ViaFect transfection reagent (Promega, Madison, WI, USA), and cells were selected by puromycin. The expression of RacGAP1-mVenus was induced by the addition of 200 ng/mL doxycycline.

4.14. In Vivo Study

One million 4T1 cells in 100 μL PBS were injected into the left thighs of six-week-old Balb/c mice subcutaneously. Eleven days after inoculation, the mouse tumors were treated with a daily fraction of 2 Gy of X-rays for four days under anesthesia. Tumors were treated with 50 mg/kg CGN or PBS 3 times by intratumoral injection one week after the first IR fraction. Tumor size was measured after radiation and normalized to the tumor size measured on the first day after radiation. All animal procedures were approved by the Institutional Animal Care and Use Committee of Hokkaido University (# 16-0137).

4.15. Immunohistochemistry

The tumor tissues were collected and placed in 4% PFA solution, fixed for 24 h, dehydrated through a gradient of ethanol, and embedded into paraffin blocks for immunohistochemistry. The paraffin blocks were cut to 4 μm sections and mounted onto microscope slides for analysis. For antigen retrieval, the slides of tumor sections were incubated with antigen unmasking solution (Vector Laboratories, Burlingame, CA, USA) at 80 °C for 1 h. Endogenous peroxidase activity was quenched using 3% H_2O_2 in 10% methanol. Each slide was incubated in 2% blocking buffer (Roche, Basel, Switzerland) for 1 h, and then incubated with a RacGAP1 primary antibody (Proteintech) overnight. Super Sensitive IHC Detection Systems (BioGenex, Fremont, CA, USA) were used to amplify the signal. Sections were stained with horseradish peroxidase (HRP) secondary antibodies. After two washes, the slides were counterstained with hematoxylin (Muto Pure Chemicals, Tokyo, Japan). Positive staining was scored using the following formula: $(r3/t) \times 3 + (r2/t) \times 2 + (r1/t) \times 1$, where t is the total area of tumor tissue for a whole tumor section, r3 is the total area of high-intensity staining (intensity 3), r2 is the total area of medium intensity staining (intensity 2), and r1 is the total area of weak intensity staining (intensity 1).

4.16. Statistical Analysis

All in vitro results were confirmed by at least three independent experiments. Data were analyzed by two-tailed Student's *t*-tests. Graphs are presented in columns as mean values ± standard error of the mean (SEM). For in vivo experiments (n = 6–7 in each group), normality of the data sets was examined by the Kolmogorov-Smirnov test, where $p > 0.05$ state in a normal distribution. After the judgment of the equality of variance by the F-test, statistical significance was examined by the two-tailed t-test for equal variance or t-test with Welch correction for unequal variances. For data with non-normal distribution, statistical significance was examined by the Mann-Whitney test after confirming equal variance by the F-test. Graphs are presented in scatter plots. Significant differences are indicated by * $p < 0.05$, ** $p < 0.01$, *** $p < 0.001$ and n.s. for not significant.

5. Conclusions

In this study, we found that RacGAP1 expression is increased after irradiation and associated with cancer cell invasion. We also found that CGN treatment following radiotherapy effectively suppressed the expression of RacGAP1 in in vitro cell culture and in vivo mouse tumor model. Based on these data, we conclude that CGN enhances the effect of radiotherapy by suppressing cancer cell survival and invasiveness through the RacGAP1 pathway. We propose the novel application of CGN as an adjuvant for radiotherapy in clinical use.

Supplementary Materials: The following are available online at http://www.mdpi.com/2072-6694/11/8/1192/s1, Figure S1: Analysis of Annexin V/PI double staining. Figure S2: The effect of CGN and/or IR in normal cell line, MCF10A. Figure S3: Caspase-3 activity measured by western blot. Figure S4: Cell viability during the Matrigel chemoinvasion assay. Figure S5: Gene expression profiles after IR and CGN treatments. Figure S6: Immunofluorescence staining of siRNA-mediated knockdown of RacGAP1 in MDA-MB-231cells. Figure S7: Uncropped scans of western blots.

Author Contributions: Conceptualization and design, P.-H.W. and J.-M.N.; Methodology, P.-H.W., F.C.R., and J.-M.N.; Acquisition of data, P.-H.W., Y.O. and J.-M.N.; Analysis and interpretation of data, P.-H.W., Y.O. and J.-M.N.; Administrative, technical, or material support, P.-H.W., Y.O., A.J.G., Q.-T.L., S.S., H.S., and J.-M.N.; Writing—original draft preparation, P.-H.W., Y.O. and J.-M.N.; Writing—review and editing, Y.O., F.C.R., A.J.G., Q.-T.L., S.S., H.S. and J.-M.N.; supervision, Y.O. and J.-M.N.

Funding: This research was supported in part by the GI-CoRE/GSQ in Hokkaido University and Grant-in-Aid from Scientific Research (C) to J.-M.N.

Acknowledgments: We thank Noriko Sasaki for the excellent assistance with animal experiments; we thank Keiko Kanno and Memu Hosoda for technical assistant; Mari Horikawa and Midori Tokuda for administrative work; Soichiro Nishioka and Chi-Che Hsieh for technical support and reading of the manuscript. We would like to thank Editage for English language editing.

Conflicts of Interest: The authors declare no conflict of interest.

References

1. Tominaga, H.; Kodama, S.; Matsuda, N.; Suzuki, K.; Watanabe, M. Involvement of reactive oxygen species (ROS) in the induction of genetic instability by radiation. *J. Radiat. Res.* **2004**, *45*, 181–188. [CrossRef] [PubMed]
2. Balcer-Kubiczek, E.K. Apoptosis in radiation therapy: A double-edged sword. *Exp. Oncol.* **2012**, *34*, 277–285. [PubMed]
3. Forastiere, A.A.; Goepfert, H.; Maor, M.; Pajak, T.F.; Weber, R.; Morrison, W.; Glisson, B.; Trotti, A.; Ridge, J.A.; Chao, C.; et al. Concurrent chemotherapy and radiotherapy for organ preservation in advanced laryngeal cancer. *New Engl. J. Med.* **2003**, *349*, 2091–2098. [CrossRef] [PubMed]
4. Mirza, A.; Choudhury, A. Bladder Preservation for Muscle Invasive Bladder Cancer. *Bladder Cancer* **2016**, *2*, 151–163. [CrossRef] [PubMed]
5. Moding, E.J.; Kastan, M.B.; Kirsch, D.G. Strategies for optimizing the response of cancer and normal tissues to radiation. *Nat. Rev. Drug Discov.* **2013**, *12*, 526–542. [CrossRef] [PubMed]
6. Aebersold, D.M.; Landt, O.; Berthou, S.; Gruber, G.; Beer, K.T.; Greiner, R.H.; Zimmer, Y. Prevalence and clinical impact of Met Y1253D-activating point mutation in radiotherapy-treated squamous cell cancer of the oropharynx. *Oncogene* **2003**, *22*, 8519–8523. [CrossRef] [PubMed]

7. Bonomo, P.; Loi, M.; Desideri, I.; Olmetto, E.; Delli Paoli, C.; Terziani, F.; Greto, D.; Mangoni, M.; Scoccianti, S.; Simontacchi, G.; et al. Incidence of skin toxicity in squamous cell carcinoma of the head and neck treated with radiotherapy and cetuximab: A systematic review. *Crit. Rev. Oncol. Hematol.* **2017**, *120*, 98–110. [CrossRef] [PubMed]
8. Higgins, G.S.; O'Cathail, S.M.; Muschel, R.J.; McKenna, W.G. Drug radiotherapy combinations: Review of previous failures and reasons for future optimism. *Cancer Treat. Rev.* **2015**, *41*, 105–113. [CrossRef] [PubMed]
9. Shanmugam, M.; Mody, K.H. Heparinoid-active sulphated polysaccharides from marine algae as potential blood anticoagulant agents. *Curr. Sci.* **2000**, *79*, 1672–1683.
10. Carlucci, M.J.; Scolaro, L.A.; Noseda, M.D.; Cerezo, A.S.; Damonte, E.B. Protective effect of a natural carrageenan on genital herpes simplex virus infection in mice. *Antivir. Res.* **2004**, *64*, 137–141. [CrossRef]
11. Luo, M.; Shao, B.; Nie, W.; Wei, X.W.; Li, Y.L.; Wang, B.L.; He, Z.Y.; Liang, X.; Ye, T.H.; Wei, Y.Q. Antitumor and Adjuvant Activity of λ-carrageenan by Stimulating Immune Response in Cancer Immunotherapy. *Sci. Rep.* **2015**, *5*, 11062. [CrossRef] [PubMed]
12. U.S. Food and Drug Administration. *Food and Drugs, 21 C.F.R. §172.620*; U.S. Food and Drug Administration: Silver Spring, MD, USA, 2018.
13. Li, J.; Aipire, A.; Zhu, H.; Wang, Y.; Guo, W.; Li, X.; Yang, J.; Liu, C. λ-Carrageenan improves the antitumor effect of dendritic cellbased vaccine. *Oncotarget* **2017**, *8*, 9996–30007. [CrossRef] [PubMed]
14. Stone, R., 2nd; Sabichi, A.L.; Gill, J.; Lee, I.L.; Adegboyega, P.; Dai, M.S.; Loganantharaj, R.; Trutschl, M.; Cvek, U.; Clifford, J.L. Identification of genes correlated with early-stage bladder cancer progression. *Cancer Prev. Res.* **2010**, *3*, 776–786. [CrossRef] [PubMed]
15. Imaoka, H.; Toiyama, Y.; Saigusa, S.; Kawamura, M.; Kawamoto, A.; Okugawa, Y.; Hiro, J.; Tanaka, K.; Inoue, Y.; Mohri, Y.; et al. RacGAP1 expression, increasing tumor malignant potential, as a predictive biomarker for lymph node metastasis and poor prognosis in colorectal cancer. *Carcinogenesis* **2015**, *36*, 346–354. [CrossRef] [PubMed]
16. Mi, S.; Lin, M.; Brouwer-Visser, J.; Heim, J.; Smotkin, D.; Hebert, T.; Gunter, M.J.; Goldberg, G.L.; Zheng, D.; Huang, G.S. RNA-seq Identification of RACGAP1 as a Metastatic Driver in Uterine Carcinosarcoma. *Clin. Cancer Res.* **2016**, *22*, 4676–4686. [CrossRef] [PubMed]
17. Toure, A.; Dorseuil, O.; Morin, L.; Timmons, P.; Jegou, B.; Reibel, L.; Gacon, G. MgcRacGAP, a new human GTPase-activating protein for Rac and Cdc42 similar to Drosophila rotundRacGAP gene product, is expressed in male germ cells. *J. Biol. Chem.* **1998**, *273*, 6019–6023. [CrossRef] [PubMed]
18. Lawson, C.D.; Der, C.J. Filling GAPs in our knowledge: ARHGAP11A and RACGAP1 act as oncogenes in basal-like breast cancers. *Small GTPases* **2018**, *9*, 290–296. [CrossRef] [PubMed]
19. Lawson, C.D.; Ridley, A.J. Rho GTPase signaling complexes in cell migration and invasion. *J. Cell Biol.* **2018**, *217*, 447–457. [CrossRef]
20. Wang, C.; Wang, W.; Liu, Y.; Yong, M.; Yang, Y.; Zhou, H. Rac GTPase activating protein 1 promotes oncogenic progression of epithelial ovarian cancer. *Cancer Sci.* **2018**, *109*, 84–93. [CrossRef]
21. Zhao, L.; Xiao, Y.; Xiao, N. Effect of lentinan combined with docetaxel and cisplatin on the proliferation and apoptosis of BGC823 cells. *Tumor Biol.* **2013**, *34*, 1531–1536. [CrossRef]
22. Zhang, Y.; Li, Q.; Wang, J.; Cheng, F.; Huang, X.; Cheng, Y.; Wang, K. Polysaccharide from Lentinus edodes combined with oxaliplatin possesses the synergy and attenuation effect in hepatocellular carcinoma. *Cancer Lett.* **2016**, *377*, 117–125. [CrossRef] [PubMed]
23. Gupta, S.C.; Hevia, D.; Patchva, S.; Park, B.; Koh, W.; Aggarwal, B.B. Upsides and downsides of reactive oxygen species for cancer: The roles of reactive oxygen species in tumorigenesis, prevention, and therapy. *Antioxid. Redox. Signal.* **2012**, *16*, 1295–1322. [CrossRef] [PubMed]
24. Bhattacharyya, S.; Borthakur, A.; Tyagi, S.; Gill, R.; Chen, M.L.; Dudeja, P.K.; Tobacman, J.K. B-cell CLL/Lymphoma 10 (BCL10) Is Required for NF-κB Production by Both Canonical and Noncanonical Pathways and for NF-κB-inducing Kinase (NIK) Phosphorylation. *J. Biol. Chem.* **2010**, *285*, 522–530. [CrossRef] [PubMed]
25. Redza-Dutordoir, M.; Averill-Bates, D.A. Activation of apoptosis signalling pathways by reactive oxygen species. *Biochim. Biophys. Acta* **2016**, *1863*, 2977–2992. [CrossRef] [PubMed]
26. Ianzini, F.; Mackey, M.A. Delayed DNA damage associated with mitotic catastrophe following X-irradiation of HeLa S3 cells. *Mutagenesis* **1998**, *13*, 337–344. [CrossRef]

27. Ianzini, F.; Mackey, M.A. Spontaneous premature chromosome condensation and mitotic catastrophe following irradiation of HeLa S3 cells. *Int. J. Radiat. Biol.* **1997**, *72*, 409–421. [CrossRef] [PubMed]
28. Eriksson, D.; Lofroth, P.O.; Johansson, L.; Riklund, K.A.; Stigbrand, T. Cell cycle disturbances and mitotic catastrophes in HeLa Hep2 cells following 2.5 to 10 Gy of ionizing radiation. *Clin. Cancer Res.* **2007**, *13*, 5501s–5508s. [CrossRef]
29. Hung, J.Y.; Wen, C.W.; Hsu, Y.L.; Lin, E.S.; Huang, M.S.; Chen, C.Y.; Kuo, P.L. Subamolide a induces mitotic catastrophe accompanied by apoptosis in human lung cancer cells. *Evid. Based Complement. Alternat. Med.* **2013**, *2013*, 828143. [CrossRef]
30. Eccles, S.A.; Box, C.; Court, W. Cell migration/invasion assays and their application in cancer drug discovery. *Biotechnol. Annu. Rev.* **2005**, *11*, 391–421.
31. Luo, H.L.; Chiang, P.H.; Chen, Y.T.; Cheng, Y.T. Lymphovascular invasion is a pathological feature related to aggressive cancer behavior and predicts early recurrence in prostate cancer. *Kaohsiung J. Med. Sci.* **2012**, *28*, 327–330. [CrossRef]
32. Camphausen, K.; Moses, M.A.; Beecken, W.D.; Khan, M.K.; Folkman, J.; O'Reilly, M.S. Radiation therapy to a primary tumor accelerates metastatic growth in mice. *Cancer Res.* **2001**, *61*, 2207–2211. [PubMed]
33. Moncharmont, C.; Levy, A.; Guy, J.B.; Falk, A.T.; Guilbert, M.; Trone, J.C.; Alphonse, G.; Gilormini, M.; Ardail, D.; Toillon, R.A.; et al. Radiation-enhanced cell migration/invasion process: A review. *Crit. Rev. Oncol. Hematol.* **2014**, *92*, 133–142. [CrossRef] [PubMed]
34. Nam, J.M.; Ahmed, K.M.; Costes, S.; Zhang, H.; Onodera, Y.; Olshen, A.B.; Hatanaka, K.C.; Kinoshita, R.; Ishikawa, M.; Sabe, H.; et al. beta1-Integrin via NF-kappaB signaling is essential for acquisition of invasiveness in a model of radiation treated in situ breast cancer. *Breast Cancer Res.* **2013**, *15*, R60. [CrossRef] [PubMed]
35. Wu, P.H.; Onodera, Y.; Ichikawa, Y.; Rankin, E.B.; Giaccia, A.J.; Watanabe, Y.; Qian, W.; Hashimoto, T.; Shirato, H.; Nam, J.M. Targeting integrins with RGD-conjugated gold nanoparticles in radiotherapy decreases the invasive activity of breast cancer cells. *Int. J. Nanomed.* **2017**, *12*, 5069–5085. [CrossRef] [PubMed]
36. Lee, G.Y.; Kenny, P.A.; Lee, E.H.; Bissell, M.J. Three-dimensional culture models of normal and malignant breast epithelial cells. *Nat. Methods* **2007**, *4*, 359–365. [CrossRef] [PubMed]
37. Onodera, Y.; Nam, J.M.; Sabe, H. Intracellular trafficking of integrins in cancer cells. *Pharm. Ther.* **2013**, *140*, 1–9. [CrossRef] [PubMed]
38. Jacquemet, G.; Green, D.M.; Bridgewater, R.E.; von Kriegsheim, A.; Humphries, M.J.; Norman, J.C.; Caswell, P.T. RCP-driven alpha5beta1 recycling suppresses Rac and promotes RhoA activity via the RacGAP1-IQGAP1 complex. *J. Cell Biol.* **2013**, *202*, 917–935. [CrossRef]
39. Saigusa, S.; Tanaka, K.; Mohri, Y.; Ohi, M.; Shimura, T.; Kitajima, T.; Kondo, S.; Okugawa, Y.; Toiyama, Y.; Inoue, Y.; et al. Clinical significance of RacGAP1 expression at the invasive front of gastric cancer. *Gastric Cancer* **2015**, *18*, 84–92. [CrossRef]
40. Yao, H.; Zeng, Z.Z.; Fay, K.S.; Veine, D.M.; Staszewski, E.D.; Morgan, M.; Wilder-Romans, K.; Williams, T.M.; Spalding, A.C.; Ben-Josef, E.; et al. Role of α5β1 Integrin Up-regulation in Radiation-Induced Invasion by Human Pancreatic Cancer Cells1. *Transl. Oncol.* **2011**, *4*, 282–292. [CrossRef]
41. Casado-Medrano, V.; Barrio-Real, L.; García-Rostán, G.; Baumann, M.; Rocks, O.; Caloca, M.J. A new role of the Rac-GAP β2-chimaerin in cell adhesion reveals opposite functions in breast cancer initiation and tumor progression. *Oncotarget* **2016**, *7*, 28301–28319. [CrossRef]
42. Yang, X.M.; Cao, X.Y.; He, P.; Li, J.; Feng, M.X.; Zhang, Y.L.; Zhang, X.L.; Wang, Y.H.; Yang, Q.; Zhu, L.; et al. Overexpression of Rac GTPase Activating Protein 1 Contributes to Proliferation of Cancer Cells by Reducing Hippo Signaling to Promote Cytokinesis. *Gastroenterology* **2018**, *155*, 1233–1249.e1222. [CrossRef] [PubMed]
43. Hazar-Rethinam, M.; de Long, L.M.; Gannon, O.M.; Boros, S.; Vargas, A.C.; Dzienis, M.; Mukhopadhyay, P.; Saenz-Ponce, N.; Dantzic, D.D.; Simpson, F.; et al. RacGAP1 Is a Novel Downstream Effector of E2F7-Dependent Resistance to Doxorubicin and Is Prognostic for Overall Survival in Squamous Cell Carcinoma. *Mol. Cancer Ther.* **2015**, *14*, 1939–1950. [CrossRef]
44. Hu, Z.Y.; Liu, Y.P.; Xie, L.Y.; Wang, X.Y.; Yang, F.; Chen, S.Y.; Li, Z.G. AKAP-9 promotes colorectal cancer development by regulating Cdc42 interacting protein 4. *Biochim. Biophys. Acta* **2016**, *1862*, 1172–1181. [CrossRef] [PubMed]
45. Frank, B.; Wiestler, M.; Kropp, S.; Hemminki, K.; Spurdle, A.B.; Sutter, C.; Wappenschmidt, B.; Chen, X.; Beesley, J.; Hopper, J.L.; et al. Association of a common AKAP9 variant with breast cancer risk: A collaborative analysis. *J. Natl. Cancer Inst.* **2008**, *100*, 437–442. [CrossRef] [PubMed]

46. Truong, T.; Sauter, W.; McKay, J.D.; Hosgood, H.D., 3rd; Gallagher, C.; Amos, C.I.; Spitz, M.; Muscat, J.; Lazarus, P.; Illig, T.; et al. International Lung Cancer Consortium: Coordinated association study of 10 potential lung cancer susceptibility variants. *Carcinogenesis* **2010**, *31*, 625–633. [CrossRef]
47. Kabbarah, O.; Nogueira, C.; Feng, B.; Nazarian, R.M.; Bosenberg, M.; Wu, M.; Scott, K.L.; Kwong, L.N.; Xiao, Y.; Cordon-Cardo, C.; et al. Integrative genome comparison of primary and metastatic melanomas. *PLoS ONE* **2010**, *5*, e10770. [CrossRef] [PubMed]
48. Caria, P.; Vanni, R. Cytogenetic and molecular events in adenoma and well-differentiated thyroid follicular-cell neoplasia. *Cancer Genet. Cytogenet.* **2010**, *203*, 21–29. [CrossRef]
49. Wood, K.W.; Lad, L.; Luo, L.; Qian, X.; Knight, S.D.; Nevins, N.; Brejc, K.; Sutton, D.; Gilmartin, A.G.; Chua, P.R.; et al. Antitumor activity of an allosteric inhibitor of centromere-associated protein-E. *Proc. Natl. Acad. Sci. USA* **2010**, *107*, 5839–5844. [CrossRef]
50. Tsang, T.Y.; Wei, W.; Itamochi, H.; Tambouret, R.; Birrer, M.J. Integrated genomic analysis of clear cell ovarian cancers identified PRKCI as a potential therapeutic target. *Oncotarget* **2017**, *8*, 96482–96495. [CrossRef]
51. Sarkar, S.; Bristow, C.A.; Dey, P.; Rai, K.; Perets, R.; Ramirez-Cardenas, A.; Malasi, S.; Huang-Hobbs, E.; Haemmerle, M.; Wu, S.Y.; et al. PRKCI promotes immune suppression in ovarian cancer. *Genes Dev.* **2017**, *31*, 1109–1121. [CrossRef]
52. Jiang, Q.H.; Wang, A.X.; Chen, Y. Radixin enhances colon cancer cell invasion by increasing MMP-7 production via Rac1-ERK pathway. *Sci. World J.* **2014**, *2014*, 340271. [CrossRef] [PubMed]
53. Bowden, A.R.; Tischkowitz, M. Clinical implications of germline mutations in breast cancer genes: RECQL. *Breast Cancer Res. Treat.* **2019**, *174*, 553–560. [CrossRef] [PubMed]
54. Jin, Y.; Dai, Z. USO1 promotes tumor progression via activating Erk pathway in multiple myeloma cells. *Biomed. Pharm.* **2016**, *78*, 264–271. [CrossRef] [PubMed]
55. Sui, J.; Li, X.; Xing, J.; Cao, F.; Wang, H.; Gong, H.; Zhang, W. Lentivirus-mediated silencing of USO1 inhibits cell proliferation and migration of human colon cancer cells. *Med. Oncol.* **2015**, *32*, 218. [CrossRef] [PubMed]
56. Erenpreisa, J.; Kalejs, M.; Ianzini, F.; Kosmacek, E.A.; Mackey, M.A.; Emzinsh, D.; Cragg, M.S.; Ivanov, A.; Illidge, T.M. Segregation of genomes in polyploid tumour cells following mitotic catastrophe. *Cell Biol. Int.* **2005**, *29*, 1005–1011. [CrossRef] [PubMed]
57. Castedo, M.; Perfettini, J.L.; Roumier, T.; Andreau, K.; Medema, R.; Kroemer, G. Cell death by mitotic catastrophe: A molecular definition. *Oncogene* **2004**, *23*, 2825–2837. [CrossRef] [PubMed]
58. Mansilla, S.; Priebe, W.; Portugal, J. Mitotic catastrophe results in cell death by caspase-dependent and caspase-independent mechanisms. *Cell Cycle* **2006**, *5*, 53–60. [CrossRef]
59. Kaur, P.; Asea, A. Radiation-induced effects and the immune system in cancer. *Front. Oncol.* **2012**, *2*, 191. [CrossRef]
60. Benard, C.; Cultrone, A.; Michel, C.; Rosales, C.; Segain, J.P.; Lahaye, M.; Galmiche, J.P.; Cherbut, C.; Blottiere, H.M. Degraded carrageenan causing colitis in rats induces TNF secretion and ICAM-1 upregulation in monocytes through NF-kappaB activation. *PLoS ONE* **2010**, *5*, e8666. [CrossRef]
61. Moyana, T.; Lalonde, J.M. Carrageenan-induced intestinal injury: Possible role of oxygen free radicals. *Ann. Clin. Lab. Sci.* **1991**, *21*, 258–263.
62. Mizokami, S.S.; Hohmann, M.S.; Staurengo-Ferrari, L.; Carvalho, T.T.; Zarpelon, A.C.; Possebon, M.I.; de Souza, A.R.; Veneziani, R.C.; Arakawa, N.S.; Casagrande, R.; et al. Pimaradienoic Acid Inhibits Carrageenan-Induced Inflammatory Leukocyte Recruitment and Edema in Mice: Inhibition of Oxidative Stress, Nitric Oxide and Cytokine Production. *PLoS ONE* **2016**, *11*, e0149656. [CrossRef] [PubMed]
63. Essel, L.B.; Obiri, D.D.; Osafo, N.; Antwi, A.O.; Duduyemi, B.M. The Ethanolic Stem-Bark Extract of Antrocaryon micraster Inhibits Carrageenan-Induced Pleurisy and Pedal Oedema in Murine Models of Inflammation. *Int. Sch. Res. Not.* **2017**, *2017*. [CrossRef] [PubMed]
64. Weiner, M.L. Food additive carrageenan: Part II: A critical review of carrageenan in vivo safety studies. *Crit. Rev. Toxicol.* **2014**, *44*, 244–269. [CrossRef] [PubMed]
65. McKim, J.M., Jr.; Wilga, P.C.; Pregenzer, J.F.; Blakemore, W.R. The common food additive carrageenan is not a ligand for Toll-Like- Receptor 4 (TLR4) in an HEK293-TLR4 reporter cell-line model. *Food Chem. Toxicol.* **2015**, *78*, 153–158. [CrossRef] [PubMed]
66. Tao, J.H.; Duan, J.A.; Zhang, W.; Jiang, S.; Guo, J.M.; Wei, D.D. Polysaccharides From Chrysanthemum morifolium Ramat Ameliorate Colitis Rats via Regulation of the Metabolic Profiling and NF-kappa B/TLR4 and IL-6/JAK2/STAT3 Signaling Pathways. *Front. Pharm.* **2018**, *9*, 746. [CrossRef] [PubMed]

67. Liu, L.Q.; Nie, S.P.; Shen, M.Y.; Hu, J.L.; Yu, Q.; Gong, D.; Xie, M.Y. Tea Polysaccharides Inhibit Colitis-Associated Colorectal Cancer via Interleukin-6/STAT3 Pathway. *J. Agric. Food Chem.* **2018**, *66*, 4384–4393. [CrossRef]
68. McKim, J.M. Food additive carrageenan: Part I: A critical review of carrageenan in vitro studies, potential pitfalls, and implications for human health and safety. *Crit. Rev. Toxicol.* **2014**, *44*, 211–243. [CrossRef] [PubMed]
69. Onodera, Y.; Nam, J.M.; Horikawa, M.; Shirato, H.; Sabe, H. Arf6-driven cell invasion is intrinsically linked to TRAK1-mediated mitochondrial anterograde trafficking to avoid oxidative catastrophe. *Nat. Commun.* **2018**, *9*, 2682. [CrossRef]

MDPI
St. Alban-Anlage 66
4052 Basel
Switzerland
Tel. +41 61 683 77 34
Fax +41 61 302 89 18
www.mdpi.com

Cancers Editorial Office
E-mail: cancers@mdpi.com
www.mdpi.com/journal/cancers

www.ingramcontent.com/pod-product-compliance
Lightning Source LLC
Chambersburg PA
CBHW061757210926
51850CB00006B/580

9783039432929